Environmental Microbiology
and Microbial Ecology

Environmental Microbiology and Microbial Ecology

Larry L. Barton and Robert J.C. McLean

WILEY Blackwell

This edition first published 2019

Registered Office
John Wiley & Sons, Inc., 111 River Street, Hoboken, NJ 07030, USA

Editorial Office
The Atrium, Southern Gate, Chichester, West Sussex, PO19 8SQ, UK

For details of our global editorial offices, customer services, and more information about Wiley products visit us at www.wiley.com.

Wiley also publishes its books in a variety of electronic formats and by print-on-demand. Some content that appears in standard print versions of this book may not be available in other formats.

Library of Congress Cataloging-in-Publication data has been applied for:

9781118966266

Cover Design: Wiley
Cover Images: Bacteria in the image were collected from an aquifer at a depth of 275 m. See Figure 2.17 and narrative discussing biofilms. Image provided by Larry L. Barton.
Deep subsurface sampling for bacteria in the Beatrix Gold Mine in South Africa. See Figure 5.5 and supporting narrative for details. Image provided by Gaetan Borgonie and Tullis C. Onstott.

Set in 10/12pt Warnock by SPi Global, Pondicherry, India

Contents

Preface

Environmental Microbiology and Microbial Ecology provides an overview and discussion of the presence of microorganisms and the significance of their interactions in numerous environments from the perspective of microbiology, environmental science, and biogeochemistry, using current publications as a resource. Diverse topics are organized and discussed in eleven chapters that include current knowledge concerning the bacteria, archaea, fungi, and viruses present in the biosphere. Concepts, results, and conclusions are presented and referenced to enable the reader to examine the original publications on which the various topics are based, with most references readily available in the Open Access literature. The extensive use of cited literature distinguishes *Environmental Microbiology and Microbial Ecology* from introductory text books.

The presentation of recent microbial studies presented in this book builds on the success of the previous text *Microbial Ecology* by Larry L. Barton and Diana E. Northup published by Wiley-Blackwell in 2011. This is an exciting time in environmental microbiology and microbial ecology, where numerous microbial interactions are being evaluated and associations are being proposed. The topics covered in this book include the following:

- Microbial formation of biofilms
- Microbial response to stress and adaption to extreme environments
- Identification of environmental sites and microbial communities relevant to exploration for life on Mars and other extraterrestrial habitats
- Structure and activities of microbial communities
- Discussion of the prokaryote–eukaryote dichotomy and the Tree of Life
- The role of viruses, lysogeny, gene transfer agents, and the CRISPR–*cas* system in horizontal gene transfer
- Microbial presence and activities in extreme environments, including deep subsurface, deserts, cold environments, and hydrothermal vents
- Extracellular electron transfer, biocorrosion, biomineralization, and bioremediation
- Biogeochemical cycles with contributions by microorganisms
- Mutualism and communication between bacteria and plants
- Mutualism between microorganisms and animals, insects, and humans, with an emphasis on intestinal microbiology.

This book is an appropriate resource for instructors of Microbial Ecology or Environmental Microbiology courses as it includes the following special features:

- Discussion questions are provided for each chapter to promote critical thinking and to instigate class debate.
- Lists of references and further reading provide students with numerous Open Access reviews and primary literature.

- Numerous tables and boxes provide extra information on specific topics related to the chapter.
- The text is supported by numerous figures that act as visual models.
- A broad range of specific environments are discussed with numerous presentations on specific sites, hosts, and interactions.

The authors have attempted to provide specific documentation for the data presented and for the generalizations made throughout the book. The extensive citation of publications enables readers to gain an insight into how the research was conducted and how any scientific conclusions were established. Currently, there is a wealth of scientific information available in the Open Access literature, which can be readily accessed online and we have identified numerous publications related to topics covered in this book. To draw attention to specific topics or clarify complex issues, numerous illustrations are presented throughout the book. The authors greatly appreciate the contributions of photographs provided by the following individuals:

Gaeton Borgonie, Ghent University, Belgium
Daniel R. Coleman, Montana State University
Stephen Giovannoni, Oregon State University
Robert Harris, University of Guelph, Canada
Gordon V. Johnson, University of New Mexico.
Roy L. Johnson, Jr, University of New Mexico
Cezar Khursigara, University of Guelph, Canada
Richard McIntosh, University of Colorado
Daniela Nicastro, University of Texas Southwestern Medical Center
Yayoi Nishiyama, Teikyo University, Japan
T.C. Onstott, Princeton University
Karsten Pedersen, Microbial Analytics, Sweden
Nathaniel L. Ritz, University of New Mexico
Helga Stan-Lotter, University of Salzburg, Austria
Xiaowei Zhao, University of Texas Southwestern Medical Center

We appreciate the numerous pictures used from the Public Domain or similar sources because they provide expertly presented colorful details. In the process of compiling this book, we received numerous encouraging comments from colleagues and for this we are most appreciative. This book is written for upper-level undergraduate and graduate students in microbiology, ecology, biology, and environmental science. Also, it is a valuable reference for professionals working in the area of microbial ecology and environmental science. This area encompasses both the microorganisms and their interactions in the environment and is rapidly evolving. It is the hope of the authors that this book will stimulate future investigations into the role of microorganisms in the biosphere.

Finally, we are indebted to the staff at Wiley for their support throughout preparation of this text and for their skill in the final presentation of this material.

Larry L. Barton
Robert J.C. McLean

1

Introduction to Microorganisms and Their Activities

1.1 Central Themes of Environmental Microbiology and Microbial Ecology

The terms "environmental microbiology" and "microbial ecology" are often used interchangeably but there are some subtle distinctions. Environmental microbiology is the study of processes in the environment mediated by microorganisms whereas microbial ecology addresses the interactions between microorganisms as well as between microorganisms and higher life forms. However, many microorganism interactions are dependent on chemicals from the environment or from other biological systems and so microbial ecology overlaps with environmental microbiology where abiotic chemistry occurs. This first chapter provides an overview of the components involved in environmental microbiology and provides a perspective on the breadth of the microbial relationships in the biosphere. The central themes of this chapter include the following:

- Discussion on the continued use of the terms prokaryote and eukaryote and on the Tree of Life
- Horizontal gene transfer and the role of viruses and gene transfer agents
- Perspective of cell size and cell shape
- Bacterial production of dormant cells

1.2 Are the Terms Prokaryotes or Eukaryotes Relevant?

Traditional microbiology classifies microorganisms into two groups: prokaryotes and eukaryotes. Several structural distinctions may be drawn between these groups of microorganisms and the major differences are listed in Table 1.1. This distinction between prokaryotic and eukaryotic life evolved from a publication by Stanier and van Niel (1962), which proved to be the stimulus to include "blue-green algae" as cyanobacteria. As stated by Stanier and van Niel, bacterial cells were unlike eukaryotic cells in that they lacked true membranes to localize the cell "nucleus" and bacteria used nuclear division by fission and not mitosis. However, after several decades of microbial phylogeny, the term "prokaryote" has become controversial because the designation of a prokaryote is based on the absence of certain characteristics (Sapp 2005; Pace 2009; Whitman 2009). It has been proposed that bacteria and archaea, unlike eukaryotes, display coupled transcription and translation where translation starts before transcription is finished (Martin and Koonin 2006; French et al. 2007). There is a concern that

Environmental Microbiology and Microbial Ecology, First Edition. Larry L. Barton and Robert J.C. McLean.

Table 1.1 Differences between prokaryotes and eukaryotes.

- Prokaryotic cells lack a true nucleus with a nuclear membrane.
- Prokaryotic cells lack histones to provide condensation of DNA into chromosomes.
- DNA in prokaryotes is circular whereas DNA in eukaryotes is linear.
- Prokaryotic cells lack organelles in the cytoplasm and cytoplasmic membranes.
- Prokaryotic cells lack carbohydrates and sterols in the plasma membrane whereas both are found in eukaryotic cells.
- In prokaryotic cells, ribosomes are 70S; in eukaryotic cells, ribosomes are 80S with 70S found in organelles.
- Cell division in prokaryotic cells is by binary fission whereas in eukaryotic cells it is by mitosis.
- Sexual recombination does not occur in prokaryotic cells except for DNA transfer. Sexual recombination in eukaryotic cells involves meiosis.
- Prokaryotic cells are 0.2–2.0 μm wide whereas eukaryotic cells are 10–100 μm wide.

bacteria and archaea themselves are sufficiently distinct and should not be united into the single group prokaryotes. The contrast between the nuclear organization and the presence of a nuclear membrane in prokaryotic and eukaryotic organisms has become blurred. Bacteria have long been considered to lack nuclear organization; however, in the bacterial phylum *Planctomycetes*, *Gemmata obscuriglobus* has a nucleoid enveloped in a membrane that forms a structure analogous to the eukaryotic nucleus (Fuerst 2005). The giant bacterium *Epulopiscium fishelsoni* has DNA highly condensed into chromosome-like structures that are physically separated from the cytoplasm (Bresler et al. 1998). In contrast, dinoflagellates, eukaryotic algae, lack histones for the condensation of DNA and lack nucleosomes (Rizzo 2003). Histone proteins are found in mesophilic, thermophilic, and hyperthermophilic archaea and the DNA interactions of archaeal histones is like that found in eukaryotes (Reeve et al. 2004).

1.2.1 Intracellular Membranes in Prokaryotes

Models of some of the intracellular membrane structures found in bacteria are presented in Figure 1.1. Bacteria that obtain energy from methane oxidization often use particulate methane monooxygenase (pMMO). which is localized in the membrane. For the greatest efficiency of methane gas oxidation, multiple membrane structures are present in the cytoplasm and, based on internal membrane structure, methanotrophic bacteria may have either stacked (type I) or paired concentric (type II) cytoplasmic membranes (Davies and Whittenbury 1970). A recently isolated filamentous bacterium *Crenothrix fusca* Roze 1896 also has stacked membranes; however, they only extend part way across the diameter of the cell (Vigliotta et al. 2007). *Nitrosomonas*, *Nitrosococcus*, and *Nitrosocystis* are genera of nitrifying bacteria that have internal membranes to achieve the oxidation of ammonia (NH_3) to nitrite (NO_2^-). The configuration of the internal membranes appears to be specific for each genus: *Nitrosomonas* has membranes along the periphery of the cell, *Nitrosococcus* has laminar membranes through the central region of the cell, and *Nitrosocystis* has a complex membrane structure of small vesicles along the exterior of the cell (Murray and Watson 1965). For almost 100 years, acidocalcisomes have been known to occur in bacteria; however, they have only been investigated recently. In bacteria and protists, acidocalcisomes store inorganic phosphates and calcium ions, and participate in maintaining intracellular pH and in osmoregulation (Docampo and Moreno 2011). Members of the *Planctomycetes* have a structure called the anammoxosome, which is used for the anaerobic oxidation of ammonia (annamox) (van Nifrik et al. 2004). Magnetotactic bacteria produce cytoplasmic magnetosomes with a surrounding membrane that originates at the plasma membrane (Lefèvre and Bazylinski 2013). In addition, phototrophic bacteria

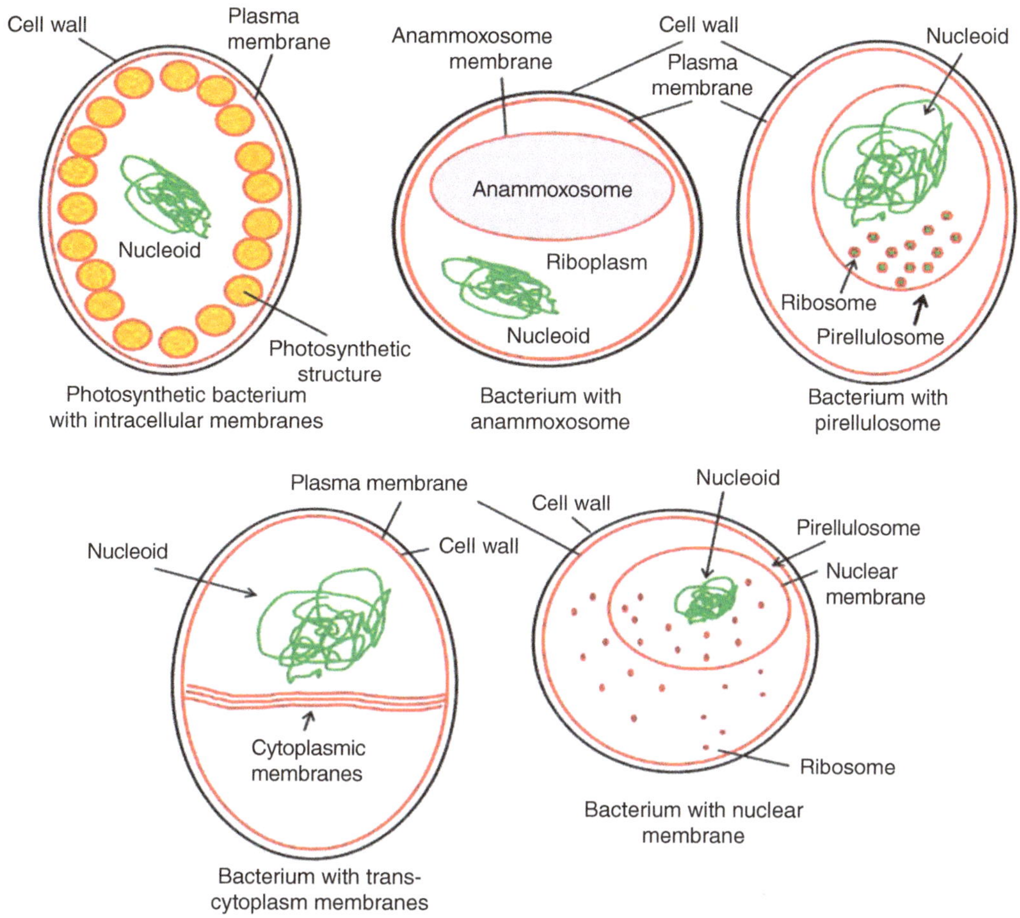

Figure 1.1 Examples of internal structures observed in bacteria.

contain photosynthetic units within cytoplasmic membrane structures (chromatophores) (Willey et al. 2014; Saier 2014). Thus, in this book the term prokaryotes is used sparingly and includes both bacteria and archaea, with considerations as stated earlier.

1.2.2 Compartmentalized Heterotrophic Bacterial Cells

The presence of a compartment inside a heterotrophic bacterium was reported initially for cells of *G. obscuriglobus* (Fuerst 2005). However, recent developments indicate that the compartmentalized bacteria, members of the *Planctomycetes-Verrucomicrobia-Chlamydiae* (PVC) superphylum, have a homolog of a eukaryotic protein that occurs in eukaryotic membranes (Santarella-Mellwig et al. 2010). Ladderanes (Figure 1.2) are unusual lipids found in the anammoxosome membrane and in *Kuenenia stuttgartiensis* and *Borcadia anammoxidans* the ladderanes account for over half of the lipids present (Damsté et al. 2002). The ladderane lipids produce a dense membrane with low permeability, which is needed to retain potentially toxic intermediates in the anammox reaction. Unlike the anaerobic ammonia-oxidizing *Planctomycetes*, members of the phylum *Verrucomicrobia* (i.e. *Verrucomicrobium spinosum,*

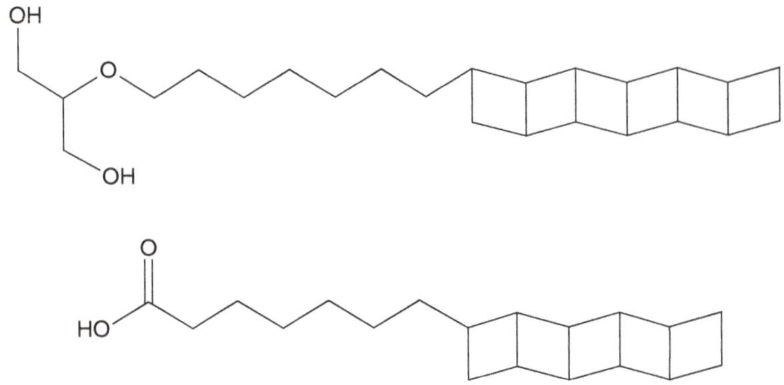

Figure 1.2 Ladderane lipids present in anammoxosome. About half of the lipid in the anammoxosome membrane consists of ladderane lipids; the fused five cyclobutane groups enable the lipids to be closely packed, which contributes to the impermeable character of this membrane.

Prosthecobacter dejongeii, and *Chthonibacter flavus*) have a compartmentalized unit in the cell but do not have the enzymes to carry out the anammox reaction (Lee et al. 2009). The membrane enclosure contains nucleoid and ribosome-like particles and is like the pirellulosome in some *Planctomycetes* (i.e. *Pirellula staleyi* and *Blastopirellula marina*) (Lindsay et al. 2001).

1.2.3 The Universal Tree of Life: Rooted or Unrooted

Microorganisms addressed here are associated with the phylogenetic domains of Bacteria, Archaea, and Eukarya. Although there have been several other systems used to arrange life forms, the classification system dealing with phylogenetic relationships or evolutionary aspects of cells or individuals that has been favored by microbiologists involves the domains of Bacteria, Archaea, and Eukarya. Salient distinctions between these groups of organisms are indicated in Table 1.2.

Tree configurations have long been used to organize physical characteristics into different categories and the Tree of Life, as proposed by Woese et al. (1990), relies on the DNA segment encoding for ribosomal RNA (rRNA). While numerous other genes could be used to express relationships between microorganisms, the genes encoding for rRNA seem to be stable and not subject to evolutionary changes. The hope was that through evaluation of the many genes found in microorganisms, a universal common ancestor for life could be determined. However, it has been difficult to construct a valid Tree of Life for organisms on Earth due to horizontal (lateral) gene transfer throughout the life span of microorganisms. Doolittle (2015) has summarized the various concerns in construction of a Tree of Life and suggested organizing the three domains using an unrooted tree (see Figure 1.3). Even though gene trees have stimulated a great amount of research, it is worth recalling Whitman's (2009) statement that "gene trees are not equivalent to organismal trees." For a thoughtful presentation of the status of the universal Tree of Life the reader is encouraged to consult the review by Forterre (2015).

1.2.4 What About the Giant Viruses?

In the past few years there have been many double stranded DNA (dsDNA) viruses isolated that are gigantic. These viruses make up the Nucleocytoplasmic Large DNA Viruses (NCLDV). which infect protozoa, algae, and other aquatic eukaryotes. The family *Phycodnaviridae* are

Table 1.2 Distinctions between members of the Tree of Life.

Characteristic	Bacteria domain	Archaea domain	Eukarya domain
Histones associated with DNA	Absent	Present in some	Present
Chromosome or nucleoid	Most circular, few linear	Circular	Linear
Nuclear membrane	Absent[a]	Absent	Present
Introns	Absent	Present in a few	Present
RNA polymerase	One type	Several types	Several types
Peptidoglycan in cell wall	Present	Absent	Absent
Amino acid initiating protein synthesis	*N*-formyl methionine	Methionine	Methionine
Ribosomes sensitive to antibiotics	70S inhibited	70S not inhibited	80S not inhibited
Cholesterol in membrane	Absent[b]	Absent	Present
Membrane lipids	Fatty acids esterified to glycerol	Phytanols ether linked to glycerol	Fatty acids esterified to glycerol
Heat-resistant endospores	Present	Absent	Absent

[a] Present in only one bacterium.
[b] An exception is that cholesterol is found in a few species of bacteria of the genus *Mycoplasma*.

large icosahedral dsDNA viruses with genomes of 160–560 kb that infect eukaryotic algae (Wilson et al. 2009). Mimiviruses are large dsDNA viruses that infect amoeba and have been associated with many human pneumonia cases; due to their size, they were initially considered to be gram-positive intracellular bacteria (Raoult et al. 2007). Faustavirus is an icosahedral virus ~200 nm that was isolated from *Acanthamoeba* spp. and appears closely related to the pathogenic African swine fever virus (Reteno et al. 2015). *Pithovirus sibericum*, an ancestral large virus that infects amoeba, was isolated from a 30 000-year-old Siberian permafrost sample (Legendre et al. 2014). The giant virus *Pandoravirus salinus*, which attacks amoeba, was isolated from seawater near the coast of central Chile (Philippe et al. 2013). Several giant viruses have genome sizes greater than some bacteria and the genomes of mimiviruses, pithoviruses, pandoraviruses, and Faustovirus are 1.2, 0.6, 2.5, and 0.466 mbp, respectively. When these viruses were discovered, some microbiologists proposed that the giant viruses represented a fourth form of cellular life or a fourth branch on the Tree of Life (Sharma et al. 2015). Although the discussion concerning the origin of these viruses continues, a strong argument has been presented that suggests giant viruses arose from smaller DNA viruses that have acquired genes from their eukaryotic hosts (Yutin et al. 2014).

1.3 Major Approach to Study Microorganisms

There are several major approaches for the study of microorganisms, especially bacteria, from an environmental or ecological perspective. Analysis of specific environmental sites includes traditional cultivation procedures as well as the use of molecular techniques to identify bacteria or archaea and to characterize their response to environmental conditions. The molecular study of environmental microorganisms is discussed in further detail in Chapter 2.

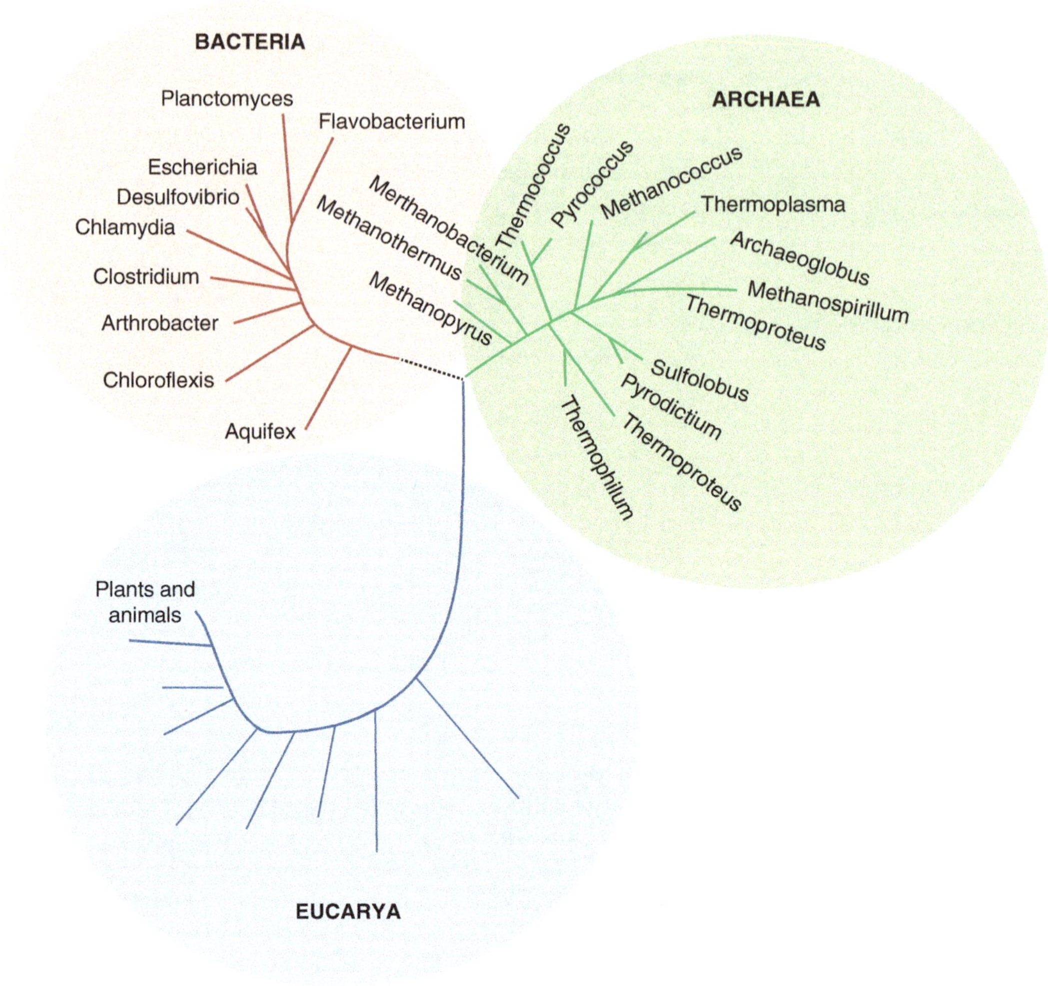

Figure 1.3 An unrooted tree showing the relationship of the three main branches of life based on the small subunit rRNA sequences.

1.3.1 Application of Genomics, Metagenomics, and Proteomics

Scientists may ask "What are these microorganisms and what specific genes do they possess?" To answer this question, scientists may use genomics, metagenomics, or proteomics. Microbiologists examining a specific site may use 16S rRNA from bacteria that were isolated and grown in the laboratory and this use of the bacterial genome to identify the bacteria present is genomics. Since not all prokaryotes at a site can be grown in the laboratory, extraction and analysis of all DNA at a site (metagenomics) provides the identity of the bacteria and archaea present, whether or not they can be cultivated. It is generally considered that less than 1% of bacteria and archaea in an environmental sample can be grown in the laboratory. The inability to count colonies of bacteria from diluted soil samples as a means of enumerating the bacteria present is referred to as the "great plate count anomaly." With the number of bacteria on Earth estimated to be 1.7×10^{30} cells (Table 1.3). it would be a daunting task to work out cultivation schemes for all bacteria and then analyze their genes. The identification of genes to

Table 1.3 Estimation of prokaryotes distributed throughout the biosphere.

Location	Number of prokaryotic cells[a]
Aquatic environment	1.2×10^{29}
Subsurface of oceans	3.5×10^{29}
Terrestrial subsurface	2.5×10^{29}
Soil	2.6×10^{27}
Plant surfaces	1.0×10^{26}

[a] Estimated number of prokaryotic cells produced each year is 1.7×10^{30} cells (Whitman et al. 1998).
Source: Reproduced with permission from Whitman et al. (1998), Lindow and Brandl (2003), and Kallmeyer et al. (2012).

explain specific activities may be obtained by sequencing DNA isolated from a site or by in situ hybridization techniques using probes targeting specific genes. Proteomics can also provide an excellent analysis of bacteria present (Dworzanski and Snyder 2005). For proteomics, protein produced by bacteria is isolated and nuclear magnetic resonance (NMR) analysis is used to match it to protein from a known gene, so the specific gene can be identified. Using genomic analysis, bacterial diversity and community structure at a site can be established. From this molecular analysis, relatively sound proposals can be made concerning the contribution of specific bacteria to the overall biological activity at a specific site.

1.3.2 Biochemical and Physiological Analysis

Another approach is to ask "How do microorganisms function in a specific process?" In this case, the activities of specific bacteria are examined by transcriptomics. The bacteria involved in the active synthesis of a specific protein can be identified by isolating mRNA from a site and its subsequent conversion to DNA by reverse transcriptase,. This is important to demonstrate which gene is being expressed because the presence of a gene in bacteria does not mean that gene is being decoded. For example, the presence of a gene for nitrogen fixation can be determined by genomic analysis but RNA isolation would indicate the actual expression of the nitrogen-fixing gene. With bacteria isolated from a site, the biochemical processes can be studied to understand how bacteria are capable of accomplishing a given activity. Included in this type of investigation would be chemical analysis to assess biotic and abiotic activities as well as processes resulting from mixed populations. This is an important approach for the study of processes such as biogeochemical cycles, bacterial symbiosis, and bioremediation.

1.4 The Impact of Horizontal Gene Transfer Between Microorganisms

The vertical transmission of genes refers to the transfer of DNA from the parent to the offspring, whereas the horizontal transmission of genes describes the movement of DNA between biological systems. The distinction between horizontal gene transfer and vertical inheritance (Andam et al. 2010) and the frequency at which DNA transfer occurs between phylogenic groups has been reviewed elsewhere (Kloesges et al. 2011). The movement of genes between bacteria has been extensively studied and DNA is transferred by conjugation, transduction,

Box 1.1 Characteristics of Bacterial DNA Exchange

Conjugation – DNA exchange between bacteria that requires cell-to-cell contact is referred to as conjugation. As described in most introductory microbiology text books (Willey et al. 2014). DNA is moved from a donor cell to a recipient cell as a plasmid or as "chromosomal" genetic material. In *Escherichia coli*, a special plasmid referred to as the F-factor enhances this horizontal gene transfer. Conjugation is assumed to occur in most bacteria and new examples are being discovered as is the case with *Mycobacterium smegmatis* (Gray et al. 2013). Often the DNA transferred to the recipient cell provides new catabolic genes, resistance to toxic metals or resistance to antibiotics. Conjugation between bacteria and plant cells accounts for tumor induction in plants by introduction of Ti plasmids from *Agrobacterium tumefaciens* or Ri plasmids from *Agrobacterium rhizogenes* (Pan et al. 1995).

Transduction – In bacteria this is the transfer of DNA into a recipient host by a bacteriophage (bacterial virus). While generalized transduction carries any gene from a donor, specialized transduction moves only a gene near the provirus in the "chromosome" of the donor. Both processes are described in introductory microbiology text books such as Willey et al. (2014).

Transformation – Bacteria have the capability of taking up DNA from outside of the cell and DNA becomes part of its genome (Chen and Dubnau 2004). The DNA acquired may be as a plasmid or a disrupted bacterial chromosome. Cells of a specific competent state are capable of acquiring exogenous DNA and there is the potential for acquiring DNA from species different from the host cell.

Gene transfer agents – The mechanism of gene transfer between prokaryotes involves a particle that is similar to a small bacterial virus. These gene transfer agents (GTAs) are released from prokaryotes by cell lysis and transduce random genomic segments to a recipient prokaryote. Enhanced gene transfer is attributed to the GTA particle (Maxmen 2010). In addition to the microorganisms listed in the text, bacteria with GTAs receiving attention include *Rhodovulum sulfidophilum, Bartonella* spp.*, and Bacillus* spp. (Nagao et al. 2015; Lang et al. 2012).

transformation, or gene transfer agents (GTAs). Characteristics of these four processes for genetic exchange are summarized in Box 1.1 and while DNA exchanges are observed in the environment, the frequency that each mechanism is used may not be apparent. It is well documented from several different environments that considerable movement of DNA between microorganisms occurs. It has been estimated that there may be 10^{13} prokaryotic gene transfers per year in the Mediterranean Sea (McDaniel et al. 2010). 1.3×10^{14} transduction events each year in Tampa Bay (Jiang and Paul 1998). and 10^{24} genes transferred by transduction each year in the oceans of the world (Rohwer and Vega Thurber 2009). The extensive horizontal exchange of DNA between microorganisms contributes to the difficulties of finding the Last Universal Common Ancestor and the rooting of the Tree of Life. Note that once DNA is transferred to a bacterial or archaeal cell, coevolution processes will determine if the host cell incorporates the DNA into its genome and if the newly acquired DNA is expressed.

Since bacteria and archaea do not have "sexual stages" in their growth processes, the mixing of the gene pool is accomplished by asexual horizontal gene transfer. There are many examples of horizontal gene transfer between microorganisms and higher biological forms and a few documented cases are given in Table 1.4. Potentially any gene in the donor cell can be mobilized but at least three classes of genes transferred between bacteria have been identified. The most common genes associated with horizontal transfer include those associated with the replication, translocation, and integration of mobile genetic elements and viruses. Additionally, genes encoding for antibiotic resistance, pathogenicity, and host–pathogen interaction are

Table 1.4 Examples of interspecies horizontal gene transfer involving microorganisms.

- From bacteria to the yeast *Saccharomyces cerevisiae* (Hall et al. 2005).
- Adzuki bean beetle, filarial nematodes, and arthropods have acquired DNA from their endosymbiont *Wolbachia* (Kondo et al. 2002).
- Pea aphids (*Acyrthosiphon pisum*) contain multiple genes from fungi (Moran and Jarvik 2010).
- *Plasmodium vivax*, a malaria pathogen, has acquired DNA from its human host and this enables it to escape defenses of host (Bar 2011).
- Genes for Shiga toxin moved from *Shigella sonnei* to *Escherichia coli* (Strauch et al. 2001).
- Genes for dissimilatory sulfite reductase have been transferred between bacteria and from bacteria to archaea (Klein et al. 2001).
- Several genes transferred from relatives of endosymbionts *Buchnera* and *Wolbachia* to pea aphid, *Acyrthosiphon pisum* (Nikoh et al. 2010).

moved between bacteria at the highest rate. At the intermediate level of mobilization are genes encoding for metabolic or structural development. Genes transferred at the lowest frequency include those dealing with transcription, translocation, or other informational processes (Keese 2008). Bacteria that have received genes enabling them to survive and grow in the presence of antibiotics or to engage in catabolism of toxic materials in the environment are often referred to as "super bugs." Horizontal gene exchanges occur between different bacteria, bacteria and archaea, and most likely between archaeal cells.

The magnitude of horizontal gene exchange between bacteria can be evaluated by comparing the core genome to the Pan-genome for bacteria (Willey et al. 2014). The core genome refers to the genes present in a species that would reflect the minimum number of genes required to enable that species to grow. Included in the core genome are the genes for transcription, translation, and replication. The Pan-genome refers the total of all genes found among all taxa and not limited to a single taxon. If the number of core genes is subtracted from the number of Pan-genes in a strain, the difference would represent the number of genes acquired to enable bacteria to colonize new niches. The examples provided by Willey et al. (2014) include the following: *Bacillus anthracis* has 3600 and 3800 in the core genome and Pan-genome, respectively. *Streptococcus agalactiae* has 1800 and 2700 in the core genome and 2700 in the Pan-genome. *Escherichia coli* has 2800 in the core genome and 6000 in the Pan-genome. This would indicate that *B. anthracis* would be limited to a few habitats whereas *E. coli* may expand into numerous habitats. Although horizontal gene exchange is a reality for bacteria, the amount of DNA exchanged varies considerably with the individual species.

1.4.1 Genetic Islands

The genes associated with horizontal gene transfer include a large segment of DNA that is 10–200 kb and which is often referred to as a genetic island (Juhas et al. 2009). Encoded in this gene cluster there are usually insertion elements or plasmid conjugation factors that mobilize this genetic information. Antibiotic resistance and virulence bacterial genes are some of the best-known genes moved by horizontal gene transfer and these discrete DNA elements are referred to as pathogenicity islands. Other discrete DNA segments may be symbiosis islands, which enhance bacterial interactions with nodulation in plants or associations with animals. Metabolic islands may describe genes for catabolism or mineral metabolism. Geochemical or geomicrobiology islands refer to a unit of genes that enable mineral transformations. Resistance or fitness islands describe those discrete DNA segments that enable bacteria to grow in toxic environments or under chemical stress.

1.4.2 Risks from Genetically Modified Organisms

With the reality that horizontal gene transfer is an ongoing process in nature, there is concern that new segments of DNA will be introduced into the environment by genetically modified organisms (GMOs). In agriculture, transgenic plants are grown and seeds from these plants are used for human consumption. The transfer of genes from these transgenic plants to humans or bacteria in the human gut has been of concern but it is highly likely that the only genes that could be transferred would be the genes of bacterial origin that were used to construct the genes for the development of GMOs. The bacterial genes of interest would be the antibiotic resistance genes used to insert the genes into the plant cells. The most common antibiotic resistance genes used for the wide transgenic development of plants are the genes encoding for kanamycin and hygromycin B. Both antibiotics are rarely used in therapeutic situations. Additionally, although there may be reports of gene transfer from transgenic plants to environmental bacteria, these are carefully controlled laboratory evaluations that have not been demonstrated in field settings (Keese 2008). Recently, transgenic salmon have been approved for human consumption and studies are needed to evaluate the potential for gene transfer from GMOs to humans. With transgenic animals, the greatest potential for gene transfer is from the viruses that were used to construct the donor DNA (Keese 2008).

1.4.3 Microbial Viruses and Gene Transfer Agents

Bacterial viruses, also known as "bacteriophages" or simply "phages," are dispersed throughout the environment and it has been estimated that there are about 10^{31-32} bacterial viruses in the biosphere. The presence of microbial viruses in various environments is indicated in Table 1.5.

Each bacterial and archaeal species is subject to phage attack and excellent reviews by Weinbauer (2004) and Suttle (2007) characterize bacterial viruses in the environment. The abundance of microbial viruses in aquatic and soil environments is determined by electron microscopy and epifluorescence microscopy using uranyl acetate staining and fluorochromes for staining nucleic acid, respectively. Electron microscopy has revealed the presence of virus-like structures in the environment with the most common structure being viruses with a head and a tail. Some bacterial phages have a long flexible tail, some have a short tail, and most

Table 1.5 Abundance of bacterial virus in the environment.

Estimates of viruses	Site of collection	Reference
10^6–10^9 ml^{-1}	Seawater	Suttle et al. 1990
10^5–10^7 ml^{-1}	Sewage effluent	Bitton 1987
10^8 mg^{-1}	Soil	Ashelford et al. 2003; Williamson et al. 2007
10^8 mg^{-1}	Cold deserts Ice-free region of Antarctica	Zablocki et al. 2016
10^5 ml^{-1}	117–292 m beneath ocean Juan de Fuca Ridge, Pacific Ocean	Nigro et al. 2017
10^4–10^9 ml^{-1}	320 m below sea floor South Pacific Gyre	Engelhardt et al. 2016
10^6–10^9 ml^{-1}	Arctic sea ice	Maranger et al. 1994
6.5×10^{10} ml^{-1}	Algal floc (marine snow)	Peduzzi and Weinbauer 1993

marine phages have a contractile tail. The phage head or capsid size is 30–60 nm with DNA-containing phages significantly more abundant than those containing RNA. As transformation and conjugation are not effective in natural settings, the process of transduction is a major mechanism for genetic transfer in most environments.

In addition to bacterial viruses, the virosphere also includes infection particles targeting cells of the Archaea domain and algal cells of the Eukarya domain (Prangishvili 2013). It has been suggested that viruses against archaeal cells should not be called "phages", the term "phage" should be reserved for bacterial viruses (Abedon and Murray 2013) due to the differences between bacterial and archaeal cells, and the domain Akamara (Greek for without chamber) should be used for archaeal viruses (Hurst 2000). As reviewed by Pietilä et al. (2013, 2014), there are over 6000 known phages but only about 100 archaeal viruses have been reported. Archaeal cells targeted by viruses include methanogens, hyperthermophiles, and extreme halophiles of the Crenarchaeota and Euryarchaeota phyla. The shapes of these archaeal viruses range from lemon, bottle, droplet to icosahedral heads, and many of the forms have tails.

Eukaryotic algae are important to provide nutrients to aquatic systems and, therefore, algae constitute an important segment of the food chain. In an unbalanced aquatic system, algal blooms occur and the decline of these blooms may be attributed to algal viruses. The virus-mediated lysis of algal cells releases nutrients into the water for use by bacteria, and the algal viruses join the phages in nutrient cycling known as the viral loop. A historical review of algal viruses is available (Van Etten et al. 1991). Recently, considerable interest has been given to algal viruses because one of these DNA-containing viruses, *Chlorovirus ATCV-1*, has been found to be part of the human virome and has been implicated in causing changes in cognitive functions in humans and mice (Yolken et al. 2014).

Viruses associated with fungi are called mycoviruses and were discovered in mushrooms over 50 years ago. Mycoviruses are broadly distributed and have been detected in Ascomycota, Basidiomycota, Chytridiomycota, Deuteromycota, and Zymomycota phyla of fungi (Son et al. 2015). Currently, most of the mycoviruses are characterized as containing dsRNA, about 30% of the mycoviruses contain positive-sense, single stranded RNA (ssRNA) and a few mycoviruses contain negative-sense ssRNA and ssDNA. Most mycoviruses do not have an extracellular route for infection and transmission of a mycovirus is dependent on cell fusion, cell division, and spore production. Therefore, to detect and identify mycoviruses, scientists rely on purification of dsRNA from fungal cultures because some mycoviruses have a genome of dsRNA but other mycoviruses produce dsRNA intermediates in their hosts. Fungi infected with mycoviruses display changes in cell pigmentation, reduced growth rate, altered sexual reproduction, and most importantly, a reduction of virulence with reduced production of mycotoxin. Considerable success has been obtained in using specific mycoviruses to control the chestnut blight fungus, *Cryphonectria parasitica*, and rapeseed stem rot by *Sclerotinia sclerotiorum*. Not all mycoviruses reduce fungal virulence but some are latent mycoviruses that have a unique effect on the host. There is one report where the symbiotic activity of an endophytic mycorrhizal fungus, mycovirus, and a grass plant enables survival at elevated temperatures when the three partners are present (Son et al. 2015). In some instances, the mycovirus is destructive to cells, as in yeast where host death occurs when the mycovirus unites with a DNA virus-like particle. Additional research is needed to clarify the role of mycoviruses in the biosphere.

The gene transfer agent (GTA) was discovered by Barry Marrs in 1994 using *Rhodobacter capsulatus*. The GTA particle consists of protein that is encoded on 15–17 genes located on the bacterial chromosome and moves dsDNA randomly acquired from the host to a recipient bacterial cell (see reviews by Lang et al. 2012; McDaniel et al. 2012). Unlike phages, GTAs do not contain DNA that encodes for GTA production but the GTA uses genes from the host cell.

Reported GTAs have a head size of about 40 nm and tail length varies from 7 to 190 nm. Unlike bacterial viruses, which carry their own DNA in the head of the infectious particle, GTAs carry 4.4–14 kb of DNA randomly collected from the host cell (McDaniel et al. 2010, 2012). GTAs have been found in several microbial strains: RcGTA from *R. capsulatus* (formerly *Rhodopseudomonas capsulatus*). Dd1 from *Desulfovibrio desulfuricans*, VSH-1 from *Brachyspira hyodysenteriae* (formerly *Serpulina hyodysenteriae*). and VAT from the archaeon *Methanococcus voltae* (Lang et al. 2012). *Roseovarius nubinhibens* ISM and *Nitratireductor* 44B9s isolated from a costal environment near Georgia, USA, were mutagenized with Tn5 (which encodes for both kanamycin and streptomycin resistance) and were found to effectively transfer dual antibiotic resistance to bacteria in environmental samples. Additional studies indicated that *R. nubinhibens*, *R. capsulatus*, *Oceanicola granulosus*, and *Ruegeria* (formerly *Silicibacter*) *pomeroyi* carried genes for GTA proteins and these were correlated with the production of 10^5–10^9 virus-like particles per milliliter (McDaniel et al. 2012). Examination of genome reports indicates that genes similar to those for the production of RcGAT are broadly distributed throughout the *Alphaproteobacteria* and that the enhanced exchange of DNA between bacteria of different species occurs (McDaniel et al. 2010, 2012; Biers et al. 2008; Lang and Beatty 2007). Using the estimated abundance of bacteria in marine waters as $1.6 \times 10^6\,ml^{-1}$ and 30% of the marine bacteria as *Alphaproteobacteria* with the potential for GTA production, suggests that the GTAs could represent 0.05% of the marine viruses. As a point of reference, the bacterial viruses in the oceans are estimated to be $4.4 \times 10^7\,ml^{-1}$ (McDaniel et al. 2012).

1.5 What Determines Which Microorganisms are Present?

When microbiologists sample different environments, the microorganisms isolated reflect the physical and chemical nature of the environment. That is, agricultural soil has different microorganisms than hot water springs and acid runoff from mines has different microbes than fresh water. However, when bacteria from agricultural soil at different sites around the world are isolated, similar types of bacteria would be expected just as bacteria from different marine settings may have similar phenotypes. The question arises "What determines which microorganisms are present in the environment?" In response to this question, Lourens Gerhard Marinus Bass Becking in 1934 proposed that "Everything is everywhere *but* the environment selects" (de Wit and Bouvier 2006). This hypothesis has been a basic component of environmental microbiology for many years. The idea is that microorganisms become airborne due to some type of physical activity and winds distribute the suspended cells over the entire Earth. Also, in fresh water or marine aquatic environments, the aquatic currents contribute to the mixing of suspended microorganisms. The physical and chemical environments where the cells are deposited select for organisms that are able to grow at those sites. However, Bass Becking's hypothesis does not explain the extent of genetic variation between the *Synechococcus* spp., a cyanobacterium, in hot springs (Pake et al. 2003) or the archaeon *Sulfolobus* spp. (Whitaker et al. 2003) that are geographically separated. Although the phenotypic characteristics of microbial communities in surface ocean waters are correlated with environmental features and not physical distance (Raes et al. 2011; Jiang et al. 2012). microbial biogeochemistry in extreme environments may be influenced by microbial dispersion with genetic selection as well as geographical isolation (Lau et al. 2014).

There are many chemical and physical environments that are highly restrictive and select for a unique type of microorganism. In environments containing high levels of acid, alkali, or salt, or areas containing toxic metals, the absence of one or more nutrients selects for a specific type of microorganism. At the bottom of oceans, the hydrostatic pressures are extreme and only barophilic bacteria can grow. Recognition sites and molecular interactions select for host-bacteria in plant symbiosis systems. From ecological studies of microbes associated with humans, it is apparent that there are several selective pressures determining which bacteria are present on the skin, ears, mouth, and intestine. And so, one could identify many selective pressures in the environment and, rather than provide an exhaustive list of them, a few examples of how microorganisms specifically respond to these selective events are given here:

- Temperatures of 80–113 °C select for hyperthermophilic archaea that have chromatin proteins (architectural histones and DNA topoisomerases) that stabilize DNA against denaturation due to heat (Visone et al. 2014).
- Acidophilic species of archaea are found in the genera *Sulfolobus*, *Thermoplasma*, and *Acidianus* and resist acid denaturation of proteins.
- Extremely acidic environments (pH 1) select for bacterial species of *Acidithiobacillus* that have unique membranes and special cellular processes for pH homeostasis (Mangold et al. 2013).
- Environments containing 3–5 M NaCl or concentrations approaching saturation levels for salt, select for halophilic archaea that actively pump K^+ and Cl^- into the cell whereas other halophiles, including the moderately halophilic bacteria, use compatible solutes to stabilize osmotic pressures. Additionally, the presence of a unique phospholipid in extremely halophilic archaea (for example *Halobacterium halobium*) helps to maintain membrane stability (Tenchov et al. 2006).

In those areas where there is a rich source of nutrients, the selective pressures are low and hundreds of different species of microorganisms abound. For example, rich agricultural soil contains an abundance of bacteria and fungi representing numerous species whereas nutrient-limited soils have relatively fewer microorganisms.

1.5.1 Metabolism as a Basis of Selection

An abundance of a specific microorganism may reflect the rate of metabolism that is characteristic of the cells present. As bacteria grow and divide faster that fungi, bacteria outcompete fungi for sugars and soluble nutrients in the environment. For this reason, bacteria are the primary microorganisms in the decomposition of mixed organic materials. Systems that regulate gene expression in bacteria and account for their dominance in an environment are reviewed in Box 1.2. Since nutrient uptake is based on solution chemistry, where soluble nutrients are transported into the cell, insoluble or particulate substrates are used slowly once they have broken down into soluble components. Bacteria such as *Mycobacteria* have generation times of many hours and they can persist in the environment because they have wax-containing cell walls that are extremely resistive to enzymatic attack. Lichen colonies growing on rock faces may increase in diameter by only a few millimeters, but they are not in danger of being overgrown because they are the only microorganism growing there. In sites contaminated with toxic heavy metals, the only bacteria that grow there are those that carry the genes and express resistance to toxic metals. In some environments, the predominant organisms present are the ones that have the metabolic capability of exploiting the toxic compounds to the benefit of the bacteria. Those bacteria with the capability of growing either aerobically or anaerobically have an advantage in environments where nutrients are not limiting. Initially, bacterial metabolism is driven by aerobic respiration but due to the rapid consumption of oxygen and the slow replenishing of oxygen from the air, the environment becomes anaerobic with anaerobic metabolism favored. Selection of bacteria in an environment is, in part, dependent on the metabolic capabilities of the bacteria present and the graphic in Figure 1.4 reflects some of these metabolic capabilities. The physical and chemical conditions of a site may be extremely changeable so metabolic flexibility (see Box 1.2) enables bacteria to persist and grow.

Box 1.2 Metabolic Quick Response by Bacteria Conserves Energy and Promotes Cell Dominance

Years of basic research with heterotrophic bacteria have revealed important features of genetic expression. The expression of many bacterial genes is regulated, which is in contrast to constitutive gene expression resulting in protein and enzyme production. Regulated gene expression may be subject to either induction or repression. In a nutrient-limited environment, induction of gene expression enables bacterial cells to overcome the deficiency of an essential nutrient. Most commonly this induction of hydrolytic enzymes occurs when an organic carbon substrate is added to a carbon-limited environment. However, limitation of soluble iron, phosphate, or nitrogen may result in induction of genes to produce siderophores, phosphatases, and nitrogenase, respectively. In bacteria, the time required for induction of gene expression is only a few minutes. When bacteria growing in a dilute nutrient environment are provided with amino acids and vitamins, the genes encoding for biosynthetic pathways of amino acid and vitamins are immediately repressed. Thus, induction and repression of gene expression regulates the synthesis of enzymes to accommodate the nutritional needs of the bacterial cell. As about 50% of the energy expended by a growing bacterial cell is used for protein synthesis, the regulation of gene expression enables the cell to become energy efficient and divert cellular energy into growth. A mechanism for this extensive regulation of induction and repression is the polycistronic character of the bacterial mRNA. Induction and repression activities impact transcription and are often referred to as the "coarse controls" of protein synthesis. "Global regulation", another mechanism for regulating genetic expression in bacteria, involves alternate sigma factors for the expression of genes that are normally silent. As a response to chemical or physical signals, several genes can be induced by a single environmental stimulus. This control is highly efficient and enables a bacterial cell to respond quickly to changes in the environment. Examples of global regulation in bacteria include metabolic adjustments for aerobic respiration, anaerobic respiration, heat shock, and nitrogen utilization. Through extensive regulation of gene expression, bacteria are capable of outcompeting fungi and yeast in the environment, which helps to explain why certain bacteria are the dominant microorganisms in a specific environment.

1.5.2 Is Persistence of Microorganisms Dependent Only on Spore Production?

It is important for bacteria and archaea to have a process that enables them to persist in environmental conditions that are not conducive to their growth. The maintenance of bacterial strains has long been dependent on laboratory manipulations, such as slow drying in which a drop of a bacterial culture is placed on top of sterile dry soil or sand to produce dried cells that remain viable for a long time. In hospital wards, it is known that *Staphylococcus aureus* infections are passed on to other hospital patients when bacteria in body fluids are dried onto bed linen. *Mycobacterium tuberculosis* can be cultivated several days to weeks after sputum samples are deposited on a sidewalk by a person with tuberculosis. Bacteria that are slowly dried and associated with proteins or other biomaterial may retain their viability for some time.

Resistive structures are associated with bacteria that are able to survive harsh and stressful environments not conducive for growth. For example, the unique bacterial endospore is remarkable as it can resist extremely high temperature and is found only in bacteria. These structures have been designated as endospores because they are formed inside the mother cell. Although most introductory microbiology classes include aerobic *Bacillus* and anaerobic *Clostridium* as sporing genera of bacteria, there are over 90 genera of bacteria reported to produce an endospore. The outer layers and coats of endospores are important in protecting them from reactive and toxic chemicals. The hallmark characteristic of the

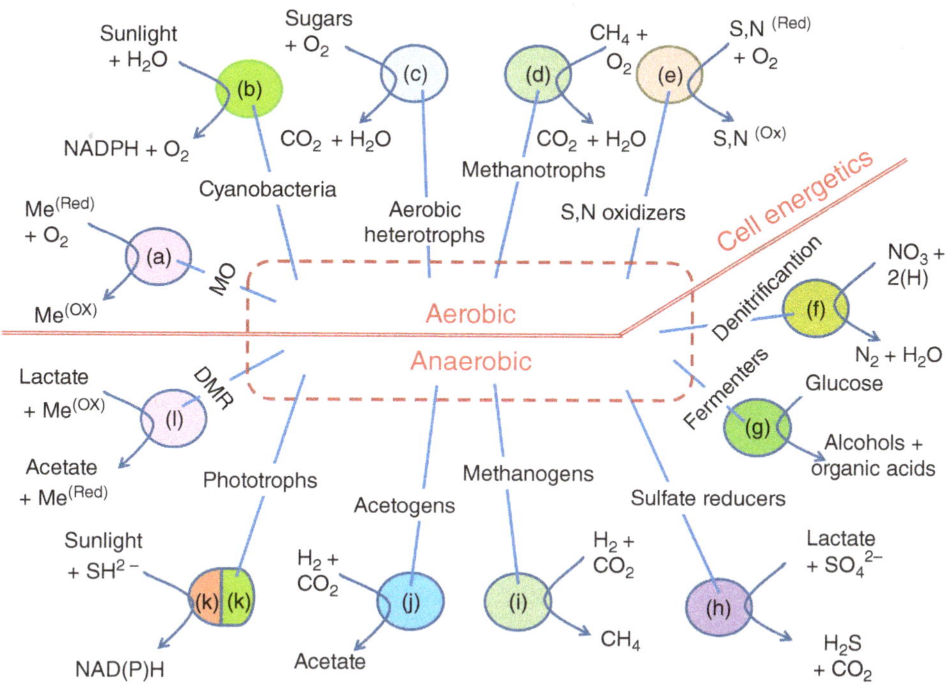

Figure 1.4 Distinctive features of several bacteria: (a) MO = aerobic metal oxidizers, most commonly includes Fe and Mn; (b) chlorophyll-mediated oxygenic respiration by cyanobacteria; (c) aerobic respiration used by microbes for oxidation of sugars, amino acids, and other organic compounds; (d) aerobic oxidation of methane by methanotrophs and methanol by methylotrophs; (e) aerobic oxidation of reduced sulfur or reduced nitrogen compounds; (f) dissimilatory nitrate reduction using anaerobic respiration; (g) microbes lacking cytochromes that use glucose and other organic compounds as substrates and obtain energy by substrate-level phosphorylation; (h) dissimilatory sulfate reduction with lactate and other electron donors with the production of small molecular weight organic compounds as endproducts; (i) unique anaerobic metabolic pathway for reduction of carbon dioxide or acetate to methane and coupling energy production to this process; (j) production of acetic acid by the Wood–Ljungdahl pathway; (k) anaerobic photosynthetic bacteria using reduced sulfur compounds as the electron donor using red or green bacteriochlorophyll molecules to harvest light; (l) DMR = dissimilatory metal reduction where electrons from a donor are directed to oxidized metal ions, which may be Fe, Mn, As, or metalloids of Se and Te.

endospore is a large layer (the cortex) filled with calcium ions (Ca^{2+}) chelated to dipicolinic acid, which withdraws all free water from the interior of the spore to make it heat resistant. Without water in the endospore's protoplasm, heat cannot damage the spore. Spores are found in other forms of life; however, they do not have a cortex containing calcium dipicolinate and are denatured by heat. Thus, it is worth restating that the bacterial endospore is the most resistive living unit.

The statement often made is that one cell produces one spore and this is not a multiplication process. However, there are at least two exceptions to this statement. *Anaerobacter polyendosporus*, an unusual anaerobic bacterium, produces up to five oblong endospores in each cell (Duda et al. 1987). A second unusual bacterium, *Metabacterium polyspora*, does not usually use cell division for multiplication but it produces four endospores per cell (Angert and Losick 1998). In *M. polyspora*, a prespore develops at the ends of the rod-shaped cell, and each prespore develops into two spores, so at least two endospores are produced from a single cell but as many as eight or nine endospores have been recorded (Angert and Losick 1998).

Endospore formation is not known to occur in archaea, but it is found in both gram-positive and gram-negative bacterial cells: however, there are only two examples of gram-negative bacteria producing endospores. *Crenothrix polyspora*, a sheathed bacterium that was isolated from water pipes, stains gram-negative and ultrastructure examination confirms the gram-negative cell wall (Völker et al. 1977). Another highly interesting bacterium that tolerates ethanol and produces endospores is the gram-negative *Brevibacillus* sp. (Thomas 2006). The formation of a bacterial endospore is not always the ideal event for persistence of a species because in nutrient-limited environments, endospore production may be an end-of-line event. In deep subsurface environments (see Chapter 3). bacterial endospores are unable to obtain sufficient nutrients for germination and, therefore, the long-term survival of bacteria is attributed to the ability of dormant vegetative cells to become active when nutrients become available.

Many bacteria produce restive structures which are differentiated cells that enable bacteria to survive winter frost or hot and dry environments. These specialized resistive cells do not contain calcium-dipicolinate and, therefore, are not resistant to high temperatures. The specific structures and bacteria that produce them are as follows: basidiospores, myxobacteria with fruiting bodies; akinetes, cyanobacteria; conidia (also called spores). streptomyces; and cysts, gram-negative soil bacteria. Cellular developments leading to the production of spores by *Streptomyces griseus* and *Myxococcus xanthus* are described in Box 1.3 and Figure 1.5. The development of resistive structures in cyanobacteria and *Azotobacter vinelandii* are described in Box 1.4. The formation of endospores by *Bacillus subtilis* is described in Box 1.5 and Figure 1.6. All these resistive structures characteristically display reduced metabolic activity and do not require a continuous supply of nutrients.

Box 1.3 Developmental Morphogenesis in *Streptomyces* and *Myxococcus*

Bacteria	Characteristics
Streptomyces griseus	Vegetative cells are branched and grow at the tip of the cell forming dense subsurface mycelial growth. *Streptomycetes griseus* produces the signaling molecule known as A-factor (γ-butyrolactone) which activates the AdpA regulon. A regulatory cascade proceeds with inductions which leads to the expression of a new sigma factor and this new need sigma factor reads silent genes and this results in morphological differentiation. The surfactant peptide SapB is produced which lowers the surface tension of the air-water interface and enables the emergence of aerial hyphae from the surface mycelium (see Figure 1.5). At the tip of the aerial hyphae is an elongated sporogenic cell that produces over 50 copies of genomic DNA. The accumulated genomic DNA is partitioned into prespore foci and septum formation synchronized with cell wall synthesis finalizing the development of the spore (Flärdh and Buttner 2009).
Myxococcus xanthus	The myxobacteria are found in the soil and a cluster of vegetative cells move by gliding across the solid surface. The cells obtain nutrients from the environment by a collective process that releases extracellular enzymes to generate nutrients from polymeric materials in the environment. Cells of *M. xanthus* are known to move as a group and digest bacterial microcolonies in their path (see YouTube cited at the end of this chapter.) With the depletion of nutrients in the environment, the stringent response induces fruiting body formation through a series of complex signal transduction systems. The intercellular signals have an influence on the fate of the aggregate of bacterial cells, with 10% becoming differentiated to produce spores, 30% remaining at the base of the fruiting body, and the remainder being lysed to provide nutrients to support the differentiation process (Claessen et al. 2014). Some refer to this use of bacterial cells for nutrients as a cannibalization process and others liken this to programed cell death. The myxospores are released from the fruiting body as the structure dries and are distributed to new environments on air currents.

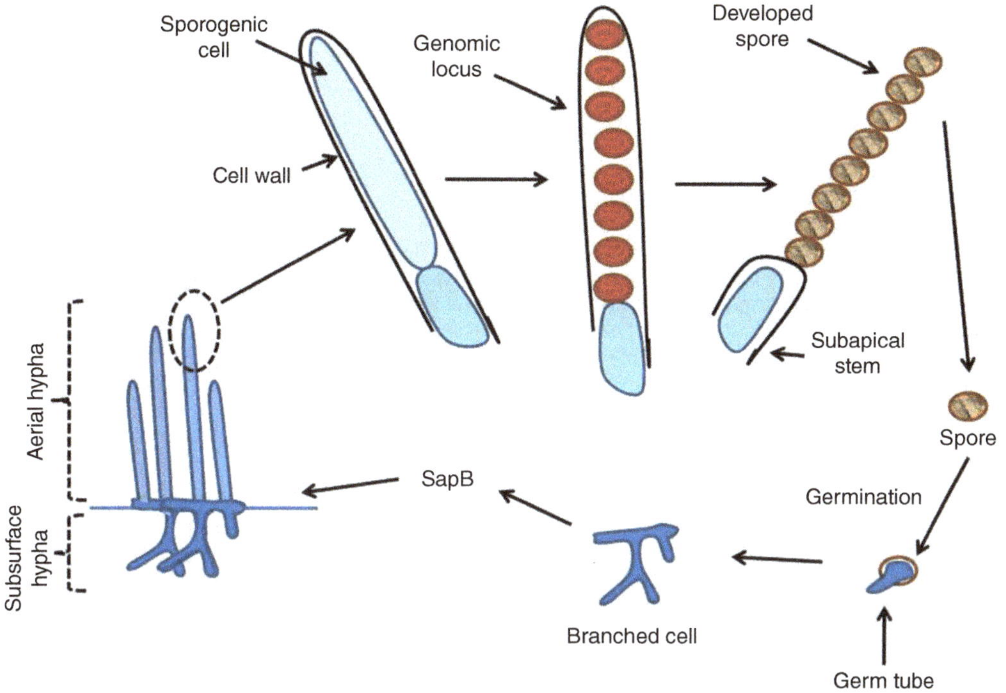

Figure 1.5 The spore cycle of *Streptomyces griseus* involves anatomical changes of aerial hyphae supported from subsurface mycelium. For additional information see Flärdh and Buttner (2009).

Box 1.4 Developmental Morphogenesis in Cyanobacteria and *Azotobacter*

Bacteria	Characteristics
Cyanobacteria	*Anabaena*, *Aphanizomenon*, *Gloeotrichia*, and *Nostoc* are the genera of oxygenic phototrophic cyanobacteria commonly encountered in the environment. The location of the akinetes in the trichome of cells varies with the species of cyanobacteria but all akinetes have a thickened cell wall and a multilayered extracellular envelope. The goal of akinete formation is to maintain the species through harsh environmental conditions. Akinetes may be located at the end of the trichome, adjacent to a heterocyst or randomly placed in the trichome. Akinetes produced by several different genera contain glycogen and cyanophycin, which serve as carbon and nitrogen sources, respectively, for germination. The mechanisms triggering akinete formation may be linked to metabolic events and is usually linked to diminished light intensity. However, sometimes formation of akinetes has been attributed to limiting levels of phosphate, inorganic carbon, inorganic nitrogen, or iron (Kaplan-Levy et al. 2010).
Azotobacter vinelandii	Several soil bacteria including *A. vinelandii* produce resting cells known as bacterial cysts. These cysts resist desiccation and remain viable for decades; however, cysts are not as resistant as bacterial endospores. AlgU and RpoS are alternate sigma factors which initiate cell differentiation resulting in cyst formation in *A. vinelandii* (Cocoti-Yañez et al. 2011). When growth becomes nutrient limited, cytoplasmic signals induce AlgUY to promote alginate synthesis and repress genes needed for flagellum synthesis. RpoS is an additional signal for transcription of the gene for alginate synthesis and activates one of the promoters for poly-β-hydroxybutyrate synthesis. The cell accumulates a considerable quantity of poly-β-hydroxybutyrate before differentiation and phospholipids in the cell membranes are replaced with 5'alkylresorcinols, which moderate oxidation of membranes and fatty acids (Funa et al. 2006). The cyst has thick layers consisting of lipopolysaccharides and lipoproteins, which bind toxic chemicals and prevent them from entering the cyst.

Box 1.5 Developmental Morphogenesis and Endospore Production by *Bacillus subtilis*

Bacillus subtilis

Endospore formation occurs in a few bacterial genera including: *Bacillus*, *Desulfotomaculum*, *Sporolactobacillus*, *Clostridium*, *Thermoactiomycetes*, and *Sporosarcina*. Most research on sporulation mechanisms has been with *B. subtilis*. Generally, nutrient limitation and production of quorum-sensing molecules triggers sporulation. Differentiation from a cell to an endospore proceeds through several morphologically distinct stages and requires a considerable metabolic effort. An overview of the sporulation cycle is given in Figure 1.6. The free endospore germinates with the production of a vegetative cell. In unfavorable environmental conditions, chemical signals are processed by the vegetative cell and DNA is segregated at one pole of the cell and the forespore develops. The Z-ring is placed near the forespore region and asymmetrical cell division occurs to produce a prespore and a mother cell. Development of the spore proceeds with the establishment of the cortex, which consists of calcium ions chelated to dipicolinic acid. The high concentration of calcium surrounding the developing spore removes water from the spore and accounts for thermal resistance of the endospore. A protein coat is added and the mother cell is lysed to release the spore.

Many genes that are not read under standard growth conditions are expressed by alternate sigma factors and allow the sporulation process to proceed (Higgins and Dworkin 2012). The first alternate sigma factor (σ^F) is activated in the forespore and accounts for the expression of about 50 genes, including *SpoIIR*. The SpoIIR protein crosses the membranes and stimulates the production of σ^E in the mother cell. The activated sigma factor σ^E induces σ^G in the mother cell and the active σ^G induces production of σ^K in the mother cell. Each alternate sigma factor has a specific region where gene expression is stimulated and their locations are presented in Figure 1.6. With a reassignment of nutrients in the mother cell to the developing spore, cell death occurs and the endospore is released following lysis of the mother cell.

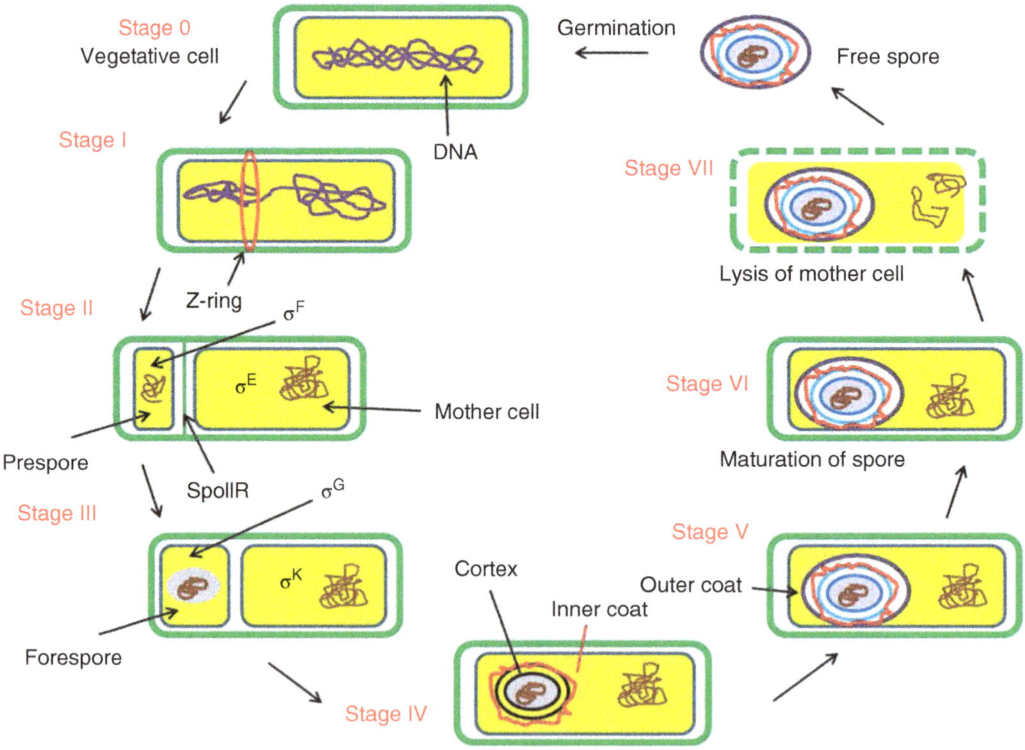

Figure 1.6 Activity associated with the different stages of sporulation in *Bacillus subtilis*.

1.6 Is the Size and Shape of a Prokaryotic Cell Important?

Much of our understanding of bacteria in the environment comes from bacteria that are isolated and grown in the laboratory. It is apparent that the microorganisms grown in the laboratory represent only 0.1–1% of those present in the environment. Although cells of bacteria and archaea have various shapes, the rod, coccus, and spiral shapes are most commonly discussed. Examples of bacteria with different shapes are given in Figure 1.7. Although there are no definite answers about why bacteria have a specific shape, it is a thought provoking question. Young (2006, 2010) has considered the shape of bacterial cells and has proposed that it is not accidental but provides the species with an advantage for survival. He reasons that cell form is an important physiological character because: (i) cell shape is consistent within a species (ii) cell shape may change with the development of specialized structures, and (iii) changes in cell shape are consistent with evolutionary progression. Cell size and shape has long been used for identification of bacteria and temporary changes in cell size are reproducible. The length of bacterial cells changes through the growth phase, but cell length is constant in the stationary phase. Cells may display dimorphic or polymorphic changes due to the environment or the available nutrients. Change in cell size is observed with stalked bacteria (*Caulobacter*) or fruiting bacteria (*Myxococcus*) and in cell shape with the development of heterocysts by cyanobacteria and of conidia by actinomycetes. Several laboratories have evaluated the evolutionary progression of cell shape and have concluded that the rod form may be the first cell form. The coccus is a degenerate form that has developed independently several times from rod-shaped cells (Young 2006). Although it is not always clear what environmental advantage the various cellular forms of bacteria (Table 1.6) and archaea gain, it is clear that cell shape is under genetic control.

1.6.1 Nanobacteria

The general perception is that prokaryotic cells are a few millimeters wide with spherical, linear, or curved shapes. However, prokaryotic cells vary greatly in size and structure. Prokaryotic cell diameters range from 0.2 to 750 μm, with a major factor limiting cell size being

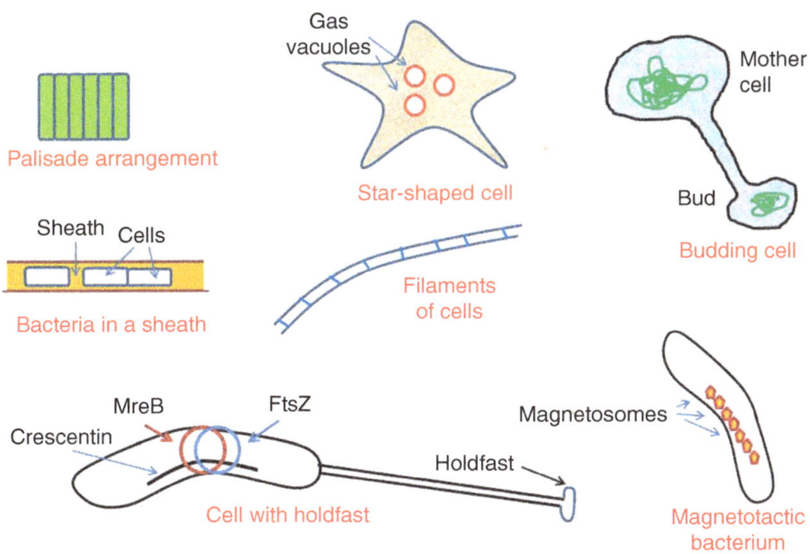

Figure 1.7 Characteristic cellular features of selected bacteria.

Table 1.6 Unusual shapes characteristic of some bacterial and archaeal cells.

Cell forms	Prokaryote		Features
Square	*Haloarcula quadrata* *Haloquadratum walsbyi*	(A)[a] (A)	Flat halophilic cells 2–5 μm wide and 0.1–0.5 μm thick. Floats on water
Triangle	*Haloarcula quadrata*	(A)	Halophilic cell that may grow in sheets of 65 cells forming triangles or squares
Star	*Stella vacuolata*is	(B)[a]	A flat Gram-negative cell with 5 or 6 points found in water and soil
	Prosthecomicrobium	(B)	Numerous cellular appendages per cell; isolated from fresh water
Pleomorphic	*Ancalomicrobium adetum* *Mycoplasma pneumonia* Numerous soil bacteria	(B) (B)	Size, shape, and arrangement of cells is not fixed but varies. May be characteristic of species or induced by changes in nutrition
Bifid	*Bradyrhizobium japonicum* *Roseibium denhamense* *Bifidobacterium* spp.	(B) (B) (B)	Polar localized complexes may lead to Y-shaped cell
Budding	*Hyphomicrobium* spp.	(B)	Aquatic bacteria that have a fibrillar filament extending from a rod cell and a bud develops at the end of the filament which is some distance from the cell
Branched bacteria	*Actinomyces israelii* *Arthrobacter globiforms*	(B) (B)	Pseudo or false branching of cells often induced by nutrient limitation
Fruiting body bacteria	*Myxococcus xanthus* *Sorangium cellulosum*	(B) (B)	Cells aggregate to form aerial structure containing myxospores

[a] A = Archaea, B = Bacteria.

the amount of space required for DNA plus related protein synthesis. It has been calculated that minimal cell size would be about 0.2 μm; these small cells are often called nanobacteria or nanoarchaea. Nanobacteria have been reported on meteorites and in various human tissues (McLean and Kirkland 2014) but the biological versus abiotic nature of these nanostructures is highly controversial (Young and Martel 2010). DNA or some other biomarker must be extracted from nanostructures before they can be designated as nanobacteria. An alternative plausible explanation of nanostructures is that they are fragments of microorganisms or aggregates of organic and inorganic minerals (McLean and Kirkland 2014).

Some small microorganisms that have been cultured in the laboratory could be in the category of nano-microorganisms. Some of the smallest bacteria include *Mycoplasma genitalium* (Daley et al. 2014) and *Buchnera aphidicola* (Pérez-Brocal et al. 2006). which have a cell size of 0.05 and 0.3 μm, respectively. *Pelagibacter ubique* is the most abundant bacterium on the Earth with 2.4×10^{28} cells. *Pelagibacter ubique* is a heterotrophic organism found in the ocean waters and its cells are 0.37–0.89 μm long and 0.12–0.20 μm wide (Rappé et al. 2002). There are several archaea including *Thermodiscus*, *Pyrodictium*, and *Nanoarchaeum equitans* (Jahn et al. 2008) with cell diameters of 0.4 μm that could be classified as nanomicrobes. *Mycoplasma genitalium*, *B. aphidicola*, and *N. equitans* are examples of nanomicrobes that have insufficient DNA to code for the enzymes needed to grow independently and so rely on nutrients from a host cell for growth.

1.6.2 Ultramicroscopic Bacteria

Another group of small bacteria are the ultramicroscopic bacteria (UMB). or "dwarf bacteria," which may be transiently small and so they should not be considered as nanobacteria. Gram-positive and gram-negative types of bacteria may decrease their cell size when undergoing nutrient starvation. These bacteria are smaller than normal and will pass through bacteriological filters with a pore diameter of 0.2 μm (Haller et al. 2000). When nutrients are provided to the UMB, the individual cells return to their original size.

Ultramicroscopic bacteria are widely distributed in soil as well as in marine waters. While bacteria are traditionally retained on filters with pore diameters of 0.45 μm, much of the heterotrophic metabolism in the deep sea is attributed to bacteria that pass through a 0.45 μm pore. When deep-sea waters from various regions were examined, species of *Alcaligenes*, *Flavobacterium*, *Pseudomonas*, and *Vibrio* were grown from water that flowed through filters with a 0.45 μm pore diameter (Tabou et al. 1981). Microorganisms from prairie, forest, or agricultural soils passed through filters with a 0.45 μm pore diameter and were identified as species of *Acidobacteria*, *Gemmatimonadetes*, *Crenarchaeota*, *Verrucomicrobia*, and *Elusimicrobia* (Portillo et al. 2013). The cell volumes of average-sized cells of *E. coli* and *Vibrio natriegens* are 0.9–3.5 μm^3, respectively, whereas the mean cell volumes of bacteria in the North Sea or Resurrection Bay, Alaska are 0.042 and 0.074 μm^3 (Schut et al. 1993). Similarly, the cell volumes of bacteria in New Port River estuary, North Carolina, and Gulf of Mexico were less than 0.1 μm^3 (Palumbo et al. 1984). As nutrients become limiting, bacterial cell diameter in many environmental settings becomes smaller and UMB become common.

1.6.3 Very Large Bacteria

A few gram-negative bacteria have a markedly greater size than *E. coli*, which is about 1.8 μm wide and about 7 μm long (Schulz and Jorgensen 2001). The larger bacteria include *Achromatium oxaliferium*, a rounded cell found in fresh water and marine waters, with a diameter of about 100 μm and *Spirochaeta plicatilis*, a spirochete found in brackish water, that is 0.75 μm wide and 250 μm long. An extremely large bacterium *E. fishelsoni* is found only in the intestine of brown surgeonfish in the Red Sea and is 80 μm wide and 600 μm long. The largest of all bacteria is *Thiomargarita namibiensis* which is 300 μm wide and 750 μm long. Smaller bacterial cells have a high surface to volume ratio, which favors diffusion across the cytoplasm. A very large cell lacking internal organization would have difficulty maintaining a distribution of nutrients throughout the cell because the diffusion rate is slow; however, *T. namibiensis* has a large central vacuole and the metabolically active segment of the cytoplasm is close to the periphery of the cell, so diffusion is not a problem. *Epulopiscium fishelsoni* has numerous intracellular membrane structures that provide some cellular organization (Bresler et al. 1998).

Several sulfur-oxidizing filamentous bacteria are exceedingly large. A working distinction is made with macrobacteria that are 1–10 μm wide with filaments that can be 10 μm to several hundred micrometers long. Megabacteria have a diameter greater than 10 μm and the filaments may be hundreds of micrometers to a millimeter long (Fonseca et al. 2017). Representatives of the megabacteria include *Candidatus* Maribeggiatoa vulgaris, *Candidatus* Isobeggiatoa divolgata, *Candidatus* Parabeggiatoa communis, *Thioploca ingrica*, *Candidatus* Thiomargarita nelsonii Thio 36, *Candidatus* Thiomargarita nelsonii Bud S10 and *Candidatus* Marithrix sp. Examples of macrobacteria would include *Beggiatoa alba*, *Beggiatoa leptomitiformis*, *Beggiatoa* sp. 35Flor, *Candidatus* Thiolava veneris and *Candidatus* Venteria ishoeyi (Fonseca et al. 2017). If these sulfur oxidizers use nitrate as the electron acceptor, a large vacuole containing nitrate occupies most of the cytoplasm.

1.6.4 Influence of Diffusion on Bacterial Cell Form

For decades it has been held that cell size was a function of surface to volume ratio with small cells having the greatest rate of diffusion. However, bacterial cells do not acquire nutrients exclusively by diffusion but employ numerous energy-driven transporters in the plasma membrane for nutrient uptake. Prokaryotic cells do rely on diffusion inside the cell for proteins and substrates to interact. Based on theoretical calculations, a sphere of 250–300 nm is required to contain the nucleic acids, proteins, and various cellular constituents required for a free-living prokaryotic cell (National Research Council Space Studies Board 1999). For diffusion, it has been calculated that it would take 0.5 s for a 50 kDa protein to move from the surface into the center of a cell of a rod shaped cell (0.8 μm by 4.8 μm) and 5 s for a protein to move from pole to pole (Beveridge 1988). Schulz and Jørgensen (2001) calculated that in one millisecond a small molecule could be distributed throughout a spherical cell (1 μm in diameter) and proteins could be dispersed in 10 ms. When rod shaped cells with a fixed width elongate before division, only a small change in the surface to volume ratio occurs, and it is only when they form filaments that there is a substantive increase in the surface to volume ratio. Curved shaped cells have the same surface to volume ratio of rod forms. Young (2006) maintains that if the surface to volume ratio was the determining factor for microbial cell shape then they would all have a thin disk-like shape or long appendages. A few bacteria and archaea have appendages and, except for a few archaea, flat disk-like cells are not present in the prokaryotic world. Thus, the shape of the prokaryotic cell may influence the interaction between the cell and the nutrients and this may be more important than the rate of nutrients uptake (Young 2006).

1.6.5 Features of a Specific Cell Form

1.6.5.1 Coccus Form

Spherical cells are found in all environments and are broadly dispersed throughout the phylogenic trees of bacteria and archaea. Some microbiologists have proposed that the coccus form is an evolutionary "dead end" and it was derived from cells with a rod shape. The rate of diffusion of nutrient uptake for a coccus-shaped cell is less than for a rod-shaped cell of the same diameter. If the precursor rod had a flagellum, the loss of flagella on coccus cells probably also occurred at an early point in their evolution. Flagella on cocci would not be effective for migration of cells from one region to another because a flagellated coccus would have a circular motion. Spherical cells are subject to sheer forces (current of flowing liquid) in aquatic environments and would be quickly dislodged from exposed surfaces. Additionally, spherical cells are unlikely to glide as a raft of cells. In nature, coccus forms are likely to penetrate small spaces and grow in regions where movement is not an issue. *Escherichia coli*, *Arthrobacter*, *Acinetobacter*, *Rhodococcus*, and other bacteria with a rod form often assume a coccus form after prolonged incubation. Bacteria such as *Bifidobacterium* may have asymmetric cell division to produce a coccus and a rod (Young 2006).

1.6.5.2 Rod Form

Cells with rod morphology are capable of directional movement using flagella in low viscosity media or of gliding in rafts on moist surfaces. Sheer forces have a minimum effect on a rod-shaped cell attached to a surface because of the flow dynamics associated with a rod shape and the numerous points of attachment between the linear cell and the surface. Rod-shaped cells may grow as filaments when cell division is incomplete so several cells remain attached at the poles to produce a long cellular unit. Filament formation has the advantage of expanding the number of sites where nutrients can be acquired as well as providing a defense against protozoan attack. Rod-shaped cells may align themselves side-by-side to produce a raft or a palisade of cells that can move across a surface. Gliding rod cells are important in the development of the fruiting structure of *Myxococcus* and swarming of *B. subtilis* is displayed with cells as a palisade and not with a single cell. Often cells of a specific length can participate in swarming or gliding, and this is a selective feature that promotes uniformity of cell length in a bacterial species. To accommodate bacterial specificity for attachment, rods have adhesive structures localized at one pole (Young 2006).

1.6.5.3 Curved Rod or Spirochete Form

There are many examples of bacteria that have a curved form. The difference between a linear rod and a curved rod may be the presence of a protein aligned along the inner side of the curvature. In *Caulobacter*, the presence of crescentin prevents the insertion of new peptidoglycan units into the cell wall, which results in a curved cell. Curved rods propelled by flagella can move through highly viscous muds or mucus environments. The helical path of curved cells is superior to that pursued by motile cells of rod form. Many marine cells are curved or vibroid – including the curved *P. ubique*, the most abundant bacterium in the sea – although the reason for this remains unclear. Spirochetes are propelled by an endoflagella system, which accounts for their corkscrew-like rotation through viscous environments. Spirochetes are common in aquatic systems and these cells can flex back and forth, which seems important to deter parasitic protozoa.

1.6.5.4 Unusual Forms

Some of the unusual shapes displayed by species of bacteria and archaea are presented in Table 1.6. These unusual forms are more commonly associated with gram-negative bacteria than with gram-positive bacterial cells and special advantages may be associated with some of these cell configurations. While some microbiologists consider bacterial form to be an accidental occurrence, others are interested in assessing what possible environmental factors could have influenced bacterial and archaeal cell shape. Although most microbial cells adhere to a specific form, some species are pleomorphic, changing shape in different media and environments. Additionally, various species have surface structures, many of which are for acquiring nutrients or for enabling cells to remain at a specific site (Table 1.7).

1.7 Microbial Predation

1.7.1 Bacteria as Prey

Forces important for driving bacterial diversity, cell size, distribution, and dominance include nutrient restrictions, protozoan predation, and bacterial viruses (Thingstad 2000). Protozoa are the nemesis for bacteria and the protozoan feeding on bacteria has been termed bacterivory. About 50% of marine bacteria are killed by grazing protists and bacteriophages; however, this activity is highly dynamic because grazing protozoa are also being preyed upon by metazooplankton. Crustaceans, rotifers, and ciliates ingest bacteria but do not appear to significantly decrease the bacterial biomass in freshwater ecosystems. Protozoa preying on bacteria may be suspension feeders, raptoral feeders, or grasping feeders and they all seem to follow the "Goldilocks approach" (Young 2006). In the story, Goldilocks found some choices were "too large" or "too small" but some were "just right." As with protozoa, each predatory species tends to select bacteria of a certain size (the "just right" size) often leaving a bimodal population of bacteria consisting of cells that are "too large" or "too small". The response of bacteria to avoid becoming prey has been the formation of biofilms, irregular surfaces, or long cellular filaments (prosthecae). It is clear that bacterivory is a selective pressure for microbial cell size and form.

Table 1.7 Structures associated with surfaces of microbial cells.

External structures	Genus		Features
Sheath from viruses	*Leptothrix discophora* *Sphaerotilus natans*	(B)[a] (B)	Straw-like structure with cells inside, collects nutrients from water, protects bacteria and grazing protozoa
Prosthecae	*Ancalomicrobium adetum* *Pedomicrobium* sp.	(B) (B)	Appendages extending from aquatic Gram-negative cells that expand cell surface for nutrient uptake
Holdfast	*Caulobacter crescentus* *Asticcacaulis* spp.	(B)	Stalks extending from cell that can bind to surfaces in the environment
Cannula	*Pyrodictum abyssi*	(A)	Extracellular matrix of hollow tubules connecting cells; function unknown but may help in suspending cells
Nanowires	*Shewanella oneidensis*	(B)	Conduit for extracellular electron transfer between cells or between cell and environment
Cellulosomes	*Acetivibrio cellulolyticus* *Ruminococcus flavefaciens* *Clostridium thermocellum*	(B) (B) (B)	Multienzyme complexes on surface of bacterial cells for digestion of cellulose
Pili	Most Gram-negative bacteria		Bind bacterial cell to target tissue or environmental structures, sex pilus for DNA exchange in conjugation, Type IV pili for twitching movement
	Streptococcus agalactiae *Corynebacterium diphtheria* Other Gram-positive bacteria	(B) (B)	"Sortase" for colonization and stabilization of biofilm

[a] A = Archaea, B = Bacteria.

1.7.2 Bacteria as Trackers and Predators

Some bacteria actively pursue algae for nutrients. The gram-negative bacteria *Pseudoalteromonas haloplanktis* and *Shewanella putrefaciens* can track the motile algae *Pavlova lutheri* in an apparent attempt to acquire nutrients released from the algae (Barbara and Mitchell 2003). When not exposed to algae, these two marine bacterial cells move at a rate of $126\,\mu m\,s^{-1}$ but accelerate to $445\,\mu m\,s^{-1}$ when tracking *P. lutheri* (Barbara and Mitchell 2003).

Some bacteria are facultative predators that do not search out specific microorganism but secrete huge quantities of hydrolytic enzymes which digest microorganisms in their immediate environment. These are gliding bacteria in the soil that may encounter bacteria or fungi in a process that is random but lethal. Swarming cells of *Myxobacteria*, often referred to as wolf packs, chemically sense the prey and move over colonies and obtain nutrients from the bacteria they digest (see the film released by JE Berleman – *M. xanthus* preying on an *E. coli* colony You Tube.) There are about two dozen species of *Lysobacter* that produce extracellular hydrolytic enzymes including multiple β-1,3-glucanases and chitinases, which are ideal for the hydrolysis of fungal cell walls. Since *Lysobacter* has been identified as a bacterium capable of digesting phytopathogenic fungi, there has been an interest in developing a biocontrol system that uses *Lysobacter* sp.

Predatory bacteria are grouped according to the region of the cell that they attack and include epiphytic, periplasmic, and cytoplasmic pathogens. The epiphytic predators are bacteria that attach to the surface of a host bacterium obtaining nutrients from the host cell with ultimate death to the host. Currently, *Vampirococcus* is the only example of an epiphytic predatory bacterium. *Vampirococcus* is an anaerobic gram-negative bacterium that was isolated by Esteve et al. (1983) from Lake Estanya and Lake Cisó in northeastern Spain, which have high concentrations of calcium sulfate and hydrogen sulfide. *Vampirococcus* (0.6 μm wide) attacks large anaerobic photosynthetic *Chromatium* cells which are 1–1.3 μm wide and 2–6 μm long. *Vampirococcus* remains on the surface of host cells and releases hydrolytic enzymes into *Chromatium* cells. Since *Vampirococcus* appears to "suck" nutrients from the host cell, it has been called the bacterial vampire.

The first bacterial predator discovered was *Bdellovibrio bacteriovorus,* which targets the periplasm of heterotrophic gram-negative bacteria (Guerrero et al. 1986). *Bdellovibrio bacteriovorus* is a small curved rod (0.35 × 1.2 μm). propelled by a single flagellum and is broadly distributed in aquatic and soil environments. The predator chases the host, attaches to the surface of the cell, and penetrates the periplasm. As *B. bacteriovorus* grows in the periplasm and acquires nutrients from the host, the host cell undergoes lysis and releases newly produced cells of the predator. Bacteria with a lifestyle similar to *Bdellovibrio* appear to be numerous in many environments and constitute a group known as *Bdellovibrio*-and-like organisms (BALOs). In addition to *Bdellovibrio,* BALOs include *Bacteriolyticum, Bacteriovorax,* and *Peredibacter.*

Daptobacter is a cytoplasmic predator that attaches to *Chromatium* cells and using hydrolytic enzymes makes its way into the cytoplasm where it reproduces (Guerrero et al. 1986). Often electron micrographs of bacterial samples from the environment contain what appears to be an intracellular bacterium and this suggests that many cytoplasmic predators remain to be discovered.

1.8 Summary

For several years microorganisms have been characterized as prokaryotic or eukaryotic; however, further studies have shown the presence of internal membranes in some bacteria, making the distinction between prokaryotic and eukaryotic organisms less distinct. Additionally, some of the large viruses have many genes, so the distinction of virus versus cell requires reassessment. It is recognized that the rooted Tree of Life has not provided clues to the universal common ancestor for life and the use of other types of phylogenic relationships are being considered. As more genomes are characterized, it is apparent that horizontal gene transfer is extensive. Species of bacteria vary greatly in size and internal structures indicating that specialization in bacteria occurs across a broad spectrum of environments. Microbial predation and host-dependent bacteria reveal nutritional specializations that have developed from bacterial interaction.

Discussion Questions

1 Discuss the advantages and disadvantages of continuing the use of the term "prokaryote" to describe bacteria.

2 State some types of systems or activities that bacteria may have acquired as defense against predators?

3 Discuss the selective pressures involved in bacterial colonization of an area following volcanic action.

4 Some internal structures of bacteria have membrane enclosures whereas others do not have membranes. Why is it beneficial for some structures to have membranes enclosing them although other internal granules do not have external membranes?

5 Consider the various shapes of bacteria. What type of environment would select for bacteria of a specific shape?

References

Abedon, S.T. and Murray, K.L. (2013). Archaeal viruses, nor archaeal phages: an archaeological dig. *Archaea* 2013: 251245, 10 pages, https://doi.org/10.1155/2013/251245.

Andam, C.P., Williams, D., and Gogarten, J.P. (2010). Biased gene transfer mimics patterns created through shared ancestry. *Proc. Natl. Acad. Sci. U.S.A.* 107: 10679–10684.

Angert, E.R. and Losick, R.M. (1998). Propagation by sporulation in the Guinea pig symbiont *Metabacterium polyspora. Proc. Natl. Acad. Sci. U.S.A.* 95: 10218–10223.

Ashelford, K.E., Day, M.J., and Fry, J.C. (2003). Elevated abundance of bacteriophage infecting bacteria in soil. *Appl. Environ. Microbiol.* 69: 285–289.

Bar, D.Z. (2011). Evidence of massive horizontal gene transfer between humans and *Plasmodium vivax. Nat. Proc.* https://doi.org/10.1038/npre.2011.5690.1.

Barbara, G.M. and Mitchell, J.G. (2003). Bacteria tracking of motile algae. *FEMS Microbiol. Ecol.* 44: 79–87.

Beveridge, T.J. (1988). The bacterial surface: general considerations towards design and function. *Can. J. Microbiol.* 34: 363–372.

Biers, E.J., Wang, K., Pennington, C. et al. (2008). Occurrence and expression of gene transfer agent genes in marine bacterioplankton. *Appl. Environ. Microbiol.* 74: 2933–2939.

Bresler, V., Montgomery, W.L., Fishelson, L. et al. (1998). Gigantism in a bacterium, *Epulopiscium fishelsoni*, correlates with complex patterns in arrangement, quantity and segregation of DNA. *J. Bacteriol.* 180: 5601–5611.

Bitton, G. (1987). Fate of bacteriophages in water and wastewater treatment plants. In: *Phage Ecology* (ed. S.M. Goyal, C.P. Gerba and G. Bitton), 181–194. New York: Wiley.

Chen, I. and Dubnau, D. (2004). DNA uptake during bacterial transformation. *Nat. Rev. Microbiol.* 2: 241–249. https://doi.org/10.1038/nrmicro844. PMID 15083159.

Claessen, D., Rozen, D.E., Kuipers, O.P. et al. (2014). Bacterial solutions to multicellularity: a tale of biofilms, filaments and fruiting bodies. *Nat. Rev. Microbiol.* 12: 115–124.

Cocotl-Yañez, M., Sampieri, A., Moreno, S. et al. (2011). Roles of RpoS and PsrA in cyst formation and alkylresorcinol synthesis in *Azotobacter vinelandii. Microbiology* 157: 1685–1963.

Daley, G., Russell, D., Tabrizi, S. et al. (2014). *Mycoplasma genitalium*: a review. *Int. J. STD AIDS* 25: 475–487.

Damsté, J.S.S., Strous, M., Rijpstra, W.I.C. et al. (2002). Linearly concatenated cyclobutane lipids form a dense bacterial membrane. *Nature* 419: 708–712.

Davies, S.L. and Whittenbury, R. (1970). The structure of methane and other hydrocarbon-utilizing bacteria. *J. Gen. Microbiol.* 61: 227–232.

de Wit, R. and Bouvier, T. (2006). '*Everything is everywhere*, but, *the environment selects*'; what did Baas Becking and Beijerinck really say? *Environ. Microbiol.* 8: 755–756.

Docampo, R. and Moreno, S.N. (2011). Acidocalcisomes. *Cell Calcium* 50: 113–119.

Doolittle, W.F. (2015). Rethinking the tree of life. *Microbe* 10: 319–323.

Duda, V.I., Lebedinsky, A.V., Mushegjan, M.S. et al. (1987). A new anaerobic bacterium, forming up to five endospores per cell – *Anaerobacter polyendosporus* gen. et spec. nov. *Arch. Microbiol.* 148: 121–127.

Dworzanski, J.P. and Snyder, A.P. (2005). Classification and identification of bacteria using mass spectrometry-based proteomics. *Expert Rev. Proteomics* 2: 863–878.

Engelhardt, T., Kallmeyer, J., Cypionka, H., and Engelen, B. (2016). High virus-to-cell ratios indicate ongoing production of viruses in deep subsurface sediments. *ISME J.* 8: 1503–1509.

Esteve, I., Guerrero, R., Montesinos, E. et al. (1983). Electron microscopy study of the interaction of epibionic bacteria with *Chromatium minus* in natural waters. *Microb. Ecol.* 9: 57–64.

Flärdh, K. and Buttner, M.J. (2009). Streptomyces morphogenetics: dissecting differentiation in a filamentous bacterium. *Nat. Rev. Microbiol.* 7: 36–49.

Fonseca, A., Ishoey, T., Espinoza, C. et al. (2017). Genomic features of "*Candidatus* venteria ishoeyi" a new sulfur oxidizing macrobacterium from the Humbolt Sulfureturn off Chile. *PLoS One* 12 (12): e0188371. https://doi.org/10.1371/journal.pone0188371.

French, S.L., Santangelo, T.J., Beyer, A.L. et al. (2007). Transcription and translation are coupled in archaea. *Mol. Biol. Evol.* 24: 893–885.

Forterre, P. (2015). The universal tree of life: an update. *Front. Microbiol.* 6: 717. https://doi.org/10.3389/fmicb.2015.00717.

Fuerst, J.A. (2005). Intracellular compartmentation in *Planctomyces*. *Annu. Rev. Microbiol.* 59: 299–328.

Funa, N., Ozawa, H., Hirata, A. et al. (2006). Phenolic lipid synthesis by type III polyketide synthesis is essential for cyst formation in *Azotobacter vinelandii*. *Proc. Natl. Acad. Sci. U.S.A.* 1103: 6356–6361.

Gray, T.A., Krywy, J.A., Harold, J. et al. (2013). Distributive conjugal transfer in mycobacteria generates progeny with meiotic-like genome-wide mosaicism, allowing mapping of a mating identity locus. *PLoS Biol.* 11 (7): e1001602. https://doi.org/10.1371/journal.pbio.1001602.PMC 3706393. PMID 23874149.

Guerrero, R., Pedrós-Alio, C., Esteve, I. et al. (1986). Predatory prokaryotes: predation and primary consumption evolved in bacteria. *Proc. Natl. Acad. Sci. U.S.A.* 83: 2138–2142.

Hall, C., Brachat, S., and Dietrich, F.S. (2005). Contribution of horizontal gene transfer to the evolution of *Saccharomyces cerevisiae*. *Eukaryot. Cell* 4: 1102–1115.

Haller, C.M., Rölleke, S., Vybiral, D. et al. (2000). Investigation of 0.2 μm filterable bacteria from the western Mediterranean Sea using a molecular approach: dominance of potential starvation forms. *FEMS Microbiol. Ecol.* 31: 153–161.

Higgins, D. and Dworkin, J. (2012). Recent progress in *Bacillus subtilis* sporulation. *FEMS Microbiol. Rev.* 36: 131–146.

Hurst, C. (2000). An introduction to viral taxonomy and the proposal of *Akamara*, a potential domain for the genomic acellular agents. In: *Viral Ecology* (ed. C. Hurst), 41–62. San Diego: Academic Press.

Jahn, U., Gallenberger, M., Paper, W. et al. (2008). *Nanoarchaeum equitans* and *Ignicoccus hospitalis*: new insights into a unique, intimate association of two archaea. *J. Bacteriol.* 190: 1743–1750.

Juhas, M., van der Meer, J.R., Gaillard, M. et al. (2009). Genomic islands: tools of bacterial horizontal gene transfer and evolution. *FEMS Microbiol. Rev.* 33: 376–393.

Jiang, S.C. and Paul, J.H. (1998). Significance of lysogeny in the marine environments: studies with isolates and a model of lysogenic phage production. *Microb. Ecol.* 35: 235–243.

Jiang, X., Langille, M.G.I., Neches, R.Y. et al. (2012). Functional biogeography of ocean microbes revealed through non-negative matrix factorization. *PLoS One* 7: e43866. https://doi.org/10.1371/journal.pone.004.3866.

Kallmeyer, J., Pockalny, R., Adhikari, R.R. et al. (2012). Global distribution of microbial abundance and biomass in subseafloor sediment. *Proc. Natl. Acad. Sci. U.S.A.* 109: 16213–16216.

Kaplan-Levy, R.N., Hadas, O., Summers, M.L. et al. (2010). Akinetes: dormant cells of cyanobacteria. In: *Dormancy and Resistance in Harsh Environments, Topics in Current Genetics*, vol. 21 (ed. E. Lubzens et al.), 5–27. Heidelberg: Springer-Verlag.

Keese, P. (2008). Risks from GMOs due to horizontal gene transfer. *Environ. Biosaf. Res.* 7: 123–149.

Klein, M., Friedrich, M., Roger, A. et al. (2001). Multiple lateral transfers of dissimilatory sulfite reductase genes between major lineages of sulfate-reducing prokaryotes. *J. Bacteriol.* 183: 6028–6035.

Kloesges, T., Popa, O., Martin, W. et al. (2011). Networks of gene sharing among 329 proteobacterial bacterial genomes reveal differences in lateral gene transfer frequency at different phylogenetic depths. *Mol. Biol. Evol.* 28: 1057–1074.

Kondo, N., Nikoh, N., Ijichi, N. et al. (2002). Genome fragment of *Wolbachia* endosymbiont transferred to X chromosome of host insect. *Proc. Natl. Acad. Sci. U.S.A.* 99: 14280–14285.

Lang, A.S. and Beatty, J.T. (2007). Importance of widespread gene transfer agent genes in alpha-proteobacteria. *Trends Microbiol.* 15: 54–62.

Lang, A.S., Zhaxybayeva, O., and Beatty, J.T. (2012). Gene transfer agents: phage-like elements of genetic exchange. *Nat. Rev. Microbiol.* 10: 472–482.

Lau, M.C.Y., Cameron, C., Magnabosco, C. et al. (2014). Phylogeny and phylogeography of functional genes shared among seven terrestrial subsurface metagenomes reveal N-cycling and microbial evolutionary relationships. *Front. Microbiol.* 5: 531. https://doi.org/10.3389/fmicb.2014.00531.

Lee, K.-C., Webb, R.I., Janssen, P.H. et al. (2009). Phylum *Verrucomicrobia* representatives share a compartmentalized cell plan with members of bacterial phylum *Planctomycetes*. *BMC Microbiol.* 2009: 5. https://doi.org/10.1186/1471-2180-9-5.

Lefèvre, C.T. and Bazylinski, D.A. (2013). Ecology, diversity and evolution of magnetotactic bacteria. *Microbiol. Mol. Biol. Rev.* 77: 497–526.

Legendre, M., Bartoli, J., Shmakova, L. et al. (2014). Thirty-thousand-year-old distant relative of giant icosahedral DNA viruses with a pandoravirus morphology. *Proc. Natl. Acad. Sci. U.S.A.* 111: 4274–4279.

Lindow, S.E. and Brandl, M.T. (2003). Microbiology of the phyllosphere. *Appl. Environ. Microbiol.* 69: 1875–1883.

Lindsay, M.R., Webb, R.I., Strous, M. et al. (2001). Cell compartmentalization in planctomycetes: novel types of structural organization for the bacterial cell. *Arch. Microbiol.* 175: 413–429.

Mangold, S., rao Jonna, V., and Dopson, M. (2013). Response of *Acidithiobacillus caldus* toward suboptimal pH conditions. *Extremophiles* 17: 689–669.

Maranger, R., Bird, D.F., and Juniper, S.K. (1994). Viral and bacterial dynamics in Arctic Sea ice during the spring algal bloom near resolute, N.W.T., Canada. *Mar. Ecol. Prog. Ser.* 111: 121–127.

Martin, W. and Koonin, E.V. (2006). A positive definition of prokaryotes. *Nature* 442: 868.

Maxmen, A. (2010). Virus-like particles speed bacterial evolution. *Nature* https://doi.org/10.1038/news.2010.507.

McDaniel, L.D., Young, E., Delaney, J. et al. (2010). High frequency of horizontal gene transfer in the oceans. *Science* 330: 50. https://doi.org/10.1126/science.1192243.

McDaniel, L.D., Young, E., Ritchie, K.B. et al. (2012). Environmental factors influencing gene transfer agent (GTA) mediated transduction in the subtropical ocean. *PLoS One* 7 (8): e43506. https://doi.org/10.1371/journal.pone.0043506.

McLean, R.J.C. and Kirkland, B.L. (2014). Nanostructures and nanobacteria. In: *Nanomicrobiology* (ed. L.L. Barton, D.A. Bazylinski and H. Xu), 1–10. New York: Springer Science + Business Media.

Moran, N.A. and Jarvik, T. (2010). Lateral transfer of genes from fungi underlies carotenoid production in aphids. *Science* 328: 624–627.

Murray, R.G.E. and Watson, S.W. (1965). Structure of *Nitrosocystis oceanus* and comparison with *Nitrosomonas* and *Nitrobacter*. *J. Bacteriol.* 89: 1594–1609.

Nagao, N., Yamamoto, J., Komatsu, H. et al. (2015). The gene transfer agent-like particle of the marine phototrophic bacterium *Rhodovulum sulfidophilum*. *Biochem. Biophys. Rep.* 4: 369–374.

National Research Council Space Studies Board (1999). *Size Limits of Very Small Microorganisms: Proceedings of a Workshop*. Washington, DC: National Academy Press.

Nigro, O.D., Jungbluth, S.P., Lin, H.T. et al. (2017). Viruses in the oceanic basement. *mBio* 8 (2): pii: e02129–16. https://doi.org/10.1128/mBio.02129-16.

Nikoh, N., McCutcheon, J.P., Kudo, T. et al. (2010). Bacterial genes in the aphid genome: absence of functional gene transfer from *Buchnera* to its host. *PLoS Genet.* 6 (2): e1000827. https://doi.org/10.1371/journal.pgen.1000827.

Pace, N.R. (2009). Problems with "Prokaryote". *J. Bacteriol.* 191: 2008–2010.

Pake, R.T., Ramsing, N.B., Bateson, M.M. et al. (2003). Geographical isolation in hot spring cyanobacteria. *Environ. Microbiol.* 5: 650–659.

Palumbo, A.V., Ferguson, R.L., and Rublee, P.A. (1984). Size of suspended bacterial cells and association of heterotrophic activity with size fractions of particles in estuary and coastal waters. *Appl. Environ. Microbiol.* 48: 157–164.

Pan, S.Q., Jin, S., Boulton, M.I. et al. (1995). An *Agrobacterium* virulence factor encoded by a Ti plasmid gene or a chromosomal gene in required for T-DNA transfer into plants. *Mol. Microbiol.* 17: 259–269.

Peduzzi, P. and Weinbauer, M.G. (1993). Effect of concentrating the virus-rich 2–200 nm size fraction of seawater on the formation of algal flocs (marine snow). *Limnol. Oceanogr.* 38: 1562–1565.

Pérez-Brocal, V., Gil, R., Ramos, S. et al. (2006). A small microbial genome: the end of a long symbiotic relationship? *Science* 314: 312–313.

Philippe, N., Legendre, M., Doutre, G. et al. (2013). Pandoraviruses: amoeba viruses with genomes up to 2.5 Mb reaching that of parasitic eukaryotes. *Science* 341: 281–286.

Pietilä, M.K., Demina, T.A., Atanasova, N.S. et al. (2014). Archaeal viruses and bacteriophages: comparisons and contrasts. *Trends Microbiol.* 22: 334–344.

Pietilä, M.K., Laurinmäki, P., Russell, D.A. et al. (2013). Insights into head-tailed viruses infecting extremely halophilic archaea. *J. Virol.* 87: 3248–3260.

Portillo, M.C., Leff, J.W., Lauber, C.L. et al. (2013). Cell size distribution of soil bacteria and archaeal taxa. *Appl. Environ. Microbiol.* 79: 7610–7617.

Prangishvili, D. (2013). The wonderful world of archaeal viruses. *Annu. Rev. Microbiol.* 67: 565–585.

Raes, J., Letunic, L., Yamada, T. et al. (2011). Toward molecular trait-based ecology through integration of biogeochemical, geographical and metagenomic data. *Mol. Syst. Biol.* 7: 473. https://doi.org/10.1038/msb.2011.6.

Raoult, D., La Scola, B., and Birtles, R. (2007). The discovery and characterization of mimivirus, the largest known virus and putative pneumonia agent. *Clin. Infect. Dis.* 45: 95–102.

Rappé, M.S., Connon, S.A., Vergin, K.L. et al. (2002). Cultivation of the ubiquitous SAR11 marine bacterioplankton clade. *Nature* 418: 630–633.

Reteno, D.G., Benamar, S., Khalil, J.B. et al. (2015). Faustovirus, an asfarvirus-related new lineage of giant viruses infecting amoebae. *J. Virol.* 89: 6585–6594.

Reeve, J.N., Bailey, K.A., Li, W.T. et al. (2004). Archaeal histones: structure, stability and DNA binding. *Biochem. Soc. Trans.* 32: 227–230.

Rizzo, P.J. (2003). Those amazing dinoflagellates chromosomes. *Cell Res.* 13: 215–217.
Rohwer, F. and Vega Thurber, R. (2009). Viruses manipulate the marine environment. *Nature* 459: 207–212.
Saier, M. (2014). Membrane-bound compartments in bacteria. *Microbe* 9: 368–372.
Santarella-Mellwig, R., Franke, J., Jaedicke, A. et al. (2010). The compartmentalized bacteria of the *Phantomycetes-Verrucomicrobia-Chlamydiae* superphylum have membrane coat-like proteins. *PLoS Biol.* https://doi.org/10.1371/journal.pbio.1000281.
Sapp, J. (2005). The prokaryote-eukaryote dicotomy: meanings and mythology. *Microbiol. Mol. Biol. Rev.* 69: 292–305.
Schulz, H.N. and Jorgensen, B.B. (2001). Big bacteria. *Annu. Rev. Microbiol.* 55: 105–137.
Schut, F., de Vries, E.J., Gottschal, J.C. et al. (1993). Isolation of typical marine bacteria by dilution culture: growth, maintenance and characteristics of isolates under laboratory conditions. *Appl. Environ. Microbiol.* 59: 2150–2160.
Sharma, V., Colson, P., Chabrol, O. et al. (2015). *Pithovirus sibericum*, a new bona fide member of the "Fourth TRUC" club. *Front. Microbiol.* 6: 722. https://doi.org/10.3389/fmicb.2015.00722.
Son, M., Yu, J., and Kim, K.-H. (2015). Five questions about mycoviruses. *PLoS Pathog.* 11 (1): e1005172. https://doi.org/10.1371/journal.ppat.1005172.
Stanier, S.Y. and van Niel, C.B. (1962). The concept of a bacterium. *Arch. Mikrobiol.* 42: 17–35.
Strauch, E., Lurz, R., and Beutin, L. (2001). Characterization of Shiga toxin-encoding temperate bacteriophage of *Shigella sonnei*. *Infect. Immun.* 69: 7588–7595.
Suttle, C.A. (2007). Marine viruses – major players in the global ecosystem. *Nat. Rev. Microbiol.* 5: 801–812.
Suttle, C.A., Chan, A.M., and Cottrell, M.T. (1990). Infection of phytoplankton by viruses and reduction of primary productivity. *Nature* 347: 467–469.
Tabou, P.S., Ohwada, K., and Colwell, R.R. (1981). Filterable marine bacteria found in the deep sea: distribution, taxonomy and response to starvation. *Microb. Ecol.* 7: 67–83.
Tenchov, B., Vescio, E.M., Sprott, G.D. et al. (2006). Salt tolerance of archaeal extremely halophilic lipid membranes. *J. Biolumin. Chemilumin.* 281: 10016–10023.
Thingstad, T.F. (2000). Elements of a theory for the mechanisms controlling abundance, diversity and biogeochemical role of lytic bacterial viruses in aquatic systems. *Limnol. Oceanogr.* 45: 1320–1328.
Thomas, P. (2006). Isolation of an ethanol-tolerant endospore-forming gram-negative *Brevibacillus* sp. as a covert contaminant in grape tissue cultures. *J. Appl. Microbiol.* 101: 764–774.
Van Etten, J.L., Lane, L.C., and Meints, R.H. (1991). Viruses and virus-like particles of eukaryotic algae. *Microbiol. Rev.* 55: 586–620.
van Nifrik, L.A., Fuerst, J.A., Sinninghe Damste, J.S. et al. (2004). The anammoxosome: an intracytoplasmic compartment in anammox bacteria. *FEMS Microbiol. Lett.* 233: 7–13.
Vigliotta, G., Nutricati, E., Carata, E. et al. (2007). *Clonothrix fusca* Roze 1896, a filamentous, sheathed, methanotrophic γ-proteobacterium. *Appl. Environ. Microbiol.* 73: 3556–3565.
Visone, V., Vettone, A., Serpe, M. et al. (2014). Chromatin structure and dynamics in hot environments: architectural proteins and DNA topoisomerases of thermophilic archaea. *Int. J. Mol. Sci.* 15: 17162–17187.
Völker, H., Schweisfurth, R., and Hirsch, P. (1977). Morphology and ultrastructure of *Crenothrix polyspora* Cohn. *J. Bacteriol.* 131: 306–313.
Weinbauer, M.G. (2004). Ecology of prokaryotic viruses. *FEMS Microbiol. Rev.* 28: 127–181.
Whitaker, R.J., Grogan, D.W., and Taylor, J.W. (2003). Geographic barriers isolate endemic populations of hyperthermophilic archaea. *Science* 301: 976–978.
Whitman, W.B. (2009). Rebuttal: problems with "Procaryote". *J. Bacteriol.* 191: 2011.

Whitman, W.B., Coleman, D.C., and Wiebe, W.J. (1998). Prokaryotes: the unseen majority. *Proc. Natl. Acad. Sci. U.S.A.* 95: 6578–6583.

Williamson, K.E., Radosevich, M., Smith, D.W. et al. (2007). Incidence of lysogeny within temperate and extreme environments. *Environ. Microbiol.* 9: 2563–2574.

Willey, J.M., Sherwood, L.M., and Woolverton, C.J. (2014). *Prescott's Microbiology*, 9e. New York: McGraw-Hill.

Wilson, W.H., Van Etten, J.L., and Allen, M.J. (2009). The Phycodnaviridae: the story of how tiny giants rule the world. *Curr. Top. Microbiol. Immunol.* 328: 1–42.

Woese, C.R., Kandler, O., and Wheelis, M.L. (1990). Towards a natural system of organisms: proposal for the domains Archaea, Bacteria and Eukarya. *Proc. Natl. Acad. Sci. U.S.A.* 104: 2043–2049.

Yolken, R.H., Jones-Brando, L., Dunigan, D.D. et al. (2014). Chlorovirus ATCV-1 is part of the human oropharyngeal virome and is associated with changes in cognitive functions in humans and mice. *Proc. Natl. Acad. Sci. U.S.A.* 111: 16106–16111.

Young, J.D. and Martel, J. (2010). The rise and fall of nanobacteria. *Sci. Am.* 302: 52–59.

Young, K.D. (2006). The selective value of bacterial shape. *Microbiol. Mol. Biol. Rev.* 70: 660–703.

Young, K.D. (2010). Bacterial shape: two-dimensional questions and possibilities. *Annu. Rev. Microbiol.* 64: 223–240.

Yutin, N., Wolf, Y.I., and Koonin, E.V. (2014). Origin of giant viruses from smaller DNA viruses not from a fourth domain of cellular life. *Virology* 466–467: 38–52.

Zablocki, O., Adriaenssens, E., and Cowan, D. (2016). Diversity and ecology of viruses in hyperarid desert soils. *Appl. Environ. Microbiol.* 82: 770–777.

Further Reading

Gyles, C. and Boerlin, P. (2014). Horizontally transferred genetic elements and their role in pathogenesis of bacterial disease. *Vet. Pathol.* 51: 328–340.

Rossello-Mora, R. and Amann, R. (2001). The species concept for prokaryotes. *FEMS Microbiol. Rev.* 25: 39–67.

Zhaxybayeva, O. and Doolittle, W.F. (2011). Lateral gene transfer. *Curr. Biol.* 21: R242–R246.

Smith, I.M., Dunez, J., Lelliot, R.A. et al. (1988). *European Handbook of Plant Diseases*. London: Blackwell Scientific Publications.

Suggested Video

Myxococcus xanthus preying on an *E. coli* colony: YouTube

2

Microbes in the Biosphere: Examination, Cultivation, and Communities

2.1 Overview and Focus

The biosphere is defined as the region on Earth wherein one might find life. More accurately, it defines the region wherein microorganisms are able to live, which is exceedingly broad because bacteria and archaea can survive in extreme environments that are inhospitable to other forms of life. One of the first experiments that many microbiology students perform is to expose agar plates to several environments within and in the vicinity of their laboratory. Through this exercise, we find microorganisms are present in the air, on inanimate objects such as shoes and door handles, water, and on our skin. Based on the diverse appearance of microorganisms (colony morphology) on Petri dishes, we see that there are many different types of microorganisms in a given environment. When microscopy is used to observe microorganisms growing in nature, bacteria are typically seen growing as surface-adherent communities (referred to as biofilms). When growth conditions are manipulated either naturally or in the laboratory, differences may be seen in the composition of microbial populations or alterations in growth, gene expression, and metabolic characteristics. In this chapter, we explore some of the techniques used to study microbial populations and the features involved in the formation and activities of microbial communities in different environments. The specific goals of this chapter include:

- Microscopy used to observe microorganisms and a tool for cellular studies
- Use of biomolecules to identify microbes in the environment
- Approaches to cultivate microorganisms in the environment including isolation chip (iChip)
- Molecular approaches to evaluate bacteria in the environment
- Assessment of community structure

2.2 Microscopy to Study Environmental Microbes

Microorganisms were first detected through microscopy by early scientists including Antonie van Leeuwenhoek. These organisms were described originally as animalcules as they were motile (in contrast to non-motile plants) yet largely invisible without magnification. In the nineteenth and early twentieth century, microbiologists including Louis Pasteur and Claude Zobell used microscopy and noticed a strong tendency for microorganisms to attach to and grow on surfaces. More recently, various types of microscopy, notably electron microscopy and confocal microscopy, were pivotal techniques used to describe surface-adherent microbial communities, now commonly referred to as biofilms (see discussion later in this chapter).

Environmental Microbiology and Microbial Ecology, First Edition. Larry L. Barton and Robert J.C. McLean.

Figure 2.1 Examples of large microorganisms: (a) black spores of the pathogenic fungus *Ustilago maydis* cover the ear of corn; (b) pigmented colonies of lichen occur on tree trunks; (c) bracket fungi growing on the trunk of a tree; (d) edible mushrooms. *Source:* Reproduced with permission from Larry L. Barton.

They noted the many different types (species) of microorganisms that were present in the different environments. In this section, we explore different approaches to studying mixed populations.

A few microbial communities can be observed with the unaided eye (Figure 2.1). Notable examples include lichens (symbiotic consortia of fungi and eukaryotic algae or fungi and cyanobacteria) on rocks and trees, and microbial mats (aggregates of algae, cyanobacteria, and bacteria on the surface of stagnant bodies of water). Fleshy forms of fungi such as mushrooms and bracket fungi on trees may vary from a few inches to a foot or more in diameter. The size of microorganisms varies ranging from 10–20 μm for eukaryotic cells (including protozoa, fungi, and algae), approximately 1 μm for many bacteria, to 20–50 nm for viruses. With rare exceptions such as *Epulopiscium* sp. and *Thiomargarita namibiensis* (Table 2.1), magnification is needed to examine individual microbial cells.

2.2.1 Light Supported Microscopy

Most students in biology are familiar with the compound light microscope, sometimes referred to as a bright-field light microscope. Here light is transmitted through a specimen and is then magnified by several lenses before being observed or photographed. Most microorganisms are transparent and difficult to see against a bright background unless they are stained (Murphy and Davidson 2013). One important advance in microbiology is the Gram stain developed in the late nineteenth century by Hans Christian Gram, which allows bacteria to be differentiated based on their cell wall structure. As microbiological stains are toxic to microorganisms, the samples observed are non-viable. At low magnification, the depth of field is relatively large and so most if not all the image will

Table 2.1 Range in size of microbial cells and viruses.

Microorganism and viruses	Cell/virus diameter (μm)	Cell length (μm)
Head of bacteriophage	0.04	
Mycoplasma genitalium	0.2	
Nanoarchaeum equitans	0.4	
Megavirus chilensis	0.44.	
Pelagibacter ubique	0.12–0.2	0.37–0.89
Staphylococcus aureus	1.1.	
Escherichia coli	0.5	2.0
Bacillus megaterium	1.5	4.0
Saccharomyces cerevisiae	4.0–6.0	
Conidia of *Alternaria* sp.	7–18	20–200
Caryophanon latum	2–4	
Achromatium oxaliferum	100	
Paramecium sp.		50–330
Spirochaeta plicatilis	0.5–0.75	50–250
Epulopiscium fishelsoni	80	300
Thiomargarita namibiensis	300	750

be in focus. Phase-contrast and dark-field microscopes are used to overcome the limitations of conventional bright-field light microscopy. Phase-contrast light microscopy takes advantage of the physical properties of light to portray biological objects as dark against a light background. With dark-field microscopy, the light source is aimed away from the objective lens and only enters when reflected from a cell. As a result, the behavior of living, unstained organisms can be observed using phase-contrast microscopy and dark-field microscopy.

2.2.2 Fluorescence Microscopy

In fluorescence microscopy, light of a specified wavelength is directed to a specimen that has been stained with a fluorescent molecule. Filters are used so that a desired wavelength from the stain is observed. In the mid twentieth century, fluorescent-labeled antibody was used to identify the agent of syphilis, serological groups of *Streptococcus,* and other bacteria. An extension of fluorescence technology resulted in the application of fluorescence to assess bacterial viability, separation of bacteria using flow cytometry, use of green fluorescent protein (GFP) to assess gene expression, and in situ DNA hybridization to identify bacteria.

Methylene blue and other dyes have long been known to stain bacteria differentially depending on cell viability. Cells with high metabolic activity were initially considered to reduce methylene blue to a colorless state whereas dead cells were stained blue. Although this difference in cell color was difficult to observe when using bacteria because of the small cell size, it was readily seen when using the larger cells of yeast. As our understanding of cell activities has progressed, it was discovered that certain dyes are excluded from living cells by the integrity of the cell membrane. This led to the development of differential fluorescence to enumerate live and dead bacterial cells. The commercially available LIVE/DEAD™ *Bac*Light™ bacterial viability kit uses propidium iodide to stain nucleic acid red in cells with damaged membranes and SYTO^{TM}9, which enters all cells and fluoresces green, to give a total cell number. Live–dead stains are important in the evaluation of viable bacteria in various foods and environmental samples, including drinking water (Berney et al. 2007; Boulos et al. 1999), and in the assessment of live–dead populations of archaea (Leuko et al. 2004). Automated calculations of live–dead populations can be quickly conducted by coupling fluorescent stain technology with flow cytometry. Combined flow fluorescence cytometry for sorting single cells is a new tool that is used to monitor and evaluate stress in non-culturable bacteria (Ferrari et al. 2012; Sachidanandham et al. 2005; Nebe-von-Caron et al. 2000).

The gene for *gfp* from the jellyfish *Aequorea victoria* is cloned into bacteria and used as a marker for a particular population or gene expression. Using green fluorescent protein (GFP) as a marker for gene expression (Chalfie et al. 1994), a fluorescent stain of *Escherichia coli* was developed to identify bacterial activity in response to a particular environment (e.g. pH, oxidation/reduction potential also known as Eh). Fusion of the *gfp* gene with the urease gene of *Proteus mirabilis* was used to demonstrate the in vivo expression of urease by *Proteus mirabilis* urinary infection in mice (Zhao et al. 1998). To study the distribution of *Desulfovibrio desulfuricans* G20 biofilms, the *gfp* gene was inserted adjacent to the dissimilatory bisulfite reductase (*Dsr*) gene (Figure 2.2a). In another example, green and red fluorescent protein vectors was used to study mixed community biofilms of *Burkholderia cepacia* and *Pseudomonas aeruginosa* (Tomlin et al. 2004). Autofluorescence is observed in some microorganisms and is attributed to the presence of fluorescent molecules in cell metabolism. Autofluorescence is displayed by *Anabaena* because there are fluorescent molecules in the photosysystem (Figure 2.2b) and by methanogens because one of the electron carriers in the methane-synthesizing bacteria is fluorescent.

Fluorescence in situ hybridization (FISH) was developed so that organisms could be identified by microscopy (Amann et al. 1990). A small DNA fragment (that will recognize and bind

(a)

(b)

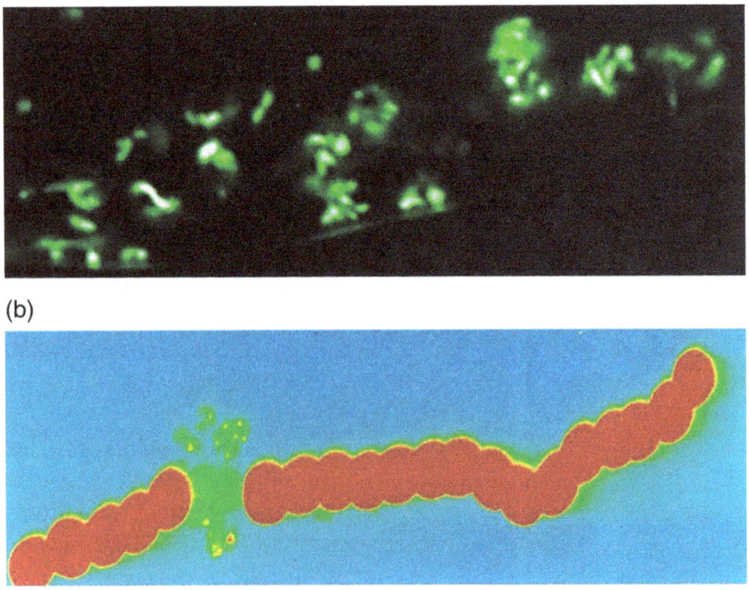

Figure 2.2 Fluorescence microscopy of bacteria: (a) fluorescence of *Desulfovibrio desulfuricans* G 20 with gfp-containing plasmid; (b) fluorescence of unstained photosynthetic bacterium *Anabaena* sp. Fluorescence is due to photosynthetic pigments in the cell. *Source:* Reproduced with permission from Larry L. Barton.

to target DNA or RNA in the target organism) is synthesized and attached to a fluorescent molecule, thus identifying a target organism. There is a wide variety of FISH probes available with specificity ranging from domain specific (Archaea, Bacteria, or Eukarya), family to specific genus. An interesting application of FISH technique is the evaluation of methanogenic, syntrophic, and sulfate-reducing bacteria in methanogenic-sulfidogenic aggregates in wastewater processing systems (Santegoeds et al. 1999).

2.2.3 Scanning Confocal Laser Microscopy

Scanning confocal laser microscopy (SCLM) is another technique that enables scientists to obtain a three-dimensional (3D) image of biological specimens and communities. A microscope is used with a computer, which is a key component of the imaging. The optics is designed so that only the portion of an image in focus is recorded. If one thinks of a specimen as having three axes (length, width, and height – these are commonly referred to as x, y, and z dimensions), then at each focal plane, the laser will scan the specimen and obtain visual data in the x–y plane. The focus will then be adjusted to the next focal plane (change in z-direction) and another scan will be done. When the scan is complete, computer reconstruction produces images in three dimensions from various angles including a side-on view (x–z or y–z projections). The specimen is placed on the stage and the focusing adjusted to that the entire height of the specimen can be recorded. As with fluorescence microscopy, various stains can be used.

Confocal laser scanning microscopy (CLSM) has been used to study biofilm growth and development including oral bacterial infections and in one example necrotic root infections. In this interesting application, bovine teeth were exposed to *Enterococcus faecalis* to evaluate the impact of this bacterial infection on dentin. The teeth exposed to bacteria were sectioned, stained with fluorescein diacetate and acridine orange or propidium iodide and examined using CLSM. The result was a 3D model of the teeth indicating the location of viable, dead, and bacterial biofilms, which were compared to in situ and ex vivo models (Zapata et al. 2008; Ordinola-Zapata et al. 2010). Other studies have used CLSM to examine bacterial biofilms in patients with different infections, as well as biofilms in industrial and other environmental situations.

2.2.4 High Resolution by Electron Microscopy

The maximum magnification commonly used in light microscopes is approximately 1000-fold. At that magnification, a bacterial cell 1 μm long would appear to be 1 mm. Viruses are too small to be seen by light microscopes. Electron microscopy uses a much higher resolution than light microscopy and so viruses and components and interactions of prokaryotic and eukaryotic cells can be seen (Beveridge (1981); Bozzola and Russell 1998). Transmission electron microscopy (TEM) uses an electron beam to provide a magnification about a thousand times greater than light microscopy and provides a two-dimensional (2D) image. Bacterial cells prepared for TEM are dried, embedded in plastic, and sectioned to produce specimens that are 50–100 nm thick. Staining is achieved using heavy metal salts (typically containing U, Os, Pb, and other heavy metals) to deflect electrons emitted from an "electron gun" that are traveling through a high vacuum to the specimen. Sectioned bacterial cells are examined by TEM to evaluate the delicate cell surface and cytoplasm (Figure 2.3), thickness of the cell wall of gram-positive bacteria (Figures 2.3 and 2.4), and multiple membranes of gram-negative bacteria (Figure 2.5).

Cryo-electron tomography (cryo-ET) is used to obtain a 3D image of bacterial cells. The cell is quick frozen within 1 ms, which preserves the cells in a native state without chemical fixation. Cryo-ET has been successfully used to examine the macromolecular organization of numerous bacteria including "*Candidatus* Pelagibacter ubique" (Zhao et al. 2017). Some of the

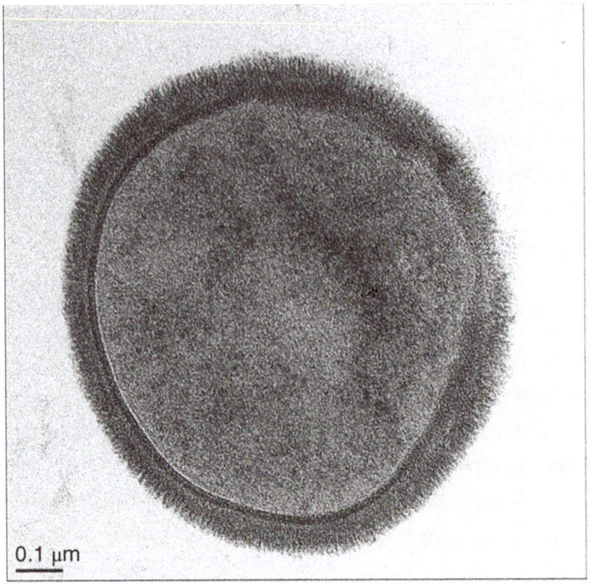

Figure 2.3 Molecular structures extending from the surface of a sectioned gram-positive cell with examination by high resolution TEM. *Source:* Courtesy of Robert Harris and Cezar Khursigara.

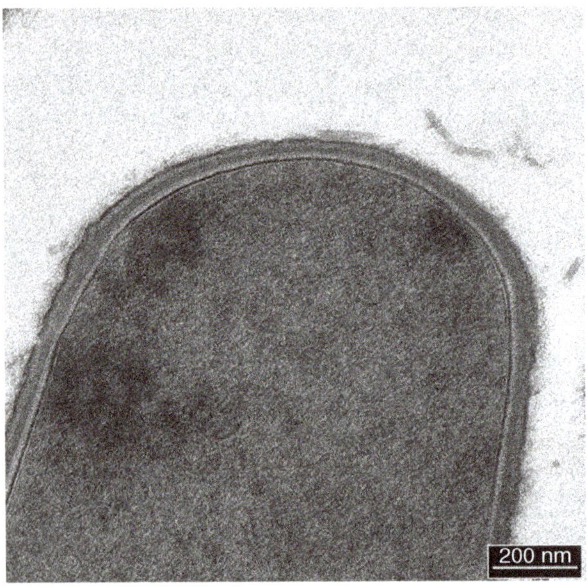

Figure 2.4 High resolution transmission electron microscopy of a cross section of a gram-positive bacterial cell. The thick cell wall is seen outside of the plasma membrane and some wall material (peptidoglycan) is sloughing from the outer edge. *Source:* Courtesy of Robert Harris and Cezar Khursigara.

structural detail of *Pelagibacter ubique* is indicated in Figure 2.6 and complete subcellular organization is presented in Zhao et al. (2017).

Another variation of electron microscopy is to operate the instrument near atmospheric conditions so that bacterial integrality is not destroyed and live bacteria can be observed. Low vacuum TEM may be referred to as "wet TEM" or "environmental TEM" (Figure 2.7a). To obtain a 3D image of bacteria, scanning electron microscopy (SEM) has been used where electrons are reflected off the surface of the specimen that has been dried by a quick freeze process (Figure 2.7b). While the SEM can produce a magnification of only 200,000, that is sufficient to give an excellent view of bacterial cell surface, morphology, and cell appendages. However, TEM is also excellent in providing cell characteristics and cell surface features.

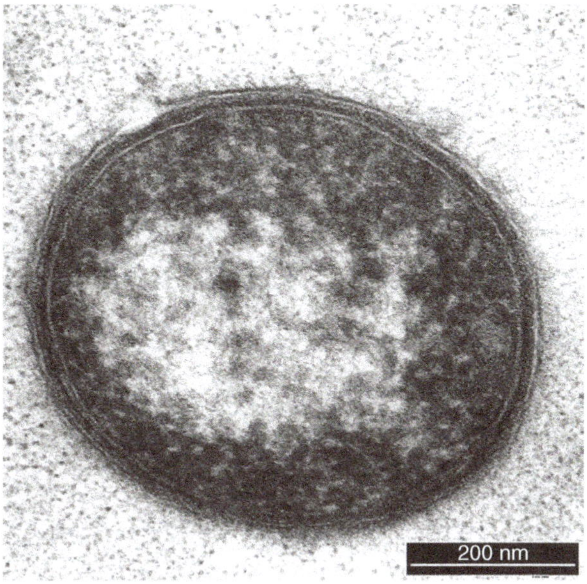

Figure 2.5 High resolution transmission electron microscopy of a gram-negative bacterial cell. The inner (plasma) membrane and the outer membrane are readily seen in the photograph. *Source:* Courtesy of Robert Harris and Cezar Khursigara.

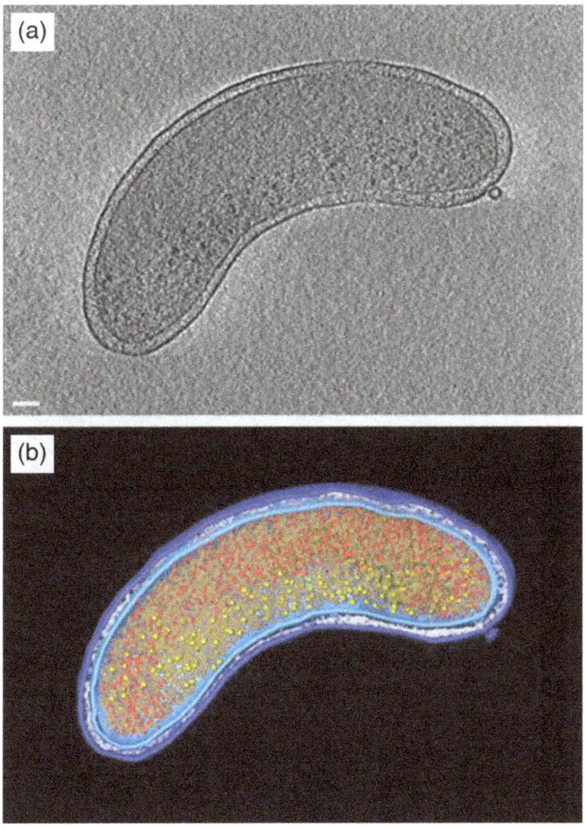

Figure 2.6 Cryo-electron tomography of an intact, frozen-hydrated *Pelagibacter ubique* cell. (a) A tomographic slice of a typical 3D reconstructed *P. ubique* cell grown in log phase. Bar = 50 nm; (b) Isosurface-rendered 3D model of the cell shown in (a). The cell was sliced open to allow viewing internal components. The outer membrane is dark blue, the plasma membrane is cyan, the peptidoglycan is white, the cytoplasm is orange, ribosome-sized particles are represented by yellow spheres and the nucleoid area is red. *Source:* Photographs of *Pelagibacter ubique* provided by Xiaowei Zhao and Daniela Nicastro at the University of Texas Southwestern Medical Center, Stephen Giovannoni at the Oregon State University, and J. Richard McIntosh at the University of Colorado. See additional pictures of cell anatomy in Zhao et al. 2017.

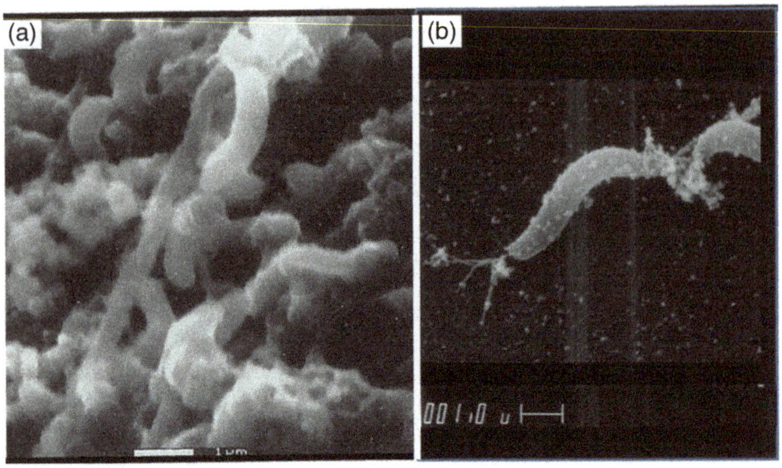

Figure 2.7 Bacteria observed using SEM and wet TEM microscopy: (a) biofilm of *Desulfovibrio desulfuricans* growing on iron wire and contributing to biocorrosion; (b) cells of *Desulfovibrio gigas* seen by SEM. *Source:* Reproduced with permission from Larry L. Barton.

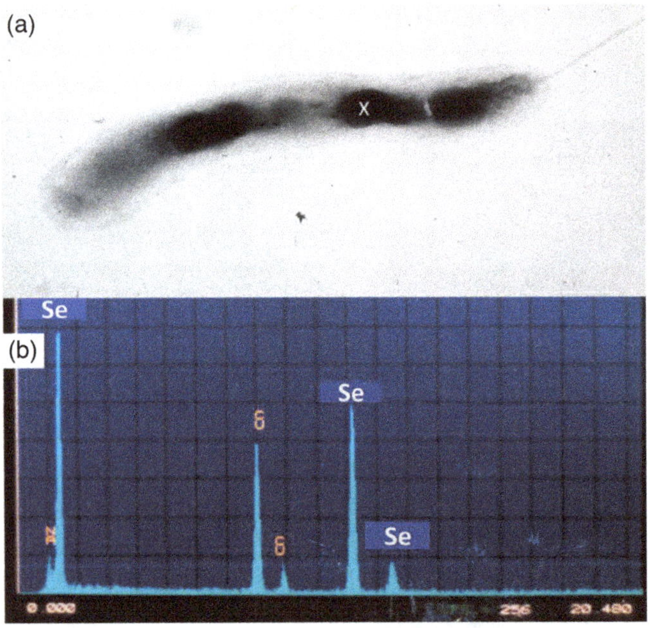

Figure 2.8 Elemental analyses of bacteria examined by EDX microscopy: (a) unstained bacterial cell from an anaerobic culture growing containing sodium selenite. The large granule at the "X" was analyzed by EDX; (b) elemental analysis by EDX of large dense bodies inside cell in (a). *Source:* Reproduced with permission from Larry L. Barton.

Energy-dispersive X-ray spectroscopy (EDX or EDXS) is a technique that is used by geomicrobiologists to follow mineral metabolism by microorganisms. The EDX uses a back-scattering detection system to identify metal elements that have been activated by electrons. The EDX detector is attached to an electron microscope and may be used to identify nanocrystals of minerals and to quantify amorphous metal precipitates (see Figure 2.8). For example, iron oxide particles have been shown to be deposited by ectosymbiotic bacteria growing on marine shrimp (Corbari et al. 2008). The identification of nanoparticles produced by bacteria (see Sections 2.3.5 and 10.8) is obtained by EDX microscopy. A futuristic application is to perform DNA sequencing using high resolution electron microscopy (Bleloch et al. 2011).

Other types of high resolution microscopes that have specific applications in microbiology include the atomic force microscope (AFM) and the scanning tunneling microscope. The AFM does not use lenses or an electron beam to illuminate the specimen but provides an image based on surface topography obtained from a laser probe scanning the specimen. As the resolution of an AFM is a fraction of a nanometer, high resolution images of DNA have been obtained (Hansma et al. 1993) and the cell wall organization of mycobacteria has been described (Alsteens et al. 2008). The scanning tunneling microscope shares several characteristics with the AFM; however, with the scanning tunneling microscope a focused beam of electrons is used to obtain an image. High resolution images of DNA have been obtained with the scanning tunneling microscope (Porath 2009). Electrically conductive pili, termed nanowires, that allow bacteria to use insoluble electron acceptors, were observed using scanning tunneling microscopy (Gorby et al., 2006).

2.3 Internal Structures in Prokaryotes

Subcellular organization is found in archaea and bacteria with the formation of inorganic or organic granules of various sizes, as well as ribosomes and the nucleoid (DNA-containing region). While a few of these structures have membranes surrounding them, most are protein structures without lipid. Some internal granules are used for nutrient storage and some for organized metabolic processes; the great diversity of these structures indicates the wide variety of metabolic processes used by prokaryotic cells. Some internal structures are discussed in the following paragraphs and other cytoplasmic structures are presented in Table 2.2.

2.3.1 Gas Vacuoles

Cylindrical structures referred to as gas vacuoles are present in the cytoplasm of some cyanobacteria and halophilic bacteria. The gas vacuoles have a glycoprotein structure that is permeable to gas and its function is to position bacteria in the environment for optimal light exposure. Cyanobacteria fixing CO_2 by photosynthesis increase the density of the cell by synthesizing carbohydrate. Dense cells sink in the water column and when light intensity is low, cell production of carbohydrates slows making the cells more buoyant. Most cyanobacterial cells grow with numerous filaments and the length of the filaments has an impact on cell buoyancy. Filament length varies with the cyanobacterial species and is 10–800 μm, but cell diameter may also vary with the species from 2 to 16 μm. Thus, the cells must regulate cell division, gas vacuole formation, and filament length to provide optimal positioning for fixation of CO_2 by photosynthesis. In some halophilic archaea, the gas vacuoles provide buoyancy and position the cells at the surface of the water. As halophilic organisms have several membrane pumps located in the plasma membrane, positioning of the cells for light exposure is important in a marine or soil environment.

2.3.2 Sulfur Globules

Several chemolithotrophic prokaryotes obtain energy from reactions coupling the oxidation or reduction of sulfur compounds. The oxidation of hydrogen sulfide by various microorganisms makes a water-insoluble sulfur globule that transiently collects in the cytoplasm or may be the metabolic end product. External sulfur globules are produced by the phototrophic bacteria that are found in the *Chlorobiacae, Ectothiorhodospiraceae,* and *Rhodospirillaceae* families and

Table 2.2 Cellular organization: internal structures and microcompartments in bacteria.

Structure	Selected bacteria and archaea	Function
Nitrate vacuole	*Thiomargartia namibiensis*	Store electron acceptor for bacteria to be used when nitrate is low concentration in the marine waters
Magnetosomes	*Magnetospirillum magneticum*	Orientation of cells to optimal O_2 concentrations in the oxic-anoxic interface
Carboxysomes	Cyanobacteria and some autotrophic bacteria	Polyhedral units with a permeable protein shell contains the RuBisCO enzyme for CO_2 fixation
Gas vacuoles	Cyanobacteria and halophilic Archaeal cells	Provide buoyancy to cells to maintain cell in optimal exposure to sunlight
Acidocalcisome[a]	*Agrobacterium tumefaciens*	Storage of cations and phosphorus, maintenance of intracellular pH
	Rhodospirillum rubrum	
Anammoxosome	*Kuenenia stuttgartiensis*	Anaerobic oxidation of ammonia
	Brocadia anammoxidans	
	Anammoxoglobus propionicus	
	Scalindua brodae	
Pirellulosome	*Gemmata obscuriglobus*	A membrane bound compartment containing nucleoid and ribosomes
	Pirellula marina	
Chlorosomes	*Chloroflexus aurantiacus*	Photosynthetic antenna complex containing thousands of BChl[b] c/d/e molecules
	Chlorobaculum tepidum	
	Chlorobium limicola	
Phycobilisomes	Cyanobacteria	Protein complexes assisting in harvesting light in photosystem II containing phycobiliproteins of phycoerythrin, phycocyanin, and allophycocyanin
Parasporal bodies	*Bacillus thuringiensis*	Crystals of protein in sporulating cells and protein are toxic to insect larvae
	Bacillus sphaericus	

[a] Some consider volutin granules to be acidocalcisomes.
[b] BChl = bacterial chlorophyll.

some thiobacilli. Cytoplasmic sulfur globules are associated with members of the *Chromatiaceae* and *Beggiatoaceae* families and species of *Thioploca, Allochromatium, Macromonas, Thiomargarita,* and *Thiovulvum*. The diameters of these sulfur globules range from 0.65 μm in *Allochromatium vinosum* to 1–3 μm in *Thiomargarita namibiensis*. The intracellular globules contain α-sulfur as elemental cyclo-octasulfur (S_8) and not as a linear polythionate chain (George et al. 2008). In *Thermococcus prieurii*, an euryarchaeon, sulfur globules are transiently in the cytoplasm and electron micrographs reveal that these sulfur globules are released from the cell by protrusions that originate from the cell envelope (Gorlas et al. 2015).

2.3.3 Polymeric Carbon Reserves

When the carbon source is abundant, microorganisms often produce intracellular structures that are nutrient reserves to be consumed when energy sources are required. Granules consisting of polyglucose, or glycogen, are present in many species of phototrophic,

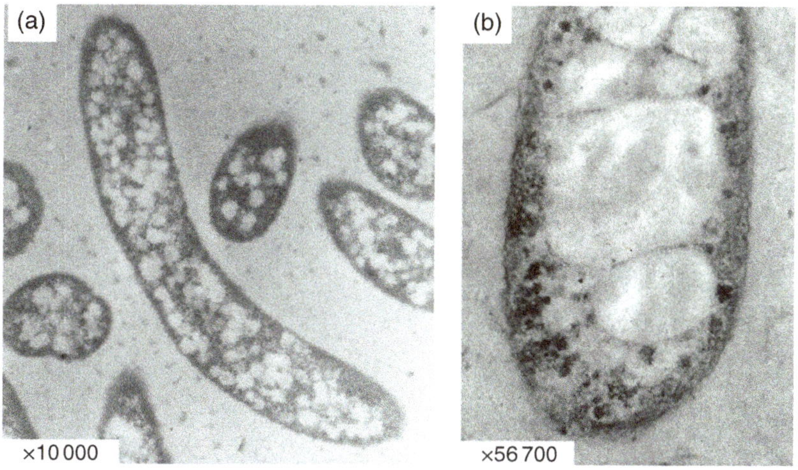

Figure 2.9 Carbon reserves in bacteria using TEM. Electron transparent regions are areas where the polymeric carbon material is collected: (a) thin section of *Desulfovibrio gigas* cells revealing polyglucose aggregates throughout the cytoplasm; (b) cell of *Azospirillum brazilense* with large granules of poly-β-hydroxybutyric acid. *Source:* Reproduced with permission from Larry Barton.

heterotrophic, or chemolithotrophic bacteria and thermoacidophilic archaea including *Sulfolobus, Thermoproteus, Desulfurococcus,* and *Thermococcus.* The presence of polyglucose materials is apparent by TEM (Figure 2.9a). In some bacteria, inclusion bodies of polyhydroxybutyric acid (PHB) constitute a considerable portion of the cell and may account for a quarter of the cell mass (Figure 2.9b). Each PHB polymer contains about 10^6 monomeric units of β-hydroxybutyrate forming an electron transparent structure with a thin protein layer. In some bacteria (i.e. *Cupriavidus necator* formerly *Ralstonia eutropus, Rhodococcus* NICMB 40126 and *Alcaligenes* SH-69), copolymers of β-hydroxybutyrate and 3-hydroxyvalerate are formed. Also, it should be noted that polymers containing 1,4 butanediol, butyrolactone, and 3-hydroxypropionate may occur inside the bacterial cell as granules and these constitute a structure called polyhydroxyalkanoates (PHA). Both PHA and PHB are electron transparent regions under the electron microscope and can be readily identified by chemical analysis.

2.3.4 Polyphosphate Granules

Bacterial cells that stain with methylene blue dye and metachromatic (blue color changing to red-violet) granules are seen inside the cell, is a good indicator that inorganic phosphate has accumulated as granules in the cytoplasm. Polyphosphate granules form in many microbial cells and function as phosphate reservoirs. In most aquatic systems, phosphorus is a limiting nutrient, so it is important for bacteria to accumulate phosphate to provide for future growth. Discrete spheres with a protein outer layer are characteristic of polyphosphate granules found in thin sections of bacteria observed under the electron microscope.

2.3.5 Metallic Nanoparticles

Bacteria inhabit environments where toxic metal and metalloid ions are present and often bacteria are able to reduce the metals and metalloids to the elemental state. Heavy metals commonly reduced to the elemental form include Au, Ag, U, Pt, and Pd with metalloids reduced including Se and Te (Barton et al. 2014). The insoluble elements accumulate as nanoparticles in the cell cytoplasm, in the periplasm, or on the surface of the cell. The presence of selenium particles (Se^0) is demonstrated inside a cell, see Figure 2.8. The native size of these dense structures may range in size from 0.2 to 1.0 ηm or as aggregates of 100–200 ηm in diameter. Several processes are associated with reduction of metals and metalloids and these vary with different bacterial species and may include electron transporting proteins such as cytochromes and hydrogenases or enzymes with dehydrogenase activities.

2.4 Strategies for Culturing Microorganisms

2.4.1 Overview

Laboratory culture, developed in the late nineteenth and early twentieth century by Louis Pasteur, Robert Koch, and several other early microbiologists, is a common strategy for studying microorganisms. For example, *Escherichia coli*, *Bacillus subtilis*, and *Staphylococcus aureus* are commonly used in introductory microbiology courses. These bacterial species grow quickly on common laboratory media, producing visible growth as turbidity in liquid media or colonies on agar plates after overnight growth. Growth of other culturable organisms may occur at a much slower pace. For example, a non-pathogenic soil organism, *Mycobacterium smegmatis*, will require 3–7 days growth to produce colonies on agar plates whereas pathogenic strains of *Mycobacterium tuberculosis* may not show growth in liquid media until 14–24 days after inoculation.

There are a variety of culturing techniques that are commonly used in microbiology and microbial ecology. During standard laboratory work, organisms are grown in liquid or solid media. Solid media contains agar, a complex polysaccharide derived from marine red algae, is the most common gelling agent used in microbiology media. Unlike other materials such as gelatin, agar is not degraded by most microorganisms and is solid at the temperatures at which most common mesophilic organisms grow. During preparation, agar solutions require temperatures close to boiling to melt, but once melted will remain liquid until cooled to 45 °C. One advantage of the lower solidifying temperature of agar is that it prevents degradation during autoclaving of media formulations containing a sugar or antibiotic that may be heat labile. In this instance, the heat-stable components of media (including agar) are autoclaved and then allowed to cool slightly (often a water bath set to ~50 °C is used to regulate temperature). The heat labile components are often filter-sterilized and then aseptically mixed with the sterile agar media and then the plates are poured. Excellent references for media preparation and techniques for isolating bacteria are found in basic laboratory manuals which are readily available at bookstores. Teske et al. (2007) have an extensive listing of media for isolation and culture of bacteria. Typical concentrations of agar in most media is 1.3–1.5% w/v. Lower concentrations (0.3–1% w/v) are used when cell motility is encouraged, whereas higher concentrations (~2% w/v) are used to restrict the motility of some bacteria. Agarose is a highly purified form of agar. Although agarose is sometimes used as a gelling agent, it is more commonly used in the analysis of DNA and other biological molecules by electrophoresis. Agar plates are commonly used in the separation of mixtures of microorganisms into pure cultures and in bacterial enumeration via dilution plating.

Liquid media lack a gelling agent and organisms grow throughout the liquid. Typically, liquid cultures are grown in test tubes or flasks. For microbial growth, organisms are supplied with nutrients including carbon, nitrogen, phosphorus, and sulfur, as well as micronutrients. For aerobic organisms, oxygen is supplied by air exchange with the liquid in the medium. Aeration can be enhanced by shaking the culture. For anaerobic culturing, oxygen must be excluded and the organisms are frequently grown in an anaerobic chamber, gas pack jars, or anaerobic tubes using anaerobic techniques developed by Hungate as he studied rumen microorganisms (Hungate 1969).

Tryptic soy agar (TSA) and tryptic soy broth (TSB) are examples of complex media that can grow several commonly encountered bacteria. TSB contains a soy bean extract that has been digested with the protein-degrading enzyme trypticase. In some instances, particularly for growing medically important agar, TSA can be supplemented with blood (typically sheep blood – 5% by volume) as the blood cells (erythrocytes) contain nutrients required by some clinical isolates. When it is important to control microbial physiology, a chemically defined medium with all chemical components specified is used. An example of a chemically defined medium used for growing *E. coli* and other related organisms can be found in the publication by Neidhardt et al. (1974).

Although there have been a number of advances in microbial culturing techniques, it is estimated that only ~1% of all microorganisms in a given environment can be cultured on lab media. Consequently, a variety of genetic approaches (described later) are being adapted to study uncultivable microorganisms.

2.4.2 Approaches for Isolation of Microorganisms

Microbiologists use a wide range of strategies for growing organisms. These often include the pure culture techniques developed largely through the work of Louis Pasteur and Robert Koch toward the end of the nineteenth century. While very useful and commonly used approach in microbiology, there are situations where desired organisms may represent only a fraction of a community. Two common approaches are used to select desired organisms in an environment, enrichment culture, and selective culture. Enrichment culture, originally developed by Martinus Beijerinck, involves creating culture conditions that promotes the growth of desired populations. As an example, one can make a growth medium that lacks any organic or inorganic nitrogen source and through repeated growth and natural selection encourages the growth of nitrogen fixing organisms. Selective culture is a situation where an inhibitory chemical is used to prevent the growth of large numbers of undesired organisms (Teske et al. 2007). An example would be MacConkey agar, containing bile salts which inhibit non-enteric bacteria, for the growth of *E. coli* and related coliform bacteria. In 1 g of fecal matter there are an estimated 10^{12} bacteria, of which 10^7–10^8 (i.e. 0.001–0.01%) are *E. coli*. In most microbial ecology studies, enrichment culture is preferred over selective culturing in that inhibitory chemicals may damage desired organisms particularly those that may be present in low numbers.

Batch culturing is the most common growth strategy used for culturing individual microorganisms. Here the organism is incubated in either liquid or solid growth media with all the ingredients present at the beginning of the culture. Another growth strategy involves continuous culture, using a device referred to as a chemostat. Here, organisms are inoculated into liquid culture and then allowed to grow (typically overnight in the case of fast-growing bacteria such as *E. coli* and *P. aeruginosa*). Once the organisms are growing, then fresh medium is added at a predetermined rate and waste medium removed. When a defined medium is used, the organisms' growth rate in a chemostat (as well as its physiology) will be determined by the rate of addition of the limiting nutrient.

2.4.3 Establishing Microbial Communities

At a larger scale, an entire microbial community can be modeled in the laboratory or even outside the laboratory. An aquarium would be a good example of a freshwater or marine aquatic ecosystem. Typically, a smaller lab-scale model environment within a laboratory is referred to as a microcosm. Microcosms have an advantage in that investigations with potentially harmful chemicals or organisms can be studied and environmental conditions manipulated, yet contained from accidental release into the environment. Larger model environments are studied outside the laboratory. Depending upon size, larger artificial environments are referred to as mesocosms or macrocosms, for example artificial streams and greenhouses. Large-scale studies and facilities are rare, for example the Experimental Lakes Area (ELA) consisting of 58 small lakes and watersheds located near Kenora, Ontario, Canada. At the ELA, ecosystem level studies were carried out to investigate the impact of various pollutants – including acidification (from acid rain), eutrophication (from phosphate and other components in municipal waste), and heavy metals – on lakes and associated terrestrial ecosystems (Schindler et al. 1985, 1996, 2008).

Aside from specialized facilities such as the ELA, field studies can also be performed. Sometimes field studies include experimental manipulation, for example shading an area to investigate the role of photosynthesis on a microbial process (such as travertine formation by cyanobacteria), or placing leaves in a porous cloth (study of leaf decomposition by microorganisms while excluding invertebrates or fish). In larger field studies, individuals from many disciplines may be involved and, in some cases (e.g. deep sea, deep drilling, or space-related experiments), the scientist may not be performing the actual experiments, but may rely on specially trained personnel or even machines. A key scientific and ethical point during field studies is to avoid introducing harmful chemicals or organisms into a natural environment.

2.4.4 The iChip and Growing Uncultured Bacteria

It has been estimated that less than 0.1% of bacteria in the environment can be cultivated in the laboratory and the number of prokaryotic cells in the environments has been estimated to be $4–6 \times 10^{30}$ (Table 1.3). As recently reviewed by Overmann (2010a, b), various novel approaches have been pursued to grow those that have become to be known as the unculturable bacteria. It is important to consider the environment where bacteria are growing and to replicate as closely as possible the chemical environment. One point of concern is the buffer employed. In bacteriological media, phosphate is used to maintain the pH as well as a phosphorus source, but very little phosphate is required to support bacterial growth and the excess may result in precipitation of Ca^{2+}, Mg^{2+}, and Fe^{3+}. Buffers such as 4-(2-hydroxyethyl)-1-piperazineethanesulfonic acid (HEPES) do not bind metals and, at low concentrations, enhance the growth of fastidious bacteria. In nature much of the buffering capacity of the environment may be attributed to carbonate. The level of soluble organic compounds is low in the environment and addition of organic compounds to $\leq 5\,mg\,l^{-1}$ is sufficient to promote growth. The levels of yeast extract or amino acids used in most bacteriological media would inhibit growth or environmental bacteria because these bacteria would be stressed by an elevated osmotic environment and significant changes in cytoplasmic signals. This phenomenon has been referred to as "substrate accelerated death." Attempts to grow bacteria from the environment include growth inside dialysis bags, use of dialysis chambers, or using capillary systems that are prepared by robotic systems to chemotactically attract bacteria (Overmann 2010a, b).

The iChip has been used to cultivate bacteria that cannot be grown by standard laboratory procedures. Basically, it is a diffusion chamber that contains water or a dilution of soil and it is used on a high-throughput platform (Nichols et al. 2010). The iChip enables more bacteria to be cultivated from the environment than can be grown using diffusion chambers of dilution plating on an agar surface. There is considerable potential to use the iChip to search for industrially useful bacteria and in situ cultivation of bacteria has recently produced the antibiotic lassomycin, which blocks a critical protease in mycobacteria. This bacterium has been called *Entotheonella* and it grows symbiotically with *Theonella swinhoei,* a marine sponge

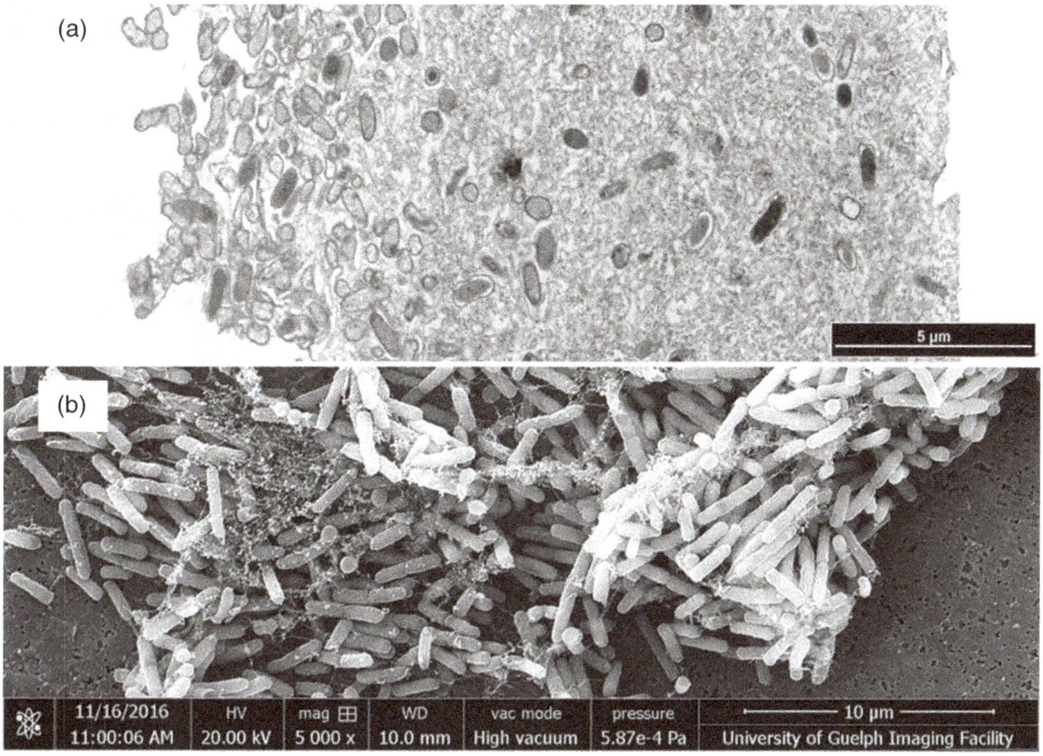

Figure 2.10 Transmission electron microscopy of a mixed biofilm: (a) section through a mixed biofilm; (b) micrograph of planktonic cells collected by centrifugation. *Source:* Courtesy of Robert Harris and Cezar Khursigara.

(Wilson et al. 2014). Using the iChip system, a new antibiotic teixobactin, which inhibits cell wall synthesis, is produced by *Eleftheria terrae*, a *Betaproteobacteria* (Ling et al. 2015). Clearly application of the iChip will enable many new secondary products of metabolism to be examined soon.

In a natural environment, bacterial cells may be in biofilms where the cells are closely packed with distance between cells being only 1–100 μm (Figure 2.10). This close relationship means that signal compounds for that species are important for growth, so gases or other end products of metabolism may need to be removed for growth to occur. Bacteria are estimated collectively to produce over a billion unique small molecules of <1 kDa (Davies 2007) and these compounds may be critical for the growth of unculturable bacteria and archaea. The addition of cAMP has been added to growth media to cultivate bacteria from limited nutrient environments because the signaling molecule, cAMP, promotes scavenging transporters to acquire nutrients from the dilute environment.

A significant discovery for the understanding of bacterial growth was the isolation of a 16–17 kDa protein from *Micrococcus luteus* that has been called a "bacterial cytokine" (Mukamolova et al. 1998). Just as the growth of vertebrate eukaryotic cells is controlled by cytokines, bacterial cytokines are signaling compounds that enable activation and multiplication of bacterial cells. Genes similar to this bacterial cytokine are present in various gram-positive bacteria including streptomyces, mycobacteria, and corynebacteria. Since this cytokine accounts for the resuscitation and promotion of growth when applied to bacteria, it is commonly referred to as a resuscitation-promoting factor (Rpf) (Mukamolova et al. 1998).

Box 2.1 Siderophores

Iron is an essential micronutrient for living systems but it is not readily available in aerobic environments because the chemistry of iron accounts for the formation of insoluble ferric iron salts. To acquire iron from these insoluble Fe^{3+} minerals, bacteria release siderophores, which have a high binding affinity for Fe^{3+} but not for Fe^{2+}. Because each species of bacteria produces a siderophore with a unique structure, there are hundreds of siderophores. A review by Hider and Kong (2010) provides characteristics and structural details of many of these iron chelating compounds. However, citrate, hydroxymate, and catechol are the three basic structures for chelation of Fe^{3+} and siderophores contain one of these moieties. Usually three molecules containing citrate, hydroxymate, or catechol moieties bind one Fe^{3+} atom and these iron-binding moieties form a "cage" with Fe^{3+} in the center.

As reviewed by Vartoukian et al. (2010), proteins with Rpf activity have been isolated from members of the *Firmicutes* including *Bacillus subtilis* and these growth-signaling compounds may be critical for the growth of unculturable bacteria.

To acquire iron from the environment, bacteria commonly use siderophores and many uncultured bacteria are dependent on other bacteria to release siderophores into the environment (Box 2.1). With the addition of different siderophores to media, the growth of uncultured isolates of *Maribacter, Hyphomonas, Reinekea, Simiduia,* and *Sulfitobacter* was achieved and *Cyclobacterium* sp. was able to grow on 26 different siderophores (D'Onofrio et al. 2010). The structure of two siderophores (i.e. enterochelin and desferrioxamine B) is given in Figure 2.11. This use of siderophores by bacteria that do not produce them has been referred

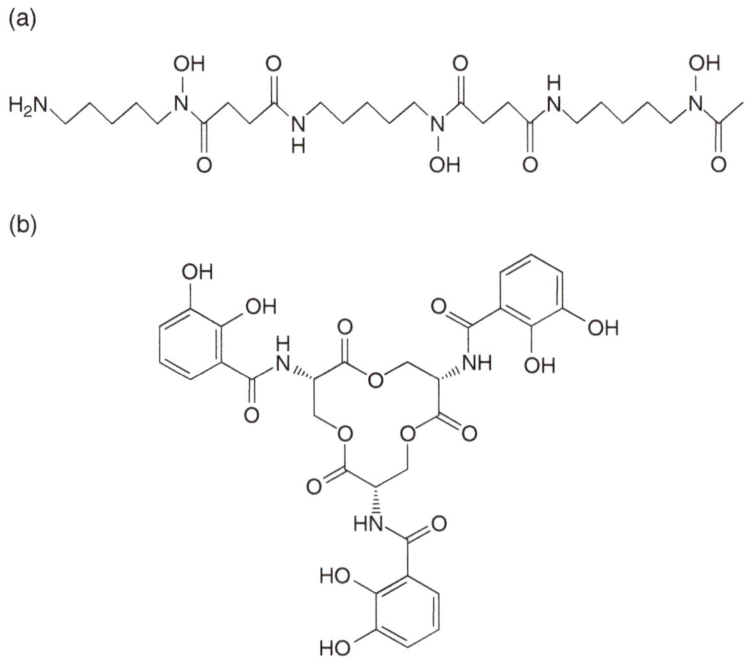

Figure 2.11 Structures of two bacterial siderophores: (a) desferrioxamine B, a hydroxamate siderophore, is produced by *Streptomyces pilosus*; (b) enterobactin, a catecholate siderophore, is produced by *E. coli*. Three of these chelators will coordinate with Fe^{3+} to provide an iron-loaded siderophore, which is transported into the cell.

to as siderophore piracy and appears to be common in nature. Bacteria that rely on helpers to produce compounds essential for their growth are sometimes considered as nutrient "cheaters" (Stewart 2012).

Rather than adding each chemical separately to a growth medium, coculture techniques may be employed. Almost 100 years ago, *Haemophilus influenza* grew as satellite colonies around colonies of *S. aureus* (Davis 1921) and we now know that *S. aureus* released heme and NAD required by *H. influenza*. The coculture of *Maribacter polysiphoniae*, a marine bacterium, with a helper microorganism (*Micrococcus luteus*, a marine sediment bacterium) has been described by D'Onofrio et al. (2010). Other cocultures reported for the growth of unculturable microorganisms include: *Lacinutrix mariniflava* supports the growth of *Verrucomicrobia* sp., heterotrophic bacteria supports growth of *Prochlorococcus* sp., *Cateellibacterium nectaripjilum* grows when the spent culture medium of a genus related to *Sphingomonas* is present, and thermophilic anaerobes related to *Clostridiaceae* grow if cell extract from *Geobacillus toebii* is present in the growth medium (Stewart 2012 and references therein). The growth of *Bacteroides forsythus* is dependent on *N*-acetylmuramic acid, which is either added to the medium or is produced by helper bacteria (Wade 2002).

2.5 Molecular Detection

In the 1980s, Rita Colwell and her colleagues (Byrd et al. 1991) noted that several normally cultivable bacteria such as *Vibrio* sp. and *E. coli* would cease to grow on common laboratory media when subjected to low nutrient conditions, yet remained viable; so they were classified as viable but non-culturable. Around the same time, other microbiologists (including David Stahl, James Tiedje, Norman Pace, and Rudolf Amann), realized that most microorganisms in various environments could not be grown on laboratory media. As a result, molecular detection strategies are commonly used to assess microbial diversity. The most commonly used molecule in prokaryotic taxonomy (Bacteria and Archaea) is the 16S rRNA gene, consisting of approximately 1500 bp (Cole et al. 2014; Woese and Fox 1977). The eukaryotic 18S rRNA gene containing approximately 1800 bp is analogous to the prokaryotic 16S rRNA gene and is also commonly used for taxonomy. All free-living organisms possess ribosomes to make proteins. Although most of the ribosome structure is highly conserved due to the natural selection pressures of having functional (i.e. protein-synthesizing) ribosomes, there are a few variable regions that are used for taxonomy. As an estimate of microbial diversity, approximately 90% of the sequences of Bacteria and Archaea represent organisms that have not been cultured in the laboratory (Cole et al. 2014).

2.5.1 Characterization of Microorganisms Using Genomics and Metagenomics

Evaluation of bacterial genomes was dependent on DNA-sequence technology established by Frederick Sanger and Walter Gilbert with recombinant DNA techniques developed by Paul Berg. These three shared the 1980 Nobel Prize in Chemistry for their efforts. Additionally, the development of molecular techniques to identify bacteria, archaea, and viruses is highly dependent on the development of computer and computer-based processing of data. An overview of applications to use DNA for characterization of bacterial isolates is given in Figure 2.12. Bacteria from an environmental sample are grown in the laboratory and pure cultures are obtained. DNA is extracted from these organisms and amplified using primers that target the 16S rRNA gene. There are several strategies for extracting and purifying DNA from microbial cultures including physical (e.g. freeze–thaw) lysis, simple chemical

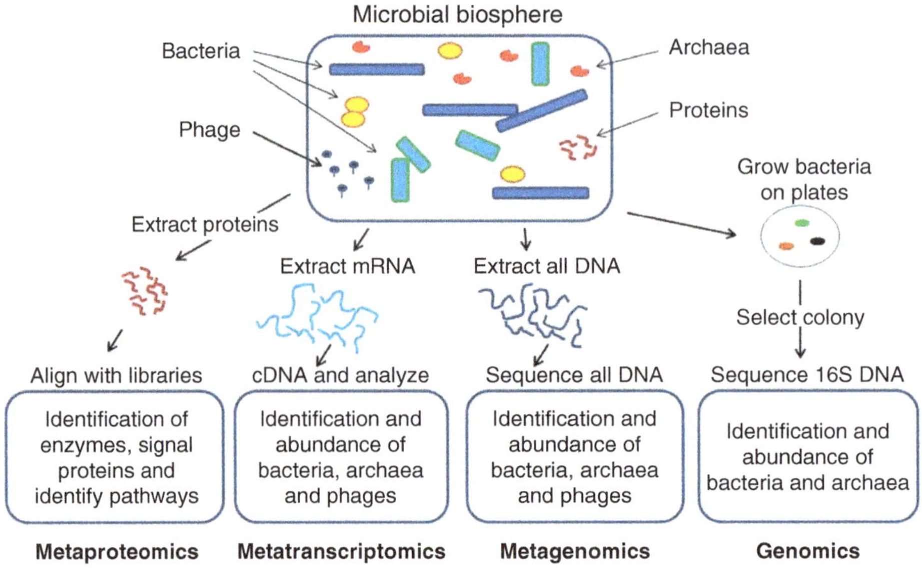

Figure 2.12 An overview of "omics" as applied to identification and characterization of microorganisms.

approaches (e.g. alkaline lysis, phenol-chloroform extraction), or the use of commercially available kits. Often primers are numbered based on their position in the *E. coli* 16S coordinates and a primer pair would include a forward and reverse primer so that both RNA-coding, DNA strands can be amplified. The specificity of primers includes domain-specific primers (targeting Archaea, Bacteria, or Eukarya) to more specific groups within the three domains. Several commonly used primers are listed in Table 2.3.

DNA sequencing can be performed using Sanger sequencing, but is increasingly being performed using high throughput, "Next Generation (NextGen)" Sequencing using technology by

Table 2.3 Representative primers used in microbial ecology.

Bacteria (general)	Sequence (5′ – 3′)[a]	Reference
8F	AGA GTT TGA TCC TGG CTC AG	Eden et al. (1991)
27F	AGA GTT TGA TCC TGG CTC AG	Frank et al. (2008)
1492R	CGG TTA CCT TGT TAC GAC TT	Eden et al. (1991)
Bacteria/Archaea (general – Earth Microbiome Project)		
515F	GTG CCA GCM GCC GCG GTA A	Caporaso et al. (2012)
806R	GGA CTA CHV GGG TWT CTA AT	Caporaso et al. (2012)
Eukaryotic (Protista – general – Earth Microbiome Project)		
1380F	GCC TCC CTC GCG CCA TCA GXX XXX CCC TGC CHT TTG TAC ACA C	Amaral-Zettler et al. (2009)
1510R	GCC TTG CCA GCC CGC TCA GCC TTC YGC AGG TTC ACC TAC	Amaral-Zettler et al. (2009)

[a] Abbreviations: A – adenine; C – cytosine; G – guanine; T – thymine; M – amino (A or C); W – weak (A or T); R – purine (A or G); Y – pyrimidine (C or T); X – region where DNA bar code can be inserted.

Table 2.4 Systems to analyze sequence information.

Name	Common use	Internet link
BLAST (Basic Local Alignment Search Tool)	DNA or protein sequence similarity	https://blast.ncbi.nlm.nih.gov/Blast.cgi
SILVA	Curated ribosomal RNA database for alignment and processing. Most currently used analysis tool for *Bacteria*, *Archaea*, and *Eukarya*	http://www.arb-silva.de
RDP (Ribosome Database Project)	Curated 16S rRNA database of Bacteria and *Archaea*, and 28S rRNA fungal database	https://rdp.cme.msu.edu/
MOTHUR	Open source software for analyzing community sequence data	https://mothur.org/
IMG/M (Integrated Microbial Genomes and Microbiomes	Annotation and analysis of microbial genomes and metagenomes	https://img.jgi.doe.gov/

manufacturers such as Illumina, Thermo-Fisher, and Roche, which is much more rapid and economical. Multiple samples can be sequenced on NextGen sequencers and to distinguish samples from each other, a short artificial DNA sequence (bar code) is included in the primer (Knapp et al. 2012). During subsequent computer analysis of the sequences (often referred to as bioinformatics), the short recognition sequences or bar codes in the primers can be recognized and used to distinguish samples. There are several strategies that can be used to analyze sequence information; several commonly used online systems are listed along with their web sites in Table 2.4.

Sequence information from 16S rDNA or other genes can be used to identify and classify organisms, especially in situations where the investigator can grow a pure culture, see Figure 2.13 for approach. There are situations where an isolate might be only distantly related to organisms in a database (e.g. ≤97% similarity) and so may represent a new species or genus. Identification of new species is relatively common in microbial ecology, whereas it is far less common in the study of higher organisms. The *International Journal of Systematic and Evolutionary Microbiology* (IJSEM) is a peer-reviewed journal and is the official journal of record for publishing new species. Author instructions with the *IJSEM* publication include the information needed for classifying a new bacterial genus and species. After the entire bacterial genome has been sequenced, a physical map for the bacterial chromosome and plasmids can be constructed. Finished genomes are annotated using Basic Local Alignment Search Tool (BLAST) or comparable programs to identify genes by finding gene homologs.

Alternatively, DNA is extracted from an environment and analysis of the bacteria present uses a process referred to as metagenomics, see Figure 2.13 for approach. This term, metagenomics, was introduced by Handelsman et al. (1998) and the first papers using metagenomics and whole genome shotgun sequencing analysis was used to characterize the biofilm in an acid drainage mine (Tyson et al. 2004) and the microbiome of the Sargasso Sea (Ventner et al. 2004). It was quickly established that new methods for isolation of DNA were required to accommodate variations in the environment. Some of the examples of environment-specific methods of isolating DNA were established for the human gut microbiota (Kennedy et al. 2014; Yang et al. 2008), human microbiome (Yuan et al. 2012; Huttenhower et al. 2012), fecal samples (Hart et al. 2015), soil and sediments (Fatima et al. 2014; Robe et al. 2003; Tsai and Olson 1991), and diverse environmental samples (Lever et al. 2015). The potential contribution of

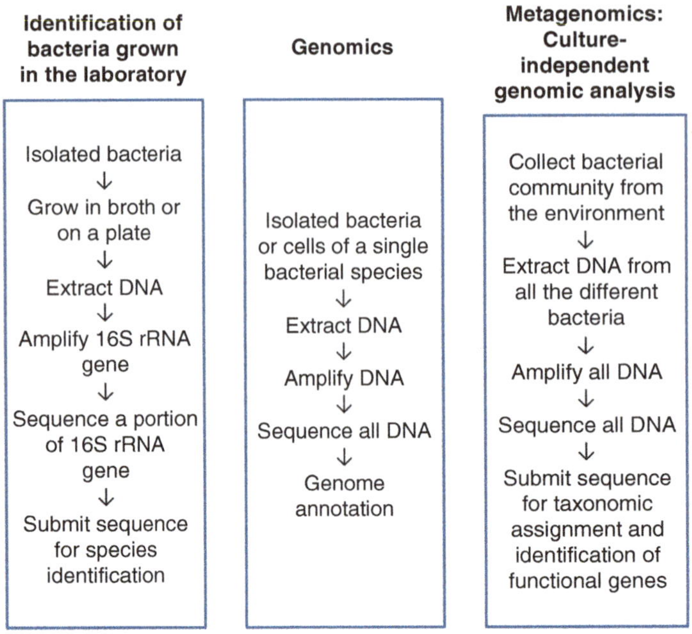

Figure 2.13 Abbreviated protocols for genomics.

metagenomics was quickly recognized and large research efforts were directed to the Human Microbiome Investigation (Human Microbiome Project Consortium 2012) and the Earth Microbiome Project (Gilbert et al. 2014), which seeks to develop a planetary database of microbial diversity.

An enormous amount of sequence information is obtained using metagenomics with DNA extracted from the environment and the organisms are not grown in the laboratory. The term operational taxonomic unit (OTU) is used generally when 16S rRNA is sequenced, to describe DNA sequences assigned to a potential organism (i.e. OTU) and some of these sequences are found in public databases such as Genbank (http://www.ncbi.nlm.nih.gov/genome). Each distinctive sequence represents a unique phylotype and many of the phylotypes from an environmental sample are uncultivated bacteria or archaea.

Metagenomic analysis of an environment enables a functional analysis of all microorganisms present and not just the ones that can be grown in pure cultures in the laboratory (Chistoserdova 2009). By using protein motifs associated with hydrolysis of polymeric carbohydrates, novel cellulases, xylanases, pectinases, amylases, esterases, and amidases have been discovered without cultivating the bacteria (de Bruijn 2011). The resistance to tetracycline, aminoglycosides, and other antibiotics has been studied using metagenomics (Boolchandani et al. 2017). Evaluation of viruses in the environment has been progressing slowly because viruses do not have a universal gene, like ribosomal 16S DNA for bacteria, but taking DNA sequences from isolated phages, metagenomics is being used to assess the distribution of bacterial and other viruses (i.e. the virome) in a specific environment.

An interesting approach to gain information about phosphorus and nitrogen metabolism was achieved using metagenomics with enrichment cultures. It is known that about half of the nitrogen lost from the oceans is attributed to anammox bacteria; however, these had not been grown in pure culture. Over 98% of the genome of *Kuenenia stuttgartiensis* was constructed from an enrichment culture using metagenomics. Gene clusters for ladderane synthesis and hydrazine metabolism, both hallmark characteristics of anammox bacteria, confirmed the identity of the genome (Strous et al. 2006). From metagenomic analysis of an enrichment culture from a sludge inoculum, nearly complete genomes of two strains of *Candidatus* Accumulibacter phosphatis, a polyphosphate-accumulating bacterium, were constructed (García Martín et al. 2006). An interesting feature of this genomic analysis of *A. phosphatis* was the discovery of a novel quinol-NAD(P)H reductase that enables the tricarboxylic cycle to function under anaerobic conditions. In another metagenomic study of an industrial digester, analysis of the genome of *Candidatus* Cloacamonas acidaminovorans predicts the use of several enzymes requiring pyrophosphate for activity (Pelletier et al. 2008).

2.5.2 Physiological Analysis Using Metatranscriptomics and Metaproteomics

More recent molecular approaches include a variety of "omics" approaches (Table 2.5) (National Research Council 2007). Advances in these broad-based approaches (transcriptomics, proteomics, and metabolomics) have been made possible by technological advances in sequencing and chemical analysis, which have greatly reduced the cost. Extremophiles may possess a number of unique biochemical pathways and so a variety of computer programs and associated algorithms are employed to complement the analysis of "omic" data. One example is the MetaGeneAlyse website (http://metagenealyse.mpimp-golm.mpg.de) through the Max Planck Society in Germany (Daub et al. 2003). Another commonly used platform is the Kyoto Encyclopedia of Genes and Genomes (KEGG, http://www.genome.jp/kegg/kegg2.html).

To address the question "How do microorganisms function in a specific process?," the activities of specific bacteria are evaluated by transcriptomics and proteomics (Figure 2.14). In a metatranscriptomic analysis of soil, mRNA was converted to cDNA by reverse transcription and direct pyrosequencing produced information about the identity and abundance of Bacteria, Archaea, and Eukarya as pertains to nitrogen metabolism (Urich et al. 2008). Metatranscriptomics of the Pacific Ocean revealed a diel (day–night) cycling with C-1 metabolism greatest in daytime and accounting for 54% of the transcripts associated with cyanobacteria, even though only 35% of the cells were cyanobacteria (Poretsky et al. 2009). A significant observation was that over half of the transcripts were unique and not observed in metagenomic studies (Frias-Lopez et al. 2008). When combining rRNA (metagenomic) and mRNA (metatranscriptomic) analysis of soil samples, quantitative information of the three domains of life can be established in a single analysis, which reduces the concern that accountability of activities may have been compromised due to the short half-life of mRNA. Additionally, it was demonstrated that the activities of *Crenarchaeota* in soil were especially important for ammonia oxidation and fixation of CO_2 (Urich et al. 2008).

Table 2.5 Broadly based "omics" molecular approaches as established by the National Research Council (2007).

Name	Molecules analyzed
Genomics	Study and analysis of the entire hereditary material (genome) in an organism (typically refers to DNA, except for RNA-viruses)
Metagenomics	Community-based DNA analysis. For microbial communities, typically 16S rRNA genes are used for identification, and other genes are used to determine metabolic potential, species interaction, etc.
Transcriptomics	RNA analysis to determine what genes are transcribed. This can be done at an organism or community level
Proteomics	Analysis of proteins at an organism or community level
Metabolomics	Analysis of all the metabolites (chemicals produced by metabolism) in an organism or community

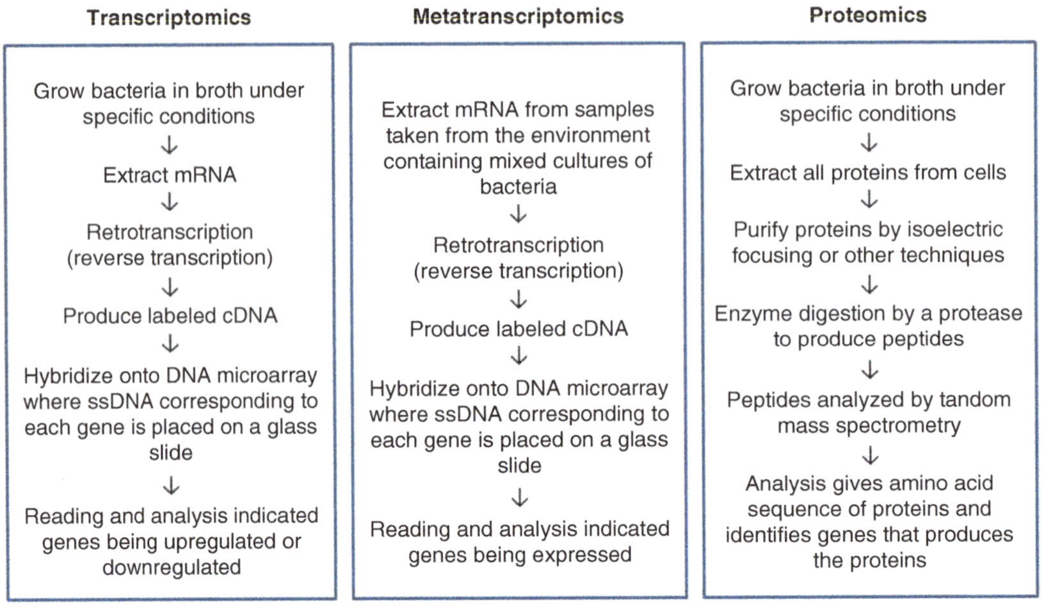

Figure 2.14 Overview for transcriptomics and proteomics.

Analysis of proteins and enzymes present in the human intestine was conducted using metaproteomics (Verberkmoes et al. 2009). In each sample analysis, 970–1340 proteins were identified and only 30–35% of these were of human origin. Approximately 20% of the proteins matched those produced by species of *Bacteroides* or *Bifidobacterium,* which correlates with the abundance of *Bacteroides* spp. in these individuals. The greatest number of bacterial proteins was associated with transcription (19%), carbohydrate metabolism (16%), and energy production (14%), with a significant number of proteins having unknown functions (Verberkmoes et al. 2009). An interesting observation was that fecal analysis could be used to predict the meal recently consumed by an individual because bacteria produced enzymes for metabolism of the food ingested.

Another application of metaproteomics is seen in the publication of Delmotte et al. (2009) which describes phyllosphere bacteria identified through protein analysis. Proteins isolated from leaves of soybean, clover, and *Arabidopsis thaliana* were digested with the products analyzed by high performance liquid chromatography coupled to mass spectrometry. The mass spectra were compared to protein sequences listed in RefSeq (ftp://ftp.ncbi.nih.gov/refseq) and the dominant bacteria were identified as *Sphingomonas* and *Methylobacterium* genera of the *Alphaproteobacteria.* Metagenomic analysis was used with metaproteomics to support the analysis based on proteins.

2.5.3 Lipid Biomarker Profiles

Over the years, biomass measurements have been based on cellular ATP, muramic acid of bacterial cell walls, and live–dead stains to enumerate bacteria; but one of the most established evaluations is based on measuring ester-linked phospholipid fatty acid (PLFA). Examination of lipids in an environment provides important information about the type of microorganisms that are present as well as the physiological status of ultramicrobacteria and dormant bacterial cells.

The PLFA pattern is sufficiently distinguishable that a commercial company (Microbial Identification System, MIDI, Newark, NJ) can identify over 650 species of bacteria. In research pioneered by D.C. White and colleagues, signature lipid biomarkers characteristic of the membranes and cell walls of bacteria have been identified to provide insight into microbial community structure and viable biomass (White and Ringelberg 1997; Balkwill et al. 1988). As indicated in the schematic of Figure 2.15, the extracted lipid holds considerable information that can be analyzed using capillary gas chromatography/mass spectrometry. Sterols including membrane cholesterol are indicative of microeukaryotic microorganisms such as algae, fungi, protozoa, and nematodes. Glycolipids such as poly β-hydroxyalkanoic acids are characteristically produced when bacteria have available carbon and electron acceptors for growth but are unable to divide. Quinones are associated with electron transport, with oxygen and nitrate as electron acceptors, while napthoquinones are used with fumarate as the electron acceptor. Quinones are not found in fermenting bacteria. By determining the quantity and types of quinones present in a sample, the amount of aerobic respiring bacteria, anaerobic respiring bacteria, and the quantity of fermenting bacteria can be estimated. The hydroxyl fatty acids from Lipid A of the lipopolysaccharide of gram-negative bacteria are quite distinctive and can be used to distinguish the various bacteria present. Archaea have alkyl ether lipids as phospholipids in the membrane whereas bacteria have an ester bond that secures the two fatty acids to glycerol with a phosphate attached to the third hydroxyl of glycerol to produce the PLFA. The PLFA is characteristic of living bacterial cells because cellular phospholipases release phosphate from PLFA when the cell dies and the diglyceride (DG) is characteristic of dead cells. Thus, by knowing the quantity of PLFA and DG, it is possible to estimate the percentage of dead cells in the population. The ratio of *trans/cis* cyclopropyl fatty acids to PLFA reflects the bacterial response to stress and toxicity. The ratio of cyclopropyl fatty acids to monounsaturated and saturated fatty acids has been used to indicate bacterial starvation when the lack of a nutrient may be growth limiting.

Lipid biomarkers are used to evaluate deep subsurface environments, and they may also be valuable to confirm the presence of bacterial life in fossilized material. Oils trapped in quartz

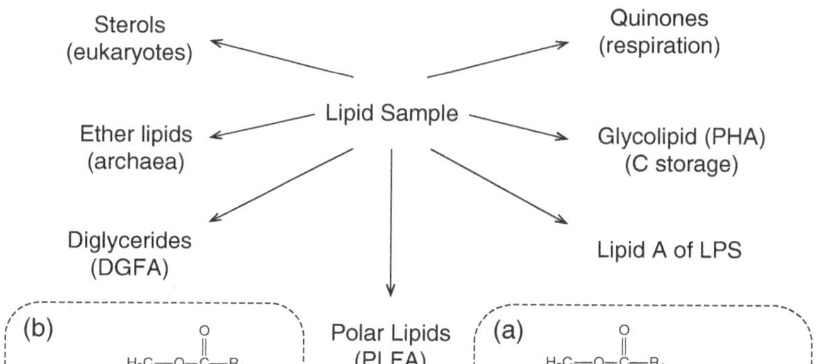

Figure 2.15 Lipid biosignals used to aid in identification of microorganisms present in a community. Structures of signature lipid biomarkers. (a) = Phosphatidyl choline, a PLFA. (b) = Diglyceride Fatty Acid (DGFA) following removal of the phosphocholine from phosphatidyl choline by a phosphoesterase. R_1 and R_2 are fatty acids.

and feldspar during diagenesis have been used to date cyanobacteria and eukaryotes in the Matinenda Formation at Elliot Lake, Canada (Dutkiewicz et al. 2006). These Huronian oil-bearing fluid inclusions suggest that oxygen-producing photosynthesis occurred before ca. 2.3 Ga (billion years before present) and before the Great Oxygenation Event.

2.6 Examining Bacteria that Do Not Grow as Pure Cultures in the Laboratory

2.6.1 Host-dependent Microorganisms

There are a few prokaryotes that have become dependent on a host to provide nutritional support. These bacteria and archaea have co-evolved with their host and as a result there is high specificity between a microorganism and the host. The inability for these microorganisms to grow independently reflects the absence of certain critical genes from the genome and reflects the small number of base pairs in the genome of host-dependent microorganism. These organisms may have resulted from traditional-like bacteria but some of the genome was lost through the process of reductive evolution. Based on the microorganisms isolated, these host-dependent microorganisms can be divided into three categories: pathogens, endosymbionts, and ecto-archaea. A few of these microorganisms are presented in Table 2.6.

2.6.1.1 Bacteria as Obligate Pathogens

A few bacteria defy the fundamental principle of Koch's Postulates used to identify agents of disease because they cannot be cultivated in the laboratory on bacteriological media. The diseases associated with these pathogens are of worldwide importance because of the number

Table 2.6 Examples of host-dependent microorganisms.

Name	Host	DNA (Mb)	Features
Bacteria as obligate pathogens			
Rickettsia typhi	Human	1.11	Agent transferred to humans by body louse
Coxiella burnetii	Human	2.10	Produces Q fever in humans and acquired by inhalation of contaminated dust
Ehrlichia chaffeensis	Horse	1.13	A tick borne zoonotic disease
Chlamydia trachomatis	Human	1.04	A sexually transmitted disease transferred by direct contact
Treponema pallidum	Human	1.13	Agent responsible for syphilis in humans
Mycobacterium leprae	Human	3.26[a]	Millions of people worldwide suffer from leprosy. Much of the DNA is nonfunctional
Bacteria as endosymbionts			
Wolbachia pipientis	Arthropods	1.26	Grows in cells of the fruit fly, *Drosophila*, and insects
Buchnera aphidicola	Aphid	0.64	Found inside special cells called bacteriocytes of most aphids
Archaea as ectosymbiont			
Nanoarchaeum equitans	*Ignicoccus hospitalis*	0.49	Acquires lipids, amino acids, cofactors, and energy is obtained from the host

[a] As a reference, *Mycobacterium tuberculosis* DNA = 4.41 Mb.

of humans infected and the high rates of mortality from these diseases. One group consists of the gram-negative bacteria of the class *Alphaproteobacteria* which are commonly referred to as rickettsia and are transmitted by arthropod (lice, fleas, mites, and ticks) vectors. Some consider that a rickettsia agent was the precursor of mitochondria and some genes of rickettsia resemble those found in mitochondria. In general, rickettsia species have a peptidoglycan cell wall, do not have genes for glycolysis or for synthesis of amino acids and nucleotides, but have genes for the tricarboxylic acid cycle and for the respiratory complex with ATP synthesis. The family *Rickettsiaceae* contains *Rickettsia* (11 species), *Ehrlichia* (2 species), *Coxiella* (1 species) and *Bartonella* (3 species). Although *Bartonella* has recently been cultivated on nutrients in the laboratory, it remains in the *Rickettsiaceae* because it shares many characteristics with the other family members.

Bacteria that are grown only in animal cell culture include the chlamydia and these pathogens are transmitted by direct contact. Currently, the 11 species of chlamydia are in the family *Chlamydiaceae* and these gram-negative bacteria constitute two different ecological groups: those producing diseases only in humans and those occurring as zoonotic diseases with humans infected occasionally. As a group, the chlamydia contain a lipopolysaccharide membrane but no peptidoglycan cell wall and they rely on the host to provide ATP.

Two other disease agents that are transferred to human hosts include *Mycobacterium leprae* and *Treponema pallidum*. Leprosy is a serious disease accounting for millions of human infections. *Mycobacterium leprae* is the agent causing the disease and it is cultivated only in foot pads of mice or in armadillos. Cultivation of *M. leprae* in animal tissues is slow due to a generation time of several weeks. The size of the *M. leprae* genome is significantly smaller than that of *M. tuberculosis* (see Table 2.1). However, the limited growth rate of *M. leprae* may be attributed to the fact that over 1000 genes are partially degraded and nonfunctional. Additionally, *M. leprae* lacks genes for energy production and for some of the genes for DNA replication. The agent of syphilis is *T. pallidum* and this spirochete is cultivated only in rabbit testicular tissue. The genome of *T. pallidum* is significantly smaller than other *Treponema* species and the genes lost due to reductive evolution include those for metabolism and nutrient uptake.

2.6.1.2 Bacteria as Endosymbionts

Wolbachia is a gram-negative bacterium that has become specialized for intracellular growth in mites, spiders, termites, mosquito, flies, crustaceans, and filarial nematodes. With *Wolbachia* infecting over a million species of insects, it is one of the most abundant bacteria on land. This bacterium is transferred from one generation to the next through eggs. To ensure an abundance of *Wolbachia* in the environment, the bacterium may kill male insect species or convert the male population to females. There is considerable interest for *W. pipientis* to serve as a vector to control insects involved in transmission of important human and animal diseases. When filarial nematodes are infected by *Wolbachia*, a symbiosis is established because if the bacterium is killed by tetracycline, the nematode also dies.

Buchnera aphidicola, a member of the *Gammaproteobacteria*, is an endosymbiont of aphids and although it has a gram-negative cell wall, it lacks lipopolysaccharides in the outer membrane. Since this symbiotic relationship was established over 150 million years ago, *B. aphidicola* has lost many of its genes by reductive evolution, and currently has only 0.64 bp of DNA. The genes lost include those for anaerobic respiration and the synthesis of fatty acids, phospholipids, and amino sugars (Nikoh et al. 2010). Additionally, this bacterium has lost control of tryptophan biosynthesis and excess tryptophan is provided to the aphid. The coexistence of *Wolbachia* and *B. aphidicola* in aphids has been reported (Gómez-Valero et al. 2004).

An interesting development is the documentation of bacteria as endosymbionts of fungal hyphae. *Rhizopus microsporus* is a fungus that is a plant pathogen especially important in rice seedling blight, which has been attributed to rhizoxin, an antimitotic virulence factor. However, the toxin is not produced by the fungus but by *Burkholderia rhizoxinica*, an endosymbiont of *R. microsporus*. This symbiosis is interesting because the fungus host is unable to produce fungal spores unless it contains the bacterial endosymbiont. Genome analysis has revealed that the bacterium lacks uptake systems for sugars but uses uptake transport systems for citrate and malate to consume organic acids in the fungal cytoplasm (Lackner et al. 2011). Bacterial endosymbionts have been detected in several arbuscular mycorrhizal fungi including *Scutellospora persica, S. castanaea*, and *Gigaspora margarita* and 16S rDNA analysis suggests that these bacteria are closely related to *Burkholderia* (Bianciotto et al. 2000). Recent evidence reveals that the arbuscular mycorrhizal fungus *G. margarita* harbors two distinct bacteria as endophytes: (i) *Candidatus* Glomeribacter gigasporarum and (ii) a coccoid *Mollicutes*-related bacterium (Desirò et al. 2014).

Marine gutless worms (*Oligochaeta*) use nutrients supplied by chemolithotrophic sulfur bacteria for growth. For *Olavius crassitunicatus*, a symbiont-bearing worm in the East Pacific Ocean, 16S rRNA analysis revealed the presence of five unidentified bacterial species: two species from *Gammaproteobacteria*, two species from *Deltaproteobacteria*, and a spirochete species (Blazejak et al. 2005). These endosymbiotic bacteria are thought to grow in the tissue of the gutless worms where some are sulfur oxidizers and some are sulfate reducers.

2.6.1.3 The *Nanoarchaeum–Ignicoccus* Relationship

Ignicoccus hospitalis is a chemolithotrophic archaeal organism living in marine hydrothermal vents and it has the archaeal parasite *Nanoarchaeum equitans* living on its surface. The genome content of this archaeon is only 0.49 bp and it is not known if it went through a strategy of gene reduction or if it represents organisms that have yet to be discovered (Das et al. 2006). Clearly, the genes for biosynthesis and metabolism have been lost, which makes it dependent on a living host for growth. Unlike bacterial endosymbionts, *N. equitans* has genes for DNA repair as well as those for expression of information (DNA replication, transcription, and translation). An extensive evaluation of the proteins present in *N. equitans* and *I. hospitalis* is available and the metabolic interchange between these two archaea can be examined (Giannone et al. 2011).

2.6.2 Molecular Analysis of Uncultivable Bacteria

By using molecular techniques, it is now possible to characterize both cultivated and uncultivated bacteria in mixed populations in seawater, hot springs, and oral microflora (Wade 2002; Giovannoni et al. 1990; Ward et al. 1990). To achieve this, DNA is extracted from the original biomass, the 16S rDNA is amplified by the polymerase chain reaction, and a mixture of all 16S rDNA segments in the biomass is produced. The 16S rDNA molecules are placed in a plasmid vector and inserted into *E. coli* to be sequenced. This will detect all the bacteria present in the sample. Routine culture procedures are used to isolate and identify the bacteria that can be cultivated in the sample. After the cultivated bacteria are subtracted from all bacteria detected by the polymerase chain reaction and cloning procedures, the uncultivated bacteria remain.

A procedure that has produced highly significant results identifies not only 16S rDNA but also all the genes present and is dependent on sequencing the uncultivated bacteria. The genome of *Candidatus* Sulcia muelleri was produced by amplification of DNA from a single bacterial cell (Woyke et al. 2010). Comparison of the genome from *C.* S. muelleri, an endosymbiont of the green sharpshooter, with the genome from *Sulcia* sp., an endosymbiont of cicada, found a unique protein in *C.* S. muelleri thought to be in the outer membrane and to account for host specificity.

2.7 Microbial Community Structures

2.7.1 Primary Production and Microbial Communities

Microorganisms are rarely found in pure cultures, but typically live in mixed populations. Biological processes, including microbial growth, require energy and so we can consider a community as having an energy source. In general ecology, sunlight and associated primary productivity through photosynthesis is considered to be the primary source of energy for the bulk of the biosphere. Although we commonly associate photosynthesis with oxygen production (oxygenic photosynthesis), there are many photosynthetic bacteria that do not produce oxygen. Detailed metabolic aspects of the oxygenic and anoxygenic photosynthetic bacteria are covered in the text by White et al. (2011) and summarized in Table 2.7. Anoxygenic photosynthesis is a feature associated with some groups of bacteria including the purple photosynthetic bacteria and the green sulfur bacteria. The major photosynthetic pigment in these organisms is bacteriochlorophyll and depending upon the organism, the electron donors are either organic compounds or sulfur compounds including H_2S and thiosulfate ($S_2O_3^{=}$). Interestingly, these bacteriochlorophyll molecules also absorb light in the far red and near infrared wavelengths (ranging from 700 to 1050 nm). Light energy (E) is inversely proportional to its wavelength (λ):

$$E = \frac{hc}{\lambda}$$

Where h represents Planck's constant and c is the speed of light. Although lower energy than visible light, this long-wavelength far red and near infrared light can penetrate turbid

Table 2.7 Characteristics of photosynthetic bacteria.

Group of bacteria	Electron donor	Major pigments[a]	Light absorption for growth	Taxonomic group
Green sulfur	S^0, HS^-, H_2	Bchl *a*, *c*, *d* or *e*	450–650 nm	*Chlorobi*[b] *Chlorobium*, *Pelodictyon*
Green nonsulfur	Organic acids, Amiono acids, Sugars	Bchl *a* and *c*,	400–700 nm	*Chloroflexi*[b] *Chloroflexus*
Purple sulfur	S^0, HS^-, H_2	Bchl *a* or *b*	400–900 nm	*α-Proteobacteria*[c] *Thiospirillum*, *Chromatium*
Purple nonsulfur	Various organic compounds	Bchl. *a* or *b*	400–950 nm	*α-Proteobacteria*[c] *Rhodobacter*, *Rhodospirillum* *β-Proteobacteria*[c] *Rhodocyclus*
Aerobic anaerobic Phototrophs	Amino acids and Sugars	Bchl *a*	400–870 nm	*α-Proteobacteria*[c] *Roseobacter* *Erythromonas*
Cyanobacteria	H_2O	Chl *a*	400–660 nm	*Cyanobacteria*[b] *Synechocystis* *Nostoc*

[a] BChl = bacterial chlorophyll; Chl = chlorophyll.
[b] Phylum.
[c] Class.

water and the upper layer of sediments, thus expanding the environments in which anoxygenic photosynthetic organisms can grow. One of the most dramatic events in evolution came with the development of oxygenic photosynthesis and photosystem II, which is estimated to have occurred 2.3 Ga (Fischer et al. 2016). Before this change (referred to as the Great Oxygenation Event), the Earth was a reducing environment. Oxygenic photosynthesis is present in cyanobacteria and in the chloroplasts of higher plants. Light absorption of a common form of chlorophyll (chlorophyll a) is in the blue (430 nm) and red (662 nm) region of the spectra, which yield considerably more energy than the far red and near infrared absorption maxima of bacteriochlorophyll. This molecule does not absorb green light but reflects it, thus explaining the green color of plants and cyanobacteria. Aquatic scientists often use the concentration of chlorophyll in water samples to assess the primary productivity of planktonic phototrophs in rivers and lakes.

Photosynthetic bacteria contain different accessory pigments including various carotenoid and phycobilin compounds. These colored pigments absorb light at different wavelengths than chlorophyll and bacteriochlorophyll but can transfer some absorbed energy to the photosynthesis complex. Consequently, the action spectra (light wavelengths at which growth can occur) is expanded beyond the narrow absorption spectra of chlorophyll and bacteriochlorophyll. Other functions of accessory pigments include protection of the organism from excess light, and for carotenoid pigments, protection from oxidative stress (Stomp et al. 2007).

Photosynthetic organisms use light energy to generate membrane potential and ATP. Lithotrophs (chemoautotrophs) are also an important source of primary production and CO_2 assimilation in some environments, particularly deep oceans or aquifers and other subsurface environments, where sunlight is not present. With chemoautotrophs, oxidation of simple chemicals such as ammonia (NH_3) and sulfide (S^{2-}) provides the energy as ATP and proton motive force. In most autotrophic microorganisms as well as plants, the Calvin cycle is used to assimilate CO_2 or bicarbonate (HCO_3^-) in aquatic environments and the key enzyme complex is ribulose bisphosphate carboxylase (RuBisCo). In some bacterial systems where carbon dioxide levels may be low, the RuBisCo enzyme complex is aggregated into a unit called a carboxysome.

Microbial communities form in a wide variety of environments. Typically, energy and nutrition (i.e. access to food) are key driving forces for community formation. The community energy source may arise from light and the corresponding primary production. As explained in Chapter 6, higher plants will promote the formation of desired microbial communities in the vicinity of their roots (rhizosphere) through a variety of organic secretions (root exudates). Alternatively, in communities where lithotrophs represent primary production, the organisms will grow in areas where there is a chemical gradient between a reduced compound and an oxidizing compound. When the energy source arises through organic matter, then heterotroph-based communities tend to disperse once the organic matter is consumed.

One notable environment in which lithotrophs provide primary production is hydrothermal vent communities (Figure 2.16), which occur near volcanoes and other regions of geothermal activity on the ocean floor. There is a difference in microbial communities associated with deep ocean (>200 m) hydrothermal vents and shallow water (<200 m) hydrothermal vents (Tarasov 2006). Gases released from vents support the growth of bacteria and archaea. Although deep vents release less gas than shallow water hydrothermal vents, the quantities of H_2, H_2S, and CH_4 are greater in the deep vents. Significant quantities of CO_2 and O_2 are released from the shallow water vents but not the deep vents. Due to the mineralization of organic materials in the shallow water vents, considerable levels of NH_4^+ are released. In waters near shallow water vents, there is an abundance of thermophilic and hyperthermophilic bacteria and archaea, with some vent waters having over 35 species of microorganisms present. The chimneys of deep

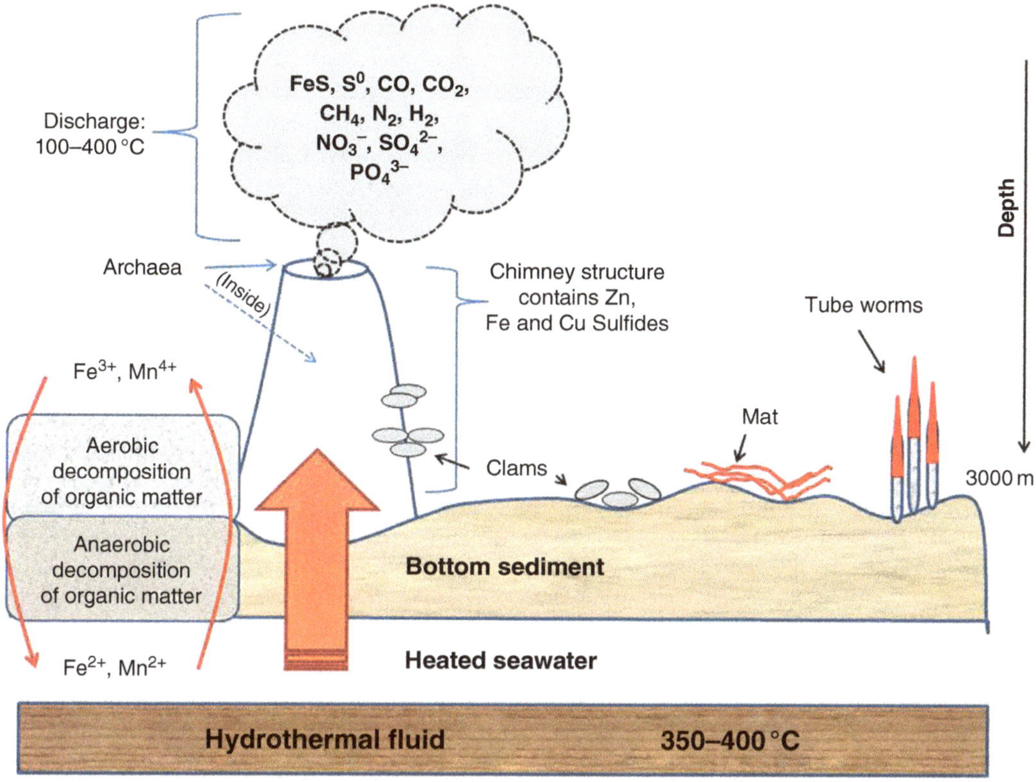

Figure 2.16 Biological communities near deep hydrothermal vents. Bacteria that may be in mats include some of the following filamentous, colorless sulfur bacteria: *Beggiatoa, Thiothrix, Thiovulvum,* and *Thiomicrospira*. Archaea at the top of the chimney are hyperthermophilic organisms whereas those inside the chimney are extreme halophiles. For additional information see Takai et al. (2001) and Tarasov et al. (2006).

hydrothermal vents are rich in archaea with hyperthermophilic archaea present on the upper surface of the chimney (Takai et al. 2001). Species of *Pyrolobus, Pyrodictium, Hyperthermus,* and *Thermococcus* are commonly found at the rim of the chimney whereas *Haloarcula* are found inside the chimney structure. Near the shallow water vents three types of mats develop: algal mats containing diatoms, mats containing mixtures of algae and bacteria, and mats consisting of bacteria only (Tarasov 2006). As indicated in Figure 2.16, bacterial mats are found on the ocean floor near deep hydrothermal vents.

In contrast to the relatively barren appearance of most of the deep ocean floor (2–3 km deep), the region close to hydrothermal vents is teaming with life including a variety of large tube worms (*Riftia pachyptila*) and clams (*Calyptogena magnifica*) (Cavanaugh et al. 1981). Within the gill-like structures of these large tube worms and clams, lithotrophic bacteria oxidize sulfide, use this energy to fix CO_2, and share the fixed carbon with their hosts. Most organisms on Earth can trace their energy source to the sun and associated photosynthesis, whereas hydrothermal vent communities are completely independent of solar energy and associated photosynthesis, instead relying on chemoautotrophic primary production. Lithotroph-based primary production can also be an important source of energy in groundwater environments.

2.7.2 Biofilms

Microbial growth on surfaces has been described by many early microbiologists including van Leeuwenhoek, Pasteur, and Zobell. In the 1980s, the term biofilm was used by J.W. Costerton to describe surface-adherent microbial communities and has since been shown by many investigators to be a prominent mode of microbial growth in most environments. Commonly encountered examples of biofilms include dental plaque and the slimy microbial growth on rocks and other surfaces in water. Bacterial colonies on agar plates are also a type of a biofilm (sometimes referred to as a colony biofilm). Microorganisms also associate with other surfaces including air–liquid interfaces (often called microbial mats) and soil surfaces, particularly in arid environments (cryptogamic crust). Suspended microorganisms in liquid environments can also aggregate into macroscopic clumps often referred to as flocs and examples of these aggregates of bacteria are shown in Figure 2.17. Marine snow is an example of microbial flocs in the pelagic oceans, see Chapter 3. In comparison to unattached (planktonic microorganisms), sessile (biofilm) organisms are much more tolerant (in some cases up to 1000-fold) to antibiotics, disinfectants, and other antimicrobial agents; so over the past 30 years there has been a tremendous amount of research on microbial growth in biofilms. Once organisms within a biofilm re-enter the planktonic population, antimicrobial susceptibility returns and so the term antibiotic tolerance is used for biofilms rather than resistance, which implies a heritable genetic change in an antibiotic target or the production of an antibiotic degrading enzyme or efflux mechanism.

Biofilms have been demonstrated for many different microorganisms including *Bacillus cereus* (Majed et al. 2016), *Desulfovibrio fairfieldensis* (Jorand et al. 2015), *Clostridium difficile* (Dapa and Unnikrishnan 2013), *Clostridium perfringens* (Charlebois and Jacques 2014; Varga et al. 2008), *Treponema denticola* and *Porphyromonas gingivalis* (Kuramitsu et al. 2005; Yamada et al. 2005), *Campylobacter jejuni* (Reuter et al. 2010), cyanobacteria (Rossi and

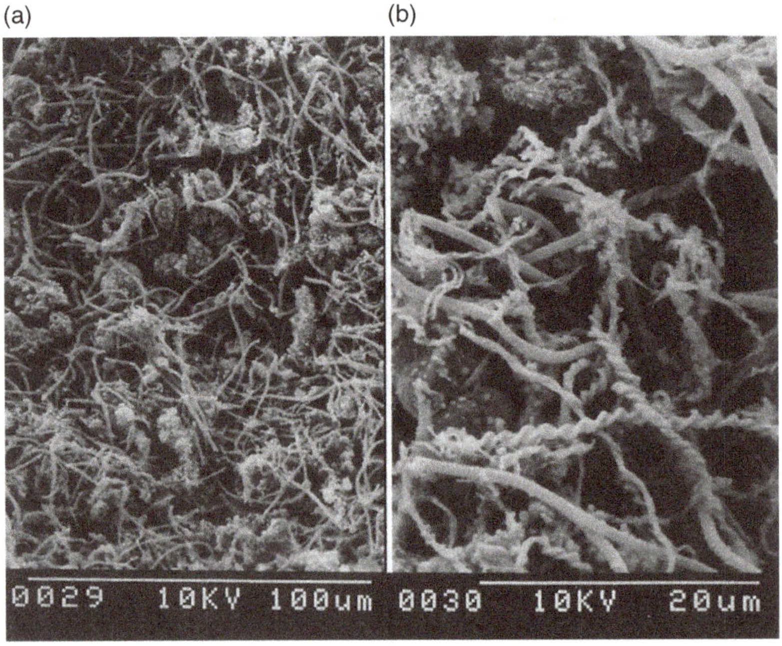

Figure 2.17 Bacteria present in an aquifer as a biofilm at a depth of 275 m. Magnification: (a) bar = 100 µm; (b) bar = 20 µm.

De Philippis 2015), *Thiobacillus* (Nikolov et al. 2002; Barahona et al. 2014), *Nitrobacter* and *Nitrosomonas* (Schramm et al. 1996), anaerobic photosynthetic bacteria (Wang et al. 2015), and archaea (Orell et al. 2013; Fröls 2013). It is quite possible that most bacteria and archaea are present in the environment as biofilms. To understand the mechanisms of biofilm formation and growth, several model organisms have been used and these include *Pseudomonas aeruginosa* (Sauer et al. 2002; Mah et al. 2003; Davies et al. 1998), *E. coli* (Schembri et al. 2003), *Vibrio cholerae* (Watnick and Kolter 1999; Yildiz and Schoolnik 1999), *Bacillus subtilis* (Cairns et al. 2014; Hamon et al. 2004), and *Candida albicans* (Chandra et al. 2001). Here, some basic concepts of biofilms are addressed, and biofilm formation is illustrated in Figure 2.18.

In aquatic and marine environments there is often a trace amount of suspended organic matter in the liquid including some proteins, polysaccharides, humic substances, or other organic molecules. This organic matter can adsorb onto abiotic surfaces and form a thin organic layer, referred to as a conditioning film. Typically, microorganisms that initially colonize a surface will adhere to the conditioning film rather than the underlying substratum. Sometimes, molecules on a conditioning film, or secreted compounds (exudates) from a biotic surface, can represent a nutrient source and motile microorganisms may be attracted to the surface via chemotaxis. In some environments, organisms that initially colonize a surface (primary colonizers) can influence subsequent population development by other species (secondary colonizers). Sequential microbial colonization on a surface is sometimes referred to as microbial succession.

The term, adhesin, is often used to describe microbial cell surface structures involved in adhesion to a surface. The two most common bacterial adhesins are pili (singular pilus, sometimes referred to as fimbria (s)/fimbriae (pl)); and exopolymeric substance (EPS). Pili are thin hair-like protein structures on the bacterial cell surface approximately 6–7 nm wide and up to 1 μm long. Flagella (used in motility) are much larger structures (~20 nm in diameter) and can extend up to 1–2 μm long. Some pili can bind to specific target molecules on a surface and contain carbohydrate-binding proteins (lectins) that recognize sugar molecules on plant root surfaces (rhizosphere) and other environments, and thus function as a target-specific adhesin.

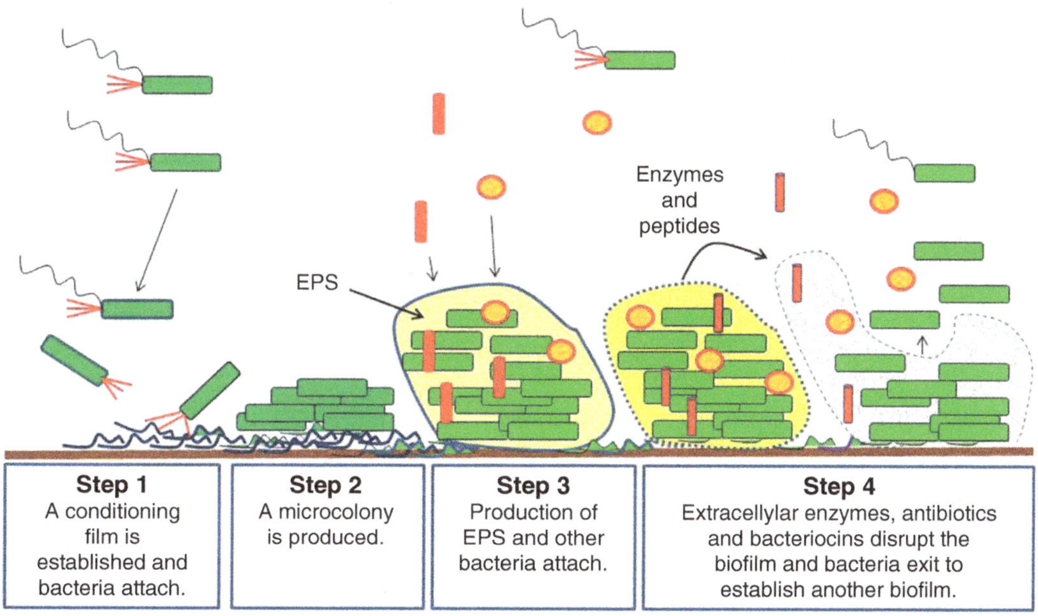

Figure 2.18 Biofilm produced by bacteria. Different shapes indicate different species. EPS = exopolymeric substance. See text for explanation.

The other major adhesin, EPS, consists of polysaccharides, extracellular DNA (eDNA), and secreted proteins. EPS is a general adhesin (i.e. attaches to all surfaces in the environment).

As shown in Figure 2.18, biofilm formation is a developmental process in which organisms attach to a surface (step 1). Once attached, organisms will aggregate together into clusters referred to as microcolonies (step 2). This aggregation involves type four pili extending from the bacterial cells. The tips of the pili attach to the surface and as the pili are retracted, the cell is pulled forward in a manner analogous to a grappling hook. Organisms within microcolonies produce copious amounts of EPS that forms the biofilm matrix (step 3). Biofilm maturation consists of continued growth of the microcolonies and EPS production. The final stage is the release of organisms from the biofilm and their return to a planktonic stage (step 4).

2.7.3 Role of Quorum Sensing

Population density-dependent gene expression (quorum signaling) is a key regulatory component of biofilm formation and becomes prominent during microcolony growth and particularly biofilm maturation, including EPS formation. The development of communities in the biosphere is dependent on the level of communication between the individuals involved. Microbial communication between organisms of different species and between plants or animals is an extension of the chemical mediated process known as quorum sensing. Cell-to-cell communication enables a population of bacteria to react to the environment, coordinate the response to achieve a common goal, and enhance the persistence of that bacterial species. A basic model of quorum sensing is given in Figure 2.19, which shows the production of small organic molecules

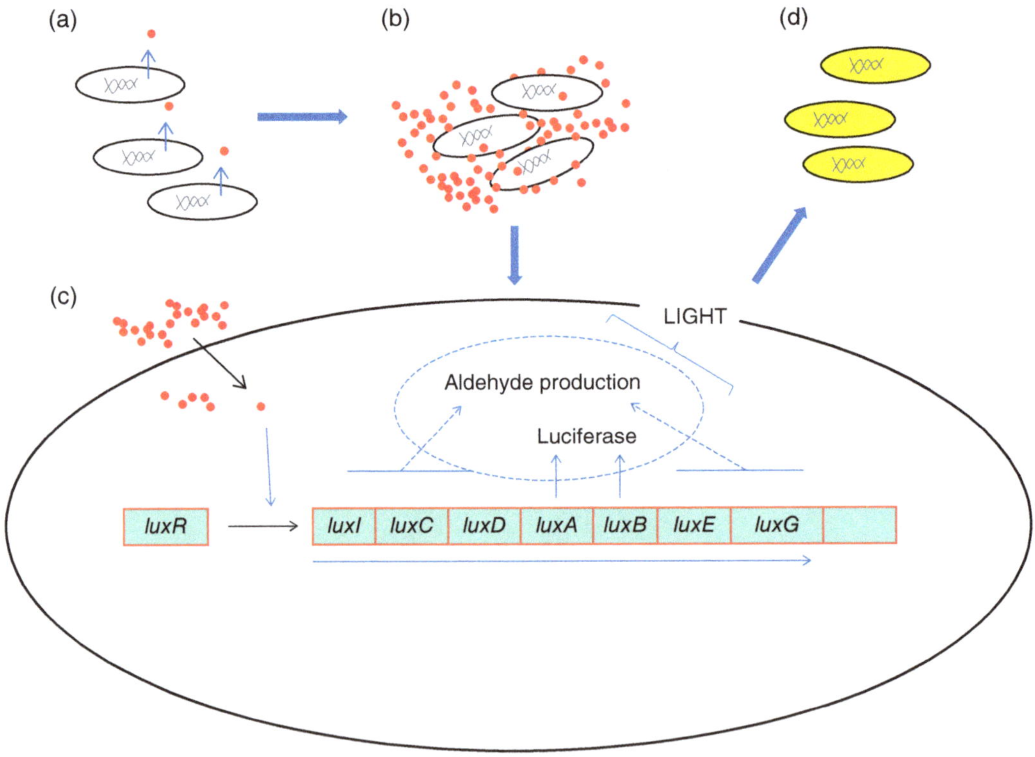

Figure 2.19 Quorum sensing in the gram negative bacterium *Vibrio fischeri* and regulation of the *lux* operon: (a) at low population levels, low concentration of AHL results in no AHL-binding to *luxR* and no *lux* expression or light production; (b) at high population levels, high density of AHL leads to AHL-binding to *luxR* which causes expression of the *lux* operon (c), and the cells are induced to produce light (d).

involved in bacterial communication. Each bacterial cell produces a limited number of signal molecules which are called autoinducers. These autoinducers can be taken into the bacterial cells to induce the production of new phenotypic expressions; however, the possibility for inducer–cell interaction is not favorable until bacterial cells density (quorum) is high enough to produce a sufficient concentration of autoinducer for interaction with cells. Usually, each species of bacteria produces a specific autoinducer structure with gram-negative bacteria using acyl-homoserine lactones (AHLs) and gram-positive bacteria using oligopeptides. Examples of autoinducers are given Figure 2.20 (Waters and Bassler 2005). Some of the activities resulting from quorum sensing are given in Table 2.8. Quorum signaling was first described in the bioluminescent organism *Vibrio fischeri* because these organisms would only produce light at high cell density. At low cell density, *V. fischeri* could be induced to produce light if exposed to spent culture (containing metabolites) from cells at high density. Initially the term, autoinducer, was used to describe this phenomenon for the genes producing (*luxI*) and responding to (*luxR*) the signal. *Vibrio fischeri* gene expression is regulated by positive feedback (Ng and Bassler 2009).

Quorum signaling is the major term used as it reflects population-density activity and is a major regulatory system contributing to global gene regulation in most prokaryotes and reflects the control attributed to quorum sensing regulons (Waters and Bassler 2005). Based on transcription profiling studies, *Streptococcus pneumonia* has over 150 genes regulated by quorum sensing, which not only results in competence (accepting DNA into recipient cell) but also impacts biofilm formation, bacteriocin production, acid tolerance, and virulence. Transcriptome analysis reveals that the quorum sensing regulon in *Pseudomonas aeruginosa* may influence as many as 616 genes, and in *V. cholera* influences over 70 virulence genes.

Quorum sensing has been observed throughout the microbial world in over 70 species of gram-positive and gram-negative bacteria as well as the archaeon *Natronococcus occultus*. Quorum sensing in bacteria is important because it is involved in several behavior activities including motility, sporulation, horizontal gene transfer, antibiotic production, conjugation virulence symbiosis, and mat formation. Gram-negative bacteria produce AHLs, which have a side chain of 4 carbons to 18 carbons in length (Figure 2.20). Quorum sensing accounts for a synchronized metabolic or physiologic response by bacteria to environmental stimuli. Different species of gram-negative bacteria produce AHLs with side chains of different lengths and this enable each species to communicate with its own species. Many of the gram-positive bacteria produce

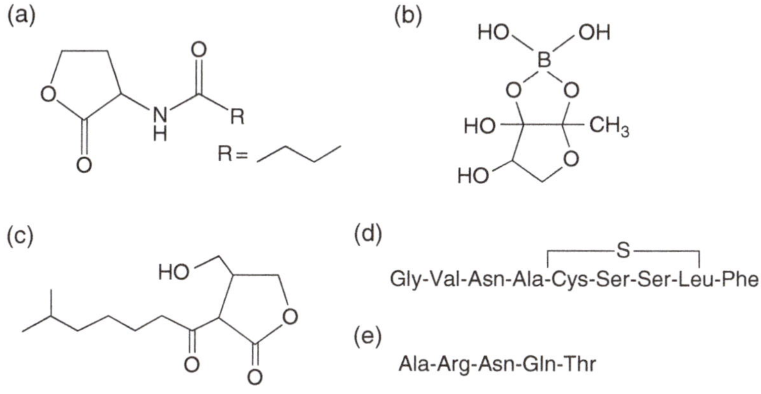

Figure 2.20 Structures of autoinducers for quorum sensing in bacteria: (a) *N*-acylhomoserine lactone, AHL, and structure presented is produced by *Serratia marcescens*; (b) furanosylborate, AI-2, produced by *Vibrio harveyi*: (c) γ-butyolactone produced by *Streptomyces griseus*; (d) peptide produced by *Staphylococcus aureus*: (e) peptide produced by *Bacillus subtilis*.

Table 2.8 Characteristics of quorum sensing systems.

Signal	Organisms found	Genes (if known)	Regulatory function in cell
N-acylated homoserine lactones (AHLs), originally called autoinducer 1 (AI-1)	Many gram-negative bacteria; individual AHLs are species-specific	*luxI* homologs (AHL synthesis); *luxR* homologs (AHL receptor)	Virulence, bioluminescence, biofilm formation; AHLs also reported in Archaea
Autoinducer 2 (AI-2) Universal signal molecule	Numerous Bacteria and Archaea	*luxS* homologs (AI-2 synthesis)	Interspecies communication
Autoinducer 3 (AI-3), cross-reacts with epinephrine and norepinephrine	Strains of *E. coli* and *Salmonella enterica*	*qseC/B*; also *qseE/F* (two-component AI-3 receptors)	Interkingdom signaling; *E. coli* intestinal infections
CAI-1	*Vibrio* sp.	*cqsA* (synthesis) *cqsS* (receptor)	Genus specific signal, genetic exchange
Competence-stimulating peptide (CSP); gram-positive QS peptides are sometimes called pheromones	*Streptococcus* sp.	*comAB* (CSP synthesis) *comDE* (two-component CSP receptors)	Competence (genetic exchange), biofilm growth
Agr – cyclic peptide pheromones	*Staphylococcus aureus*	*agrD* (encodes pheromone); *agrC/A* (two-component receptor)	Virulence

peptide autoinducers (Figure 2.20) with each species producing a unique peptide. In the environment, peptide autoinducers are found to be more stable to elevated temperatures than AHLs. Although the research with autoinducers is being initiated with extremophiles, initial reports suggest that quorum sensing occurs in the environment by halophiles, acidophiles, thermophiles, psychrophiles, piezophiles, and archaea (Montgomery et al. 2013). The stability of the autoinducer molecule is associated with the length of the R groups of AHLs, with shorter R groups enabling the AHL to have a longer half-life in the environment and to be more thermal resistant. In natural environments, such as sludge, about 10% of the bacteria are capable of producing quorum sensing molecules but approximately 60% of the bacteria were capable of quenching (enzymatically inactivating) the AHLs (Tan et al. 2015). Dense mats or granules of bacteria were formed in sludge when AHLs were produced but when the AHLs were quenched there was no biofilm produced. Thus, in an open environment dealing with mixed cultures of bacteria, the establishment of a biofilm is dependent on the extent that the quorum sensing molecules can function.

One key aspect of biofilm growth is the greatly enhanced tolerance of microorganisms to antibiotics and disinfectants as well as enhanced stress resistance and virulence. In *Pseudomonas aeruginosa* and several other organisms, quorum signaling has been shown to be an essential regulator for antimicrobial tolerance. In some organisms, notably the gram-positive genera *Streptococcus* and *Enterococcus*, quorum signaling is required for genetic exchange and the quorum signal molecules may be referred to as bacterial pheromones. This use of autoinducers as molecular signals appears to be a key component of biofilm formation. Chemical antagonists to quorum sensing molecules result in inhibition of biofilm formation (Hentzer and Givskov 2003a, b; Estrela and Abraham 2010). Several natural products serve as chemical antagonists of signal molecules and include natural products such as extracts from garlic, citrus flavonoids, halogenated furanones produced by marine alga, and small bromine-containing molecules produced by marine sponges. The addition of the fatty acid signal, 2.5 nM *cis*-2-decanoic acid

(Davies and Marques 2009) accounts for the dispersal of biofilms containing *P. aeruginosa, Escherichia coli, Klebsiella pneumonia, Proteus mirabilis, Streptococcus pyogenes, Bacillus subtilis,* and *Staphylococcus aureus.* Also, nutrient levels have been associated with detachment, with both elevated and reduced nutrient levels being associated with dispersion. Although a common mode of microbial growth, biofilms are associated with various problems for human society including ship and pipeline fouling, and many chronic infectious diseases. As a result, there is considerable interest in identifying molecules that disrupt quorum signaling (to prevent biofilm formation) and promote biofilm detachment.

In polymicrobial biofilms, distribution of individual species is impacted by several factors including attraction to or repulsion by metabolites or signals of other members. The distribution of organisms in mixed communities is not random, but generally involves organisms of a different species occupying distinct niches in association with other species. For example, biofilms containing cells of methanogens and sulfate-reducing bacteria participate in the cycling of sulfur. Another notable aspect of biofilm growth is the tendency of individual organisms to specialize and perform metabolic functions that may help the overall community. There are also instances where some organisms within a biofilm may function as "cheats" so they benefit from the activities of other organisms in the community, yet do not contribute themselves. Quorum signal cheats (able to use but not produce quorum signals) may also be present in biofilms, as are community members that do not produce EPS. In this context, biofilms may represent an evolutionary bridge between single-cell and multicellular life.

Bacterial interspecies signaling involves the signal molecule autoinducer 2 (AI-2), which was discovered by Bonnie Bassler and her laboratory. The molecule is unique because it contains boron – see structure in Figure 2.20b. Al-2 is recognized by many bacterial species and, therefore, can function as a universal bacterial signaling molecule regulating numerous activities, including biofilm formation and virulence (Pereira et al. 2013). Several bacteria have developed a quenching response to Al-2, which inhibits quorum sensing in a specific bacterium and thereby gives that species uninterrupted growth (Han et al. 2015). As reviewed by Waters and Bassaler (2005), *cis* (Davies and Marques 2009) spp., *Staphylococcus aureus, Variovorax paradoxus, Ralstonia* spp., *Escherichia coli, Salmonella typhimurium, Pseudomonas aeruginosa, Erwinia carotovora,* and *Agrobacteroum tumefaciens* produce enzymes that modify the structure of Al-2. The pharmaceutical industry has focused on developing therapies to interfere with virulence development in pathogenic bacteria by interfering with quorum sensing (Hentzer and Givskov 2003a, b; Zhu and Li 2012; Reuter et al. 2016).

2.8 Summary

Microorganisms can form intricate communities throughout the biosphere and numerous laboratory-based activities to dissect these communities are discussed in this chapter. The observation of individual cells in the community by fluorescence microscopy and electron microscopy gives useful insight into microbial types and activities in the environment. To evaluate microbial communities, individual cells have historically been cultivated in the laboratory but using genomics, information can be obtained on the presence and activities of bacteria. The application of metagenomics, metatranscriptomics, and metaproteomics provides a wealth of information about the activities of microorganism in the environment and is especially useful for uncultured microorganisms. A few microorganisms grow in close association as symbionts and obligate pathogens with a living host. A common microbial community structure involves the growth of biofilms in aquatic environments or biomats in terrestrial areas. Cell-to-cell communication in bacteria is by quorum sensing using unique chemicals as signals to optimize bacterial response to changes in the environment.

Discussion Questions

1 As molecular techniques can now be used to identify bacteria and their genes without cultivating them, why is it still important to grow environmental bacteria in the laboratory?

2 It is known that bacteria in the environment can degrade hydrocarbons, what are some of the concerns that need to be addressed so that a person could see living bacteria growing on the surface of an oil droplet?

3 Cyanobacterial mats not only contain cyanobacteria but also facultative and obligate anaerobic bacteria. Construct a series of experiments that would aid understanding of carbon and nitrogen cycles in the cyanobacterial mat.

4 Describe some of the features that required co-evolution between *Wolbachia* and an insect host.

5 Discuss some of the activities of bacteria in a biofilm that help to stabilize the bacterial community within the biofilm.

References

Alsteens, D., Verbelen, C., Dague, E. et al. (2008). Organization of the mycobacterial cell wall: a nanoscale view. *Pflügers Arch. Eur. J. Physiol.* 456: 117–125.

Amann, R.I., Krumholz, L., and Stahl, D.A. (1990). Fluorescent oligonucleotide probing of whole cells for determinative, phylogenetic, and environmental studies in microbiology. *J. Bacteriol.* 172: 762–770.

Amaral-Zettler, L.A., McCliment, E.A., Ducklow, H.W. et al. (2009). A method for studying protistan diversity using massively parallel sequencing of V9 hypervariable regions of small-subunit ribosomal RNA genes. *PLoS One* 4 (7): e6372. https://doi.org/10.1371/journal.pone.0006372.

Balkwill, D.L., Leach, F.R., Wilson, J.T. et al. (1988). Equivalence of microbial biomass measures based on membrane lipid and cell wall components, adenosine triphosphate, and direct counts in subsurface sediments. *Microb. Ecol.* 16: 73–84.

Barahona, S., Dorador, C., Zhang, R. et al. (2014). Isolation and characterization of a novel *Acidithiobacillus ferrivorans* strain from the Chilean Altiplano: attachment and biofilm formation on pyrite at low temperature. *Res. Microbiol.* 165: 782–793.

Barton, L.L., Tomei-Torres, F.A., Xu, H. et al. (2014). Nanoparticles formed by microbial metabolism of metals and minerals. In: *Nanomicrobiology* (ed. L.L. Barton, D.A. Bazylinski and H. Xu), 145–176. New York, NY: Springer.

Berney, M., Hammes, F., Bosshard, F. et al. (2007). Assessment and interpretation of bacterial viability by using the LIVE/DEAD BacLight Kit in combination with flow cytometry. *Appl. Environ. Microbiol.* 73: 3283–3290.

Beveridge, T.J. (1981). Ultrastructure, chemistry, and function of the bacterial wall. *Int. J. Cytol.* 72: 229–317. https://doi 10.1016/S0074-7696(08)61198-5.

Bianciotto, V., Lumini, E., Lanfranco, L. et al. (2000). Detection and identification of bacterial endosymbionts in arbuscular mycorrhizal fungi belonging to the family *Gigasporaceae*. *Appl. Environ. Microbiol.* 66: 4503–4509.

Blazejak, A., Erséus, C., Amann, R. et al. (2005). Coexistence of bacterial sulfide oxidizers, sulfate reducers, and spirochetes in a gutless worm *(Oligochaeta)* from the Peru Margin. *Appl. Environ. Microbiol.* 71: 1553–1561.

Bleloch, A.L., Own, C.S., Hamalainen, M. et al. (2011). DNA sequencing by electron microscopy. *Microsc. Microanal.* 17: 1274–1275.

Boolchandani, M., Patel, S., and Dantas, G. (2017). Functional metagenomics to study antibiotic resistance. In: *Antibiotics. Methods in Molecular Biology*, vol. 1520 (ed. P. Sass), 307–329. New York, NY: Humana Press.

Boulos, L., Prévost, M., Barbeau, B. et al. (1999). LIVE/DEAD BacLight: application of a new rapid staining method for direct enumeration of viable and total bacteria in drinking water. *J. Microbiol. Methods* 37: 77–86.

Bozzola, J.J. and Russell, L.D. (1998). *Electron Microscopy*, 2e. London, United Kingdom: Jones and Bartlett.

Byrd, J.J., Xu, H.S., and Colwell, R.R. (1991). Viable but nonculturable bacteria in drinking water. *Appl. Environ. Microbiol.* 57: 875–878.

Cairns, L.S., Hobley, L., and Stanley-Wall, N.R. (2014). Biofilm formation by *Bacillus subtilis*: new insights into regulatory strategies and assembly mechanisms. *Mol. Microbiol.* 93: 587–598.

Caporaso, J.G., Lauber, C.L., Walters, W.A. et al. (2012). Ultra-high-throughput microbial community analysis on the Illumina HiSeq and MiSeq platforms. *ISME J.* 6: 1621–1624.

Cavanaugh, C.M., Gardiner, S.L., Jones, M.L. et al. (1981). Prokaryotic cells in the hydrothermal vent tube worm *Riftia pachyptila* Jones: possible chemoautotrophic symbionts. *Science* 213 (4505): 340–342.

Chalfie, M., Tu, Y., Euskirchen, G. et al. (1994). Green fluorescent protein as a marker for gene expression. *Science* 263: 802–805.

Chandra, J., Kuhn, D.M., Mukherjee, P.K. et al. (2001). Biofilm formation by the fungal pathogen *Candida albicans*: development, architecture, and drug resistance. *J. Bacteriol.* 183: 5385–5394.

Charlebois, A., Jacques, M., and Archambault, M. (2014). Biofilm formation of *Clostridium perfringens* and its exposure to low-dose antimicrobials. *Front. Microbiol.* 5: 183. https://doi.org/10.3389/fmicb.2014.00183.

Chistoserdova, L. (2009). Functional metagenomics: recent advances and future challenges. *Biotechnol. Genet. Eng. Rev.* 26: 335–352.

Cole, J.R., Wang, Q., Fish, J.A. et al. (2014). Ribosomal database project: data and tools for high throughput rRNA analysis. *Nucleic Acids Res.* 42: D633–D642.

Corbari, L., Cambon-Bonavita, M.-A., Long, G.J. et al. (2008). Iron oxide deposits associated with the ectosymbiotic bacteria in the hydrothermal vent shrimp *Rimicaris exoculata*. *Biogeosciences* 5: 1295–1310.

Dapa, T. and Unnikrishnan, M. (2013). Biofilm formation by *Clostridium difficile*. *Gut Microbes* 4: 397–402.

Das, S., Paul, S., Bag, S.K. et al. (2006). Analysis of *Nanoarchaeum equitans* genome and proteome composition: indications for hyperthermophilic and parasitic adaptation. *BMC Genomics* 7: 186. https://doi.org/10.1186/1471-2164-7-186.

Daub, C.O., Kloska, S., and Selbig, J. (2003). MetaGeneAlyse: analysis of integrated transcriptional and metabolite data. *Bioinformatics* 19: 2332–2333.

Davies, D.G. and Marques, C.N.H. (2009). A fatty acid messenger is responsible for inducing dispersion in microbial biofilms. *J. Bacteriol.* 191: 1393–1403.

Davies, D.G., Parsek, M.R., Pearson, J.P. et al. (1998). The involvement of cell-to-cell signals in the development of a bacterial biofilm. *Science* 280: 295–298.

Davies, J. (2007). Small molecules: the lexicon of biodiversity. *J. Biotechnol.* 129: 3–5.

Davis, D.J. (1921). The assessor factors in bacterial growth. IV. The "satellite" or symbiosis phenomenon of Pfeiffer's *Bacillus* (*B. influenza*). *J. Infect. Dis.* 29: 178–186.

de Bruijn, F.J. (2011). *Handbook of Molecular Ecology II: Metagenomics in Different Habitats.* London: Wiley.

Delmotte, N., Knief, C., Chaffron, S. et al. (2009). Community proteogenomics reveals insights into the physiology of phyllosphere bacteria. *Proc. Natl. Acad. Sci. U.S.A.* 106: 16428–16433.

Desirò, A., Salvioli, A., Ngonkeu, E.L. et al. (2014). Detection of a novel intracellular microbiome hosted in arbuscular mycorrhizal fungi. *ISME J.* 8: 257–270.

D'Onofrio, A., Crawford, J.M., Stewart, E.J. et al. (2010). Siderophores from neighboring organisms promote the growth of uncultured bacteria. *Chem. Biol.* 17: 254–264.

Dutkiewicz, A., Volk, H., George, S.C. et al. (2006). Biomarkers from Huronian oil-bearing fluid inclusions: an uncontaminated record of life before the great oxidation event. *Geology* 34: 437–440.

Eden, P.A., Schmidt, T.M., Blakemore, R.P. et al. (1991). Phylogenetic analysis of *Aquaspirillum magnetotacticum* using polymerase chain reaction-amplified 16s rRNA-specific DNA. *Int. J. Syst. Bacteriol.* 41: 324–325.

Estrela, A.B. and Abraham, W.-R. (2010). Combining biofilm-controlling compounds and antibiotics as a promising new way to control biofilm infections. *Pharmaceuticals* 3: 1374–1393.

Fatima, F., Pathak, N., and Verma, S.R. (2014). An improved method for soil DNA extraction to study the microbial assortment within rhizospheric region. *Mol. Biol. Int.* 2014: 518960. https://doi.org/10.1155/2014/518960.

Ferrari, B.C., Winsley, T.J., Bergquist, P.L. et al. (2012). Flow cytometry in environmental microbiology: a rapid approach for the isolation of single cells for advanced molecular biology analysis. *Methods Mol. Biol.* 881: 3–26.

Fischer, W.W., Hemp, J., and Valentine, J.S. (2016). How did life survive Earth's great oxygenation? *Curr. Opin. Chem. Biol.* 31: 166–178.

Frank, J.A., Reich, C.I., Sharma, S. et al. (2008). Critical evaluation of two primers commonly used for amplification of bacterial 16S rRNA genes. *Appl. Environ. Microbiol.* 74: 2461–2470.

Frias-Lopez, J., Shi, Y., Tyson, G.W. et al. (2008). Microbial community gene expression in ocean surface waters. *Proc. Natl. Acad. Sci. U.S.A.* 105: 3805–3810.

Fröls, S. (2013). Archaeal biofilms: widespread and complex. *Biochem. Soc. Trans.* 41: 393–398.

García Martín, H., Ivanova, N., and Kunin, V. (2006). Metagenomic analysis of two enhanced biological phosphorus removal (EBPR) sludge communities. *Nat. Biotechnol.* 24: 1263–1269.

George, G.N., Gnida, M., Bazylinski, D.A. et al. (2008). X-ray absorption spectroscopy as a probe of microbial sulfur biochemistry: the nature of bacterial sulfur globules revisited. *J. Bacteriol.* 190: 6376–6383.

Giannone, R.J., Huber, H., Karpinets, T. et al. (2011). Proteomic characterization of cellular and molecular processes that enable the *Nanoarchaeum equitans-Ignicoccus hospitalis* relationship. *PLoS One* 6 (8): e22942. https://doi.org/10.1371/journal.pone.0022942.

Gilbert, J.A., Jansson, J.K., and Knight, R. (2014). The earth microbiome project: successes and aspirations. *BMC Biol.* 12 (69): https://doi.org/10.1186/s12915-014-0069-1.

Giovannoni, S.J., Britschgi, T.B., Ch, M. et al. (1990). Genetic diversity in Sargasso Sea bacterioplankton. *Nature* 345: 60–63.

Gómez-Valero, L., Soriano-Navarro, M., Pérez-Brocal, V. et al. (2004). Coexistence of *Wolbachia* with *Buchnera aphidicola* and a secondary symbiont in the aphid *Cinara cedri*. *J. Bacteriol.* 186: 6626–6633.

Gorby, Y.A., Yanina, S., McLean, J.S. et al. (2006). Electrically conductive bacterial nanowires produced by Shewanella oneidensis strain MR-1 and other microorganisms. *Proc. Natl. Acad. Sci. USA* 103: 11358–11363.

Gorlas, A., Marguet, E., Gill, S. et al. (2015). Sulfur vesicles from *Thermococcales*: a possible role in sulfur detoxifying mechanisms. *Biochime* 118: 356–364.

Hamon, M.A., Stanley, N.R., Britton, R.A. et al. (2004). Identification of AbrB-regulated genes involved in biofilm formation by *Bacillus subtilis*. *Mol. Microbiol.* 52: 847–860.

Han, C.H., Koh, K.S., Xie, C. et al. (2015). Community quorum sensing signaling and quenching microbial granular biofilm assembly. *Biofilms Microbiomes* 1 (2015): 15006. https://doi.org/10.1038/npjbiofilms.2015.6.

Handelsman, J., Rondon, M.R., Brady, S.F. et al. (1998). Molecular biological access to the chemistry of unknown soil microbes: a new frontier for natural products. *Chem. Biol.* 5: R245–R249.

Hansma, H.G., Bezanilla, M., Zenhausem, F. et al. (1993). Atomic force microscopy of DNA in aqueous solutions. *Nucleic Acids Res.* 21: 505–512.

Hart, M.L., Meyer, A., Johnson, P.J. et al. (2015). Comparative evaluation of DNA extraction methods from feces of multiple host species for downstream next-generation sequencing. *PLoS One* 10 (11): e0143334. https://doi.org/10.1371/journal.pone.0143334.

Hentzer, M. and Givskov, M. (2003a). Pharmological inhibition of quorum sensing for the treatment of chronic bacterial infections. *J. Clin. Invest.* 112: 1300–1307.

Hentzer, M. and Givskov, M. (2003b). Pharmacological inhibition of quorum sensing for the treatment of chronic bacterial infections. *J. Clin. Invest.* 112: 1300–1307.

Hider, R.C. and Kong, X. (2010). Chemistry and biology of siderophores. *Nat. Prod. Rep.* 27: 637–657.

Human Microbiome Project Consortium (2012). Structure, function and diversity of the healthy human microbiome. *Nature* 486: 207–214.

Hungate, R.E. (1969). A roll tube method for cultivation of strict anaerobes. *Methods Microbiol.* 3B: 117–132.

Huttenhower, C., Gevers, D., Knight, R. et al. (2012). Structure, function and diversity of the healthy human microbiome. *Nature* 486: 207–214.

Jorand, F.P., Debuy, S., Kamagate, S.F. et al. (2015). Evaluation of a biofilm formation by *Desulfovibrio fairfieldensis* on titanium implants. *Lett. Appl. Microbiol.* 60: 279–287.

Kennedy, N.A., Walker, A.W., Berry, S.H. et al. (2014). The impact of different DNA extraction kits and laboratories upon the assessment of human gut microbiota composition by 16S rRNA gene sequencing. *PLoS One* 9 (2): e88982. https://doi.org/10.1371/journal.pone.0088982.

Knapp, M.K., Stiller, M., and Meyer, M. (2012). Generating barcoded libraries for multiplex high-throughput sequencing. *Methods Mol. Biol.* 840: 155–170.

Kuramitsu, H.K., Chen, W., and Ikegami, A. (2005). Biofilm formation by the periodontopathic bacteria *Treponema denticola* and *Porphyromonas gingivalis*. *J. Periodontol.* 76 (11 Suppl): 2047–2051.

Lackner, G., Moebius, N., Partida-Martinez, L. et al. (2011). Complete genome sequence of *Burkholderia rhizoxinica*, an endosymbiont of *Rhizopus microspores*. *J. Bacteriol.* 193: 783–784.

Leuko, S., Legat, A., Fendrihan, S. et al. (2004). Evaluation of the LIVE/DEAD *Bac*Light kit for detection of extremophilic archaea and visualization of microorganisms in environmental hypersaline samples. *Appl. Environ. Microbiol.* 70: 6884–6886.

Lever, M.A., Torti, A., Eickenbusch, P. et al. (2015). A modular method for the extraction of DNA and RNA, and the separation of DNA pools from diverse environmental sample types. *Front. Microbiol.* https://doi.org/10.3389/fmicb.2015.00476.

Ling, L.L., Schneider, T., Peoples, A.J. et al. (2015). A new antibiotic kills pathogens without detectable resistance. *Nature* 517: 455–459.

Mah, T.F., Pitts, B., Pellock, B. et al. (2003). A genetic basis for *Pseudomonas aeruginosa* biofilm antibiotic resistance. *Nature* 426: 306–310.

Majed, R., Faille, C., Kallassy, M. et al. (2016). *Bacillus cereus* biofilms – same, only different. *Front. Microbiol.* 7: 1054. https://doi.org/10.3389/fmicb.2016.01054. eCollection 2016.

Montgomery, K., Charlesworthy, J.C., LeBard, R. et al. (2013). Quorum sensing in extreme environments. *Life* 3: 131–148.

Mukamolova, G.V., Kaprelyants, A.S., Young, D.I. et al. (1998). A bacterial cytokine. *Proc. Natl. Acad. Sci. U.S.A.* 95: 8916–8921.

Murphy, D.B. and Davidson, M.W. (2013). *Fundamentals of Light Microscopy and Electronic Imaging*. New York, NY: Wiley-Blackwell.

National Research Council (US) Committee on Metagenomics (2007). Challenges and functional applications. In: *The New Science of Metagenomics*. Washington, DC: Academic Press.

Neidhardt, F.C., Bloch, P.L., and Smith, D.F. (1974). Culture medium for enterobacteria. *J. Bacteriol.* 119: 736–747.

Nebe-von-Caron, G., Stephens, P.J., Hewitt, C.J. et al. (2000). Analysis of bacterial function by multi-colour fluorescence flow cytometry and single cell sorting. *J. Microbiol. Methods* 42: 97–114.

Nichols, D., Cahoon, N., Trakhtenberg, E.M. et al. (2010). Use of ichip for high-throughput *in situ* cultivation of "uncultivable" microbial species. *Appl. Environ. Microbiol.* 76: 2445–2450.

Nikoh, N., McCutcheon, J.P., Kudo, T. et al. (2010). Bacterial genes in the aphid genome: absence of functional gene transfer from *Buchnera* to its host. *PLoS Genet.* https://doi.org/10.1371/journal.pgen.1000827.

Nikolov, L., Karamanev, D., Mamatarkova, V. et al. (2002). Properties of the biofilm of *Thiobacillus ferrooxidans* formed in rotating biological contactor. *Biochem. Eng. J.* 12: 43–48.

Ng, W.L. and Bassler, B.L. (2009). Bacterial quorum-sensing network architectures. *Ann. Rev. Genet.* 43: 197–222.

Ordinola-Zapata, R., Monteiro Bramante, C., Gomes de Moraes, I. et al. (2010). The use of confocal laser scanning microscopy for the study of dentin infection. In: *Microscopy: Science, Technology, Applications and Education*, vol. 3 (ed. A. Méndez-Vilas and J. Díaz), 583–589. Badajoz, Spain: FORMATEX.

Orell, A., Fröls, S., and Albers, S.-V. (2013). Archaeal biofilms: the great unexplored. *Annu. Rev. Microbiol.* 67: 337–354.

Overmann, J. (2010a). Novel cultivation strategies for environmentally important microorganisms. In: *Geomicrobiology: Molecular and Environmental Perspective* (ed. L.L. Barton, M. Mandl and A. Loy), 69–89. New York, NY: Springer.

Overmann, J. (2010b). Chemotaxis and behaviorial physiology of not-yet-cultivated microbes. *Methods Enzymol.* 397: 133–147.

Pelletier, E., Kreimeyer, A., Bocs, S. et al. (2008). "*Candidatus* Cloacamonas acidaminovorans": genome sequence reconstruction provides a first glimpse of a new bacterial division. *J. Bacteriol.* 190: 2572–2579.

Pereira, C.S., Thompson, J.A., and Xavier, K.B. (2013). AI-2-mediated signalling in bacteria. *FEMS Microbiol. Rev.* 37: 156–181.

Porath, D. (2009). Scanning tunneling microscopy: a DNA sequence scanned. *Nat. Nanotechnol.* 4: 476–477.

Poretsky, R.S., Hewson, I., Sun, S. et al. (2009). Comparative day/night metatranscriptomic analysis of microbial communities in the North Pacific subtropical gyre. *Environ. Microbiol.* 11: 1358–1375.

Reuter, M., Mallett, A., Pearson, B.M. et al. (2010). Biofilm formation by *Campylobacter jejuni* is increased under aerobic conditions. *Appl. Environ. Microbiol.* 76: 2122–2128.

Reuter, K., Steinbach, A., and Helms, V. (2016). Interfering with bacterial quorum sensing. *Perspect. Med. Chem.* 8: 1–15.

Robe, P., Nalin, R., Capellano, C. et al. (2003). Extraction of DNA from soil. *Eur. J. Soil Biol.* 39: 183–190.

Rossi, F. and De Philippis, R. (2015). Role of cyanobacterial exopolysaccharides in phototrophic biofilms and in complex microbial mats. *Life* 5: 1218–1238.

Sachidanandham, R., Gin, K.Y., and Poh, C.L. (2005). Monitoring of active but non-culturable bacterial cells by flow cytometry. *Biotechnol. Bioeng.* 89: 24–31.

Santegoeds, C.M., Damgaard, L.R., Hesselink, G. et al. (1999). Distribution of sulfate-reducing and methanogenic bacteria in anaerobic aggregates determined by microsensor and molecular analyses. *Appl. Environ. Microbiol.* 65: 4618–4629.

Sauer, K., Camper, A.K., Ehrlich, G.D. et al. (2002). *Pseudomonas aeruginosa* displays multiple phenotypes during development as a biofilm. *J. Bacteriol.* 184: 1140–1154.

Schembri, M.A., Kjaergaard, K., and Klemm, P. (2003). Global gene expression in *Escherichia coli* biofilms. *Mol. Microbiol.* 48: 253–267.

Schindler, D.W., Curtis, P.J., Parker, B.R. et al. (1996). Consequences of climate warming and lake acidification for UV-B penetration in North American boreal lakes. *Nature* 379: 705–708.

Schindler, D.W., Hecky, R.E., Findlay, D.L. et al. (2008). Eutrophication of lakes cannot be controlled by reducing nitrogen input: results of a 37-year whole-ecosystem experiment. *Proc. Natl. Acad. Sci. U.S.A.* 105: 11254–11258.

Schindler, D.W., Mills, K.H., Malley, D.F. et al. (1985). Long-term ecosystem stress: the effects of years of experimental acidification on a small lake. *Science* 228: 1395–1401.

Schramm, A., Larsen, L.H., Revsbech, N.P. et al. (1996). Structure and function of a nitrifying biofilm as determined by in situ hybridization and the use of microelectrodes. *Appl. Environ. Microbiol.* 62: 4641–4647.

Stewart, E.J. (2012). Growing unculturable bacteria. *J. Bacteriol.* 194: 4151–4160.

Stomp, M., Huisman, J., Stal, L.J. et al. (2007). Colorful niches of phototrophic microorganisms shaped by vibrations of the water molecule. *ISME J.* 1: 271–282.

Strous, M., Pelletier, E., Mangenot, S. et al. (2006). Deciphering the evolution and metabolism of an anammox bacterium from a community genome. *Nature* 440: 790–794.

Takai, K., Komatsu, T., Inagaki, F. et al. (2001). Distribution of archaea in a black smoker chimney structure. *Appl. Environ. Microbiol.* 67: 3618–3629.

Tan, C.H., Koh, K.S., Xie, C. et al. (2015). Community quorum sensing signaling and quenching: microbial granular biofilm assembly. *NPJ Biofilms Microbiomes* (2015): 15006. https://doi.org/10.1038/npjbiofilms.2015.6.

Tarasov, V.G. (2006). Effects of shallow-water hydrothermal venting on biological communities of costal marine ecosystems of the western Pacific. *Adv. Mar. Biol.* 50: 267–421.

Teske, A., Cypionka, H., Holt, J.G. et al. (2007). Enrichment and isolation. In: *Methods for General and Molecular Microbiology*, 3e (ed. C.A. Reddy), 215–269. Washington, DC: American Society for Microbiology.

Tomlin, K.L., Clark, S.R., and Ceri, H. (2004). Green and red fluorescent protein vectors for use in biofilm studies of the intrinsically resistant *Burkholderia cepacia* complex. *J. Microbiol. Methods* 57: 95–106.

Tsai, Y.-L. and Olson, B.H. (1991). Rapid method for direct extraction of DNA from soil and sediments. *Appl. Environ. Microbiol.* 57: 1070–1074.

Tyson, G.W., Chapman, J., Hugenholtz, P. et al. (2004). Community structure and metabolism through reconstruction of microbial genomes from the environment. *Nature* 428: 37–43.

Urich, T., Lanzén, A., Qi, J. et al. (2008). Simultaneous assessment of soil microbial community structure and function through analysis of the meta-transcriptome. *PLoS One* 3: e2527. https://doi.org/10.1371/ journal.pone.0002527.

Varga, J.J., Therit, B., and Melville, S.B. (2008). Type IV pili and the CcpA protein are needed for maximal biofilm formation by the gram-positive anaerobic pathogen *Clostridium perfringens*. *Infect. Immun.* 76: 4944–4951.

Vartoukian, S.R., Palmer, R.M., and Wade, W.G. (2010). Strategies for culture of 'unculturable' bacteria. *FEMS Microbiol. Lett.* 309: 1–7.

Ventner, J.C., Remington, K., Heidelberg, J.F. et al. (2004). Environmental genome shotgun sequencing of the Sargasso Sea. *Science* 304: 66–74.

Verberkmoes, N.C., Russell, A.L., Shah, M. et al. (2009). Shotgun metaproteomics of the human distal gut microbiota. *ISME J.* 3: 179–189.

Wade, W. (2002). Unculturable bacteria - the uncharacterized organisms that cause oral infections. *J. R. Soc. Med.* 95: 81–83.

Wang, X., Xie, K., Cheng, X. et al. (2015). Biofilm formation of anoxygenic photosynthetic bacteria induced by phototaxis for enhancing hydrogen production. *Environ. Sci. Water Res. Technol.* 1: 383–393.

Ward, D.M., Weller, B., and Bateson, M.M. (1990). 16S rRNA sequences reveal numerous uncultured microorganisms in a natural community. *Nature* 345: 63–65.

Waters, C.M. and Bassler, B.L. (2005). Quorum sensing: cell-to-cell communication in bacteria. *Annu. Rev. Cell Dev. Biol.* 21: 319–346.

Watnick, P.I. and Kolter, R. (1999). Steps in the development of a *Vibrio cholerae* El tor biofilm. *Mol. Microbiol.* 34: 586–595.

White, D., Drummond, J., and Fuqua, C. (2011). *The Physiology and Biochemistry of Prokaryotes*, 4e. New York, NY: Oxford University Press.

White, D.C. and Ringelberg, D.B. (1997). Utility of the signature lipid biomarker analysis in determining the in situ viable biomass, community structure, and nutritional/physiologic status of deep subsurface microbiota. In: *Microbiology of the Terrestrial Deep Subsurface* (ed. P.S. Amy and D.L. Halderman), 119–136. Boca Raton, FL: CRC Press.

Wilson, M.C., Mori, T., Rückert, C. et al. (2014). An environmental bacterial taxon with a large and distinct metabolic repertoire. *Nature* 506: 58–62.

Woese, C.R. and Fox, G.E. (1977). Phylogenetic structure of the prokaryotic domain: the primary kingdoms. *Proc. Natl. Acad. Sci. U.S.A.* 74: 5088–5090.

Woyke, T., Tighe, D., Mavromatis, K. et al. (2010). One bacterial cell, one complete genome. *PLoS One* 5 (4): e10314. https://doi.org/10.1371/journal.pone.0010314.

Yildiz, F.H. and Schoolnik, G.K. (1999). *Vibrio cholerae* O1 El Tor: identification of a gene cluster required for the rugose colony type, exopolysaccharide production, chlorine resistance, and biofilm formation. *Proc. Natl. Acad. Sci. U.S.A.* 96: 4028–4033.

Yamada, M., Ikegami, A., and Kuramitsu, H.K. (2005). Synergistic biofilm formation by *Treponema denticola* and *Porphyromonas gingivalis*. *FEMS Microbiol. Lett.* 250: 271–277.

Yang, J.-L., Wang, M.-S., Cheng, A.-C. et al. (2008). A simple and rapid method for extracting bacterial DNA from intestinal microflora for ERIC-PCR detection. *World J. Gastroenterol.* 14: 2872–2876.

Yuan, S., Cohen, D.B., Ravel, J. et al. (2012). Evaluation of methods for the extraction and purification of DNA from the human microbiome. *PLoS One* 7 (3): e33865. https://doi.org/10.1371/journal.pone.0033865.

Zapata, R.O., Bramante, C.M., de Moraes, I.G. et al. (2008). Confocal laser scanning microscopy is appropriate to detect viability of *Enterococcus faecalis* in infected dentin. *J. Endod.* 34: 1198–1201.

Zhao, X., Schwartz, C.L., Pierson, J. et al. (2017). Three-dimensional structure of the ultraoligotrophic marine bacterium "*Candidatus* Pelagibacter ubique.". *Appl. Environ. Microbiol.* 83: e02807–e02816. https://doi.org/10.1128/AEM.02807-16.

Zhao, H., Thompson, R.B., Lockatell, V. et al. (1998). Use of green fluorescent protein to assess urease gene expression by uropathogenic *Proteus mirabilis* during experimental ascending urinary tract infection. *Infect. Immun.* 66: 330–335.

Zhu, P. and Li, M. (2012). Recent progresses on AI-2 bacterial quorum sensing inhibitors. *Curr. Med. Chem.* 19: 174–186.

Further Reading

Bjarnsholt, T., Ciofu, O., Molin, S. et al. (2013). Applying insights from biofilm biology to drug development – can a new approach be developed? *Nat. Rev. Drug Discov.* 12: 791–808. https://doi.org/10.1038/nrd4000.

Handelsman, J. (2004). Metagenomics: application of genomics to uncultured microorganisms. *Microbiol. Mol. Biol. Rev.* 68: 669–685.

Sangwan, N., Xia, F. and Gilbert, J.A. (2016). Recovering complete and draft population genomes from metagenome datasets. Microbiome 4: 8 https://doi.org/10.1186/s40168-016-0154-5.

Web Sites

Kyoto Encyclopedia of Genes and Genomes: http://www.genome.jp/kegg/kegg2.html
MetaGeneAlyse website: http://metagenealyse.mpimp-golm.mpg.de

3

Terrestrial Systems: Soil and Subsurface Environments

3.1 Overview and Focus

About 30% of the Earth's surface is land and of this, about 40% is arable or capable of being farmed whereas 33% is desert. Agricultural microbiology is important because 99.7% of human food based on calories comes from arable land, with the remaining 0.3% coming from oceans. The importance of agriculture was undoubtedly an important factor for the production of the initial reports by Sergei Winogradsky and Martinus Beijerinck on the role of soil bacteria in the nitrogen, sulfur, and carbon cycles (Lechevalier and Solorovsky 1974). To understand the ecology of microorganisms in arable land, this chapter initially provides the background for the environment that supports and regulates the growth of soil microorganisms.

Important segments of terrestrial microbiology are the subsurface and deep subsurface environments. As presented here, the microorganisms are found either within the rocks or in fracture water flowing through the rock environment. Since the growth of microorganisms in the subsurface regions is dependent on water, many of the samples are from shallow aquifers or fracture water. In subsurface environments, primary productivity for microbial communities is not light supported but comes from biotic and abiotic gases coupled to mineral transformation. As the topic of terrestrial microbiology is developed, this chapter focuses on the following:

- Soil microbiology: habitat and microorganisms
- Microbial ecosystems in arable land
- Distribution of microorganisms in subsurface and deep subsurface
- Microbial concerns when using deep subsurface as storage sites
- Life in deep subsurface
- Microbiology of minerals: weathering and transformations

3.2 Soil: An Environment for Microorganisms

Soil is defined as a layer of weathered rock on the Earth's surface, which overlies solid bedrock. Aside from weathered rock, soil contains minerals, nutrients, organic matter, moisture, gases, and a wide variety of microscopic and macroscopic organisms. Soil is typically characterized as supporting plant growth. Minerals within soils are characterized based on size as: sand (2.00–0.05 mm [50 μm]), silt (2–50 μm), and clays (<2 μm). Soils can be characterized based on mineral content in that sandy soils have a high proportion of sand (>50%), clay soils have a higher proportion of clay minerals (>40% clay), and loam soils have a roughly equal concentrations of sand, silt, and clays. For further information on mineral classifications, the reader is

Environmental Microbiology and Microbial Ecology, First Edition. Larry L. Barton and Robert J.C. McLean.

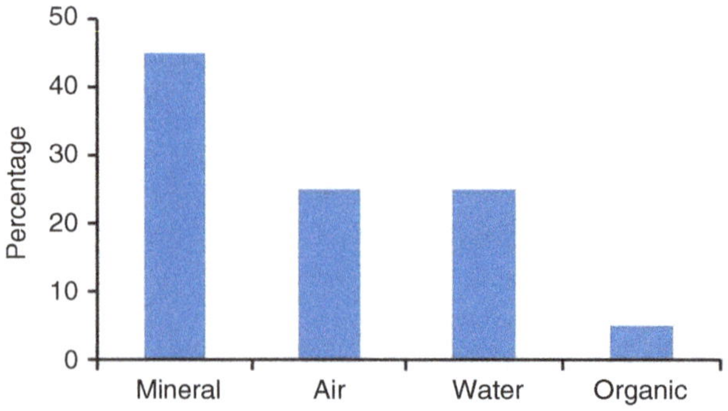

Figure 3.1 Abundance of minerals, organics, water, and air in agricultural soil.

referred to the Soil Science Society of America (SSSA) website (www.soils.org). There are different soil types established by the SSSA, which vary according to geographic region and climate (forest, grassland, wetland, alpine, arctic, desert). Examples of broad groups include: andisols (volcanic ash soils), gelisols (frozen soils, i.e. Arctic regions), vertisols (typically contain higher amounts of clays and shrink or swell during dry and moist periods), and mollisols (deep, fertile soils). The names end in –sol, derived from the Latin "solum" (soil or ground).

The soil habitat supports the growth and viability of microorganisms. Historically, agricultural soil was the model used to study microbial activities because of the economic importance of supplying human and animal food. A general distribution of minerals, organic matter, water, and air in fertile agricultural soil is presented in Figure 3.1. For production by agricultural soil, appropriate values for the following parameters are addressed: (i) atmosphere, (ii) water, (iii) oxidation–reduction activities, (iv) pH, and (v) temperature. Gas exchanges of O_2 into the soil which is needed for biotic respiration and removal of CO_2 is needed especially with decomposition of organic matter. Water in pore spaces is required to support plant growth and water potential of −0.01 MPa is optimal for soil microorganisms. Droughts produce an increase in negative water potential with a decline in growth of important soil bacteria such as *Rhizobium* and *Nitrosomonas* while the common soil fungi are not affected. Increase of water in soil due to flooding results in anaerobic conditions as water replaces air in the soil.

3.2.1 Soil Horizons

Soil is made up of a number of layers called horizons (Table 3.1). Most biological activity occurs in the A horizon, which is a surface layer. The soil profiles in Figure 3.2 show that the depth of the soil layers varies with the biological activity above it. Extensive deposition of organic material occurs at the bottom of an aquatic system and an extremely thick humic layer is present. Woodlands have less surface vegetation than prairies and, thus have fewer humic materials in the soil than prairies. Microbial abundance and activities reflect the biological activities at the surface of the soil because the humic layer contains nutrients supportive of microbial growth.

3.2.2 Soil Organic Matter (SOM)

Organic matter in soils includes typically humic and fulvic acids (Box 3.1), which are high molecular weight organic compounds (Sylvia et al. 1999). These are thought to arise from alterations in decayed plant and microbial material. Humic compounds are typically black or brown (Figure 3.3) and give most soils a characteristic dark color. In some aquatic environments (often associated with swamps) these compounds leach into the water and turn it brown. These two organic soil components can be extracted with a mild alkaline solution and separated from

Table 3.1 Soil horizons from surface to bedrock. Some soils do not contain all of these layers.

Horizon	Details
O	Layer with a considerable amount of organic matter. For example, leaves on a forest floor (in various states of decay) covering the underlying layers
A	Layer immediately below the O horizon containing organic matter mixed in with mineral components – often referred to as topsoil
E	Layer immediately below the A horizon, which has a higher concentration of sand and silt, and lower concentration of clay than the A horizon (subsoil)
B	Layer immediately below the E horizon. Different chemistry than E and C horizons
C	Layer immediately below B horizon and above bedrock. Often has similar chemistry to bedrock
R layer	Bedrock – intact underlying rock

Source: Based on data from https://www.soils.org/publications/glossary/appendix.

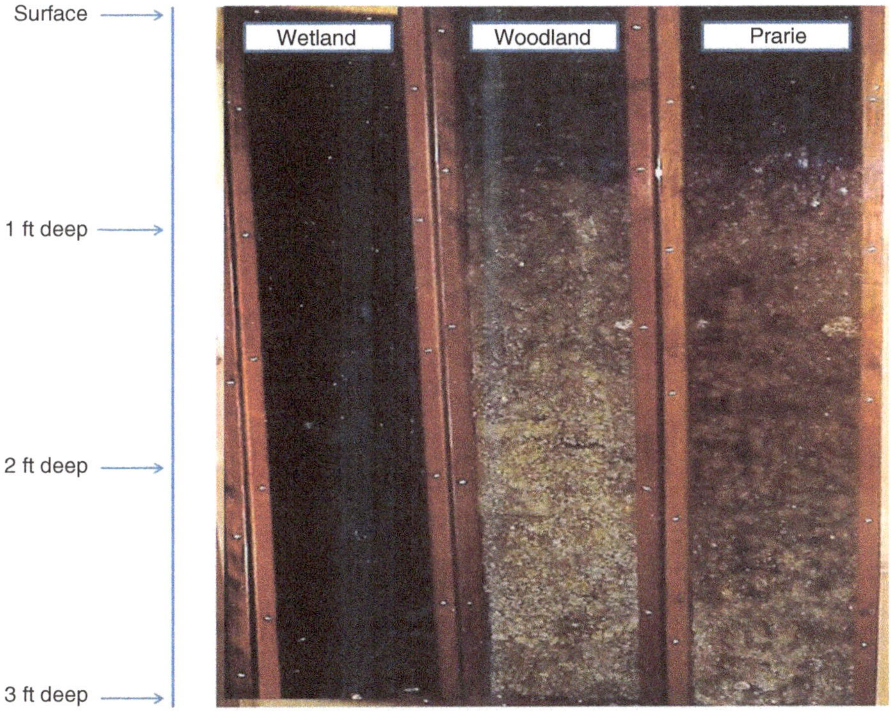

Figure 3.2 Soil profiles from different sites. High humic materials and DOM in wetland soil extend a considerable distance from the surface and account for the black soil. Relatively little organic material settles on the surface of woodland sites and accounts for a shallow layer of topsoil. With a moderate amount of biological growth in a prairie, the organic material extends down to a deeper level than in woodlands. Soil profiles displayed at Prairie Woods Environmental Learning Center, Spicer, Minnesota, USA. *Source:* Reproduced with permission from Larry L. Barton.

each other by their relative solubility at very acidic pH levels. At pH 1.0, humic acid is insoluble but fulvic acid is soluble.

There are a variety of clay minerals commonly associated with soils including kaolinite, smectite, montmorillonite, and illite. Clays are aluminum silicate minerals that typically contain

Box 3.1 Humic and Fulvic Acids

Humic and fulvic acids are weak organic acids consisting of aromatic and aliphatic (carbon chains) moieties found in rich agricultural soils. These soil compounds do not have a defined chemical structure and are best considered as a soil fraction with the molecular weight of fulvic acids being 1000–10 000 and the molecular weight of humic acids are 10 000–100 000. Both of these soil acids are important for binding trace elements that can be used by soil microorganisms. The formation of humic acids is unresolved but it has been proposed that the lignin-like structure is the result of microbial enzymes. Humic acids are highly resistant to enzyme attack and have a half-life of several hundred years.

Figure 3.3 A complex chemical structure makes up humic acids and accounts for the dark color in productive agricultural soil. For research experiments, commercially available salts of humic acids are used. The authors do not endorse any commercial source for humic acid products. *Source:* Reproduced with permission from Larry L. Barton.

a variety of other ions. Clay minerals typically have a net negative charge (i.e. are anionic) a feature that enables them to bind metal cations such as Mg^{2+} and Ca^{2+}, and nutrients such as ammonium (NH_4^+, a common dissolved form of ammonia at pH <9). The crystal structure and chemical composition varies among the various clay minerals (Sylvia et al. 1999).

Clays will swell or contract depending upon their water content, which explains the cracked appearance of clay-containing soils during dry periods (drought). In terms of water content, the upper layer of soil – the vadose zone – contains a mixture of water and air. Lower down is the phreatic zone (water table), where the soil is saturated with water. Water enters the vadose zone via rainfall and other precipitation, flood events, or through capillary action from the phreatic zone (Sylvia et al. 1999). Within the vadose zone, water, and air are both important for the support of living organisms (plants, animals, and microorganisms). The pH value of soils is determined by the minerals and organic components present. Unless completely saturated with water (i.e. waterlogged), soils are generally aerobic.

3.3 Soil Microbiology

3.3.1 Soil Prokaryotes

Microorganisms within soils are associated with various activities including nutrient cycling, nutrient availability, and decomposition (addressed in Chapter 8). Soil consists of a series of irregular aggregations of minerals, organic matter, and microcolonies of microorganisms (Figure 3.4). Due to the spatial distribution of microorganisms in soil, there is considerable differences in rates of growth and bacterial interactions as compared to solution chemistry that occurs with microorganisms growing in broth. The presence of microorganisms in the soil is not static but highly dynamic (see Box 3.2). Using broad-based approaches, including metagenomics, a significantly higher biodiversity of microorganisms in soils has been

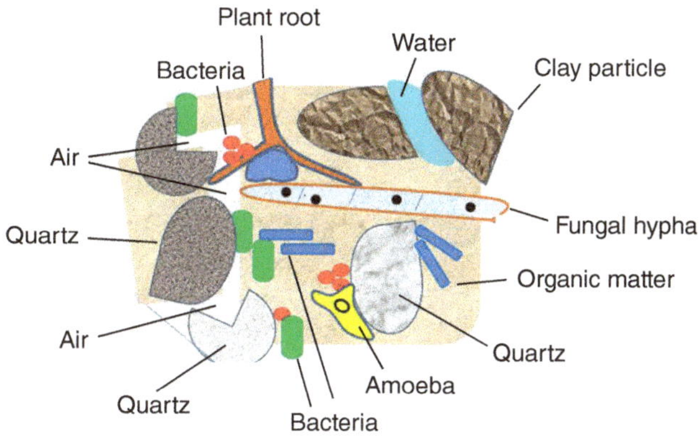

Figure 3.4 Soil aggregate showing the distribution of microorganisms, minerals, organic material, and plant root in soil.

Box 3.2 Microbial Homeostasis in Soil

Evaluations of microorganisms in a soil sample give an overall view of what is occurring in the soil at any one point in time. In reality, microbial activities in soil are highly dynamic with each fungal, bacterial, and archaeal species growing at different rates and subject to the influences of microsites within each niche. Microbial numbers may decline because of viral attacks or digestion by protozoa. Within the soil there are a myriad of microbial cycles and activities that are not apparent from a single sample as this does not include fluctuations that occur over short periods.

Box 3.3 Terrestrial Sampling is Destructive

The study of microorganisms in soil is complicated because sampling destroys the habitat and the microbial community. A challenge for the future is to evaluate microorganisms in the native state to obtain information on microbial interactions and "real time" response to chemical and physical perturbations.

observed than previously, such as many different species of unknown soil bacteria found to participate in the cycling of organic carbon matter (Long et al. 2016).

Soil provides an environment for the growth of many species of prokaryotes and a high microbial cell density. Soil sampling is not perfect because it is a one-point test in time and space (Box 3.3). Bacteria and archaea frequently encountered in soil samples are listed in Table 3.2. These samples are from bulk soil taken within 10 cm of the surface. Using high throughput DNA pyrosequencing, an average gram of soil from four different soils was determined to be about 10^{10} bacterial cells and this reflected the presence of 4×10^3–5×10^4 different species. Members of the *Bacteroidetes, Betaproteobacteria,* and *Alphaproteobacteria* were the most numerous, with the number of bacterial phyla in forest soil greatly exceeding that of agricultural soil (Roesch et al. 2007). Over 40% of all bacterial belonged to *Proteobacteria,* with and 6–12% of sequences representing uncultivated prokaryotes. The distance between bacterial cells in the soil is estimated to be about 12.5 μm, and the distance for interaction between cells is 20 μm, so each soil bacterial cell could be communicating with about 120 other cells (Raynaud and Nunan 2014).

Archaea are less abundant in soil than bacteria. DNA sequences for *Crenarchaeota* were present in 4–12% of the total number of sequences, with greater abundance in agricultural soil

Table 3.2 Prokaryotic microorganisms commonly encountered in soil.

Bacteria	Common examples
Firmicutes (low G + C gram-positive bacteria)	Gram-positive *Bacillus* and *Clostridium* spp.
High G + C gram-positive bacteria	*Actinobacter* (sometimes called actinomycetes) including *Streptomyces* spp., *Frankia* spp.
Proteobacteria (many commonly encountered gram-negative bacteria)	*Rhizobium, Agrobacterium* spp. (*Alphaproteobacteria*) *Nitrosomonas, Nitrobacter, Thiobacillus* spp. (*Betaproteobacteria*) *Pseudomonas, Shewanella* spp. (*Gammaproteobacteria*) *Desulfomonas, Bdellovibrio* spp. (*Deltaproteobacteria*) *Hydrogenimonas, Sulfurimonas* spp. (*Epsilonproteobacteria*)
Cyanobacteria	*Microcoleus, Nostoc* spp. (common in soil crusts)
Archaea	
Euryarchaeota	Includes methanogens (*Methanococcus, Methanobacterium* spp.), extreme halophiles (*Halobacterium* spp.), and some extreme thermophiles (*Thermoplasma* spp.)
Crenarchaeota	Originally thought to be extreme sulfur-oxidizing thermophiles; now at least one low temperature (28 °C) autotrophic, ammonia-oxidizing organism (*Nitrosopumilus maritimus*) has been cultured

than forest soil (Raynaud and Nunan 2014). In an extensive survey of 146 soils from different ecosystems, the greatest abundance of archaea was in soils with a low C:N ratio (Bates et al. 2011). When evaluating all the soils samples around the globe, only 2% of the 16S rRNA gene sequences were from archaea and over 70% of these were associated with two phylotypes in the *Crenarchaeota* phylum. Considerable phylogenetic diversity occurs in the soil archaea, even though some phylotypes dominate. Analysis of operational taxonomic unit (OTU) abundance in several communities revealed that in desert soils, archaea had greater OTU richness than bacteria, fungi, or viruses (Fierer et al. 2007).

3.3.2 Soil Fungi

It has been estimated that there are over 1.5 million fungal species on Earth (Gomes et al. 2003) and about 70 000 species are expected to be found in terrestrial environments. Cultivation techniques had estimated that about 3000 fungal species were present in a field site; however, small-subunit RNA analysis showed that fungal OTU richness is greater than initially suggested (Fierer et al. 2007). There are three types of soil fungi: saprophytic fungi, pathogens, and mycorrhizae. Saprophytic fungi are important because they decompose plant residues that are not readily hydrolyzed by bacteria and these fungi are abundant in prairie soils. In forest soil and soil with annual plant cropping, members of the *Basidiomycota* and *Ascomycota* are most abundant. *Phanerochaete chrysosporium,* a basidiomycete that is commonly called "white rot fungus" is important in the decomposition of lignin. Another important group of fungi include the mycorrhizal fungi that live in association with many plant roots. Mycorrhizal fungi enable plants to gain access to many nutrients within soils. In 1997, a group of scientists led by Suzanne Simard showed that carbon compounds could be directly transferred between two different tree species via shared mycorrhizal fungal hyphae, a process nicknamed the "wood wide web" (Simard et al. 1997). Activities of mycorrhizae are discussed in Chapter 7. Soil fungi adsorb small particles including bacteria and viruses on their surfaces. As the fungi grow with hyphae extensions, bacteria may be transported across air spaces in the soil and account for the distribution of plant viruses in the rhizosphere.

3.3.3 Soil Crusts

Soil crusts, commonly called cryptogamic or cryptobiotic crusts, are a common feature of many semi-arid and arid environments. These microbial communities resemble biofilms because they consist of soil surface adherent communities of cyanobacteria and other bacteria, fungi, eukaryotic algae, lichens, and mosses. Intact soil crusts are very useful in allowing soils to resist wind erosion, retain water during dry periods, and – depending upon their biological composition – can fix carbon and nitrogen. In overall ecology, soil crusts can influence the growth of plant seeds (Belnap et al. 2001) and by inherent pigmentation can absorb more energy from the sun and increase the temperature of the underlying soil. Cyanobacteria in desert crusts produce scytonemin, which is a type of sunscreen, to reduce the harmful effects of intense sunlight; however, by absorbing the sunlight, scytonemin heats the immediate environment by as much as 10 °C (Couradeau et al. 2016). This increase in temperature may account for replacement of temperature sensitive microbiota with temperature resistant strains. When soil crusts are broken up by animal traffic (grazing livestock) or human activities, the underlying soil becomes more susceptible to erosion.

3.3.4 Soil Invertebrates and Burrowing Animals

Soil microorganisms are one biological component of soil. In many soils there is a thriving invertebrate population, including insects (e.g. ants, termites, beetle larvae), and other invertebrates, including earthworms, pill bugs (terrestrial isopod crustaceans), and nematodes. Burrowing invertebrates and small vertebrate animals (e.g. gophers, moles, and mice) increase the porosity of soil and transfer microorganisms throughout soil. Animals that ingest leaf litter aid in its decomposition through digestive processes and also physical damage to the leaf litter, which increases surface area and provides access for decomposing microorganisms.

3.3.5 The Rhizosphere and Associated Bacteria

Plants have a significant role in soil microbiology. The region immediately adjacent to plant roots is called the rhizoplane and the zone around the root containing soil is the rhizosphere. The rhizosphere is one of the zones within surface soils where microorganisms are most active. One way of obtaining the rhizosphere of an individual plant is to carefully remove it from the soil (see Figure 3.5). Exudates from plant roots contain organic acids, carbohydrates, and amino acids, which attract microorganisms into the rhizosphere. The plant exudate may represent a significant contribution of the plant's photosynthetic activity, with about 10% of a plant's total carbon production released as exudate carbon. Due to the differences in exudates from the various plants, the rhizosphere varies considerably with plant species and geochemical features of the soil. Additionally, there is a difference in the microorganisms present and their activity in bulk soil compared with areas of soil adjacent to roots of plants fixing nitrogen or plants without this capability (Figure 3.6).

In some rhizospheres, bacteria produce hydrogen cyanide, oomycin A, 2,4-diacetylphloroglucinol, and phenazine, which protect plant roots from pathogens (Lareen et al. 2016). The production of phenazine by *Pseudomonas chlororaphis* was found to inhibit the growth of *Fusarium oxysporum,* and lipoproteins produced by *Pseudomonas* spp. and *Bacillus* spp. inhibit the growth of various plant pathogens. Several activities of microorganisms in the

Figure 3.5 A grass plant removed from soil shows: (a) the rhizosphere with the soil attached to the roots; (b) the rhizoplane where the soil has been washed from the roots. *Source:* Reproduced with permission from Larry L. Barton.

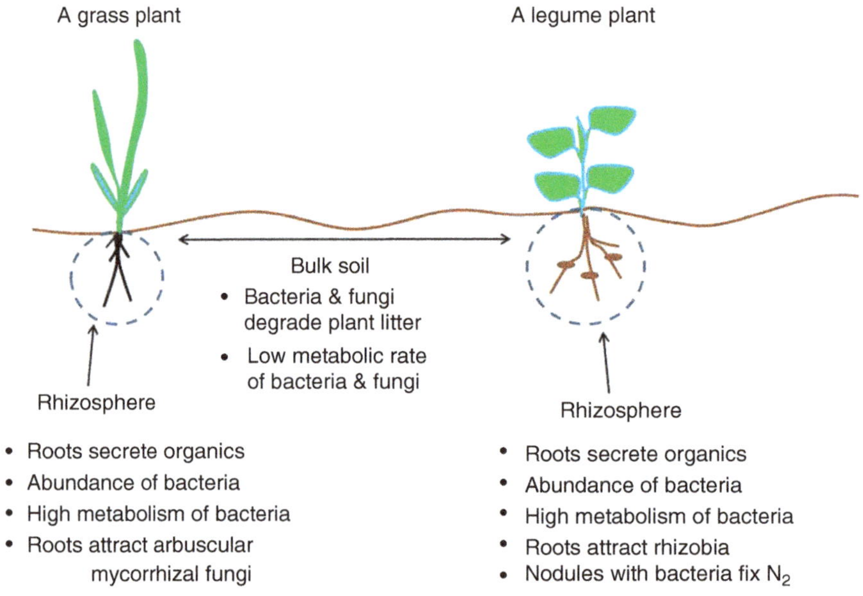

Figure 3.6 Difference in activities in bulk soil compared with rhizospheric soil.

rhizosphere require coordination and have been related to bacterial cell density. The activities of biofilm formation in the soil, adhesion of bacteria onto plant roots, production of extracellular enzymes, and expression of virulence are behavior characteristics reminiscent of quorum sensing. The production of acyl homoserine lactones (AHLs), an important quorum sensing molecule, influences the extracellular production of enzymes by soil bacteria (e.g. *Inquilinus* spp., *Rhizobium* spp., and *Agrobacterium* sp.). AHL-producing *Alphaproteobacteria* are more common in the rhizosphere than in bulk soil (see quorum sensing discussion in Chapter 2). The release of nitrogen from chitin, proteins, and nucleic acids in the soil has been attributed to the quorum-sensing-dependent production of extracellular enzymes by rhizospheric bacteria (DeAngelis et al. 2008). It has long been established that the beneficial activities of rhizospheric organisms are attributed to symbiotic nitrogen-fixing bacteria of *Frankia* spp., *Bradyrhizobium* spp., and *Rhizobium* spp. Additionally, rhizospheric microorganisms have been shown to enhance the tolerance of plants to various physical and environmental stresses, including increasing salinity and drought stress (Lau and Lennon 2012).

3.4 Understanding Soil Ecosystems

3.4.1 The Carbon : Nitrogen Ratio

One of the important determinants for assessing the growth potential of a substance is to consider the carbon:nitrogen (C:N) ratio, which compares the mass of carbon with the mass of nitrogen in plant material. Soil microorganisms have a C:N ratio of about 8:1 (Figure 3.7) and the ideal plant material added to soil should have a C:N ratio of about 24:1. Soil microorganisms require about 16 parts of carbon for energy production leaving eight parts for cell mass production. Alfalfa hay has a C:N ratio of 25–30:1, dry plant material such as wheat straw has a C:N ratio exceeding 80:1 which will result in slow microbial growth because nitrogen is limited. Grass clippings will have a C:N ratio of about 16:1 whereas animal manure is about 14:1. In general, if the C:N ratio exceeds 30, the growth of fungi is favored and if the C:N ratio is less than 20, bacteria will be dominant. Vesicular-arbuscular mycorrhizae are the dominant type of fungi in agricultural and prairie soils whereas ectomycorrhizae are dominant in forest soil. In composting, it is often best to mix dry leaves with fresh green material because if the C:N is too low, the excess nitrogen may be converted to ammonia, which would make the compost alkaline.

3.4.2 The Fungi : Bacteria Ratio

The abundance of microorganisms in the soil reflects the nutrients available and the relative quantity of fungi and bacteria may be presented as the fungi:bacteria (F:B) ratio. There are different ways to calculate the fungi and bacteria in the soil and unless specified, the F:B ratio

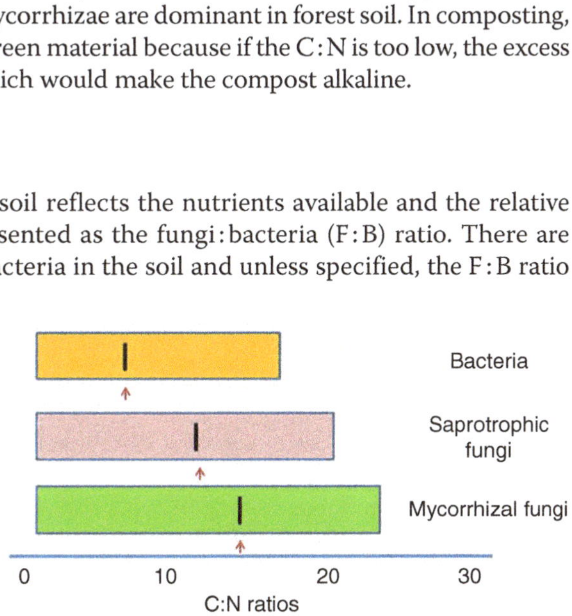

Figure 3.7 The carbon:nitrogen ratio of soil microorganisms based on mass. Arrow indicates average mass for the different microorganisms.

is based on microbial mass. Lawns and prairie soils have nearly equal quantities of fungi and bacteria whereas forest soils are dominated by fungi. Different plants prefer soils populated by a dominance of bacteria or fungi. Cabbage, lettuce, and carrots prefer bacteria dominated soil, corn, wheat, and tomatoes grow best with near equal quantities of bacteria and fungi, and trees and shrubs show optimal growth in a fungal dominated soil. The F : B in rich, highly productive agricultural soil, may approach 1, in deciduous forests it may be 5 : 1 to 10 : 1, but in a coniferous forest the value may exceed 100 : 1. The slow decay of pine needles is attributed to fungi and the decomposition releases organic acids that produce an acidic soil.

Different methods based on direct or indirect measurements have been used to calculate soil F : B ratios and correlation of the different methods may not be clear. The basic systems include (i) direct counting of fungi and bacteria in soil using microscopy, (ii) calculating fungal and bacterial biomass using precursors for cell structures, (iii) using antibiotics to determine the substrate induced respiration on a specific substrate, and (iv) using short term incubations to measure growth rates by following rate at which precursors molecules are incorporated into cellular material (Strickland and Rousk 2010). Assessment of bacteria and fungi by plating on media and counting colonies is problematic because only a fraction of the viable bacteria in soil can be cultivated and fragmentation of fungal filaments in sample preparation can give erroneously large plate-count numbers. Chemical analysis of cell wall components can be used to assess total (including dead) fungi and bacteria by measuring chitin and muramic acid, respectively. Ergosterol, a lipid limited to fungi, measurements can be used to indicate the quantity of fungal membranes present. Phospholipid fatty acids specific for fungi and bacteria can be used to estimate the mass of fungi and bacteria (Rousk et al. 2009). Fungal activity has been expressed as the rate of incorporation of [^{14}C] acetate into ergosterol, and bacterial activity has been determined by following the incorporation of [^{3}H]thymidine and [^{14}C]leucine into bacterial DNA and protein, respectively (de Vries et al. 2006). Quantitative polymerase chain reaction (qPCR), with primers specific for bacteria or fungi that target DNA for synthesis of small subunit RNA, has also been used to determine the growth rates of these microorganisms (Strickland and Rousk 2010). Since most assessments are based on cultured microorganisms, field characteristics must also be included when investigating the F : B of soil. In general, the greater the nitrogen concentration of the soil, the lower the F : B; the lower the soil pH, the greater the F : B. Soil aggregates have also been implicated in influencing the F : B; however, the F : B appears to be influenced by other physical/chemical features which make the effect of soil aggregates difficult to evaluate.

3.4.3 SOM and Soil Food Webs

Soil is the part of the terrestrial environment where plant roots and diverse forms of life are found. There are several different kinds of soil but for the discussion here, the focus will be on productive agricultural soil. As indicated in Figure 3.1, water filled pore space and air account for about half of the volume of topsoil, with about 45% being minerals. The amount of soil organic matter (SOM) is about 5%, with living organisms, including not only plant roots, but also bacteria, fungi, protozoa, nematodes, mites, insects and other arthropods, and earthworms making up <5% of SOM.

Soil contains equal amounts of decomposing organic matter (DOM; 33–50% of SOM) and humus (stable organic matter; 33–50% of SOM) (Figure 3.8), with small amounts (<10% of SOM) of fresh material consisting of roots and animal wastes.

The soil environment is highly dynamic, leading some researchers to suggest that soil should be evaluated as a living system (Harshberger 1911; Paul and Clark 1989). Humus is a complex organic matrix of aromatic polymers that is highly resistant to enzyme attack and consists of subunits containing quinones, phenols, catechols, and sugar moieties. The turnover rate for

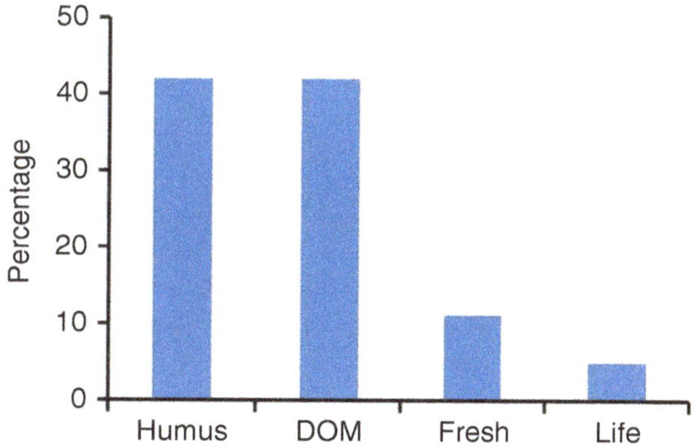

Figure 3.8 Composition of soil based on organic matter and living components. DOM = Decomposing organic matter. Fresh = fresh residues. Life = all living organisms in the soil.

humus in soil is extremely slow due to its enzymatically resistant structure and some have proposed the turnover may take several decades. Humus is not a carbon source for bacteria but serves to retain moisture and bind nutrients in the soil.

Living organisms in the topsoil are organized into a complex soil food web. The bacteria and fungi are supported by plant root exudates and decomposing plant residues. To some extent, the diversity and abundance of living organism below ground can be predicted by the quantity and type of plant growth above ground. The general model to describe the distribution of carbon in plants from CO_2 fixation in plants is that 10% of carbon is released as organic acids from plant roots, 25% goes to plant roots, and 65% to growth above ground. Each plant species secretes organic acids, which stimulate the growth of bacteria and fungi that benefit the plant. This specificity of root exudates for specific microorganisms includes nitrogen-fixing bacteria and mycorrhizal fungi for plant roots. Often the selection of microorganisms by organic compounds released from plant roots to promote the health of plants may be relatively subtle.

Bacteria and fungi are active participants in the soil microbial food web. When ^{13}C-labeled *Escherichia coli* cells were added to agricultural soil, the labeled carbon material was quickly located in *Lysobacter* spp., *Myxococcales* spp., and *Bacteroidetes,* which are bacterial predators (Lueders et al. 2006). Additionally, fungal species of the *Microascaceae* were observed to rapidly degrade the ^{13}C-labeled *E. coli* cells. This indicates that microorganisms not only degrade fresh residue in the soil but also function to control the microbial populations. Although the input of plant residue into the soil coupled with microbial mineralization is commonly recognized as an important contributor to subsurface life in the soil, organic matter from animals is also important. In soils with low plant productivity, the death of an animal or the deposit of manure provides a bioavailable source of nitrogen that stimulates microbial growth in a confined region.

Cycling of carbon, nitrogen, and sulfur compounds is a hallmark characteristic of microorganisms, with nutrient cycling as an efficient system for management of resources. Several metabolic schemes (i.e. citric acid cycle, glyoxylate cycle) in cells are based on cyclic processes and, at the global level, cycling processes for nitrogen, carbon, and sulfur occur as a summation of individual microbial enzymatic activities. CO_2 exchanges between the atmosphere and the terrestrial environment have been described and a model for carbon stored in various biological systems is indicated in Figure 3.9. The major terrestrial consumption of CO_2 is photosynthetic-driven primary productivity, and microbial decomposition of dead organic material from animals and plants releases CO_2 back into the environment. Other volatile carbon compounds released from the terrestrial environment include carbon monoxide and methane. Nitrogen cycling involves several oxidation states of nitrogen and the characteristic activities in the nitrogen cycle are shown in the model in Figure 3.10. Sulfur compounds are recycled primarily through the terrestrial environment; however, some hydrogen sulfide and sulfur dioxide

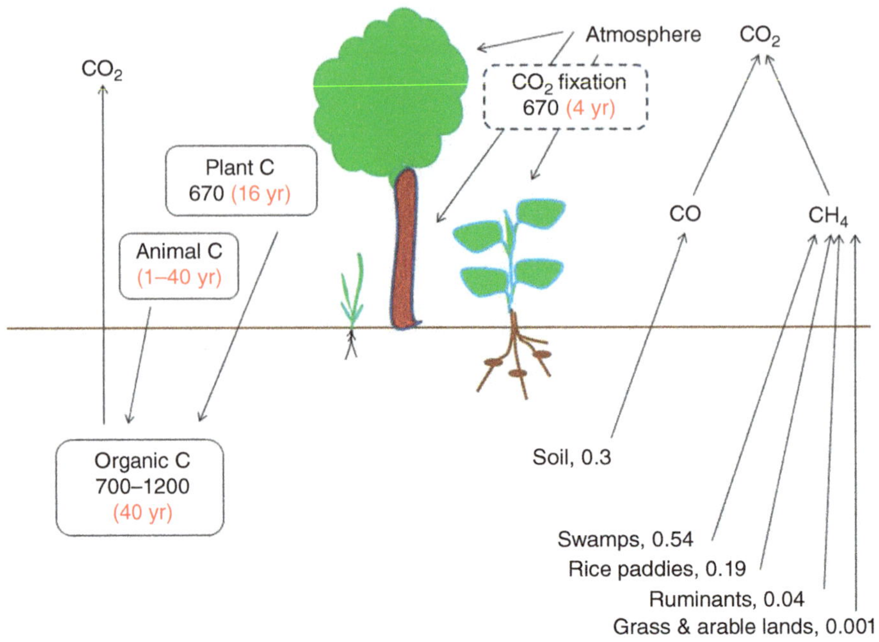

Figure 3.9 Atmospheric production of CH_4, carbon reservoirs, and residence times. Values for CO and CH_4 production are 10^{15} g per year. Values for CO_2 and C reservoirs are 10^{15} g per year. Residence time in years (yr). For additional information see Brock et al. (1984).

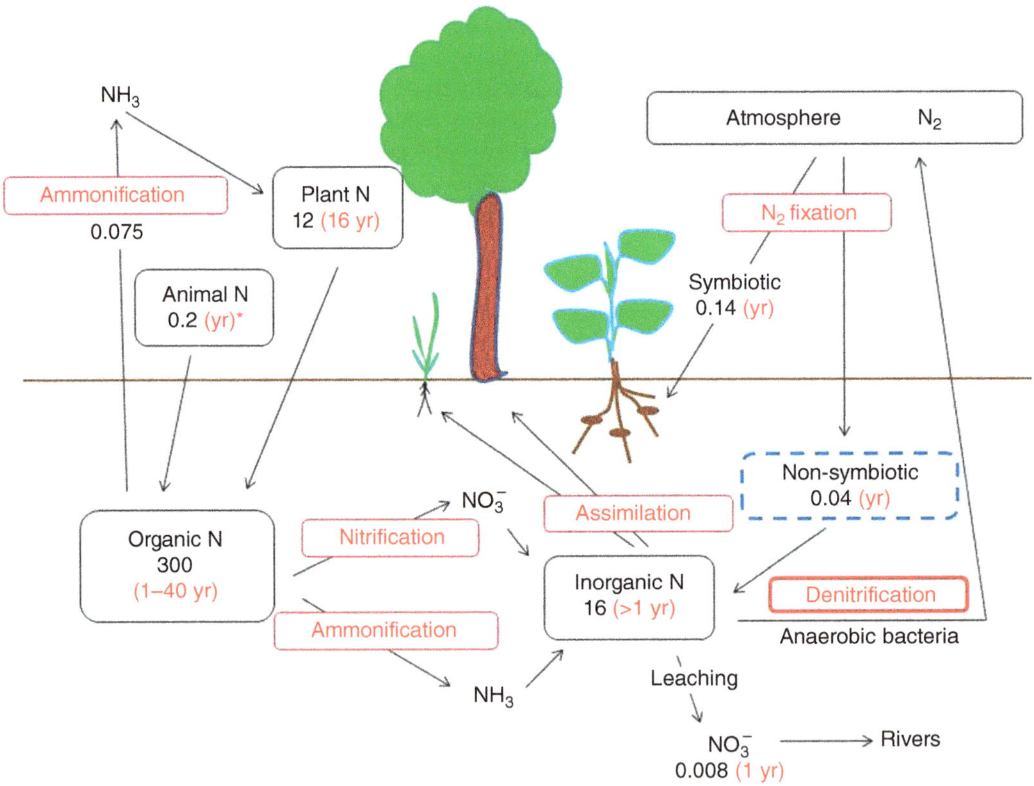

Figure 3.10 Major components of the global nitrogen (N) cycle for terrestrial processes. Values listed are 10^{15} g nitrogen. Residence time in years (yr) and time (yr)* varies considerably. For additional information see Brock et al. (1984).

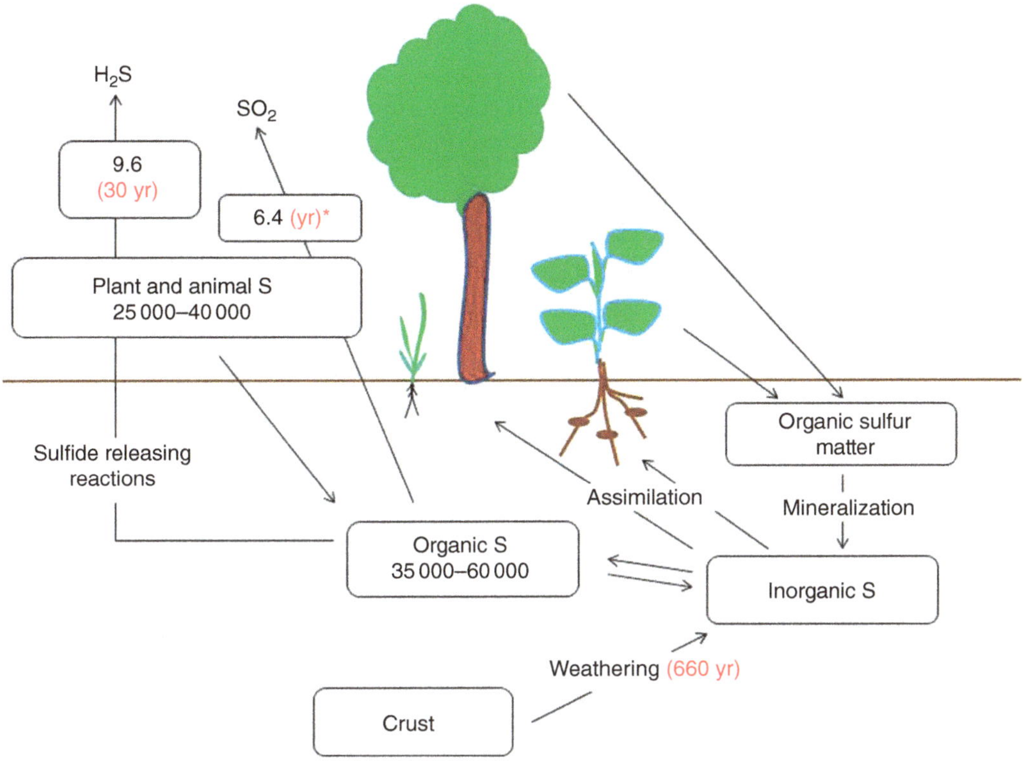

Figure 3.11 The terrestrial metabolism of sulfur compounds. Values listed are 10^{11} g sulfur. Residence time in years (yr) and time (yr)* varies considerably. Values for plants and animals reflect organic sulfur as reservoir sulfur and have no time status. For additional information see Brock et al. (1984).

may enter the atmosphere (Figure 3.11). The cycling of the major nutrients of nitrogen, carbon, and sulfur in aquatic systems is addressed in Chapter 4 and the global biogeochemical cycles are discussed in Chapter 8.

3.4.4 Influence of Agricultural Management on the Soil Microbe Community

The structure of soil microbial communities is highly dependent on land use with the most important factors driving gene expression including C:N, phosphate level, amount of available nitrogen, moisture, and the clay content of soil. When the C:N is low, many enzymes associated with decomposition of complex polysaccharides, CO_2 fixation, and metal reduction are expressed, whereas a reduction of methane oxidation, dissimilatory sulfate reduction, nitrification, and nitrogen-fixation is characteristic of fields with a high C:N ratio (Kuramae et al. 2014). For grasslands, which account for over 25% of the land surface, fertilization with nitrogen and phosphorus markedly shifts the microbial population (Leff et al. 2015). When Amazonian rainforests were converted to pasture, the gene diversity in the newly established pasture was significantly less than in the original forest land (Paula et al. 2014). In most soils, *Acidobacteria* are the most abundant groups of bacteria when the soil organic carbon is relatively high but *Actinobacteria* are more abundant in soils with a low soluble organic carbon level. Tilling of soil promotes aeration and in untilled soils the microbial populations are much lower below 7 cm deep. The numbers of denitrifying bacteria and facultative anaerobes are several times greater in untilled soils compared with tilled soils (Doran 1980). Examination of microorganisms growing in and on the surface of corn (*Zea mays* L.), the plant–root environment revealed a mixed population of *Alphaproteobacteria* and a few archaea with DNA sequences that clustered near Marine group I (*Crenarchaea*) and Marine group II (*Euryarchaea*) (Chelius and Triplett 2001).

3.4.5 Impact of Viruses on the Soil Microbiota

It follows that wherever there are bacteria and archaea, their viruses will be there also. The distribution of extracellular bacterial and archaeal viruses is important but poorly understood. The global virosphere is thought to contain about 10^{31} viral-like particles, with bacterial viruses contributing greatly to this number. Bacterial viruses are estimated infect 10^{23}–10^{25} bacterial cells daily, with 25% of the bacterial population killed daily by phages (Keen et al. 2017). Several soil environments contain about 10^{8} bacterioviruses per gram of dry soil (tailed bacterial viruses are the most abundant). Although there are major advances in the molecular biology of bacterial ecology, about 70% of viral environmental sequences have no homologs in available data banks (Zablocki et al. 2016). The gene banks containing sequences for cyanobacterial viruses need to be developed to provide a reference for metagenomic studies. Evaluation of bacterial viruses in surface soils has been conducted using sequence methodologies and other culture-independent procedures (Zablocki et al. 2016). Bacterial biofilms are dispersed by lytic bacteriophages but proviruses tend to stabilize biofilm communities. In hot desert soils, few extracellular viruses are found but a high number of bacterial cells harbor proviruses. In cold deserts, the extracellular viruses are abundant, with few cells containing proviruses (Adriaenssens et al. 2017). Although some information about the distribution and diversity of bacterial viruses is available, questions concerning the contribution of bacterial viruses to energy flux and their influence on soil nutrient cycles remains unresolved.

3.5 Subsurface Microbiology

3.5.1 Groundwater

Groundwater represents a major source of potable water in several regions of the world. The phreatic zone, described at the beginning of this chapter, represents the waterlogged region of soil or water below the water table. The study of groundwater is referred to as hydrogeology and reviews concerning geological aspects, water flow, and bioremediation of hydrology are available (Kresic 2009; Fetter 2001).

Groundwater can exist in porous rock or even sand and an underground, water-bearing region of porous rock is typically called an aquifer. In some instances, there are underground streams. If the aquifer is surrounded by a layer of impermeable rock, the aquifer is a confined aquifer (Figure 3.12). In an unconfined aquifer, the rock is relatively porous and the water within the aquifer can exchange freely with surface water through springs. Water from precipitation or surface water can enter the aquifer through seepage via porous rocks (this area is typically called the recharge zone). Well water within an aquifer can enter the surface water through natural springs or by being pumped from a well. Porous rock regions overlying an aquifer can function as recharge zones because surface precipitation can seep through the porous rock and enter the aquifer. Alternatively, water can enter the aquifer via seepage from lakes and rivers. In groundwater, bacterial distribution is relatively homogenous due to the ease

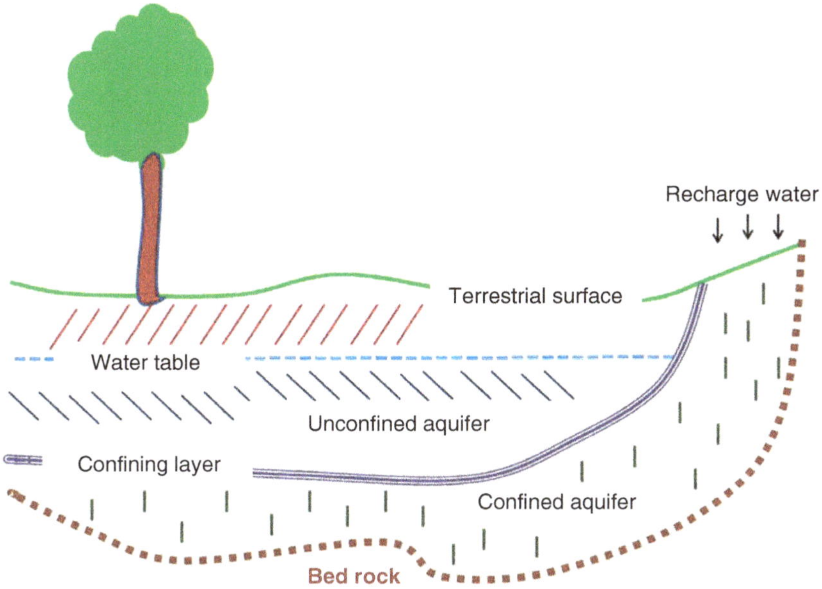

Figure 3.12 Model showing relative positions of unconfined and confined aquifer.

of microbial movement and the physical features contributing to solution mixing. However, in shallow aquifers, plantonic bacteria make up only 10% of the total bacterial number with the remaining 90% attached to minerals (Hazen et al. 1991; Griebler et al. 2002; Alfreider et al. 1997). It is generally considered, that the greatest microbial activity is at the water–mineral interface.

An area of microbiology that has not received much attention is the presence of Protista in shallow aquifers. Even though the protists are found in small numbers in aquifers, they may have an important role in the microbial ecology of some aquifers. Amoeba and ciliates may be present; however, small flagellates (in situ size is 2–3 μm diameter) are most commonly found in sandy aquifers (Novarino et al. 1997). Many of the protists are encysted and the detection of protists in aquifers relies on cultivation of protists and direct microscopic observation. The relatively fragile character of the cells and cysts requires that samples are protected against desiccation. The unique swimming pattern of the flagellates suggests that flagellates feeding on bacteria would consume different groups of bacteria within a community. One effect of bacterivorous flagellates may be to impede bioremediation by feeding on bacteria that are involved in the decomposition of organic contaminants. Aquifers containing protists may have a water connection to nearby rivers (Sinclair and Ghiorse 1987) and, indeed, many historical accounts of protists present in subsurface water concern wells associated with coastal regions (Novarino et al. 1997). With the development of primers appropriate for the Protista, a better understanding of protists in groundwater will emerge.

Groundwater, used as a potable water source, is typically tested for coliforms (indicative of fecal contamination). In rural environments, fecal contamination may arise from the outflow of septic drainage fields and represent a contamination issue for drinking water wells. Another concern with groundwater safety arises from chemical contamination via industrial activities, leaking petroleum storage tanks, and hydraulic fracturing (see Chapter 11, Section 11.9) during petroleum exploration. Groundwater bioremediation is addressed in Chapter 9.

3.5.2 Cave Water

Caves can provide direct access to groundwater and so the entry of higher organisms into caves for shelter or exploration can introduce organic matter in some aquifers. There are several prominent examples of this in caves that are used by bats and other organisms. One such location, Bracken Cave, located a few miles north of San Antonio, Texas, USA, contains an estimated 15 million Mexican free-tailed bats (*Tadarida brasiliensis*) and the considerable guano (bat excrement, approximately 85 tons per annum) contributes significant quantities of carbon, nitrogen, and phosphorous to the groundwater in the vicinity.

However, in most instances, groundwater environments are highly oligotrophic, with trace amounts of carbon entering the system from surface runoff. Due to the absence of light, primary production within aquifers relies on lithotrophic microorganisms that can use reactions such as the oxidation of iron, sulfur, ammonia, or other inorganic compounds as a source of energy to fix CO_2. The resulting microbial-based food web may support higher organisms, typically invertebrates in some groundwater environments (Barton 2006). Aquifers represent a significant source of potable water in many parts of the world, particularly in arid regions. Microorganisms within aquifers contribute significantly to the water quality and are now being increasingly studied. Except for caves that may provide direct access to groundwater, aquifer testing for microorganisms and water chemistry is typically achieved using sampling wells.

The microbiology of cave waters from epigenic caves and karst landscapes have been examined and these appear to have different microbial communities. In an alpine study (Wilhartitz et al. 2006), microorganisms in a dolomitic-limestone karst aquifer were compared with those in a limestone karst aquifer. Considerable similarity was found in the 16S DNA profiles of bacteria obtained from both aquifers. Additional similarities were detected using catalyzed reporter deposition fluorescence in situ hybridization (CARD-FISH) to evaluate the microbial communities. Because of this study it was concluded that a karst aquifer maintains a stable ecosystem, which has been designated autochthonous microbial endokarst communities (AMEC) (Wilhartitz et al. 2006). In contrast, epigenic cave streams do not have AMEC but rather display a highly dynamic microbial community that shows population successions (Brannen-Donnelly and Engel 2015).

3.5.3 Deep Subsurface Aquifers

For deep subsurface aquifers, the question is not are microorganisms present, but which bacteria and archaea are found there and how does the environment shape the microbial community? From the early publications (see review by Lovley and Chapelle 1995) concerning microorganisms isolated from aquifers (e.g. granite rock, formation waters of oil fields, limestone groundwater, alkaline basaltic water, and Atlantic coastal plain), each subsurface aquifer was found to have unique physical and chemical parameters that influence microbial growth. While habitat-based microbiological evaluations continue to be important, the microbial community is being investigated in several deep subsurface aquifers.

3.5.3.1 Aquifer in a Coal-bearing Basin

Microorganisms were characterized in a deep aquifer of a coal containing sedimentary basin in Germany and were found to actively degrade lignite coal to methane (Gründger et al. 2015). Using 16S rRNA sequence analysis, the aquifer water was found to contain a highly diverse group of bacteria and an equally varied set of archaea. Many of the bacteria were nitrate-reducing and hydrocarbon-degrading members of the *Actinobacteria, Gammaproteobacteria,* and *Deltaproteobacteria*. Some major groups of bacteria were isolated: *Pseudomonas* spp., which characteristically degrade alkane and alkene hydrocarbons; *Acinetobacter* spp., which have the capability of metabolizing short linear alkanes and aromatic hydrocarbons; *Arthrobacter* spp., which degrade polysaccharides; *Georgfuchsia* sp. which degrade benzene, toluene, ethylbenzene, and xylene (BTEX) compounds using iron or manganese as an electron acceptor; and *Desulfatiferula* sp., a sulfate reducer that oxidizes long chain alkenes. Other bacteria found in oil and coal environments include *Acidovorax, Acrobacter,* and *Methylotenera.* Some of the dominant archaeal isolates included unclassified *Crenarchaeota* members with unknown metabolic activities and methanogens (e.g. *Methanosarcina, Methanosaeta,* and *Methanoregula*), which characteristically used acetate or H_2 as electron donors. The syntrophic activity of bacteria and archaea appears to be characteristic in anaerobic aquifers that produced methane from lignite and in other hydrocarbon fed aquifers (Gründger et al. 2015 and references therein). The abundance of microbial activity at the lignite–water interface not only contributes to the degradation of lignite with the formation of low molecular weight organic carbons but may also contribute to the translocation of these bacteria into adjacent areas through the seams or fractures in the lignite (Detmers et al. 2001)

3.5.3.2 Deep Granitic Aquifer

Chemolithotrophic communities are present in granitic waters containing H_2 and methane (Kotelnikova and Pedersen 1998). Borehole samples were taken from the granitic formation on the island of Äspö on the Baltic coast of Sweden. These samples were from 45 to 446 m below sea level, which represented three types of water: shallow or modern marine, intermediate or mixed water, and old saline water. The age of the modern marine water is estimated to be <20 years while the old saline was estimated to be thousands of years old. Some dissolved organic carbon (e.g. reduced humic and fulvic acids and their decomposition products) compounds were present in the modern marine and intermediate waters; however, water at the deepest level receives no organic matter from the surface environment. Using cultivation methods, 10^3–10^6 microbial cells per milliliter were present in the water and these cells were predominately acetogens and methanogens. The energy source to support the microbes in the deep granitic aquifer is considered to be abiotic H_2, and this would agree with H_2 as the universal energy source for life (Morita 2000). With about 50 μM H_2 in the groundwater, the autotrophic methanogens would grow and produce methane whereas the autotrophic homoacetogens would produce acetate. Acetotrophic methanogens and other anaerobic bacteria using H_2 or acetate would also grow in this environment. The metabolic reaction associated with these methanogens and acetogens is:

$$\text{Autotrophic methanogens} \quad 4H_2 + CO_2 \rightarrow CH_4 + 2H_2O$$

$$\text{Homoacetogens} \quad 4H_2 + 2CO_2 \rightarrow CH_3COO^- + H^+ + 2H_2O$$

$$\text{Acetotrophic methanogens} \quad CH_3COO^- + H^+ + H_2O \rightarrow CH_4 + HCO_3^- + H^+$$

In a recent assessment of the major populations of microorganisms in the granitic aquifers, a distinction in physiological types could be seen in the three water types (Wu et al. 2016). Heterotrophic activity was the major type of metabolism seen in organisms found in the modern marine water. The greatest amount of sulfate reduction and bioproduction of hydrogen sulfide was observed in the intermediate water, whereas anaerobic ferric iron reduction was dominant in the modern marine water. Cells in the old saline water had gene homologs to hydrogen oxidation and carbon dioxide fixation suggesting that chemoautotrophic growth

Table 3.3 Microbial population in two alkaline aquifers in the Columbia River Basalt Group, Washington, USA.

Formations	Grande Ronde	Priest rapids
Characteristics of water		
Sample depth	1270 m	316 m
pH of water	9.94	8.05
Eh	−137 mV	−27
Sulfate	$142\,mg\,l^{-1}$	$<0.05\,mg\,l^{-1}$
Sulfide	$22.8\,\mu mol\,l^{-1}$	not detected
Well type	sulfidogenic	methanogenic
Abundance of microorganisms (%) based on percentage of target group probed		
Bacteria	92.0	64.4
Archaea	1.8	2.5
Eucarya	5.7	14.4
Desulfovibrionaceae	2.5	2.9
Desulfobacter group	0.3	not detected

Source: Based on data from Fry et al. (1997).

was the principle activity of these indigenous microorganisms. Small cells with a diameter of <0.22 μm had a small genome and these small bacteria were members of *Alpha-*, *Beta-*, and *Gammaproteobacteria* group. These cells of <0.22 μm accounted for about half of the cells found in each of the three water types and appeared to generate energy by fermentation of pyruvate and degrade ethanol to acetyl~CoA.

3.5.3.3 Anaerobic, Alkaline Aquifer

Several alkaline, saline aquifers are present in the Columbia River Basalt Group in the state of Washington, USA, and geochemical energy (e.g. production of H_2 by abiotic reactions) is thought to energize microorganisms in the water (Stevens and McKinley 1995). A survey of two alkaline aquifers revealed considerable differences in microbial content (Table 3.3). This difference in bacterial populations in the same geological formations is not unique because adjacent subsurface environments have been reported to contain different microbial populations (Hazen et al. 1991). Methanogens are considered to be highly represented in the Archaea group and would indicate the methanogenic character of the Priest Rapids well. About 100 times more homoacetogenic bacteria than methanogens were cultured from the Priest Rapids well, whereas similar numbers of fermentative heterotrophs, homoacetogens, and sulfate-reducing bacteria were cultivated from the Grande Ronde aquifer (Stevens and McKinley 1995). The nature of the microeukaryotes is unknown but it should be noted that eukaryotes have been reported in other deep aquifers (Fry et al. 1997 and references within).

3.5.3.4 Saline Hydrothermal Aquifer

The Naica Mine in Mexico produces giant gypsum crystals that are over 11 m long. Periodically this mine is flooded by saline hydrothermal water from a hydrothermal aquifer, and water from about 700 m deep in the mineshaft was examined for microorganisms (Ragon et al. 2013).

Although primers were used for eukaryotes, none were detected. Archaea and bacteria were present and seemed to be most similar to thermophilic microorganisms found in deep subsurface environments. Although sulfate reducers and methanogens were not present, archaea with monoamine oxidase were detected and appeared to be *Thaumarchaeota*. The cellular biomass of these microorganisms was small and no bacteria were found inside the giant quartz crystals.

3.6 Deep Subsurface Microbiology

Some of the first reports concerning microorganisms in deep subsurface environments were from Zobell's group which cultivated bacteria from marine sediments (Morita and ZoBell 1955; ZoBell and Anderson 1936). The exploration of the deep subsurface sediments by the Ocean Drilling Program (1983–2003) provided microbiologists access to mud samples taken from the deep biosphere where numerous physiological types of bacteria were obtained (Parkes et al. 2000). The report that approximately 39% of terrestrial prokaryotes and 55% of marine prokaryotes were subsurface (Whitman et al. 1998) encouraged new subsurface investigations. A deep coring expedition in 2003 provided a basis for evaluating physiological activities in all deep subsurfaces (D'Hondt et al. 2004). Not only are there numerous physiological types of bacteria in the deep subsurface but there is also an abundance of archaea (Takai et al. 2001) and a new class (*Hadesarchaea*), which is rooted close to the *Thermococci*, has been reported (Baker et al. 2016). With an expansion of 16S rRNA primers (Teske and Sørensen 2008), there are sure to be numerous other archaea discovered in the deep terrestrial environment.

Since the latter part of the twentieth century, microbiologists have explored the boundaries of life (i.e. defining the scope of the biosphere). The deep subsurface (broadly defined as microbiology within bedrock and extending into the Earth's crust), has been a component of these studies as scientists are curious about the depths below the Earth's surface in which life exists. Deep subsurface microbiology can have an economic impact on petroleum and gas reservoirs, deep groundwater sources, geochemical processes, and minerals. Access to deep subsurface environments and their associated microorganisms can occur via mine shafts, drilling, or via a combination of these two processes. The regions in proximity to human activity (i.e. mine shafts, petroleum wells, etc.) can obtain nutrients and oxygen from the surface, in contrast to microorganisms in the deep subsurface, which use subsurface nutrients to support their anaerobic growth.

Another aspect of the terrestrial subsurface is the constant yet increasing temperature as one descends below the Earth's surface. The temperature just below the soil surface reflects the average temperature of the area; then the average increase on descent into the Earth is 3 °C per 100 m depth (Zlotnikov 2012). Pressure also increases in the deep terrestrial regions and bacteria that grow optimally at 0.1 MPa are called piezophiles and those few that require >60 MPa for growth are known as hyperpiezophiles (Barlett 2002). The upper pressure limit for microbial growth appears to be 70–130 MPa because at this pressure, protein and DNA interactions are greatly disrupted. Proteins are not denatured until pressures reach 400–800 MPa.

Deep subsurface environments are extremely low in the nutrients needed to support microbial growth and as a consequence the starved microorganisms have a generation time projected to be ten years or even hundreds of years (Frederickson and Onstott 1996). Others have suggested that the turnover rate for microorganisms in the deep biosphere may be hundreds to thousands of years (Hoehler and Jorgensen 2013). It is a challenge to understand the physiological status of the deep subsurface microorganisms because it appears that over 75% of these cells may be viable (Morono et al. 2011) and capable of acquiring carbon nutrients. This prompted Jørgensen (2011) to suggest that deep subsurface cells are on "physiological standby." It should also be recognized that subsurface environments have an abundance of bacterial viruses with 10^4–$10^9\,cm^{-3}$ bacteriophage observed by electron microscopy and the ratio of viruses : bacteria may exceed 200 : 1 in the marine deep biosphere (Engelhardt et al. 2014).

Technical difficulties and the expense of studying subsurface microbiology are considerable and require the collaboration of microbiologists with geologists, engineers, and other professions. It has been estimated that 40–69% of bacteria in the Earth's biosphere, which is about 25×10^{29} bacterial cells (Whitman et al. 1998; McMahon and Parnell 2014), are found in the terrestrial subsurface. As discussed at a recent workshop on the deep terrestrial biosphere (Wilkins et al. 2014), the collection, processing, and measuring of deep terrestrial samples is indeed challenging and requires special consideration. The relevant questions raised by scientists pertaining to bacteria in the deep terrestrial region (Wilkins et al. 2014) can be summarized as follows:

- What are the limiting factors that influence the abundance of bacteria and archaea?
- How are bacteria able to survive with variable, limiting nutrient levels and what physiological adjustments enable bacteria to grow with extremely slow doubling times?
- What is the taxonomic diversity of microorganisms present and what is their community structure?
- Why do sub-terrestrial sites that are physically isolated have microorganisms with similar taxonomic diversity?

Drilling to obtain rock cores is often referred to as diamond drilling as the drill bits contain industrial diamonds. The drilling process involves friction as the drill bit (end of the drill) cuts through rock and so lubricant must be added. Obviously the lubricant is not sterile so during microbial analysis the contribution of the drilling lubricant must be considered. One approach would be to add some type of chemical or physical tracer material to the lubricant (such as bacteria-sized fluorescent spheres) and consider the region of a rock core containing the tracer to be contaminated, whereas the region lacking the tracer to be pristine. Inagaki et al. (2015) provides a detailed description of the precautions used to avoid contamination in this context. In spite of the difficulties, we are finding some fascinating things about subsurface microorganisms and will explore some examples. Also, microorganisms in the deep subsurface and elsewhere have an impact on geology and that is discussed at the end of this chapter. Based on metagenomics studies, the diversity of organisms in the subsurface is remarkable.

3.6.1 Marine Sediment Microbiology

The amount of ocean sediment varies with over 10 km near the continents and less than 0.1 km in the open ocean (Amend and LaRowe 2016). The total number of microbial cells present in marine sediment has been estimated to be 3×10^{29} with an approximately equal abundance of bacteria and archaea (Kallmeyer et al. 2012). Commonly, anaerobic microorganisms are dominant in ocean sediments with sulfate-reducing microorganisms having a prominent role in the conversion of organic carbon materials to carbon dioxide. The physiological types of microorganisms present in marine sediments reflects the temperature of the environment and includes psychrophiles, mesophiles, and thermophiles but not hyperthermophiles. The deep sea sediments lack the nutrients to support microbial growth but it appears that many of these microbial cells remain viable for hundreds of years. Endospores may be produced by appropriate bacteria but it is unlikely that these endospores would be induced to germinate because the required nutrients are not available.

3.6.2 Deep Mines and Boreholes

Assessing the types of microorganisms present in the terrestrial deep subsurface is a challenge and requires making drill holes that are hundreds of meters to a few kilometers deep through rocks, mines, and caves to collect water. The water collected may be from deep subsurface aquifers or from water seeping through fractures in the rocks. To establish the presence of microorganisms with specific activities in the carbon, nitrogen, and sulfur cycles, specific gene markers are used (Table 3.4). Although there were numerous investigations into terrestrial deep subsurface regions, a few sites will be discussed here and additional sites are covered in other sections of this book. Deep granitic aquifers and Columbia River basaltic aquifer were addressed in Sections 3.5.3.2 and 3.5.3.3. The abundance of bacteria, archaea, and microscopic nematodes in the Beatrix gold mine in South Africa is addressed in Section 3.6.4.

Considerable diversity of archaea occurred in water samples from several deep gold mines in South Africa (Takai et al. 2001). Based on 16S rDNA analysis, the archaea were more abundant than bacteria in acid mine waters but in groundwater samples, bacteria were more abundant than archaea. Microorganisms associated with the *Euryarchaeota* were dominant in fissure water, whereas microorganisms most similar to *Crenarchaeota* were present in service water. Many of the archaeal DNA sequences were dissimilar to other archaeal DNA sequences; however, clones similar to *Methanosarcina barkerii* and *Methanosaeta concilli* were present. Fissure water from the Kloof Mine contained a clone with 99.4% similarity to *Pyrococcus abyssi*, a hyperthermophilic archaea that had previously been found only in marine hydrothermal vents. It is possible that *P. abyssi* may have been growing at a depth of about 5 km, where the environment would have a temperature appropriate for this hyperthermophile, and with time it migrated upward into the water of this mine.

The water in the Mponeng Mine near Carletonville, South Africa was found to be unique because 99.9% of the genomic DNA was from a single species (Chivian et al. 2008). This bacterium was collected 2.8 km deep in the mine and it has been designated *Candidatus* Desulforudis audaxviator, even though it has not yet been cultivated in the laboratory. It has a 2.35 megabase pair genome which encodes for synthesis of flagella, pili, and endospore in addition to having a complete metabolic capability for N_2 and CO_2 fixation. Since this initial report, DNA for this bacterium has been isolated from other deep subsurface environments. Future studies are needed to determine why this bacterium is so suited for growth in this environment.

Table 3.4 Marker genes used in anaerobic deep terrestrial biosphere research.

Cycle	Gene	Characteristic
Carbon	*mcrA*	Methyl coenzyme M reductase, the final step in methane production
	pmoA, mmo	Methane monooxygenase, oxidation of methane to methanol
Sulfur	*dsrB*	sulfite reductase, dissimilatory sulfate reducers use this enzyme to produce sulfide from sulfite
Nitrogen	*amoA*	Ammonia monooxygenase, oxidation of ammonia to hydroxylamine in nitrification
	nifH	Nitrogenase, fixes nitrogen by producing ammonia from N_2
	nirK, nirS	dissimilatory nitrite reduction, reduction of nitrite to NO
	napA, narG	dissimilatory nitrate reduction, denitrification with nitrate reduced to nitrite

Source: Based on data from Itävaara et al. (2016).

3.6.3 Deep Subsea Floor

The ocean basin consists of rocks such as basalt, which consist of dense iron and manganese silicates. One notable recent study (Inagaki et al. 2015) showed that microbes could be detected at the deepest level drilled (2458 m below the sea floor) and based on 16S rRNA sequencing, the microbes associated with the lignite (coal) layer were similar to those associated with forest soils, suggesting that the microorganisms in lignite were likely buried along with the plant material (from which coal was derived) during the late Paleogene approximately 23 million years ago. These organisms differed notably from organisms associated with the surrounding rock.

Several interesting reports have been produced that indicate the diversity of microorganisms in the deep subsurface marine sediments. Aerobic microorganisms have been found several meters below the sediment surface or material deposited 86 million years ago where the rate of oxygenation in the sediment was greater than microbial metabolism (Røy et al. 2012). In another report, $10–10^4$ cells per cubic cm were found in lignite coal beds beneath the seafloor in the Pacific Ocean and these microorganisms resembled microbial communities in forest soils and not microorganisms in adjacent subseafloor communities (Inagaki et al. 2015). Methane production was reported in black shale deposited over 100 million years ago in sediment hundreds of meters beneath the sea floor. The oxidation of methane to carbon dioxide was attributed to an anaerobic consortia consisting of sulfate-reducing bacteria and methane oxidizing bacteria (Arndt et al. 2006). Modeling systems are useful for evaluating microbial communities in deep subsurface marine sediments.

3.6.4 Deep Subsurface Storage Sites

Currently, deep subsurface storage is is used for the containment of high level nuclear material and the long-term storage of gases. This storage is of great scientific interest and considerable economic, social, and political concern. As the sites for storing these two materials are significantly different, they will be discussed separately.

3.6.4.1 Storage of Nuclear Fuel Waste

The containment of high levels of nuclear fuels is a concern of several governments and depositing of these nuclear wastes in deep underground environments is one of the options being considered. With approximately 11% of the world's electricity coming from nuclear facilities, nuclear waste is continuously produced. A report on the Canadian nuclear fuel waste management program provides information concerning microbial activity in a nuclear fuel storage site (Stroes-Gascoyne and West 1997). The site for Canadian nuclear fuel waste was a vault constructed in granitic rock 500–1000 m below the surface. The containment followed the guidelines established by the Atomic Energy of Canada Limited proposal, which calls for fuel wastes to be placed in corrosion resistant containers placed on the floor of the vault. The backfill surrounding the nuclear waste containers consists of a bentonite-silica sand mixture which is overlaid with granitic rock and clay. The area surrounding the waste containers reaches a maximum temperature of 95 °C, is nutrient poor, has a pH range of 7–9, and an Eh range of −300 to +200 mV. In a pilot study with gamma radiation from ^{60}Co, the death rate (D_{10} value) for bacteria was influenced by temperature and moisture content. Microbial death in the contained waste would be immediate whereas on the container surface it would take days to weeks for the bacteria to be killed and in time, the backfill would be sterilized up to a distance of 50 cm from the container. A recent report of microbiological activities in borehole samples taken from the nuclear fuel waste site (Stroes-Gascoyne et al. 1997)

Figure 3.13 Underground laboratory used by Karsten Pedersen and his group as they assessed subsurface microorganisms. *Source:* Courtesy of Karsten Pedersen, Microbial Analytics Sweden.

suggests that the dramatic reduction of heterotrophic bacteria, sulfate reducers, and methanogens is attributed to the water activity (a_W). With water levels of <15%, bacteria and archaea are not capable of growing. The observation that sulfate-reducing bacteria are not capable of growing in this environment is important because the hydrogen sulfide produced by these bacteria is highly corrosive. After 2.5 years of in situ placement, the abundance of bacterial cells per gram of backfill was 10^1–10^5 for aerobe and anaerobic heterotrophs and 10^2 for sulfate-reducing bacteria. Most of the bacteria isolated were members of *Beta-*, *Gamma-*, and *Deltaproteobacteria* and gram-positive bacteria while about 20% of the bacteria isolated were gram-negative. The most frequently isolated genera of gram-positive bacteria were *Amycolatopsis*, *Bacillus*, and *Nocardia* while *Acinetobacter* and *Pseudomonas* were frequently found as gram-negative bacteria (Stroes-Gascoyne et al. 1997).

There has been considerable interest in establishing a repository for spent nuclear fuel in deep regions of the Fennoscandian Shield in southeastern Sweden. To study microorganisms under in situ pressures, Karsten Pedersen constructed the Äspö Hard Rock Laboratory (HRL) (see Figure 3.13) 450 m underground and evaluated microbial activities in granitic groundwaters (Pedersen 2012). A diverse population of subterranean microorganisms with $<10^5$ cells ml^{-1} were reported in groundwater including aerobes, acetogens, methanogens, and bacteria capable of dissimilatory reduction of nitrate, iron, manganese, and sulfate. Pedersen and colleagues have isolated numerous novel organisms from the Äspö HRL aquifer, including *Methanobacterium subterraneum*, *Desulfovibrio aespoeensis*, and a bacterial virus specific for *D. aespoeensis* (Pedersen 2012). With an abundance of phages present in the groundwater, the viral shunt is important in recycling carbon compounds for bacterial growth, with H_2 from the underground environment used as the energy source. The bacterial population in the Äspö HRL aquifer of $<10^5$ cells ml^{-1} is maintained by limiting nutrients and the lethal activity of phages. With about

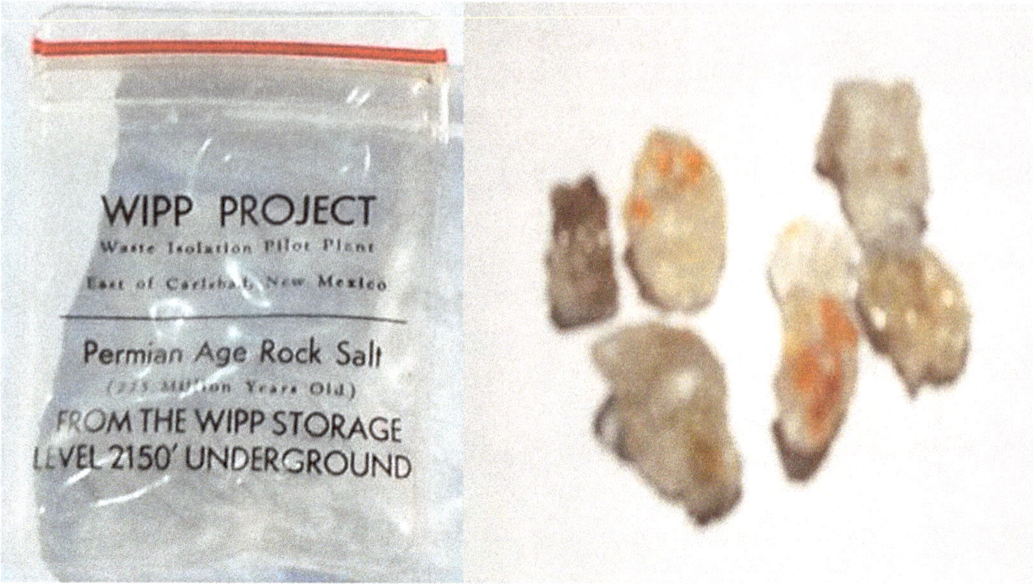

Figure 3.14 Salt crystals from the Waste Isolation Pilot Plant in New Mexico for the storage of radioactive wastes. These crystals are typical of those present 2150 ft. below the surface. Source Reproduced with permission from Larry L. Barton.

half the bacteria in the aquifer immobilized in a biofilm, the bacterial viruses would have limited access to bacteria in a biofilm and would be very effective in attacking the planktonic bacteria.

In the USA, possible sites for the subsurface storage of nuclear cycle wastes have included repositories at Yucca Mountain, Nevada, and Carlsbad, New Mexico. The Yucca Mountain site is in heat-consolidated volcanic rock which is about 300 m below the surface and 275 m above groundwater. Based on plate counts, the number of heterotrophic bacteria was 3×10^4–2×10^5 bacteria/g dry wt. The limiting factors for bacterial growth were water and organic carbon (Kieft et al. 1997). The Waste Isolation Pilot Plant (WIPP) near Carlsbad is 300 m underground in a 600 m thick salt formation. The radionuclides are put in specified containers, placed in the salt caves at WIPP (typical salt crystals are in Figure 3.14), and the salt very slowly “creeps” over the storage material to envelope it. The microbiological studies pertaining to WIPP are model systems evaluating the rate of anaerobic cellulose decomposition by bacteria.

Low level radioactive liquid waste has been injected into the subsurface horizon at the Severnyi radioisotope disposal site in Eastern Siberia, Russia. Microorganisms in samples taken 160–400 m below sea level revealed robust activity compared with samples outside the injection area (Nazina et al. 2004). Heterotrophic aerobes, sulfate-reducing bacteria, methanogens, and anaerobic denitrifying bacteria were more abundant in the zone where the radioactive waste was introduced. The stimulation of microbial growth was clearly attributed to the various chemicals present in the liquid waste, which were used as nutrients for resident microorganisms.

3.6.4.2 Underground Storage for H_2 and CH_4

Energy companies store town gas (a flammable fuel made from coal), H_2 and CH_4 in over 600 underground sites around the world (Gniese et al. 2013). Most commonly, these storage sites are aquifers, salt caverns, and old oil reservoirs. The relevant questions of concern are: (i) Will the introduction of these gases impact indigenous microorganisms present in the reservoir sites? (ii) What effect will the microorganisms have on the stored gas? It can be assumed that there are 10^3–10^6 microorganisms/ml of reservoir water and that the growth limiting factors are phosphorus and nitrogen (Gniese et al. 2013). Methane and H_2 are suitable electron donors for the indigenous bacteria, while sulfate, manganese, iron, and CO_2 are suitable electron acceptors for microbial metabolism. While methane oxidation occurs quickest under aerobic conditions, anaerobic consumption of methane could proceed by methanotrophic archaea, and anaerobic bacterial consortium involving sulfate-reducing, manganese-reducing or iron-reducing bacteria (Beal et al. 2009; Knittel and Boitius 2009; Milucka et al. 2012). The depletion of H_2 can be attributed to indigenous bacteria and archaea and in many instances H_2 would be converted to methane. Obviously, long term sampling would be a part of the management plan for the underground storage of H_2 and CH_4.

3.6.4.3 Underground Storage for CO_2

As increasing levels of atmospheric CO_2 may potentially be caused by industrial emissions, a new technology has evolved that focuses on collecting CO_2 and storing it in underground vaults. The carbon capture and storage (CCS) program focuses on CO_2 since it is the greatest contributor to the greenhouse effect even though fluorocarbons and methane are also important emissions to the atmosphere. The physical and chemical technology associated with the capture of CO_2 from flue gas is discussed in published reports (Gniese et al. 2013) and the storage of CO_2 is of interest to microbiologists. It is proposed to contain 200–2000 Gt of CO_2 in saline aquifers (Frio Brine Pilot Test, USA; CO_2SINK, Ketzin, Germany; Sleipner, Norway; Nagaoka, Japan), in gas fields (In Salah Gas, Algeria; CO_2-CRC Otway Basin, Australia), and in an enhanced oil recovery site (Weyburn CO_2-EOR, Canada) (Gniese et al. 2013).

While the role of microorganisms in the storage of CO_2 in subsurface regions is an ongoing study, some generalizations based on in situ studies can be made. Subsurface biofilms are important because their presence diminishes the rate at which minerals dissolve and reduces the rate of carbonate mineral deposition. In areas without biofilms, elevated CO_2 levels promote dissolving of minerals, perhaps due to localized acidification reactions, but after many months, the pores in subsurface rocks become blocked due to carbonate precipitates. Additionally, microorganisms in biofilms are more resistant to CO_2 exposure compared with planktonic microorganisms. Sulfate-reducing bacteria and methanogens are less sensitive to CO_2 injections compared with other microorganisms.

One of the best analyzed sites for the impact of CO_2 on microorganisms was conducted at the saline aquifer in Ketzin, Germany (Morozova et al. 2010, 2011). Prior to injection of CO_2, the number of bacterial cells determined by DAPI staining was $2–6 \times 10^6\,ml^{-1}$. After CO_2 was injected, the number of bacteria was about $10^5\,ml^{-1}$ and five months after injection of CO_2 the number of bacterial cells returned to $2 \times 10^6\,ml^{-1}$. Using 16S rRNA molecular methods, the dominant bacteria were identified as anaerobic halophilic *Halanaerobium* sp. and *Halobacteroidaceae* and sulfate reducers of the *Desulfohalobium* and *Desulfotomaculum* genera. After CO_2 injection, the chemolithotrophic bacteria were dominant over the chemoorganotrophs but over time the numbers of these bacteria returned to that of preinjection of CO_2. FISH probes containing sequences for specific microorganisms, revealed that numerous species of methanogens and sulfate reducers were present after CO_2 injections. Also present post injection were bacteria from the *Alpha-, Beta-, Gammaproteobacteria* as well as *Firmicutes,* indicating a rich diversity of microorganisms tolerant to elevated levels of CO_2.

3.6.4.4 Geothermal Energy Production

Geothermal systems consist of a closed loop system where hot water is extracted from an underground area, passed through a heat exchanger, and reintroduced back into the subsurface. Costa Rica, El Salvador, Iceland, Kenya, and the Philippines obtain 20% of their energy from sites that may be 4000 m below the surface. Private households may use shallow underground water recycling systems where the depth is less than 400 m. In these subsurface regions, most of the microorganisms are attached to the rock environment and with the introduction of water they become dislodged and circulated throughout the system. From various subsurface studies, it is apparent that there is a diverse array of thermophilic microorganisms present in subsurface environments. Any stimulation of sulfate-reducing bacteria and archaea would be detrimental to the geothermal process because of the production of hydrogen sulfide by these microorganisms. Not only is this biogas (hydrogen sulfide) corrosive but it also produces metal sulfide precipitates that restrict subsurface water flow. Basically, there are two questions concerning microorganisms and geothermal energy (Gniese et al. 2013): (i) Do microorganisms have an influence on use of geothermal energy? (ii) What is the impact of geothermal energy waters on subsurface microbiology? Future studies are required to provide answers to these questions.

3.6.5 Endolithic Microorganisms

Organisms can associate with pore water and live within various crevices in rocks. At the Earth's surface, the association of microorganisms within sandstone and other porous rocks is known as an endolithic (i.e. within rocks) community and represents a mechanism whereby organisms can survive an extremely hostile environment. Although first described in desert environments such as the McMurdo Dry Valleys in Antarctica and the Negev Desert in Israel in the 1980s by Friedmann (1982), endolithic activity is a common feature of microbial growth in non-extreme environments. The McMurdo endolithic microbial communities are most prominent in the initial millimeter depth in the rock, being limited on the outside by the extreme weather conditions, and on the inside by the depth to which light can penetrate.

Considerable research interest has been shown in the McMurdo Dry Valleys. Examination of rRNA sequences of over 1100 clones from two cryptoendolithic communities (one lichen dominated and one cyanobacterium dominated) revealed the presence of 46 bacterial phylotypes and five eucaryl phylotypes (de la Torre et al. 2003). From samples taken by drilling into black, white, green, and red layers in granite and sandstone, it was determined that cyanobacterial communities were most abundant in granite and sandstone while different types of lichen were found in weathered sandstone (Archer et al. 2017). The production of extracellular polymeric materials appears to account for persistence in this harsh environment. It is proposed that these endolithotrophic microorganisms in the McMurdo Dry Valleys resulted from isolation attributed to evolution of the continent and not from dispersed microorganisms that became adapted to this environment (De Los Ríos et al. 2014).

Endolithic microbial communities are of relevance to the relatively new field of astrobiology (study of life outside of Earth) as they are one strategy whereby microorganisms might survive in an otherwise hostile extraterrestrial environment such as Mars. Students are undoubtedly familiar with rapidly growing bacteria in microbiology laboratories such as *E. coli* and *P. aeruginosa*, which can have generation times of 30 minutes or less. In contrast, endolithic and other subsurface bacteria may largely exist in a state of dormancy due to a lack of water or other growth requirements, and have generation times measured in months, years, or longer (Reith 2011).

3.7 Life in Deep Subsurfaces

3.7.1 Adjusting to a Subsurface Diet

A number of organisms have been isolated and grown from subsurface environments and several have shown some fascinating properties. At least two subsurface organisms, *Geobacter sulfurreducens* and *Shewanella oneidensis* MR-1, are of considerable interest in research involving microbial fuel cells. In terms of physiology, microbiologists are familiar with organisms that use oxygen or other soluble materials (e.g. NO_3^-, $SO_4^=$, CO_2, and small organic molecules) as terminal electron acceptors. In contrast, if *G. sulfurreducens* or *S. oneidensis* MR-1 are grown anaerobically without a soluble electron acceptor, they produce electrically conductive pilus-like structures from the cell. These structures (nanowires) are able to transfer electrons to insoluble minerals that can serve as electron acceptors. This process is currently referred to as extracellular electron transfer. Nanowires conduct electrons via cytochrome-like elements or alternatively flavin-like molecules. In contrast, regular pili, which are involved in bacterial adhesion to surfaces, do not conduct electrons. In other organisms, electron transfer between cells and insoluble materials can be mediated by compounds that act as electron shuttles. One example of a naturally occurring compound that can function in this manner is pyocyanin. Pyocyanin is produced by *P. aeruginosa* and as it is secreted, it is reduced at the cell surface and diffuses out from the cell. When it encounters an appropriate material in the environment, it can function as an anode. Electrons from the reduced pyocyanin are transferred to the anode (Lovley 2012) (see discussion of extracellular electron transport in Chapter 11).

Of considerable interest is the anaerobic oxidation of methane by a microbial aggregation consisting of methane-oxidizing archaea and sulfate-reducing bacteria. This co-aggregation of different microbial species is hypothesized to enhance oxidation–reduction reactions through direct electron transfer using unique electric conductive biomolecules (McGlynn et al. 2015). In support of this proposed syntrophic activity is the observation that the methane-oxidizing archaea overexpress genes for extracellular cytochrome synthesis and sulfate-reducing bacteria overexpress genes for pili production. This pili formation may be nanowires that would facilitate extracellular electron movement (Wegener et al. 2015).

Other organisms that have been isolated and cultured show some interesting morphological characteristics. Under prolonged oligotrophic conditions (such as in most subsurface environments), bacteria can become reduced in size and show little evidence of extracellular polymeric substance, which is needed for biofilm formation (Lever et al. 2015). However, under appropriate nutrient conditions, many of these organisms can regain normal growth and can be induced to form biofilms. However, given the difficulties of culturing these bacteria, much of our knowledge of subsurface microbiology is based on molecular data.

3.7.2 Energy Sources in the Deep Biosphere

To support life, cells must have an electron donor and an electron acceptor. In the deep dark biosphere the common energy source are H_2 and CH_4 which may be formed by abiotic activities or originate from cellular processes. As reviewed recently (Itävaara et al. 2016; Kieft 2016), abiotic processes release CO_2 from minerals and abiotic H_2 chemically reacts to produce CH_4, CO, and short chain hydrocarbons. The reactions contributing to the formation of abiotic H_2 and CH_4 are presented in Figure 3.15. These abiotic gases are often referred to as "geogas." With H_2 attaining a subsurface concentration of 7.4 mM, there is sufficient electron donor to account for considerable microbial growth and many microorganisms in the deep subsurface have hydrogenase enzymes to use molecular hydrogen.

Abiotic reactions producing H_2 and CH_4

1. Radiolysis of water due to U^{238} and Th^{232} present in rocks

$$\text{water} \rightarrow H_2$$

2. Hydration of iron silicate minerals

$$3FeO + H_2O \rightarrow Fe_3O_4 \text{ (magnetite)} + H_2$$

3. From iron olivine

$$3Fe_2SiO_4 + 2H_2O \rightarrow Fe_3O_4 + 3SiO_2 + 2H_2$$

4. From magnesium-iron olivine

$$(Fe,Mg)_2SiO_4 + H_2O + CO_2 \rightarrow Mg_3Si_2O_5(OH)_4 + Fe_3O_4 + CH_4$$

Figure 3.15 Abiotic reactions generating electron donors for bacteria and archaea.

Methane is commonly produced by abiotic processes and is found in many deep subsurface drilling sites (Kietäväinen and Purkamo 2015). While there are numerous aerobes that oxidize methane, anaerobic methane oxidation is generally associated with microorganisms. One anaerobic process is reverse methanogenesis where anaerobic methanotrophic archaea oxidize methane with coupled reduction of sulfate, nitrate, nitrite, iron, and manganese (Kietäväinen and Purkamo 2015). It has been suggested that certain methanogenic archaea oxidize methane with reduction of sulfate without the presence of sulfate-reducing microorganisms (Milucka et al. 2012). The bacterium, *Methanomirabilis oxyfera,* couples methane oxidation to nitrite reduction (Ettwig et al. 2010). Methane cycling may result from acetogenic methanotrophy where acetate and H_2 are produced from two molecules of CH_4.·Some of the reactions involving terminal electron acceptors in anaerobic methane oxidation have been enumerated by Itävaara et al. (2016) and are given here:

$$5CH_4 + 8NO_3^- + 8H^+ \rightarrow 5CO_2 + 4N_2 + 14H_2O$$

$$3CH_3 + 8NO_2^- + 8H^+ \rightarrow 3CO_2 + 4N_2 + 10H_2O$$

$$CH_4 + 8Fe(OH)_3 + 15H^+ \rightarrow HCO_3^- + 8Fe^{2+} + 21H_2O$$

$$CH_4 + 4MnO_2 + 7H^+ \rightarrow HCO_3^- + 4Mn^{2+} + 5H_2O$$

$$CH_4 + SO_4^{2-} \rightarrow HS^- + HCO_3^- + H_2O$$

Several inorganic compounds can serve as electron donors in the subsurface regions where H_2 or methane is not available. Disproportionation reactions are present in various environments throughout the biosphere and if bacteria and archaea are present to catalyze these reactions, alternate sources of energy become available. In disproportionation, a single chemical is both the electron donor and the electron acceptor. *Desulfocapsa sulfoexigens* and *Desulfocapsa thiozymogenes,* two marine bacteria, obtain energy from the disproportionation of elemental sulfur, sulfite, or thiosulfate (Amend and Teskeb 2005). Some anaerobic methanotropic archaea obtain energy by dissimilatory reduction of sulfate with the production of hydrogen sulfide and elemental sulfur (Itävaara et al. 2016). These disproportionation reactions are:

$$4S^0 + 4H_2O \rightarrow SO_4{}^{2-} + 3H_2S + 2H^+ \quad \text{disproportionation of elemental sulfur}$$

$$4SO_3{}^{2-} + 2H^+ \rightarrow 3SO_4{}^{2-} + H_2S \quad \text{disproportionation of sulfite}$$

$$S_2O_3{}^{2-} + H_2O \rightarrow SO_4{}^{2-} + H_2S \quad \text{disproportionation of thiosulfate}$$

Another disproportionation reaction involves acetate where two volatile end products are produced (Pedersen 1997):

$$CH_3COO^- + H^+ \rightarrow CH_4 + CO_2 \quad \text{disproportionation of acetate}$$

Disproportionation reactions provide useful electron donors and electron acceptors for other physiologically active microorganisms and also release sufficient energy to support growth by anaerobic respiration. The metabolism of redox-active elements found in minerals (e.g. Fe, Mn, U, As) may fuel the energetics of these anaerobic cells by serving as either electron donors or electron acceptors in anaerobic respiration. Additionally, syntrophic growth involving anaerobic methanotrophic (ANME) archaea and sulfate reducers involves electron passage between microbial taxa.

Soluble organic carbon material may be present in some subsurface areas in sufficient quantity to support microbial growth. At some sites, all the organic carbon is tied up in microbial cells and cell death releases organic compounds used by microorganisms for growth. In the review by Itävaara et al. (2016), the role of viruses in cell lysis is important for cycling deep subsurface carbon. In some instances, such as the Outokumpu formation in Finland, fermenting microorganisms have been reported (Nyyssönen et al. 2014).

3.7.3 The Benefit of Living Together

In an environment where nutrients are abundant, the microbial consumption of nutrients is frenetic and has been likened to a street fight for food. While some microbial environments showing abundance of a physiological group there is an activity which seems to target the dominant microorganisms, and this has often been termed "kill the winner." In nutrient limited environments such as the oligotrophic subsurface, microorganisms appear to work together. This activity of cooperation is referred to as syntrophy, with one species releasing an end product of metabolism that becomes the substrate for another species or where two or more species cooperate to metabolize a specific organic material. Syntrophism occurs for the anaerobic oxidation of methane coupled to sulfate reduction. The oxidizing microorganisms are ANME archaea and the sulfate reduction is attributed to *Desulfosarcina, Desulfococcus,* and *Desulfobulbus*-type bacteria. An anaerobic bacterium, *Methylomirabilis oxyfera*, couples methane oxidation to nitrite reduction and in the process produces O_2, which is used by aerobic methane oxidizers. In some lithoautotrophic ecosystems, a highly diverse collection of microorganism abound because H_2 supports the growth of methanogens that in turn supports the growth of other bacteria, including acetogens and sulfate reducers.

In an interesting study, water from the Witwatersrand Basin, South Africa was obtained from a fault 1.3 km deep and the microbial DNA was examined (Lau et al. 2016). Major microbial groups in the subsurface community contained anaerobic sulfur oxidizers, methanogens, anaerobic methane oxidizers, denitrifying microorganism, N_2 fixing microorganisms, and sulfate reducers. Using protein-encoding genes and ribosomal RNA transcripts, members of 39 microbial phyla were detected with *Proteobacteria* being the most abundant. Less than 2% of the organisms were archaea. About 28% of the microbial community – represented by taxa similar to *Sulfuritalea, Tauera, Thiobacillus,* and *Sulfuricella* – coupled the oxidization of sulfur (HS^-, S^0, $S_2O_3^{2-}$, HSO_3^-) to denitrification. Also, microorganisms carrying out the anaerobic ammonium oxidation (anammox) coupled to nitrate reduction with N_2 production were detected. The production of nitrate may be attributed to radiolytic oxidation of NH_3/NH_4^+ in rock formation water. Sulfur oxidation and sulfate reduction occurred at this site by microorganisms which grew autotrophically fixing CO_2. The syntrophic activity was attributed to three partners (Lau et al. 2016): (i) methane cycling attributed to anaerobic methane oxidizers and methanogens; (ii) CO_2 from methane oxidation is cycled to denitrifying sulfur oxidizers; (iii) CO_2 from methane oxidation is cycled to autotrophic sulfate reducers. The syntrophic activity required for survival in deep subsurface environments is in stark contrast to the presence of a single species (*Candidatus* Desulforudis audaxviator; see Section 3.5.1.1) accounting for almost the entire microbiome.

3.8 Geomicrobiology

Microorganisms, including those in the subsurface, can have a significant impact on geology. A commonly observed example is lichen growth on rocks and stone structures (monuments, tombstones, etc.), where microbial growth causes weathering. In terms of microbial succession, degradation of rocks by lichen and the deposition of associated organic material ultimately results in the formation of soil. Overall, the impact of microorganisms on geological processes can be summarized in three general concepts: (i) rock and mineral weathering, (ii) mineral transformations, and (iii) rock and mineral formation. We address a couple of these concepts in the following sections.

3.8.1 Rock and Mineral Weathering

A number of microbial processes result in the formation of acid molecules. Typically these include a variety of organic acids that can be produced by lichens and other microbial communities. When growing on calcium carbonate-containing rocks such as limestone and travertine, organic acids can cause dissolution of Ca^{2+} from $CaCO_3$ and the release of CO_2. A common demonstration of this process involves adding some vinegar (which typically contains approximately 5–7% (v/v) acetic acid) to a limestone fragment and observing the bubbling that results from the release of CO_2. The acid content of vinegar is much higher than in most microbial excretions, but gives a good visualization of the dissolution process. Calcium carbonate-containing minerals have various crystalline forms (e.g. calcite, aragonite, and vaterite), often with trace amounts of other elements. The rate of weathering due to microbial activity depends upon the amount of acid produced, chemical buffering from other materials present, and the chemistry and crystalline nature of the rocks. Rock weathering and dissolution can occur on the Earth's surface, but can also occur in subsurface environments, a process that can lead to cave formation.

Other microbial activities can enhance weathering and rock dissolution. One of the most extreme examples arises from sulfur-oxidizing bacteria (sulfur oxidation is described in Chapter 8). In aerobic regions with reduced sulfur present, such as hydrogen sulfide emissions or sulfide-containing minerals, *Thiobacillus* spp. and other sulfur-oxidizing bacteria will oxidize elemental sulfur (S^0) or sulfide (S^{2-}) to sulfate (SO_4^{2-}), which becomes sulfuric acid (H_2SO_4) in the presence of water. Sulfuric acid and other strong acids can readily dissolve carbonate and some other types of rocks. Acid production can also release metals that are bound to clays and other soil components. Aside from acid production, microorganisms can selectively extract elements and compounds from soils and rock through specific enzymes and transport mechanisms. One notable example of this is the production of iron-chelating siderophores by most soil bacteria. Iron is an essential element for growth as it is a cofactor for many enzymes and a component of cytochromes. In aerobic, non-acidic environments, Fe is rapidly oxidized to a variety of iron oxide minerals (rust) and becomes biologically unavailable. Siderophores chelate iron with a very high affinity (the dissociation constant is typically $<10^{-30}$ M) and the iron–siderophore complex can be transported into the cell, after which the iron is used by the cell.

Mineral solubilization occurs as a surface event and also in weathering of material deep in the marine environment. Iron-oxidizing bacteria are important contributors to the weathering of the numerous inactive sulfide chimneys in the Main Endeavor Field of the Juan de Fuca Ridge in the Pacific Ocean (Li et al. 2017). *Gallionella* spp. and *Mariprofundus* spp. are the major iron oxidizers present and other bacteria include members of the *Alpha-*, *Epsilon-*, and *Gammaproteobacteria*. The microbial composition associated with chimney weathering is distinct from the microbes present in active sulfide chimneys; those bacteria involved in mineralization participate in nitrogen fixation, ammonia oxidation, and denitrification.

3.8.2 Mineral Transformations

Nutrient (biogeochemical) cycling is a key aspect of microbial ecology and will certainly impact geology. Although addressed in detail in Chapter 8, some aspects of nutrient cycling are dealt with here. In general, various elements can be modified directly by microorganisms, and this process may affect solubility. For Mn oxidation, the more soluble form of manganese is Mn^{2+} (prevalent in reducing environments) and when oxidized to Mn^{3+}, this ion often precipitates as an oxide or may interact with other ions. Similarly, if used as an electron acceptor, Mn^{3+} can be reduced to Mn^{2+} with enhanced solubility. Typically, these mineralizing bacteria will live in chemical conditions (pH, Eh conditions) where both oxidized and reduced forms of the metal are favorable. Iron oxidizing and reducing bacteria will have a similar effect on iron.

Most microbial influences on geology are indirect, with microbial activities altering a chemical environment (pH, Eh, presence of various ions or elements) due to their metabolic activities. One example is the microbial oxidation of sulfur-containing waste rocks from mining activities (acid mine waste). In regions where nitrogen-containing wastes are present (such as urea), microbial urease activity in which urea is hydrolyzed to ammonia and carbon dioxide, will cause an elevation of pH by the following reactions:

$$(NH_2)_2CO(\text{urea}) + H_2O\_\text{urease} \rightarrow 2NH_3 + CO_2$$

In an aqueous environment, the ammonia combines with water, resulting in a pH increase:

$$NH_3 + H_2O \rightarrow NH_4^+(\text{ammonium}) + OH^-(\text{pH increases})$$

At alkaline pH (typically between 7.2 and 9.0), Ca^{2+} and Mg^{2+} will precipitate as a variety of minerals. This and other examples of mineral-forming features of microorganisms is being explored as a mechanism to fix small cracks in a variety of concrete- and rock-containing structures and buildings in a process called biocementation.

Figure 3.16 Travertine is an interesting stone formed when calcium-containing waters flow through a layered mineral matrix containing bacteria. Small cavities are observed that result from oxygen bubbles produced by photosynthetic cyanobacteria. *Source:* Reproduced with permission from Larry L. Barton.

Another example of a mineral-forming reaction arises from photosynthesis in cyanobacteria. In some hot spring environments (examples are at Mammoth hot springs in Yellowstone Park, USA, and several hot springs in Italy), the genus *Synechococcus* and other photosynthetic cyanobacteria are associated with the formation of the calcium carbonate rock, travertine (Figure 3.16). The associated microorganisms grow as a biofilm and the extracellular polymers within the biofilm matrix enable the aggregation of the various calcite and aragonite crystals that form as a consequence of microbial photosynthesis. This matrix also entraps oxygen bubbles that form. In one travertine forming area that the author visited (Le Zitelle hot springs, Viterbo, Italy) the water temperature was 51 °C and the pH was 7.2. In some parts of this hot spring, travertine could form up to 1-cm thickness per day, although in most areas it was 1-mm per day. At pH 7.2, dissolved CO_2 is present as both bicarbonate (HCO_3^-) and carbonate (CO_3^{2-}). The bicarbonate is assimilated into organic carbon by the microorganisms, resulting in a pH elevation, and the free carbonate precipitates with Ca^{2+} as calcium carbonate, thus forming travertine. Several notable buildings have been constructed of travertine including the columns at St. Peter's Basilica, Vatican City. One might argue that microorganisms and biofilms are responsible for the construction of St. Peter's although people like Michelangelo may have also contributed (Chafetz and Folk 1984).

In the Earth's history, arguably the most significant microbial contribution to geology and life in general is the Great Oxygenation Event, which was mediated by the advent of oxygen-producing photosystem 2 in cyanobacteria and chloroplasts (evolutionarily related to cyanobacteria) in higher plants. The Great Oxygenation Event and photosynthesis since that time is responsible for aerobic life on Earth including humans. Rocks that formed earlier than 2.3 billion years ago (2.3 Ga) show evidence of reducing conditions and are considered to reflect the lack of free oxygen and other chemical conditions on the Earth at that time. While there is evidence of localized oxygen production prior to the great oxygenation event (Dutkiewicz et al. 2006), most of the free oxygen would have combined with iron and other elements. Once the chemical buffering occurred, then O_2 accumulated in the atmosphere (present day levels are 21%) and made possible aerobic life in prehistoric and present-day Earth.

3.8.3 Microbial Metal Binding

Microbial cell surface components are anionic due to negatively charged carboxylate and phosphate moieties on extracellular polysaccharides, peptidoglycan, and other cell wall components. As a result they can bind metal cations. Ca^{2+} binding can benefit the cells because some extracellular enzymes, such as peptidoglycan autolysins (which recycle peptidoglycan components during cell growth and elongation), require calcium. However, other metals are bound as well. One notable example of this is the rust-colored ferrihydrite (amorphous iron oxide), which may form in biofilms that can grow on the inner surfaces of water pipes. When this biofilm becomes detached, the water briefly appears rusty until the detached biofilm is washed away. Another common example is calcium deposition (containing calcium phosphates and other minerals) on dental plaque (a biofilm that forms on teeth). This calcified plaque is called tartar or dental calculus.

Metal binding and concentration by individual bacterial cells and biofilms can be an important feature of mineral formation. The bacterial cell components can nucleate (i.e. promote – this is a more geologically precise term than catalyze, as catalyze implies promotion of a chemical reaction) the formation of various minerals, with the exact type of mineral formed influenced by local physical and chemical conditions. A number of economically important mineral deposits (e.g. copper, gold, etc.) are associated with sedimentary rocks that were formed in association with a prehistoric river. One possible contribution of bacteria in this mineralization process is that biofilms on river sediments may have concentrated trace quantities of dissolved metal ions that, over geological time and rock formation (including elevated temperature and pressures), resulted in the formation of ore deposits.

3.8.4 Microbiota of Subsurface Crystals

An interesting observation has been made concerning organisms associated with stalactites in deep subsurface mines. Stalactites of calcium carbonate consisting of calcite and aragonite were collected along water-bearing faults in the Beatnix gold mine in South Africa at 1.4 km below ground. Following careful dissection of the stalactites, DNA extractions yielded sequence similarities to marine bacteria (*Bowmanella* sp. and *Alcanivoraz* sp.) and archaea (*Halobacteriales*). Also present inside the stalactite was a marine nematode, *Monhystrella parvella* (Borgonie et al. 2015). *Alicyclobacillus* sp., a spore-forming sulfur and ferrous iron oxidizer was found associated with *M. parvella*. This nematode is found in brackish water/marine environments and uses bacteria as food. Previously, nematodes including the new nematode species *Halicephalobus mephisto* were found in fracture water from a 0.9–3.6 km deep mine in Africa (Borgonie et al. 2011). It is presumed, but not yet fully confirmed, that the marine bacteria and bacterivorous nematode(s) have moved through the fractures in the rocks and that stalactites with biological specimens were formed within a few years after the drilling in the mine.

3.9 Summary

The abundance and type of microorganisms in the soil depends to a great extent on the abundance and type of chemicals present. The chemicals of importance include those that are required for microbial nutrition (i.e. C, N, S) as well as humic complexes and clay particles, which bind nutrients and water. Rich agricultural soil supports an abundance of growing bacteria, archaea, and fungi while nutrient deficient desert soils harbor metabolic inactive microorganisms in soil crusts that are waiting for the supply of water and appropriate nutrients. Plant roots secrete organic acids, which are nutrients that stimulate the growth of prokaryotes and fungi to produce a greater abundance of microorganisms in the rhizosphere than are present in barren bulk soil. Soil invertebrates and burrowing animals contribute to the aeration and mixing of nutrients in the soil. Carbon to nitrogen ratios (C:N) of about 24:1 are ideal for bacterial growth, and when there is an abundance of carbon material, nitrogen is limiting for bacteria and fungi will dominate. Thus, the fungi:bacteria ratio (F:B) varies with the above ground vegetation with fungi being highly dominant in forest soil and nearly equal ratios of fungi and bacteria in productive agricultural soil. While bacterial viruses are highly abundant in soil and viruses are suspected of transferring antibiotic resistance in the natural environment, documentation to support the impact of viruses on bacterial communities in soil needs to be presented.

An approach to assess the presence and activities of microorganisms in subsurface and deep subsurface environments is to examine water from the site of interest. Since microorganisms in subsurface environments resist cultivation in the laboratory, there is a high reliance on metagenomics to assess the bacteria and archaea present. In shallow subsurfaces, microorganisms are carried into the underground waters while deep subsurface environments are unlikely to receive surface microorganisms. Unlike soil microorganisms, which are highly dependent on organic carbon compounds attributed to photosynthetic activity, the keystone for microbial life in deep subsurface environments are the anaerobic chemolithotrophic microorganisms. Geochemical formation of molecular hydrogen and carbon dioxide enable acetogenic microorganisms and methanogens to grow, while the cellular organic compounds released upon microbial death enable other microorganisms to grow. The anaerobes are the primary producers in the deep subsurface while aerobic chemolithotrophs grow at the anoxic–oxic interface. Biologically produced methane and geologically produced hydrogen sulfide are oxidized by bacteria and the production of carbon dioxide and sulfate is used to support the growth of sulfate reducers. With this limited source of energy-yielding nutrients, microorganisms in the deep subsurface are highly dependent on the metabolic activities of other microorganism present and antagonistic processes appear to be minimized. The low nutrient state of the deep subsurface environment does not favor the presence of spore-forming bacteria because the nutrients required for germination are not present and sporulation would be a terminal event. Due to the relatively low metabolic activities of microorganisms, several deep subsurface storage vaults have been used or are under consideration for storage of nuclear fuel wastes, molecular hydrogen, methane, and carbon dioxide.

Not only do microorganisms live in the terrestrial environment but some are capable of attacking mineral formations. Weathering of rocks is a characteristic attributed to mild acidic products from lichen and other microorganisms growing on their surface and inside the cracks within rock surface. Bacterial metabolism leading to the oxidation/reduction of iron and manganese minerals provides energy for chemolithotrophic bacteria. Microorganisms use iron as a micronutrient and most bacteria and fungi produce siderophores which can extract Fe^{3+} from insoluble minerals. The microbial release of carbonate under alkaline conditions may result in the production of mineral deposits and, for water containing high levels of calcium, biofilms may lead to travertine formations.

Discussion Questions

1 Why would a significant quantity of water need to be removed from a sample well for drinking water prior to testing for *E. coli*?

2 In areas where logs are cut into boards, large piles of sawdust accumulate. Would it be a good idea to package this sawdust and distribute it to stores to be sold to individuals and companies for distribution on lawns, golf courses, and parks to enhance plant growth?

3 As evidenced from deep subsurface testing, bacteria, and viruses are found thousands of meters below the surface. How did the bacteria get there and are new bacteria being continuously added to that environment?

4 With the great abundance of viruses in the terrestrial environment, why are bacteria so abundant in the soil even though viruses that attack them are also present there?

5 Many countries are using underground sites to store nuclear wastes. Since bacteria are present everywhere, what are the possibilities that bacterial activities in these sites may compromise the long term storage of low-level nuclear wastes in these areas?

References

Adriaenssens, E.M., Zyl, L.V., De Maayer, P. et al. (2017). Metagenomic analysis of the viral community in Namib Desert hypoliths. *Environ. Microbiol.* 17: 480–495.

Alfreider, A., Krössbacher, M., and Psenner, R. (1997). Groundwater samples do not reflect bacterial densities and activity in subsurface systems. *Water Res.* 31: 832–840.

Amend, J.P. and LaRowe, D.E. (2016). Ocean sediments – an enormous but underappreciated microbial habitat. *Microbe* 11: 427–432.

Amenda, J.P. and Teskeb, A. (2005). Expanding frontiers in deep subsurface microbiology. *Palaeogeogr. Palaeoclimatol. Palaeoecol.* 219: 131–155.

Archer, S.D.J., de los Ríos, A., Lee, K.C. et al. (2017). Endolithic microbial diversity in sandstone and granite from the McMurdo Dry Valleys, Antarctica. *Polar Biol.* 40: 997–1006.

Arndt, S., Brumsack, H.-J., Kai, W. et al. (2006). Cretaceous black shales as active bioreactors: a biogeochemical model for the deep biosphere encountered during ODP Leg 207 (emerara rise). *Geochem. Cosmochim. Acta* 70: 408–425.

Baker, B.J., Saw, J.H., Lind, A.E. et al. (2016). Genomic inference of the metabolism of cosmopolitan subsurface Archaea, *Hadesarchaea*. *Nat. Microbiol.* 16002. https://doi.org/10.1038/NMICROBIOL.2016.2.

Barlett, H.D. (2002). Pressure effects on in vivo microbial processes. *Biochim. Biophys. Acta* 1595: 367–381.

Barton, H.A. (2006). Introduction to cave microbiology: a review for the non-specialist. *J. Cave Karst Studies* 68: 43–54.

Bates, S.T., Berg-Lyons, D., Caporaso, J.G. et al. (2011). Examining the global distribution of dominant archaeal populations in soil. *ISEM J.* 5: 908–917.

Beal, E.J., House, C.H., and Orphan, V.J. (2009). Manganese- and iron-dependent marine methane oxidation. *Science* 325: 184–187.

Belnap J, Kaltenecker JH, Rosentreter R, et al., (2001). Biological soil crusts: Ecology and management. US Department of the Interior, Denver, Colorado. Technical Reference 1730-2

Borgonie, G., García-Moyano, A., Litthauer, D. et al. (2011). Nematoda from the terrestrial deep subsurface of South Africa. *Nature* 474: 79–82.

Borgonie, G., Linage-Alvarez, B., Ojo, A. et al. (2015). Deep subsurface mine stalactites trap endemic fissure fluid *Archaea, Bacteria*, and *Nematoda* possibly originating from ancient seas. *Front. Microbiol.* https://doi.org/10.3389/fmicb.2015.00833.

Brannen-Donnelly, K. and Engel, A.S. (2015). Bacterial diversity differences along an epigenic cave stream reveal evidence of community dynamics, succession, and stability. *Front. Microbiol.* 6: 729. https://doi.org/10.3389/fmicb.2015.00729.

Brock, T.D., Smith, D.W., and Madigan, M.T. (1984). *Biology of Microorganisms*. Englewood Cliffs, NJ: Prentice-Hall, Inc.

Chafetz, H.S. and Folk, R.L. (1984). Travertines: depositional morphology and bacterially constructed constituents. *J. Sed. Petrol.* 54: 289–316.

Chelius, M. and Triplett, E. (2001). The diversity of archaea and bacteria in association with the roots of *Zea mays* L. *Microbiol. Ecol.* 41: 252–263.

Chivian, D., Brodie, E.L., Alm, E.J. et al. (2008). Environmental genomics reveals a single-species ecosystem deep within earth. *Science* 322: 275–278.

Couradeau, E., Karaoz, U., Lim, H.C. et al. (2016). Bacteria increase arid-land soil surface temperature through the production of sunscreens. *Nat. Commun.* 7: 10373. https://doi.org/10.1038/ncomms10373.

De Los Ríos, A., Wierzchos, J., and Ascaso, C. (2014). The lithic microbial ecosystems of Antarctica's McMurdo Dry Valleys. *Antarctic Sci.* 26: 459–477.

DeAngelis, K.M., Lindow, S.E., and Firestone, M.K. (2008). Bacterial quorum sensing and nitrogen cycling in rhizosphere soil. *FEMS Microbiol. Ecol.* 66: 167–207.

Detmers, J., Schulte, U., Strauss, H. et al. (2001). Sulfate reduction at a lignite seam: microbial abundance and activity. *Microbiol. Ecol.* 42: 238–247.

D'Hondt, S., Jørgensen, B.B., Miller, D.J. et al. (2004). Distribution of microbial activities in deep subseafloor sediments. *Science* 306: 2216–2221.

Doran, J.W. (1980). Soil microbial and biochemical changes associated with reduced tillage. *Soil Sci. Soc. Am. J.* 44: 765–771.

Dutkiewicz, A., Volk, H., George, S.C. et al. (2006). Biomarkers from Huronian oil-bearing fluid inclusions: an uncontaminated record of life before the great oxidation event. *Geology* 34: 437–440.

Engelhardt, T., Kallmeyer, J., Cypionka, H. et al. (2014). High virus-to-cell ratios indicate ongoing production of viruses in deep subsurface sediments. *ISME J.* 8: 1503–1509.

Ettwig, K.F., Butler, M.K., Le Paslier, D. et al. (2010). Nitrite-driven anaerobic methane oxidation by oxygenic bacteria. *Nature* 464: 543–548.

Fetter, C.W. Jr. (2001). *Applied Hydrogeology*, 4e. Essex, England: Pearson Education Limited.

Fierer, N., Breitbart, M., Nulton, J. et al. (2007). Metagenomic and small-subunit rRNA analyses reveal the genetic diversity of bacteria, archaea, fungi and viruses in soil. *Appl. Environ. Microbiol.* 73: 7059–7066.

Fredrickson, J. and Onstott, T. (1996). Microbes deep inside the earth. *Sci. Am.* 68–73.

Friedmann, E.I. (1982). Endolithic microorganisms in the Antarctic cold desert. *Science* 215: 1045–1053.

Fry, N.K., Fredrickson, J.K., Fishbain, S. et al. (1997). Microbiology population structure of microbial communities associated with two deep, anaerobic, alkaline aquifers. *Appl. Environ. Microbiol.* 63: 1498–1504.

Gniese, C., Bombach, P., Rakoczy, J. et al. (2013). Relevance of deep-subsurface microbiology for underground gas storage and geothermal energy production. *Adv. Biochem. Eng. Biotechnol.* 142: 95–121.

Gomes, N.C.M., Fagbola, O., Costa, R. et al. (2003). Dynamics of fungal communities in bulk and maize rhizosphere soil in the tropics. *Appl. Environ. Microbiol.* 69: 3758–3766.

Griebler, C., Mindl, B., Slezak, D. et al. (2002). Distribution patterns of attached and suspended bacteria in pristine and contaminated shallow aquifers studied with an in situ sediment exposure microcosm. *Aquat. Microbiol. Ecol.* 28: 117–129.

Gründger, F., Jiménez, N., Thielemann, T. et al. (2015). Microbial methane formation in deep aquifers of a coal-bearing sedimentary basin, Germany. *Front. Microbiol.* https://doi.org/10.3389/fmicb.2015.00200.

Hazen, T., Jiménez, L., López de Victoria, G. et al. (1991). Comparison of bacteria from deep subsurface sediment and adjacent groundwater. *Microbiol. Ecol.* 22: 293–304.

Harshberger, J.W. (1911). The soil, a living thing. *Science* 33: 791–744.

Hoehler, T.M. and Jorgensen, B.B. (2013). Microbial life under extreme energy limitation. *Nat. Rev. Microbiol.* 11: 83–94.

Inagaki, F., Hinrichs, K.-U., Kubo, Y. et al. (2015). Exploring deep microbial life in coal-bearing sediment down to ~2.5 km below the ocean floor. *Science* 349: 420–424.

Itävaara, M., Salavirta, H., Marjamaa, K. et al. (2016). Geomicrobiology and metagenomics of terrestrial deep subsurface microbiomes. *Adv. Appl. Microbiol.* 94: 3–77.

Jørgensen, B.B. (2011). Deepsubseafloor microbial cells on physiological standby. *Proc. Natl. Acad. Sci. U.S.A.* 108: 18193–18194.

Kallmeyer, J., Pockalny, R., Adhikari, R.R. et al. (2012). Global distribution of microbial abundance and biomass in subseafloor sediments. *Proc. Natl. Acad. Sci. U.S.A.* 109: 16213–16216.

Keen, E.C., Bliskovsky, V.V., Malagon, F. et al. (2017). Novel "Superspreader" bacteriophages promote horizontal gene transfer by transformation. *mBio* 8 (1): e02115–e02116. https://doi.org/10.1128/mBio.02115-1617.

Kieft, T.L. (2016). Microbiology of the deep continental biosphere, in their world: a diversity of microbial environments. In: *Advances in Environmental Microbiology* (ed. C.J. Hurst), 225–249. Cham: Springer International Publishing.

Kieft, T.L., Kovacik, W.P. Jr., Ringelberg, D.B. et al. (1997). Factors limiting microbial growth and activity at a proposed high-level nuclear repository, Yucca Mountain, Nevada. *Appl. Environ. Microbiol.* 63: 3128–3133.

Kietäväinen, R. and Purkamo, L. (2015). The origin, source, and cycling of methane in deep crystalline rock biosphere. *Front. Microbiol.* 6: 725. https://doi.org/10.3389/fmicb.2015.00725.

Knittel, K. and Boitius, A. (2009). Anaerobic oxidation of methane: progress with an unknown process. *Annu. Rev. Microbiol.* 63: 311–334.

Kotelnikova, S. and Pedersen, K. (1998). Distribution and activity of methanogens and homoacetogens in deep granitic aquifers at Äspö Hard Rock Laboratory, Sweden. *FEMS Microbiol. Ecol.* 26: 121–134.

Kresic, N. (2009). *Groundwater Resources: Sustainability, Management, and Restoration*. New York, NY: McGraw-Hill Publishing Companies, Inc.

Kuramae, E.E., Zhou, J.Z., Kowalchuk, G.A. et al. (2014). Soil-borne microbial functional structure across different land uses. *Sci. World J.* 2014: 216071, 8 pages. https://doi.org/10.1155/2014/216071.

de la Torre, J.R., Goebel, B.M., Friedmann, E.I. et al. (2003). Microbial diversity of cryptoendolithic communities from the McMurdo Dry Valleys, Antarctica. *Appl. Environ. Microbiol.* 69: 3858–3867.

Lareen, A., Burton, F., and Schäfer, P. (2016). Plant root-microbe communication in shaping root microbiomes. *Plant. Mol. Biol.* 90: 575–587.

Lau, J.A. and Lennon, J.T. (2012). Rapid responses of soil microorganisms improves plant fitness in novel environments. *Proc. Natl. Acad. Sci. U.S.A.* 109: 14058–14062.

Lau, M.C.Y., Kieft, T.L., Kuloyo, O. et al. (2016). An oligotrophic deep-subsurface community dependent on syntrophy is dominated by sulfur-driven autotrophic denitrifiers. *Proc. Natl. Acad. Sci. U.S.A.* 113: E7927–E7936.

Lechevalier, H.A. and Solotorovsky, M. (1974). *Three Centuries of Microbiology*. New York: Dover Publications.

Leff, J.W., Jones, S.E., Prober, S.M. et al. (2015). Consistent responses of soil microbial communities to elevated nutrient inputs in grasslands across the globe. *Proc. Natl. Acad. Sci. U.S.A.* 112: 10967–10972.

Lever, M.A., Rogers, K.L., Lloyd, K.G. et al. (2015). Life under extreme energy limitation: a synthesis of laboratory- and field-based investigations. *FEMS Microbiol. Rev.* 39: 688–728.

Li, J., Cui, J., Yang, Q. et al. (2017). Oxidative weathering and microbial diversity of an inactive seafloor hydrothermal sulfide chimney. *Front. Microbiol.* 8: 1378. https://doi.org/10.3389/fmicb.2017.01378.

Long, P.E., Williams, K.H., Hubbard, S.S. et al. (2016). Microbial metagenomics reveals climate-relevant subsurface biogeochemical processes. *Trends Microbiol.* 24: 600–610.

Lovley, D.R. (2012). Electromicrobiology. *Annu. Rev. Microbiol.* 66: 391–409.

Lovley, D.R. and Chapelle, F.H. (1995). Deep subsurface microbial processes. *Rev. Geophys.* 33: 365–381.

Lueders, T., Kindler, R., Miltner, A. et al. (2006). Identification of bacterial micropredators distinctively active in a soil microbial food web. *Appl. Environ. Microbiol.* 72: 5342–5348.

McGlynn, S.E., Chadwick, G.L., Kempes, C.P. et al. (2015). Single cell activity reveals direct electron transfer in methanotrophic consortia. *Nature* 526: 531–535.

McMahon, S. and Parnell, J. (2014). Weighing the deep continental biosphere. *FEMS Microbiol. Ecol.* 87: 113–120.

Milucka, J., Ferdelman, T.G., Polerecky, L. et al. (2012). Zero-valent sulphur is a key intermediate in marine methane oxidation. *Nature* 491: 541–546.

Morita, R.Y. (2000). Is H_2 the universal energy source for long-term survival? *Microbiol. Ecol.* 38: 307–320.

Morita, R.Y. and ZoBell, C.F. (1955). Occurrence of bacteria in pelagic sediments collected during the Mid-Pacific Expedition. *Deep-Sea Res.* 3: 66–73.

Morono, Y., Terada, T., Nishizawa, M. et al. (2011). Carbon and nitrogen assimilation in deep subseafloor microbial cells. *Proc. Natl. Acad. Sci. U.S.A.* 108: 18295–18300.

Morozova, D., Wandrey, M., Alawi, M. et al. (2010). Monitoring of the microbial community composition in saline aquifers during CO_2 storage by fluorescence in situ hybridization. *Int. J. Greenh. Gas. Conf.* 4: 981–989.

Morozova, D., Zettlitzer, M., Let, D. et al. (2011). Monitoring in the microbial community composition in deep subsurface saline aquifers during CO_2 storage in Ketzin, Germany. *Energy Procedia* 4: 4362–4370.

Nazina, T.N., Kosareva, I.M., Petrunyaka, V.V. et al. (2004). Microbiology of formation waters from the deep repository of liquid radioactive wastes Severnyi. *FEMS Microbiol. Ecol.* 49: 97–107. https://doi.org/10.1016/j.femsec.2004.02.017.

Novarino, G., Warren, A., Butler, H. et al. (1997). Protistan communities in aquifers: a review. *FEMS Microbiol. Rev.* 20: 261–275.

Nyyssönen, M., Hultman, J., Ahonen, L. et al. (2014). Taxonomically and functionally diverse microbial communities in deep crystalline rocks of the Fennoscandian shield. *ISME J.* 8: 126–138.

Parkes, R.J., Cragg, B.A., and Wellsbury, P. (2000). Recent studies on bacterial populations and processes in subseafloor sediments: a review. *Hydrogeol. J.* 8: 11–28.

Paul, E.A. and Clark, F.E. (1989). *Soil Microbiology and Biochemistry*. San Diego, CA: Academic Press.

Paula, F.S., Rodrigues, J.L.M., Zhou, J. et al. (2014). Land use change alters functional gene diversity, composition and abundance in Amazon forest soil microbial communities. *Mol. Ecol.* 23: 2988–2999.
Pedersen, K. (1997). Microbial life in deep granitic rock. *FEMS Microbiol. Rev.* 20: 399–414.
Pedersen, K. (2012). Subterranean microbial populations metabolize hydrogen and acetate under in situ conditions in granitic groundwater at 450 m depth in the Äspö Hard Rock Laboratory, Sweden. *FEMS Microbiol. Ecol.* 81: 217–229.
Ragon, M., Van Driessche, A.E.S., García-Ruíz, J.M. et al. (2013). Microbial diversity in the deep-subsurface hydrothermal aquifer feeding the giant gypsum crystal-bearing Naica Mine, Mexico. *Front. Microbiol.* 4: 37. https://doi.org/10.3389/fmicb.2013.00037.
Raynaud, X. and Nunan, N. (2014). Spatial ecology of bacteria at the microscale in soil. *PLoS One* 9 (1): e87217. https://doi.org/10.1371/journal.pone.0087217.
Reith, F. (2011). Life in the deep subsurface. *Geology* 39: 287–288.
Roesch, L.F.W., Fulthorpe, R.R., Riva, A. et al. (2007). Pyrosequencing enumerates and contrasts soil microbial diversity. *ISME J.* 1: 283–290.
Rousk, J., Brookes, P.C., and Bååth, E. (2009). Contrasting soil pH effects on fungal and bacterial growth suggest functional redundancy in carbon mineralization. *Appl. Environ. Microbiol.* 75: 1589–1596.
Røy, H., Kallmeyer, J., Adhikari, R.R. et al. (2012). Aerobic microbial respiration in 86-million-year-old Deep-Sea red clay. *Science* 336: 922–925.
Simard, S.W., Perry, D.A., Jones, M.D. et al. (1997). Net transfer of carbon between ectomycorrhizal tree species in the field. *Nature* 388: 579–582.
Sinclair, J.L. and Ghiorse, W.C. (1987). Distribution of protozoa in subsurface sediments of a pristine groundwater study in Oklahoma. *Appl. Environ. Microbiol.* 53: 1157–1163.
Stevens, T.O. and McKinley, J.P. (1995). Lithoautotrophic microbial ecosystem in deep basalt aquifers. *Science* 270: 450–454.
Strickland, M.S. and Rousk, J. (2010). Considering fungal:bacterial dominance in soils – methods, controls and ecosystem implications. *Soil Biol. Biochem.* 42: 1385–1395.
Stroes-Gascoyne, S. and West, J.M. (1997). Microbial studies in the Canadian nuclear fuel waste management program. *FEMS Microbiol. Rev.* 20: 573–590.
Stroes-Gascoyne, S., Haveman, S.A., Hamon, C.J. et al. (1997). Occurrence and identification of microorganisms in compacted clay-based buffer material designed for use in a nuclear fuel waste disposal vault. *Can. J. Microbiol.* 43: 1133–1146.
Sylvia, D.M., Fuhrmann, J.J., Hartel, P.G. et al. (1999). *Principles and Applications of Soil Microbiology*. Upper Saddle River, NJ: Prentice-Hall.
Takai, K., Moser, D.P., DeFlaun, M. et al. (2001). Archival diversity in waters from deep south African gold mines. *Appl. Environ. Microbiol.* 67: 5750–5760.
Teske, A. and Sørensen, K.B. (2008). Uncultured archaea in deep marine subsurface sediments: have we caught them all? *ISEM J.* 2: 3–18.
de Vries, F.T., Hoffland, E., van Eekeren, N. et al. (2006). Fungal/bacterial ratios in grasslands with contrasting nitrogen management. *Soil Biol. Biochem.* 38: 2092–2103.
Wegener, G., Krukenberg, V., Riedel, D. et al. (2015). Intercellular wiring enables electron transfer between methanotrophic archaea and bacteria. *Nature* 526: 587–590.
Whitman, W.B., Coleman, D.C., and Wiebe, W.J. (1998). Prokaryotes: the unseen majority. *Proc. Natl. Acad. Sci. U.S.A.* 95: 6578–6583.
Wilhartitz I, Mach RL, Ryzinska G, et al., (2006). Stable autochthonous microbial endokarst communities in spring water of two different alpine karst aquifers, *Geophysical Research Abstracts*, Vol. 8,05941,2006 SRef-ID:1607-7962/gra/EGU06-A-05941
Wilkins, M.J., Daly, R.A., Mouser, P.J. et al. (2014). Trends and future challenges in sampling the deep terrestrial biosphere. *Front Microbiol.* 5: 481.

Wu, X., Holmfeldt, K., Hubalek, V. et al. (2016). Microbial metagenomes from three aquifers in the Fennoscandian shield terrestrial deep biosphere reveal metabolic partitioning among populations. *ISME J.* 10: 1192–1203.

Zablocki, O., Adriaenssens, E.M., and Cowan, D. (2016). Diversity and ecology of viruses in hyperarid desert soils. *Appl. Environ. Microbiol.* 82: 3770–3777.

Zlotnikov D (2012). Mining in the Extreme, *Canadian Institute of Mining Magazine*, August 2012.

ZoBell, C.F. and Anderson, Q.A. (1936). Vertical distribution of bacteria in marine sediments. *Bull. Am. Assoc. Petrol. Geol.* 20: 258–269.

Further Reading

Adriaenssens, E.M., Kramer, R., Van Goethem, M.W. et al. (2017, 2017). Environmental drivers of viral community composition in Antarctic soils identified by viromics. *Microbiome* 5: 83. https://doi.org/10.1186/s40168-017-0301-7.

Colwell, F.S. and D'Hondt, S. (2013). Nature and extent of the deep biosphere. *Rev. Minerol. Geochem.* 75: 547–574.

Hurst, C.J. (2016). *Their World: A Diversity of Microbial Environments*. New York, NY: Springer.

Edwards, K.J., Becker, K., and Colwell, F. (2012). The deep, dark energy biosphere. Interrestrial life on earth. *Ann. Rev. Earth Planetary Sci.* 40: 551–568.

Miettinen, H., Kietäväinen, R., Sohlberg, E. et al. (2015). Microbiome composition and geochemical characteristics of deep subsurface high-pressure environment, Pyhäsalmi mine Finland. *Front. Microbiol.* 6: 1203. https://doi.org/10.3389/fmicb.2015.01203.

Nigro, L.M., Harris, K., Orcutt, B.N. et al. (2012). Microbial communities at the borehole observatory on the Costa Rica Rift flank (Ocean Drilling Program Hole 896A). *Front. Microbiol.* https://doi.org/10.3389/fmicb.2012.00232.

Sohlberg, E., Bomberg, M., Miettinen, H. et al. (2015). Revealing the unexplored fungal communities in deep groundwater of crystalline bedrock fracture zones in Olkiluoto, Finland. *Front. Microbiol.* 6: 573. https://doi.org/10.3389/fmicb.2015.00573.

Yulia, A., Frank, Y.A., Kadnikov, V.V. et al. (2016). Stable and variable parts of microbial community in Siberian deep subsurface thermal aquifer system revealed in a long-term monitoring study. *Front. Microbiol* 7: 2101. https://doi.org/10.3389/fmicb.2016.02101.

Web Sites

www.darkenergybiosphere.org (Center for Dark Energy Biosphere Investigations).
https://dco.tw.rpi.edu/ (Deep Carbon Observatory).
https://www.soils.org/publications/glossary/appendix (Soil horizons).

4

Aquatic Surface Environments: Freshwater, Marine, and Wastewater

4.1 Overview and Focus

The Earth's surface (>70%) is covered in water, with oceans making up the bulk of this. The remaining surface water is found in lakes, rivers, and wetlands, or may be frozen in snow and ice. Water vapor is also found in the atmosphere, where it is present as humidity, clouds, and precipitation. Groundwater is a notable reservoir of water and regardless of the physical environment, water is an essential requirement for life. In this chapter, we address the microbiome of some of the surface aquatic environments. Some of the microbial communities in these aquatic environments are explored.

- Bacterial viruses in water
- Bacteria in snow and glacial ice
- Wetland microbiome
- Microorganisms in cold and hot springs
- Primary productivity in surface and deep waters

4.2 Water as Relevant to Microbial Growth

Water has a relatively simple chemical formula (H_2O), but its chemical and physical properties have an essential impact on life. Water is a polar molecule with the oxygen atom having a slight negative charge and the hydrogen atoms having a positive charge. Hydrogen bonding occurs when the slight positive charge of a hydrogen atom in one water molecule is attracted to an electronegative atom such as oxygen in an adjacent water molecule. Hydrogen bonding controls several features of water, including surface tension and density. The density of most substances increases as the temperature becomes lower. In contrast, water achieves its maximum density ($1\,g\,ml^{-1}$) at 4 °C, while at higher or lower temperatures its density becomes reduced due to alterations in hydrogen bonding. This explains why ice floats on water, rather than becoming submerged. Another aspect of hydrogen bonding relates to surface tension at the water–air interface. From a biological perspective, insects such as water striders (family *Gerridae*) can walk on water. In microbial ecology, the air–water interface is referred to as the neuston, and there are microbial communities, notably microbial mats, that colonize this environment (see Chapter 2). Hydrogen bonding is not unique to water but is a common feature in many biological molecules in which a hydrogen atom is covalently bonded to an electronegative atom such as O, S, or N, thus forming a dipole. Adjacent electronegative atoms attract the bound hydrogen, so that it becomes shared between two electronegative atoms.

Environmental Microbiology and Microbial Ecology, First Edition. Larry L. Barton and Robert J.C. McLean.

DNA provides a notable example of the importance of hydrogen bonding in biological molecules because the various nucleotide bases are joined by hydrogen bonding – a feature that allows DNA to be a double stranded molecule.

4.2.1 Water Activity

Water is a universal solvent (capable of dissolving a wide variety of chemicals). Substances that dissolve readily in water will form ions (Na^+ and Cl^- for common salt, NaCl), or have polar characteristics of their own (e.g. sugars such as glucose have polar hydroxyl groups that readily form hydrogen bonds with water). Water chemistry is influenced by the solids and gases dissolved in it. Oceans contain an average of 3.5% (w/v) salt, most of which is NaCl. Marine and brackish environments are addressed in Section 4.3. The solute concentration controls osmotic concentration and thus influences the availability of water. One common term is water activity (A_W), which is defined by the vapor pressure of water from a solution, divided by the vapor pressure of pure water. Examples of water activity are listed in Table 4.1.

As seen in Table 4.1, organisms such as *Caulobacter crescentus* and *Pseudomonas fluorescens* live in freshwater environments, whereas *Vibrio* spp. live in brackish and marine environments. The archaeon *Halobacterium* is an extreme halophile and is found in environments such as Great Salt Lake and the Dead Sea where the water is saturated with salt. The fungus *Xeromyces bisporus* is associated with spoilage in foods that have a high sugar content. Microorganisms have developed several mechanisms to adapt to changing osmotic conditions while still allowing for growth. Several examples of chemicals that serve as osmoprotectants in bacteria are given in Figure 4.1. The mechanisms involved include the synthesis or degradation of osmoprotectant molecules and altered membrane transport of Na^+ and K^+ ions. Organisms can become dormant when growth conditions are not suitable. Spore development by fungi, cyst formation by protozoa, and endospore production by gram-positive bacteria are classic examples of dormant forms of life.

Table 4.1 Common water activities required for growth of various microorganisms.

Water activity (A_W)	Bacteria	Fungi
1.00 (distilled water)	*Caulobacter, Sprillum*	
0.97	*Pseudomonas fluorescens*	
0.95	*Escherichia coli* *Salmonella* spp. *Vibrio cholerae*	
0.94	*Vibrio parahaemolyticus*	*Stachybotrys atra*
0.90	*Lactobacillus* spp. *Bacillus* spp. *Staphylococcus aureus* (anaerobic growth)	*Fusarium* *Mucor*
0.85	*Staphylococcus aureus* (aerobic growth)	*Debaronmyces*
0.80		*Penicillium*
0.75 (saturated salt solution)	*Halobacterium* spp.	*Aspergillus* *Saccharomyces rouxii*
0.60		*Xeromyces bisporus*

Source: Based on data from Fontana (2001, 2008) and Atlas and Bartha (1997).

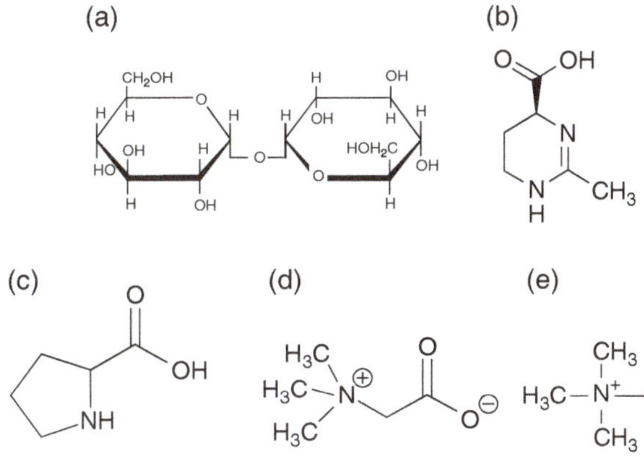

Figure 4.1 The structures of molecules produced by microorganisms as a response to osmotic pressure: (a) trehalose (a glucose dimer); (b) ectoine; (c) proline; (d) glycine betaine; (e) choline. *Source:* Reproduced with permission from Barton (2005).

4.3 Marine Environments and Associated Microbiomes

The marine environment consists of oceans, coral reefs, and estuaries. Oceans have a considerable diversity of species that are reflective of four different zones. The abyssal zone is the deepest ocean region, where chemosynthetic bacteria are primary producers that grow on the ocean floor and by hydrothermal vents. The benthic zone is above the abyssal zone. The pelagic zone includes the open ocean is above the abyssal zone. The zone at the ocean surface includes the intertidal zone, which interfaces with land. Salinity is often expressed in parts per thousand (per mille, which uses the symbol ‰), so the average salinity of 3.5% is equal to 35‰ and typically ranges from 32‰ to 37‰, with variations arising from enhanced freshwater input from rivers and precipitation (low salinity) to areas of high evaporation and limited freshwater input (higher salinity). Water temperatures at the ocean surface range from 28 to 30 °C in tropical regions to 0 °C and colder in polar regions. Higher temperatures may be encountered in tropical regions by the shore. Although the freezing point of pure water is 0 °C, the freezing point of ocean water is −2 °C due to salinity. Water temperature in the deeper parts of the ocean is approximately 3 °C as colder and dense polar water displaces warmer surface water. There are localized regions of much warmer water in areas with geothermal activity. At the ocean surface, wind and waves mix the water and the temperature is relatively constant. Below this surface layer (called the epipelagic layer), the temperature begin to drop noticeably (the transition zone is called the thermocline).

Two other features of oceans are tides and currents. Tides are caused primarily by the gravitational attraction of the moon and to a lesser extent the sun. The net result of tides is that the ocean water will rise (high tide) and fall (low tide) twice in a lunar day (approximately 24 hours 50 minutes). The water level fluctuation between high and low tides vary around the world. In some locations in the Arctic Ocean, tidal water levels fluctuate under 0.5 m; at San Diego, the fluctuation may be up to 2 m. The highest tides are seen in the Bay of Fundy on the east coast of Canada with fluctuations being 15 m (~50 ft.). The portion of the shore that is between the high and low tide levels– the intertidal zone or eulittoral zone – is submerged at high tide and exposed to air during low tide. Above the intertidal zone is the supralittoral zone or splash zone, which gets water from wave splashes and precipitation. Below the intertidal zone at the edge of the ocean is the sublittoral zone, which is permanently covered by water.

Ocean currents involve the continuous movement of large quantities of water (Figure 4.2). Surface ocean currents are caused by a combination of solar heating, prevailing winds, gravity, and the Coriolis effect (influenced by the Earth's rotation around its axis). Aside from

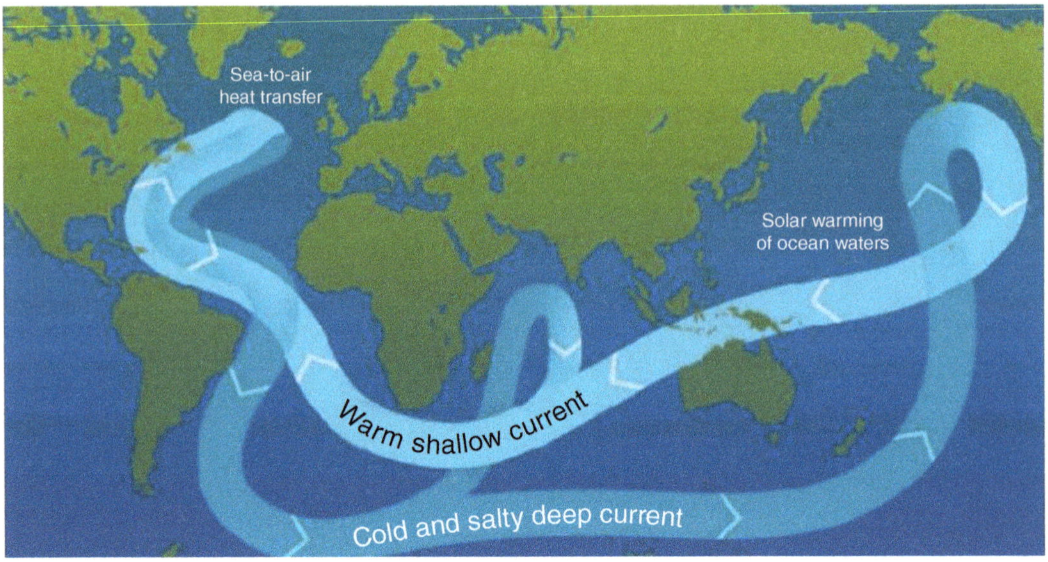

Figure 4.2 The global conveyor belt moves water all around the world. *Source:* https://commons.wikimedia.org/wiki/File:Ocean_circulation_conveyor_belt.jpg.

influencing conditions such as water temperature, ocean currents can influence climate. An example of this is the Gulf Stream in which warm water from sub-tropical regions by the Gulf of Mexico, moves along the east coast of North America and across the northern Atlantic to Europe. As a result, portions of northern Europe benefiting from the Gulf Stream proximity including Dublin, Ireland (latitude 53°N) is much warmer (average winter daily temperature 4.5 °C) in comparison to an inland city at the same latitude such as Edmonton, Alberta, Canada (average winter daily temperature – 12.5 °C). There are also currents in the deeper ocean and in some places, upwelling can occur where deeper nutrient-containing waters rise to the surface and the influx of nutrients results in increased primary production and associated flourishing marine communities. For further information, readers are referred to Trujillo and Thurman (2017).

4.3.1 Marine Primary Productivity

Net primary productivity in the oceans is estimated at 44–67 Gt ($4.4–6.7 \times 10^{13}$ kg) per year, which accounts for approximately half the global primary productivity. The bulk of the primary productivity in marine environments occurs in shallower regions at or near the coast. Here, the influx of nutrients occurs from rivers and other runoff from terrestrial environments. In ocean gyres (open ocean far from land), there are few nutrients available and so primary productivity is low (Details are provided in Moore et al. 2013; Sigman and Hain 2012). Some coastal regions are rocky and the land ends abruptly at the coast. In areas where the land slopes gently into the ocean, there may be extensive coastal wetlands that contain a wide variety of plant, animal, and microbial life. Primary productivity in coastal wetlands is high (estimates range from 600 to 2000 g carbon fixed per m^2 per year), with productivity higher in some tropical salt marshes (Mitsch and Gosselink 2007).

Primary productivity in the wetlands comes from the plants, trees, and photosynthetic microorganisms. In coastal wetlands, the levels of primary productivity are comparable to coral reefs and rainforests. The vegetation in coastal wetlands can resist wave action and

Table 4.2 Nitrogen input into the ocean expressed as Gmol/yr.

Year	Fluvial (river)	Atmospheric (precipitation)
1860	1000	1000
2000	2100	4800
2050 (estimate)	2300	5500

Source: Based on data from Moore et al. (2013).

erosion. Coral reefs, which also have high rates of primary productivity, are found in shallow tropical and sub-tropical oceans. In the open ocean, the bulk of primary productivity arises from photosynthetic microorganisms (phytoplankton) and their populations are controlled by nitrogen and to a lesser extent, phosphorous limitation (Moore et al. 2013). Nutrients enter the oceans from rivers (Table 4.2), atmospheric precipitation, and upwelling of nutrient-containing water from deeper regions of the ocean. Nutrients, present in atmospheric particulates can also enter the ocean and in the absence of upwelling or terrestrial runoff, can represent a source of nutrients.

Estimates of nitrogen input have increased since the onset of the industrial revolution, along with associated primary productivity. One extreme example of eutrophication has arisen in the Gulf of Mexico due to nutrient-laden river waters entering the Gulf. Although the growth of algae and cyanobacteria has been enhanced, the decomposition of these organisms results in a localized depletion of oxygen (the hypoxic zone, more commonly known as a dead zone). Nitrogen and other nutrient cycling will be influenced at various depths in the oceans, and thus affect microbial growth and activity. Further details are presented in Moore et al. (2013).

Nitrogen is not the only limiting nutrient for primary production and other biological activities, as phosphorus limitation is also common in many areas. The photic zone (region of the ocean in which light can penetrate) is affected by turbidity. In very turbid water, light may penetrate only a few millimeters, whereas in the open ocean, the euphotic zone (where photosynthesis is possible) extends to a depth of approximately 200 m. From 200 to 1000 m (the dysphotic zone or twilight zone), light intensity rapidly decreases as it is absorbed by water, and no light is present below 1000 m in the aphotic zone (https://costalscience.noaa.gov website).

Our knowledge of the organisms associated with marine primary productivity has been greatly enhanced by metagenomics and other culture-independent, molecular-based approaches to microbial ecology. While macrophytes such as kelp, mangroves, sea grasses, and *Sargassum* spp. (a macroalga in the Sargasso Sea) are locally important in some environments, the bulk of primary production is carried out by microorganisms. One key study in this context was carried out by J.C. Venter and associates (Venter et al. 2004). The Sargasso Sea is a region in the Atlantic Ocean off the Florida Coast and is within the North Atlantic Sub-tropical Gyre (rotating ocean current) (Figure 4.2). Its boundaries are four ocean currents: the Gulf Stream is the western boundary, the North Atlantic Current is the north boundary, Canary Current is the eastern boundary, and the North Atlantic Equatorial Current is the south boundary. Based on surface water analysis, the major representative of the cyanobacteria clade was *Prochlorococcus* spp., with *Synechococcus* spp. accounting for approximately 10% of sequences. Delong et al. (2006) conducted a similar study near Hawaii in the Pacific and examined the changes in microbial community composition at various depths (shown in Table 4.3).

In Arctic and Antarctic waters, primary productivity is affected by the seasons. In summer, there is prolonged daylight, whereas in winter there is prolonged darkness and waters have an ice cover. A recent metagenomic study showed that primary production in the summer

Table 4.3 Representative microorganisms associated with marine primary productivity in the Pacific Ocean.

Ocean depth	Phylum	Organism
10-70 m	*Cyanobacteria*	*Prochlorococcus marinus* (most prominent at 10 m, declines below that depth) Uncultivated cyanobacteria
	Alphaproteobacteria	*Rhodobacteriaceae* (photoheterotroph)
130 m	*Cyanobacteria*	*Prochlorococcus marinus* Uncultivated cyanobacteria
200 m	*Chloroflexi*	Uncultured *Chloroflexi*-related clade SAR202 (found down to 3000–4000 m depth)

Source: Based on data from DeLong et al. (2006).

Box 4.1 Aerobic Anoxygenic Phototrophic Bacteria (AAPBs)

While "aerobic anoxygenic phototrophic" sounds like a contradiction, AAPBs are obligate aerobic bacteria that are energized by anaerobic photosynthesis processes using bacteriochlorophyll *a*. Water is not used as the electron donor for photosynthesis and, therefore, O_2 is not produced. These bacteria are found in the *Alphaproteobacteria* and *Betaproteobacteria* groups and are photoheterotrophs that are unable to use carbon dioxide as the only carbon source.

was dominated by photoheterotrophs and aerobic anoxygenic photosynthetic organisms (see Box 4.1) including *Roseobacter* spp. In the winter, chemoautotrophs (both bacteria and archaea) became more dominant with energy likely arising from sulfur and ammonia oxidation (Grzymski et al. 2012).

Like green, brown (kelp), and red algae, diatoms are most abundant in water along the coastline, where nutrients are plentiful. Diatoms have been estimated to account for as much as 20% of total primary productivity on Earth and the most dominant genera include *Chaetoceros, Fragilariopsis, Thalassiosira*, and *Corethron* (Malviya et al. 2016). Cyanobacteria account for 30–80% of primary productivity in oligotrophic waters between 40°N and 40°S (Price et al. 2004). *Prochlorococcus* and *Synechococcus* are the major genera of cyanobacteria in marine waters and it has been estimated these phototrophs reach densities of 10^8 cells l^{-1} with an annual production of *Prochlorococcus* and *Synechococcus* at 2.9×10^{27} and 7×10^{26} cells, respectively (Flombaum et al. 2013). *Synechococcus* possesses chlorophyll *a* and phycobiliproteins as accessory pigments, while *Prochlorococcus* has chlorophyll *b* and no accessory pigments.

4.3.2 Marine Heterotrophs

A wide variety of microorganisms including bacteria, archaea, protists, and fungi have been identified in marine environments using traditional culturing and molecular based techniques. Bacteria commonly encountered in coastal waters include several gram-negative bacteria such as *Vibrio* spp., *Aeromonas* spp., and *Pseudoalteromonas* spp. gram-positive coastal isolates include *Bacillus* spp., *Halobacillus* spp., and *Streptomyces* spp. A variety of coliforms and other fecal indicator organisms (addressed below) are also encountered in coastal regions near cities and areas with notable agricultural runoff.

The SAR11 clade of the *Alphaproteobacteria* (now identified as *Candidatus* Pelagibacter ubique) is prominent in most ocean waters and based on the frequency of its sequences in many regions (30–50% of sequences in some locations), it may be the most prominent organism in the world. Pictures of C. Pelagibacter ubique are in Chapter 2, Figure 2.6). Based on sequence information, the unusual nutrient requirements of one SAR11 strain, HTCC1062, were determined by Steve Giovanonni and colleagues (Carini et al. 2011) and include an obligate requirement for reduced sulfur compounds, and serine or glycine. Generation time for this organism in the lab ranges from 25 to 50 hours. which indicates that this is a slow-growing microorganism. Another interesting aspect of this clade is the presence of a light-driven proton pump, which features proteorhodopsin. Therefore, organisms within this clade can be classified as photoheterotrophs because they gain energy from the sun, yet require organic carbon. This organism has one of the smallest genomes (1.3 Mb) when compared with other organisms such as *E. coli* (4.6 Mb), *Bacillus subtilis* (4.2 Mb), *Pseudomonas aeruginosa* (5.9 Mb), and *Methanococcus voltae* (1.9 Mb). Genomic analysis has shown that members of this clade (but not HTCC1062) are able to release methane, possibly under conditions of phosphate limitation. On a global scale, the contribution of the SAR11 clade to carbon cycling is quite notable. This organism has not been formally assigned a species name but is called *Candidatus* Pelagibacter ubique ("C. P. ubique") (Steindler et al. 2011).

4.3.3 Bacterial Symbionts and Marine Hosts

A range of invertebrates in oligotrophic regions of the marine environment have bacteria as symbionts. These bacteria are located as either endosymbionts or epibionts that function as primary producers by converting CO_2 into cellular organic material using inorganic metabolism for energy production (Ruehland and Dubilier 2010). In several instances, hydrogen sulfide from the environment is oxidized to sulfate by oxygen-dependent respiration and the energy from the oxidation reaction is coupled to ATP production. Alternately, H_2 may be used as an energy source (electron donor) by the symbiotic bacteria. Epibiotic bacteria may form long filaments that collect the minerals and nutrients needed to support their nutrition and these nutrients are frequently at low concentrations in the environment. The symbiotic bacteria have efficient recycling processes to maximize the use of C, N, and P elements, including high affinity systems to acquire nutrients from the environment. The bacteria benefit by having the host organism move them throughout the environment and especially to environments supportive of bacterial metabolism. Some examples of endosymbiotic and epibiotic bacteria are discussed as follows.

The symbiotic activity between bacteria and small marine organisms as host has focused on marine ciliates and oligochaete worms (Box 4.2). The ciliate *Euplotes harpa* is found in brackish water and it was initially found to harbor the symbiotic bacterium *Polynucleobacter* of the *Betaproteobacteria* group. This is an obligate relationship on the part of this protest ciliate because when the bacterial symbiont is removed by the use of antibiotics, the ciliate is unable to divide. The *Polynucleobacter* bacterium has also been found in three additional species of *Euplotes*. More recently, two strains of the *Alphaprotobacter* group of the "*Candidatus* Midichloria" clade have been identified in the cytoplasm of *Euplotes harpa* (Vannini et al. 2010). These secondary symbionts have been designated "*Candidatus* Anadelfobacter veles" and "*Candidatus* Cyrtobacter comes" in the order *Rickettsiales*. The presence of endosymbionts may be common for ciliated protists, with "*Candidatus* Cryptoprodotis polytropus" reported for the common ciliate *Pseudomicrothorax dubius* (Ferrantini et al. 2009).

The gutless oligochaete *Olavius algarvensis* found in the oligotrophic environment of the Mediterranean has chemosynthetic endosymbionts in the body wall. These endosymbionts include: symbionts 1 and 2, members of the *Gammaproteobacteria*; symbionts 3 and 4, members of the *Deltaproteobacteria;* and the fifth symbiont is a spirochete. Using metaproteomics

Box 4.2 Oligochaete Worms

Oligochaete worms are members of the phylum *Annelida* and are found in both aquatic and terrestrial regions. There are thousands of species of oligochaete worms and while they are broadly distributed throughout the biosphere, they are not found in the desert. Marine oligochaetes are found primarily in the vegetation of shallow tidal zones and terrestrial forms are commonly found in muddy areas between the terrestrial and aquatic zones. They respire through their skin and aquatic oligochaetes secrete ammonia while terrestrial species secrete urea. Some oligochaete worms have a mouth cavity, a highly developed pharynx for ingesting food, and a primitive intestine but several marine oligochaete worms lack a mouth and are gutless. Commonly, oligochaete worms are hosts for symbiotic bacteria.

and metabolomics, the physiological activities of the endosymbionts of *O. algarvensis* have been found to include: (i) the capability of using CO and H_2 as energy sources, (ii) recycling metabolic end products of acetate, malate, succinate and propionate from the host, (iii) using highly efficient uptake transporters for nutrients including amino acids, and (iv) coupling pyrophosphate hydrolysis with ATP production (Kleiner et al. 2012). A model depicting the proposed metabolic activities of the endosymbionts in *O. algarvensis* is given in Figure 4.3.

A number of epibiotic bacteria have been observed on marine invertebrates that harbor specific bacteria. The Yeti crab (*Kiwa hirsute*) has bacteria attached to the chitin layer of their setae. These bacteria represented a spectrum of species with 56% of associated organisms from the group of *Epsilonproteobacteria*, 25% were *Gammaproteobacteria*, and 10% were *Bacteroidetes* (Goffredi et al. 2008). From the hydrothermal vents in the Indian Ocean, the foot of the scaley snail (genus and species unknown) has epibiotic bacteria from the *Deltaproteobacteria* attached, while bacteria attached to the snail shell are from the *Epsilonproteobacteria* group, and the esophageal gland contains a single species of endosymbiotic bacteria from the *Gammaproteobacteria* group (Goffredi et al. 2004). The sulfate-reducing bacteria of the *Deltaproteobacteria* group on the foot of the scaley snail produce hydrogen sulfide, which precipitates iron sulfides (pyrite and greigite), while other bacterial activities remain unresolved. The stalked barnacle, *Vulcanolepas osheal*, in the Kermadec Arc, New Zealand, has epibiotic filamentous bacteria of the *Epsilonproteobacteria* group on their elongated cirral setae and showed considerable similarity to epibiotic bacteria of the bresiliid shrimp (Suzuki et al. 2009). The shrimp *Rimicaris exoculata* is found in dense swarms of 3000 shrimp per m^2 on chimney vents in the Mid-Atlantic Ridge has two novel symbionts growing on the setae in the mouth parts of the gill chamber. There two species of filamentous bacteria that are shrimp symbionts with one species from the *Epsilonproteobacteria* and the other from the *Gammaproteobacteria* (Petersen et al. 2010). Chemoautotrophic sulfide – oxidizing *Thiothrix* spp. grow on the exoskeleton of the freshwater cave amphipod (*Niphargus ictus*) (Dattagupta et al. 2009) and *Thiothrix* spp. are found attached to setae or spines of the marine amphipod *Urothoe poseidonis* (Gillan and Dubilier 2004).

The worm *Tubificoides benedii* is a marine oligochaete found in sediments of shallow marine water where sulfide concentrations are appreciable and oxygen levels are low. There is a bacterial community that lives attached to the worm's tail that consists of two species. One bacterial ectosymbiont is a member of the *Leucothrix mucor* clade (*Gammaproteobacteria*) and the other belongs to the *Thiovulgaceae* (*Epsilonproteobacteria*) (Ruehland and Dubilier 2010). This bacterial symbiont of the *L. mucor* clade is phylogenetically related to symbiont 3 of the gutless oligochaete *O. algarvensis*. This feature of a bacterial species serving as an endosymbiont with one host and an ectobiont with another may be common in the marine environment.

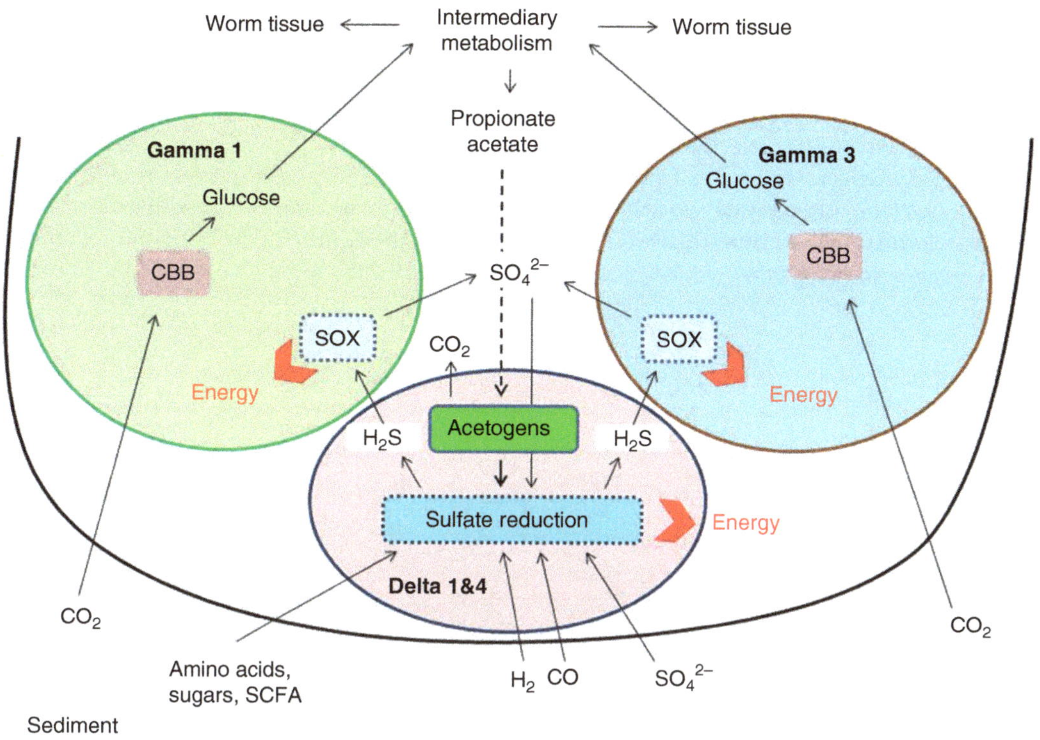

Figure 4.3 The metabolic contribution of endosymbionts in the gutless oligochaete *O. algarvensis*. Gamma 1 and Gamma 2 are endosymbionts of *Gammaproteobacteria* that can oxidize sulfur. Delta 1 & 4 are endosymbionts of *Deltaproteobacteria* where one symbiont is a sulfate reducer and another is an acetogen. The colored spheres represent the biomass of bacterial endosymbionts: Gamma 1 and Gamma 2 are assigned to different units while Delta 1 & 4 are combined into one unit. External energy sources are carbon monoxide (CO), carbon dioxide (CO_2), and hydrogen (H_2). CBB = Calvin–Benson–Bassham cycle, SOX = sulfur oxidizing, SFCA = short chain fatty acids. Gamma 1 and gamma 2 symbionts provide the initial productivity by converting CO_2 to organic material including propionate, which energizes acetogens and ultimately sulfate reducers. *Source:* Based on data from Kleiner et al. (2012).

4.3.4 Microbial EPSs, Marine Snow, and Marine Gel Particles

Marine microorganisms produce extracellular polymeric substances (EPSs), which consist of polysaccharides, proteins, nucleic acids, and lipids with molecules ranging from <10 kDa to >100 kDa. The aggregates of colloidal EPS consist of a mixture of dissolved organic matter (DOM) and particulate organic matter (POM) called a marine gel matrix (Decho and Gutierrez 2017), which contributes to the composition of marine snow (Box 4.3). A model indicating the distribution of EPSs in marine waters is presented in Figure 4.4. Transparent colloidal gels are >0.4 μm wide and contain a small amount of amino acids, which enables them to be stained by Coomassie Brilliant Blue. The number of colloidal gels in a liter of ocean water may reach 10^6–10^8 and based on observations off the coast of Santa Barbara, California, USA. These EPSs appear to have a unique microbial content and the Santa Barbara colloidal gels contained *Cytophaga*, *Planctomyces*, and *Gammaproteobacteria*, while the planktonic bacteria around these gels contained organisms of the *Alphaproteobacteria* group (DeLong et al. 1993). Bacterial EPS along with dissolved organic carbon from phytoplankton collect at the water–atmosphere

Box 4.3 Marine Snow

Particulate biological material in the upper water column falls slowly to the ocean floor and is frequently referred to as marine snow. The particulate material is a collection of animal wastes and complex biological molecules that are resistant to rapid degradation. Organic particulates have various sizes and some small particles fall so slowly that it takes them weeks to reach the ocean floor. When this material was first observed in the lights of a marine submersible vehicle, it appeared to be similar to snow observed in a street light or the lights of a car. Thus, this material has been called marine snow.

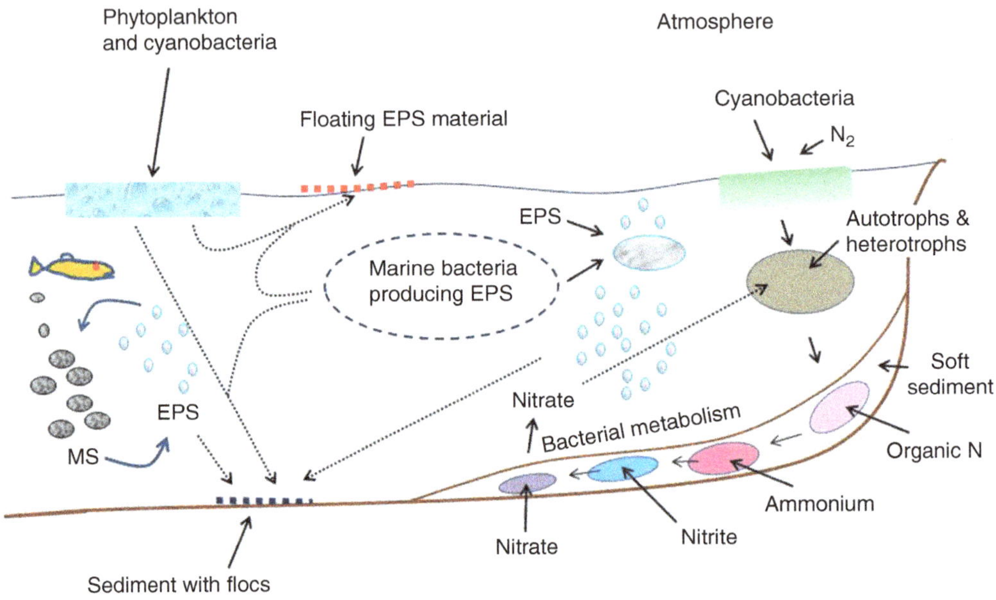

Figure 4.4 Extracellular polymeric substances (EPS) and marine snow (MS) in marine environments. *Source:* Based on data from Decho and Gutierrez (2017).

interface and bubbles or foam may result. As the marine bubbles are carried into the atmosphere, they burst releasing bacteria and various organic materials associated with the EPS into the atmosphere.

Recently, it has been suggested that gel particles in aquatic environments have an important role in the absorption of nutrients and subsequent bacterial colonization. Gel particles have been found in fresh water (Simon et al. 2002) as well as in marine settings (Busch et al. 2017). These transparent exopolymer particles (TEPs) are 10–30 μm wide and appear to be products of the phytoplankton. The TEPs are present throughout the water column and represent organic carbon material that reflects a chemical state between dissolved and particulate material. Bacteria colonizing these TEPs hydrolyze the carbon material and consume organic material attached to the TEP, thereby contributing to the biogeochemical marine carbon cycle. These TEPs are readily demonstrated by staining with Alcain Blue, which binds to acidic groups, and recently Coomassie Blue has been used, which binds to basic units of amino acids.

These TEPs and EPSs can be distinguished from "marine snow", which has a diameter of >500 μm and consists of fecal matter from various small aquatic animals. As for TEP and EPS, marine snow is a source of nutrients for bacteria and its metabolism by microbes contributes to the consumption of oxygen in the sea (Bochdanskya et al. 2010). When oil spills occur, the combination of oil with marine snow produces marine oil snow (MOS), in which oil droplets are embedded in the mucilaginous material of marine snow. Following the Deepwater Horizon blowout in the Gulf of Mexico, marine snow carried about 14% of the oil spill to the bottom of the Gulf of Mexico (Valentine et al. 2014). Research indicated that bacteria associated with the MOS were unlike bacteria in the surrounding water and consisted of organisms associated with the *Gammaproteobacteria* (*Cycloclasticus, Congregibacter, Haliela, Halomonas,* and *Marinobacter*), *Alphaproteobacteria* (*Roseobacter* clade), *Bacteroidetes,* and *Planctomyces* (Arnosti et al. 2015).

4.3.5 Brackish Water and Intertidal Zones

Brackish water (salt concentrations between 5‰ and 35‰) occurs in the transition zone between freshwater and marine systems. Ocean tidal influences are prominent in the estuary (the transition zone between fresh water and marine water), but are otherwise absent. In some coastal areas, mud carried by rivers is deposited as the water flows into the sea and forms mud flats. The region between the terrestrial and the marine ecosystem is the salt marsh zone (see Figure 4.5), which is a region of high microbial activity that is important for both marine and terrestrial ecosystems.

The cyanobacteria and microalgae in the littoral zone provide organic material that enables many animals (i.e. snails, crabs, clams, and fish) to grow. Photosynthetic process produce oxygen and promote an aerobic environment in the littoral zone as the oxygen does not readily escape into the atmosphere, as happens with large marsh plants. Only a small percentage of marsh vegetation is consumed by herbivores and most of the marsh plants are converted into detritus by bacteria and fungi to be used as a food source. Root exudates are often high in

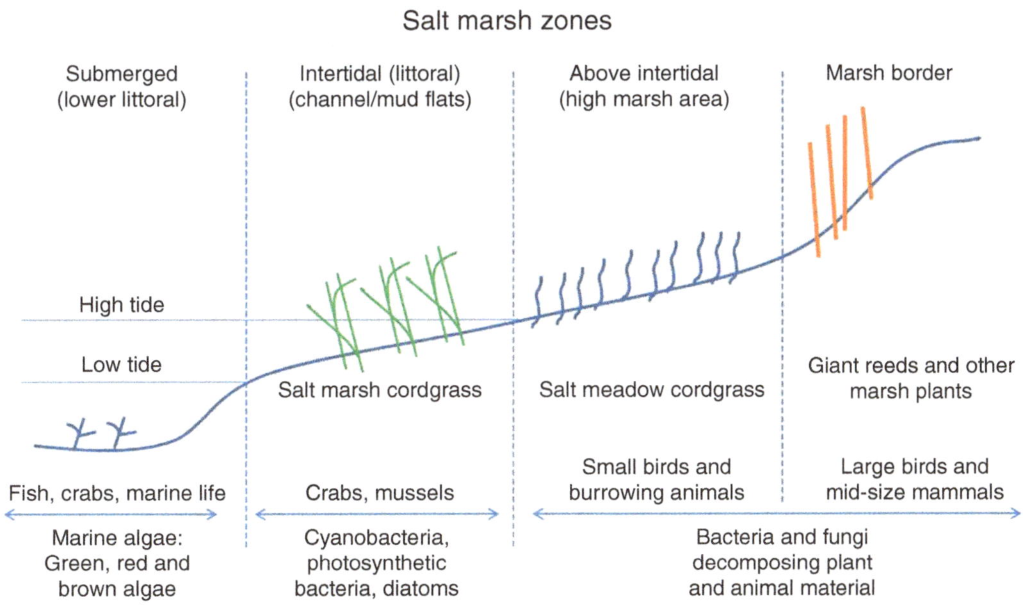

Figure 4.5 A model of a salt marsh showing the environmental zones. This costal ecosystem, also known as a costal marsh or tidal marsh, is located between the land and the ocean and is subject to flooding by tides.

organic acids and enhance the growth of bacteria, with bacterial abundance attributed to the type of plants in the littoral zone. The decomposition and mineralization of plant matter by bacteria may be sufficiently vigorous to make some regions in the salt marsh anaerobic and the smell of rotten eggs (due to hydrogen sulfide produced by sulfate-reducing bacteria) may be highly pronounced.

Nitrogen enters the salt marshes from watershed runoff and by bacterial nitrogen fixation. Over half of this nitrogen is released to the sea and the remainder is either stored as a "sink" (i.e. insoluble nitrogen compounds) in the salt marsh or lost to atmospheric N2 due to bacterial denitrification (Hofmann et al. 2008). Plants growing in the salt marsh are often limited by nitrogen and bacteria are commonly limited by phosphorus (Sundareschwar et al. 2003). This limitation of phosphorus for bacteria has an impact on the bacterial transformation of nitrogen compounds in the salt marsh. The grazing activities of snow geese at salt marshes along the Hudson Bay were compared with grazing at freshwater marshes at La Pérouse Bay, Manitoba, to assess goose fitness (Ngai and Jefferies 2004). Grasses in the salt marshes were limited by both nitrogen and phosphorus, while the freshwater marsh was limited by nitrogen. Even though feces from the geese enriched the marshes, geese foraging on low-phosphorus plants had feces low in phosphorus, which therefore indicated the difficulty of plant recovery without the addition of phosphorus.

4.3.6 Coral Reefs

Marine environments contain a broad array of coral reefs, which are constructed from algae and animal polyps with a frame constructed from calcium carbonate. Frequently, corals are collected for their esthetic effects (see Figure 4.6). Corals can be considered complex metaorganisms often called the coral holobiont because their composition includes photosynthetic

Figure 4.6 Microscopic marine life solidifies bicarbonate and calcium ions in water to produce calcium carbonate, which functions to protect the coral polyps and coralline algae. When these life forms die, the skeleton of the coral remains. Coral colors include blue, green, red, and golden brown. When corals are severely stressed they become white. Reproduced with permission from Larry Barton.

partners, bacteria, archaea, fungi, and viruses. Shallow water corals grow in nutrient poor environments with nutritional support from photosynthetic organisms and numerous bacteria. The bacteria making up the coral microbiome are distinct from bacteria in seawater. A hallmark feature of shallow water (<60 m) corals is the presence of zooxanthellae, which are photosynthetic dinoflagellates growing in the polyp tissue. A recent metagenomic study has identified a green sulfur phototroph (*Candidatus* Prosthecochloris korallensis) that is also a coral symbiont (Cai et al. 2017). Some corals are found in deep waters and live at 80–3000 m depths in nutrient poor environments where the polyps obtain nutrients from dead phytoplankton and zooplankton that descend from the surface following spring blooms.

A considerable number of bacteria are associated with coral communities. Research on the deep sea coral *Lophelia pertusa* reveals a conserved set of microbiomes. From samples collected in the western Atlantic, *Proteobacteria* represented 75–96% of the bacterial taxa with almost all these sequences assigned to *Pirellulaceae*, *Pseudonocardiaceae*, *Rhodobacteriaceae*, *Sphingomonacaceae*, and *Oceanospirillales* (Kellogg et al. 2017). The core bacterial community consisted of *Novosphingobium*, *Pseudonocardia*, and *Propionobacterium*. Species of *Novosphingobium* and *Pseudonocardia* are hydrocarbon degraders and have been enriched due to the seeping of oil through the ocean floor. While *Propionobacterium* spp. are located inside the photosynthetic dinoflagellates in tropical corals, which are dependent on photosynthesis, the great depths where *Lophelia* grow precluded the presence of photosynthetic dinoflagellates. Species of *Pseudonocardia* are recognized for their production of antibiotics and these bacteria may contribute this to the coral community. Other bacteria associated with *Lophelia* colonies vary depending on the availability of nutrients and geochemical features. Evaluation of another deep sea coral (*Paramuricea placomus*) revealed that the core microbiome included 70–90% of bacterial DNA sequences that belonged to the phyla *Proteobacteria* (>50%), *Firmicutes* (10%), *Planctomyces* (10%), and *Tenericutes* (1–20%) (Kellogg et al. 2016).

An evaluation of bacteria associated with two coral species (*Montipora aequituberculata* and *Acropora millepora*) from the Great Barrier Reef, Australia, was conducted using direct cultivation and identification in library sequences. The genera of bacteria accounting for 30–60% sequences included *Roseobacter*, *Spongiobacter*, *Vibrio*, and *Alteromonas* (Raina et al. 2009). *Roseobacter* is a well-known marine organism that is dominant in phytoplankton blooms and polar waters as well as being a symbiont in corals which have zooxanthellae. *Roseobacter* produces antibiotics, such as thiotropocin, and appears to have an important role in the health of corals. The presence of *Spongiobacter* in corals has been proposed to indicate healthy coral, although how it contributes is not yet ascertained.

Commonly, the isolation of bacteria from coral structures includes samples containing mucus. Bacterial cells are appreciably more abundant in the surface mucous layer which coats the coral. There are numerous possible roles for coral mucus and they include defense against pathogens, invasion by other corals, UV damage, and pollutants (Brown and Bythell 2005). The mucous matrix is sloughed from the coral surface by ocean currents and degraded by coral-associated bacteria. Coral mucus is known to contain carbohydrates, proteins, DOM, and lipids (Brown and Bythell 2005), which are excellent substrates for microorganisms. Mucus from *L. pertusa* contains 14 different amino acids, while bacteria and fungi from this coral are able to cycle nitrogen and carbon (Kellogg et al. 2017). Bacteria associated with shallow water coral metabolize dimethylsulfoniopropionate (DMSP) and dimethylsufide (DMS) (Table 4.4), which indicates these bacteria also participate in the sulfur cycle (Raina et al. 2009). Since corals grow in a low nutrient environment, the recycling of nutrients by microorganisms is an important feature for coral growth. Figure 4.7 illustrates the bacteria associated with corals that are involved in recycling nitrogen (Table 4.5) and sulfur (Table 4.6).

Table 4.4 Bacteria associated with coral that can grow on sulfur compounds produced in the coral biosphere.

Reactions: Zooxanthellae→ DMSP → DMS + Acrylic acid	
Bacteria using DMSP[a]:	**Bacteria using DMS[b]:**
Alteromonas	*Alteromonas*
Arhodomonas	*Idiomarina*
Idiomarina	*Photobacterium*
Oseudomonas	*Pseudomonas*
Roseobacter	*Vibrio*
Spongiobacter	

Source: Reproduced with permission from Raina et al. (2009).
[a] Dimethylsulfoniopropionate.
[b] Dimethylsulfide.

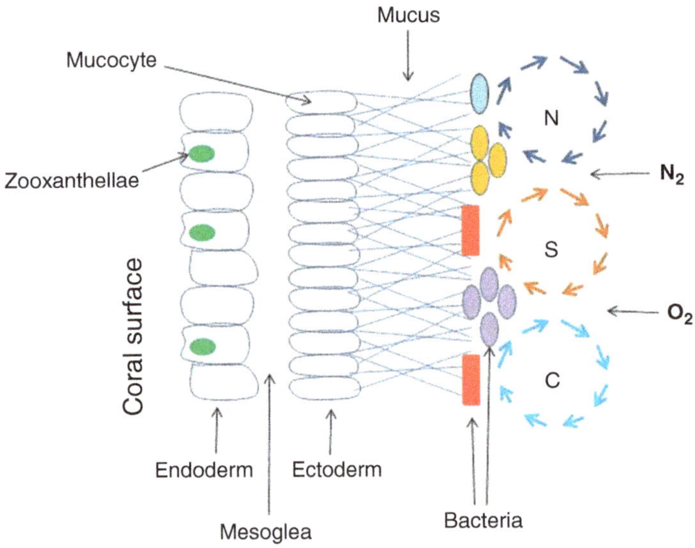

Figure 4.7 A model indicating the surface of a shallow water coral. The mucous layer protects the coral surface from microbial destruction and microcolonies of bacteria form at the mucus–water interface. Nutrient cycling by the core bacteria that make up the coral microbiome conserves essential elements. *Source:* Based on data from Brown and Bythell (2005).

Table 4.5 The capability of the coral biosphere to recycle nitrogen based on the presence of bacteria associated with deep sea coral.

Nitrogen transformation	Reactions	Bacteria
Nitrogen fixation	$N_2 \rightarrow NH_3$	*Bradyrhizobium, Magnetospirillum*
Nitrification	$NH_4^+ \rightarrow NO_3^-$	*Pirellulaceae*
Ammonification	$NO_3^- \rightarrow NH_4^+$	*Bacillus*
Nitrate reduction	$NO_3^- \rightarrow NO_2^-$	*Photobacterium, Propionibacterium*
Denitrification	$NO_3^- \rightarrow N_2$	*Bacillus*

Source: Based on data from Kellog et al. (2016).

Table 4.6 Potential for cycling of sulfur in a hot spring mat in Yellowstone National Park based on bacteria present. Mat community was determined from metagenomic analysis.

Sulfur transformation	Reactions	Bacteria
Sulfur oxidation	$S^0 \rightarrow SO_4^{2-}$	*Thermocrinis*
Dissimilatory sulfate reduction	$SO_4^{2-} \rightarrow H_2S$	*Thermodesulfovibrio*
Sulfide oxidation	$H_2S \rightarrow S^0$	*Chloroflexi, Thermodesulfovibrio, Pseudothermotoga, Synechococcus, Aramatimonadetes*
Sulfur disproportionation	$S^0 \rightarrow H_2S + SO_4^{2-}$	*Thermodesulfobacterium*

Source: Based on data from Thiel et al. (2017).

The most abundant bacterium in the common shallow-water corals *Stylophora pistillata* and *Pocillopora verrucosa* is *Endozoicomonas* spp. of the *Betaproteobacteria* group (Neave et al. 2017). The *Endozoicomonas* genus is an endosymbiont present in the animal polyps and genetically distinct *S. pistillata* host different strains of *Endozoicomonas.* Neave et al. (2017) compare this species specificity to the difference of symbionts with bees. The honey bee (*Apis* spp.) has the bacterium *Snodgrassella alvi* as an endosymbiont but colonization of the gut by this bacterium is limited to a specific species (Kwong et al. 2014). Clearly the role of bacteria as endosymbionts in coral animal tissue is an interesting area for further research.

Diseases leading to the loss of corals are difficult to control because several factors could contribute to the demise of corals. In fact, it is unknown which area of the coral holobiont is subject to attack. Corals grow exceedingly slowly, with photosynthetic-supported corals growing 1–3 cm a year, while corals in the deep oceans grow only about 1 mm annually.

4.4 Freshwater Environments and Associated Microbiomes

There are a diverse number of freshwater environments throughout the world, ranging from small intermittent streams and ponds, wetlands in which water flow is limited or negligible, to large rivers and lakes. Most inland water is fresh water, which is defined as having a salt concentration of less than 1% (10‰). Surface water can move into groundwater via porous rock. Conversely, groundwater can also enter surface water via springs or through human activity (being pumped from wells). Water chemistry is influenced by various features including the geography of the region, climate, and human activities. Here, we address some common features. The study of inland waters is typically referred to as limnology and some excellent resources are available for additional reading (Cole and Weihe 2016).

A watershed or drainage basin is defined as the region of land in which precipitation collects and drains into a common outlet (such as a river or bay). Usually, the original source of water for rivers and lakes is precipitation and corresponding runoff and drainage from the watershed. In alpine and other colder (e.g. polar and sub-polar) regions, the original source of water may be melting snow or ice (including meltwater from glaciers). The other major source of water are artesian springs and discharge from groundwater. In alpine regions, particularly above the tree line, there is very little plant life adjacent to or within the streams, and so the water in these streams is markedly oligotrophic and consequently has very little organic matter present. A high level of dissolved oxygen in these alpine streams is attributed to the rapid flow rate and the presence of waterfalls. Conversely, in stagnant water the aerobic portion may be restricted to the first few millimeters or centimeters and may be partially or completely blocked if a microbial mat (a biofilm at an air–water interface) forms.

Primary productivity in inland waters is influenced by various factors including light penetration, available nutrients, water chemistry, and climate. Contributing factors to light penetration include silt and other particulate matter from soil runoff or the wave erosion of sediments; suspended algae and cyanobacteria; shade from trees or floating plants; and humic acids from soils. In very turbid waters, primary production may be limited to adjacent terrestrial plants, or alternatively to the initial few millimeters water depth.

4.4.1 Lakes and Rivers

Reservoirs and most lakes receive water from the surrounding environment, with the input of nutrients greatly influencing the abundance and type of microorganisms present. If nitrogen and phosphorus are available in fresh waters, *Chrysophyta* (golden-brown algae including diatoms), *Chlorophyta* (green algae), and *Pyrrhophyta* or *Dinophyta* (including dinoflagellates), and *Cryptophyta* (unicellular biflagellate cells) are commonly present. If phosphorus is not a limiting nutrient but nitrogen is limiting, nitrogen-fixing cyanobacteria (Table 4.7) may be found in the water. The dense growth of cyanobacteria – known as a cyanobacterial bloom (Figure 4.8) – produces undesirable odors, makes the water taste bad, and disrupts normal food webs (Paerl et al. 2001). As CO_2 is removed from the water through photosynthesis-driven metabolism, the alkalinity of the water increases and the cyanobacterial bloom is halted by cyanophages (Yoshida et al. 2008). In late winter and early spring, algal blooms are attributed to *Cryptomonas,* while the dominant organisms in the water in summer and fall blooms include diatoms, green algae, dinoflagellates, and cyanobacteria. The criteria used to assess aquatic-based microbial activities include the measurement of chlorophyll *a* ($\mu g\,l^{-1}$), the number of bacteria present (cells l^{-1}), primary production and microbial respiration (μg C/l/day) (Defore et al. 2016), and nitrogen fixed using acetylene reduction (Box 4.4).

As organic matter and other nutrients (notably nitrogen and phosphorous) enter the water, primary productivity can increase and microbial growth makes the water turbid. Although photosynthesis (primary productivity) and associated O_2 production is a consequence of oxygenic photosynthesis, heterotrophic aerobic bacteria grow on the organic material released

Table 4.7 Characteristics of bloom-forming cyanobacteria.

Location and genus	Nitrogen (N_2) fixing	Toxin producing
Forms a scum at surface		
Anabaena	Yes	Anatoxins
Aphanizomenon	Yes	Saxatoxins
Nodularua	Yes	Nodulins
Microcystis	No	Microcystins
Bloom forms subsurface		
Cylindrospermopsis	Yes	Saxatoxins
Lyngbya[a]	No	Lyngbyatoxin A
Osacillatoria	No	Microcystins

Source: Reproduced with permission from Pearl et al. (2001).
[a] Forms dense subsurface mat but releases and floats on surface.

Figure 4.8 Cyanobacteria (blue-green algae) bloom on the shore of Catawaba Island, Ohio, in Lake Erie. Summer, 2009. *Source:* https://landsat.gsfc.nasa.gov/nasa-noaa-epa-and-usgs-join-forces-to-put-satellite-eyes-on-threat-to-u-s-freshwater, NASA.

Box 4.4 Acetylene Reduction

The acetylene reduction assay has often been used to measure N_2 fixation. The enzyme nitrogenase recognizes the triple bond between $HC\equiv CH$ (acetylene) as if it was a triple bond between $N\equiv N$ (N_2). The product of N_2 fixation is NH_3 and the product of acetylene reduction is $H_2C=CH_2$ (ethylene). The microliters or micromoles of ethylene produced is measured using a gas chromatograph. The acetylene reduction assay is simple to perform and it has high reproducibility.

from the dying cyanobacteria. Concomitant with the rapid bacterial growth is the rapid consumption of O_2 and the low level of O_2 is detrimental for fish and other animals. Lakes and reservoirs containing minimal levels of photosynthetic microorganisms due to nitrogen and phosphorus limitation are oligotrophic. In contrast, a body of water with dense algal or cyanobacterial growth is designated as eutrophic. The trophic status of a water body can be determined by the amount of algal/cyanobacterial activity related to the amount of chlorophyll present or the amount O_2 fixed using $^{14}CO_2$ incorporated into the water sample or by chemically measuring carbon directly. If nutrients (P and N) are introduced into the water from a specific site, the contamination of the water is at a point source but if the nutrient input is from a general area, pollution is from a non-point source.

Deep mountain lakes are highly stratified in the summer with two photo zones. The upper photosynthetic region is for oxygenic cyanobacteria and algae, while the bottom segment of the lake is anaerobic. As indicated in Figure 4.9, anoxygenic photosynthetic bacteria are beneath the oxic–anoxyic interface. Methane oxidizing bacteria and sulfide oxidizing bacteria are active at the aerobic/anaerobic interface while sulfate-reducing bacteria and methanogenic archaea are present in the lake bottom. In temperate climates with prominent summer and winter

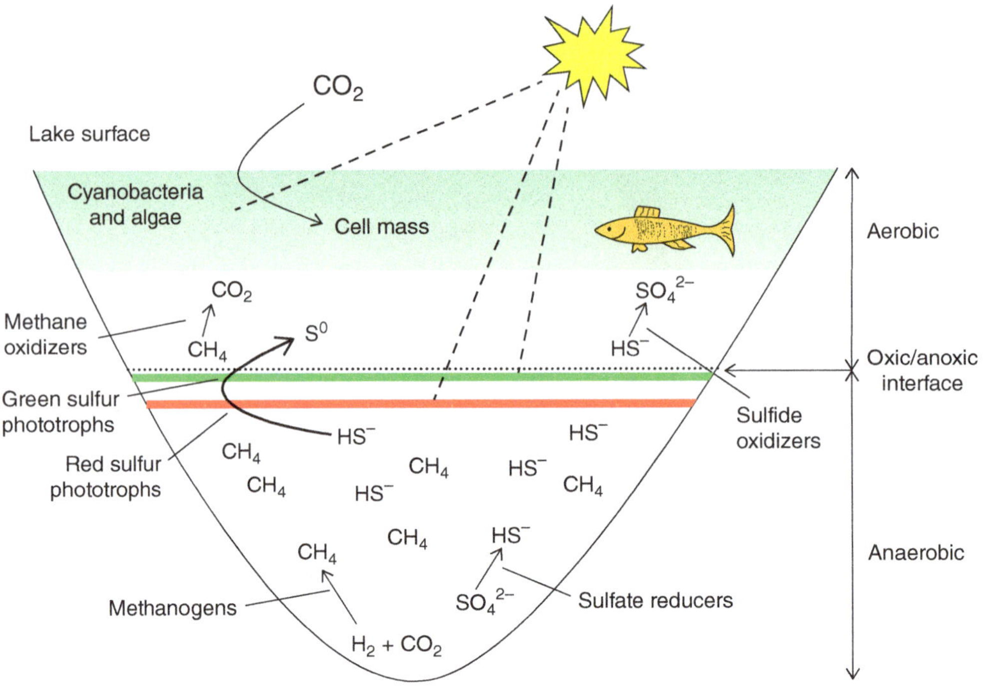

Figure 4.9 A deep lake showing the stratified microbial processes. Cyanobacteria grow on the surface while anaerobic phototrophic bacteria are found in the sulfide rich environment. The position of the red and green photosynthetic bacteria indicates light penetration and the type of baceteriochlorophyll of the phototrophs. Hydrogen sulfide is produced in the anaerobic zone while methane and sulfide oxidation occurs at the oxic/anoxic interface.

seasons and where lake water is at least 25 m deep, the surface and deeper waters will interchange, referred to as a turnover (see Figure 4.10). Deep lakes characteristically have a spring and a fall turnover. This turnover may take only a few days and serves to mix sediment nutrients with surface waters and introduce O_2 into the subsurface water. Note that in some frozen lakes, microbial degradation of plant materials may occur in water under the ice, with photosynthetic blooms reported in some under-ice waters (Bertilsson et al. 2013).

Photosynthesis is used by bacteria but not archaea, and for both bacteria and green plants the capture of light energy is due to the unique properties of chlorophyll. The chlorophyll molecule (Figure 4.11) has a characteristic magnesium-containing tetrapyrrole ring that is found in both eukaryotic and bacterial phototrophs; however, bacteriochlorophyll has numerous substitutions at the sites indicated in Figure 4.11. Because of these changes in the bacteriochlorophyll molecule, the wavelength for adsorption of light also varies (see Table 4.8). Thus, this difference in the wavelength of adsorption of light enables the photosynthetic bacteria to avoid competition for light energy, so vertical stacking of photosynthetic bacteria in water columns occurs (Figure 4.9).

The Laurentian Great Lakes system includes the five lakes (Michigan, Superior, Huron, Ontario, and Erie) that border Canada and the USA. Lake Superior is the largest and due to the limited human impact is extremely oligotrophic with $<2 \times 10^6$ cells per liter (Reed and Hicks 2011). Similar to other oligotrophic lakes, sediments of Lake Superior have a greater abundance of ammonia oxidizing archaea (AOA) than ammonia oxidizing bacteria (AOB) and perhaps this is attributed to AOA having a greater K_m for ammonium/ammonia than AOB.

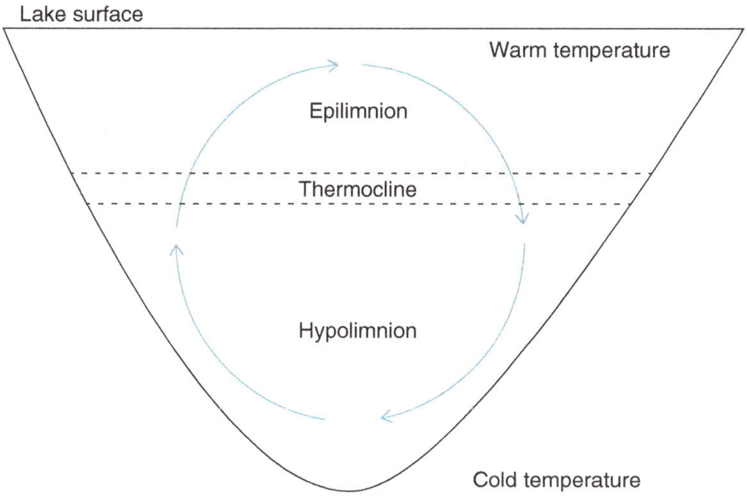

Figure 4.10 Turnover of a deep lake typically occurs in the spring and fall and this mixing allows nutrient cycling in lakes. Winds promote mixing of the surface waters (epilimnion) throughout the year and the turnover process is due to temperature changes, with cold water being denser than warm water. The decomposition of algal and cyanobacterial matter occurs slowly in the hypolimnion zone and in conjunction with lake turnover, provides nutrients for the surface waters.

Figure 4.11 Structure of the chlorophyll *a* molecule. Different side groups are added at the encircled regions to produce different bacteriochlorophylls.

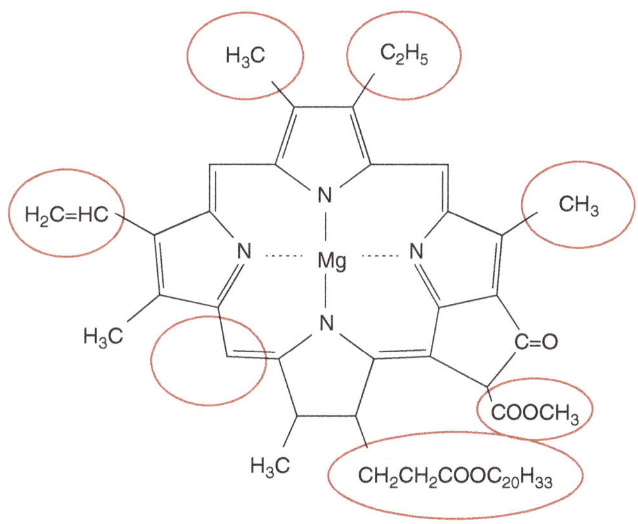

From DNA sequence analysis, the most abundant archaea in Lake Superior were with the *Nitrosopumilus* clusters 1 and 15 and the *Nitrosotalea* cluster (Reed and Hicks 2011). Lake Erie is mesotrophic to eutrophic and the trophic status of this shallow lake reflects the large amount of urban and agricultural runoff that it receives. The quantity of *amoA* genes from AOA were comparable to that for AOB and while the AOB in Lake Erie were like those found in other freshwater environments, the AOA were primarily of the *Nitrosphaera* cluster (Reed and Hicks 2011). For over 200 years, the algal blooms and water status of Lake Ontario has been recorded (Estepp and Reavie 2015) and for much of the time it has had a mesotrophic status with heavy spring growth of the diatom *Cyclotella* and summer proliferation of the cyanobacteria *Aphanocapsa* and *Oscillatoria*. Nutrient input from urban systems, agricultural land, and

Table 4.8 Selected photosynthetic bacteria and their photocharacteristics.

Phylum	Genus	Chlorophyll[a]	Photo-system[b]	Wavelength of adsorption for purified chlorophyll (nm)
Cyanobacteria	*Anabaena*	Chl *a/b*	PS I and PSII	680, 700
Chloroflexi	*Chloroflexus* (green non-sulfur)[c]	Bchl *a*	PS II	800–810
Chlorobi	*Chlorobium* (green sulfur)	Bchl *a*	PS I	800–810
Betaprotobacteria	*Rhodospirillum* (purple non-sulfur)	Bchl *a*	PS I	800–810
Alphaproteobacteria	*Blastochloris*	Bchl *b*	PS II	835–850
Betaproteobacteria	*Chromatium* (purple sulfur)	Bchl *a/b*	PS II	800–810 835–950
Firmicutes	*Heliobacterium*	Bchl *g*	PS I	670, 788

Source: Reproduced with permission from Slonczewski and Foster (2009).
[a] Chl = chlorophyll; Bchl = bacteriochlorophyll.
[b] Photosystem which most closely describes the bacteria.
[c] Color refers to appearance of colony and non-sulfur indicated inability to use sulfur compounds as electron source for anaerobic photosynthesis. Sulfur refers to the use of sulfur compounds as an electron donor in photosynthesis.

industrial organizations have promoted algal development in spring and cyanobacterial growth in summer. Wastewater from urban environments carries bacteria with diverse metabolism into the lakes; however, these microorganisms from the urbanized community do not become established in oligotrophic lakes such as Lake Michigan (Newton and McLellan 2015).

The microorganisms present in rivers and streams have been extensively examined to assess water quality and safety. It is well recognized that sediments in rivers and streams harbor many bacteria and that the number of microorganisms in flowing water is only a fraction of those present. For some time, it was thought that the microorganisms in rivers resembled those in lake waters and that typical freshwater bacteria were present in all surface waters; however, the presence of core microbial communities in rivers as suggested by Staley et al. (2013) is now attracting considerable attention. Seasonal changes influence river microbiota, with the Amazon River having different seasonal microbial communities associated with the high flow levels in May and the low flow rates in December (Doherty et al. 2017). Temperate and arctic rivers have three seasonal microbial communities, which are influenced by environmental factors associated with discharges into the river. In the Danube and Changjiang Rivers (and presumably many other rivers), the abundance and type of microorganisms reflect activity in the different regions of inflow, middle segment, and plume. The dominance of *Actinobacteria* and *Proteobacteria* (especially *Betaproteobacteria*) in the Upper Mississippi River, Yenisei River, and the River Thames (see discussion by Savio et al. 2015) reflects the input of soil and groundwater into the rivers. Additionally, major shifts in the microbial populations in the Mississippi River occur because of confluent tributaries and urban point sources, but further downriver, microbial diversity changes gradually (Payne et al. 2017). The influx of water into rivers brings nutrients and diverse microorganisms that metabolize the DOM, leaving complex biomolecules slowly degraded by a new community of bacteria. Thus, long stretches of the Mississippi, Danube, and Amazon rivers (Savio et al. 2015; Doherty et al. 2017; Payne et al. 2017) have only a small change in microbial diversity. As these three rivers discharge water into marine environments, a change in microbial communities occurs due to the increase in salinity. Importantly, these plumes enrich the oceans with nutrients, including bacterioplankton and other particulate biological compounds (Mason et al. 2016).

4.4.2 Wetlands

The two categories for wetlands are coastal wetlands and inland wetlands (Figure 4.12). Coastal wetlands vary in salinity because fresh water from rivers mixes with seawater and they are subject to mixing by tides. Inland wetlands are flooded regions along rivers or lakes and in various regions where the soil is continuously underwater. Frequently an inland wetland site is used for wildlife refuge sites and the Bosque del Apache wildlife refuge in New Mexico is along the Rio Grande. Migrating birds and especially geese (Figure 4.13) feed on alfalfa and corn adjacent to the Bosque del Apache wetland refuge have an increase of 40% total nitrogen and 75% of total phosphorus (Kitchell et al. 1999).

Constructed inland wetlands (Figure 4.14) are often used to remove nitrogen and phosphorus from agricultural sources, waste water treatment systems, and treatment of acid mine drainage. To manage the acid waters containing numerous metals from the Wheal Jane mine in England, a three-tiered system was constructed and monitored over a 30 month period (Hallberg and Johnson 2004). Initially, the dominant bacterium in the constructed wetland was an organism related to *Halothiobacillus neapolitarus* but this was replaced by moderately and

Figure 4.12 Wetlands in central Minnesota. These regions are especially important sites for wildlife. *Source:* Reproduced with permission from Larry Barton.

Figure 4.13 Geese in the bosque along the Rio Grande in New Mexico. *Source:* Reproduced with permission from Larry Barton.

Figure 4.14 Constructed wetlands being used to remove excess fertilizer from fields. Note the presence of water lilies and other floating plants. *Source:* Reproduced with permission from Larry Barton.

extremely acidophilic sulfur- and iron-oxidizing organisms closely related to *Thiomonas intermedia*. The acidophiles eventually accounted for up to 25% of the microbes present and many of these were not previously reported for acid mine drainage streams.

4.4.3 The Snow and Glacier Ice Ecosystems

Since glacier ice is produced from the compacting of snow over hundreds of years, the biospheres of these two environments are interrelated. While 35% of the Earth's terrestrial surface is covered permanently or in the winter with snow, 10% is covered with the Greenland and Antarctic ice sheets. These ice sheets have been estimated to contain 9.6×10^{25} bacterial cells (Miteva 2008). Bacteria are found in the thin films of liquid around particulate minerals and in liquid veins running through glaciers. Nutrients are most abundant at the bottom of the glacier, so bacteria are also abundant there. The surface of the Greenland ice has $\sim 10^3$ bacterial cells ml^{-1} of water from ice but the bottom of the glacier has 10^7 cells ml^{-1}. Another habitat for bacteria in glaciers is in the cryoconite holes, which are cylindrical holes in the ice attributed to dark debris established on the ice surface. The dark material heats with sun exposure, the ice melts and water fills the bottom of the hole. Cultivation procedures have grown cold-tolerant (psychrotolerant and psychrophilic) bacteria from these areas and the amount of nutrients present in the liquid between the ice veins could support >100 bacterial cells ml^{-1} for 400 000 years (Miteva 2008). Bacteria have been isolated from 420 000-year-old Vostok accretion ice and from the 750 000-year-old Guliya ice cap (Christner et al. 2001,2003). It is likely that the presence of these bacteria in ancient glacier ice is attributed to long periods of dormancy alternating with slow growth. When comparing the 16S rDNA sequences from glacial ice, most of the bacteria were from the High G + C Gram Positive group followed by *Proteobacteria*, Low G + C Gram Positive, *Cytophaga-Flavobacterium-Bacteroidetes*, listed in decreasing abundance. Bacteria of the following genera were most frequently isolated: *Acinetobacter*, *Arthrobacter*, *Chyrseobacterium*, *Exiguobacterium*, *Frogoribacterium*, *Janthinobacterium*, *Methylobacterium*, *Rhodococcus*, and *Sphingomonas* (Miteva 2008). A few of the novel bacteria isolated from glacial ice are listed in Table 4.9. The bacteria isolated had physiological capabilities that enabled them to survive the harsh environment.

The number of bacteria in mountain and polar snow is estimated at 10^3–10^5 cells ml^{-1} of melted snow. Based on ribosomal RNA sequence analysis, bacteria in snow samples were psychrotolerant or psychrophilic and resistant to desiccation and high radiation. Red snow was

Table 4.9 New strains of bacteria isolated from glacial ice.

Bacteria	Location of glacial ice	Reference
Dyadobacter hamtensis	India	Chaturvedi et al. (2005)
Exiguobacterium indicum	India	Chaturvedi and Shivaji (2006)
Flavobacterium glaciei	China	Zhang et al. (2006)
Flavobacterium omnivorum	China	Zhu et al. (2003)
Flavobacterium xinjiangense	China	Zhu et al. (2003)
Nocardiopsis antarcticus	Antarctica	Abyzov et al. (1983)
Pedobacter cryoconitis	Alpine	Margesin et al. (2003)
Pedobacter himalayesis	India	Shivaji et al. (2005)

Source: Based on data from Miteva (2008).

reported initially by Aristotle and reports of red snow and pigmented snow from around the world on high altitude mountains are reported annually. Red snow is attributed to a bloom of the algae *Chlamydomonas nivalis*, green snow to growth of the algae *Chloromonas*, and orange snow to pigmented cysts of the *Chlamydomonas polyptera* (Terashima et al. 2017). In spring and summer when snow melts, algal snow is common. These algae are the primary producers that support the growth of various cold tolerant heterotrophic bacteria. In the snow fields of Svalbard and Arctic Sweden, *Betaproteobacteria* are the most abundant bacteria associated with green algae, while *Bacteroidetes* are associated with algae in red snow. The reasons for the alignment of these classes of bacteria with algae producing green or red snow is unresolved. It is known that neither of the algae have an association with a specific bacterial species to form a symbiotic partnership. Algae are the primary producers in this low nutrient environment, with bacteria using compounds released from living and dead algal cells. The bacterial strains of *Glaciimonas*, *Polaromonas*, and *Callimonas* isolated with the green algae may provide vitamins, siderophores, or other growth factors for the algae as well as producing chemicals to protect *Chloromonas* from the parasitic *Chrytridomycota*. Members of the *Bacteroides* isolated along with *Chlamydomonas* produce proteins that promote adhesion to the algal cell and that optimize the acquisition of nutrients released from dead cells of *Chlamydomonas* (Terashima et al. 2017).

Numerous bacteria have been detected in snow using 16S rDNA sequence analysis; however, it is important to distinguish endogenous viable bacterial strains from transient bacteria distributed across the snow from the atmosphere. Snow samples from the Tateyama Mountains, Japan, have a broad spectrum of bacterial DNA sequences including those from enteric bacteria. The snow environment is laden with terrestrial bacteria brought in by wind currents and the fecal deposits of birds (Segawa et al. 2005). Based on the growth of bacteria isolated from Tateyama Mountain snow, resident bacteria include: *Cryobacterium psychrophilum*, a psychrophile, the psychrotrophic *Variovorax paradoxus*, and *Janthinobacterium lividum*. Since these three bacteria are also present in glacier ice cores from the Antarctic and Greenland, they represent members of the snow microbiota (Segawa et al. 2005). Cold tolerant bacteria, identified from operational taxonomic units, were found near the Russian Arctic stations and included members of *Variovorax*, *Janthinobacterium*, *Pseudomonas*, and *Sphingomonas* (Lopatina et al. 2013).

4.4.4 Microbiota of Cold and Hot Springs

Terrestrial springs are distributed through the Earth and have be characterized by different parameters including temperature, salinity, and sulfur content. Each spring reflects an extreme environment and for cold springs, the microorganisms are psychrophilic or at least tolerant to cold environments. The following discussion shows that each cold environment has its own unique microbial community. The Gypsum Hill spring of the Canadian Arctic arises from permafrost where the mean temperature is −15 °C, the water contains 7.5–15% brine, and the oxidation/reduction potential is −325 mV. Based on 16S rRNA gene analysis, this site has a rich diversity of bacteria with sequences affiliated with *Actinobacteria*, *Bacteroidetes*, *Firmicutes*, *Gemmatimonadetes*, *Proteobacteria*, *Spirochaetes*, and *Verrucomicrobia* phyla (Perreault et al. 2007). Half the bacterial sequences align with *Deltaproteobacteria* and *Gammaproteobacteria*. Of the archaeal sequences, 79% were related to *Crenarchaeota*. An examination of cold spring water (4 °C) in the Qinghai-Tibetan Plateau revealed the alignment of DNA sequences to members of the *Proteobacteria*, *Acidobacteria*, *Deinococci*, *Sphingobacteria*, *Flavobacteria*, *Nitrospirae*, *Actinobacteria*, and *Gemmatimonadetes* phyla (Li et al. 2011), while the archaeal sequences indicated the presence of *Crenarchaeota* and *Thaumarchaeota*.

A second habitat dealing with cold spring environments contains sulfide and this environment may support a microbial community reflecting sulfur cycling and the use of sulfur compounds for energy production. The cold spring water along Lake Erie is anaerobic and contains H_2 (5.4 nM), CH_4 (2.7 μM), HS^- (0.03 mM), SO_4^{2-} (14.8 mM), and HCO_3^- (4.1 mM), which supports the growth of sulfur metabolizing and chemolithotrophic bacteria commonly associated with sulfide environments (Chaudhary et al. 2009). This spring near Lake Erie contains *Epsilonproteobacteria* and *Cyanobacteria,* which are present in other sulfidogenic waters. An unusual community consisting of bacteria and archaea were found in a cold spring in the Sippenauer Moor, Germany (Rudolph et al. 2001). The water contained 1.2 $mg\,l^{-1}$ sulfide and filaments of numerous types of microorganisms were found in the 10 °C environment. White globules, 0.5–3.0 mm in diameter, were connected by white threads of unknown chemical composition to produce a "string-of-pearls" morphology that formed a dense network when compacted. The white globules contained about 10^7 cells of archaea and had a covering of filamentous bacteria of unknown taxa. The 16S rRNA gene analysis indicated that the unknown archaea are associated with the *Euryarchaeota*. In Lake Huron, groundwater seeps or sink holes that are fed by groundwater seeps have localized *Beggiatoa* mats and an adjacent sulfide-rich artesian fountain dominated by bacteria related to *Thiothrix lacustris* and a symbiotic euryarchaeon called "SM1" (Sharrar et al. 2017).

The acidic Yellowstone hot springs are a favorite site for microbiologists to isolate unique bacteria and archaea. The isolation of *Bacillus stearothermophilus* by Marsh and Larsen (1952) from several hot springs in Yellowstone National Park (YNP) provided an impetus for subsequent microbiological examinations of hot spring water. *Thermus aquaticus*, the thermophilic microbe capable of growing near 100 °C, was isolated by Thomas Brock from Yellowstone hot springs and this gram-negative bacterium produces Taq polymerase, which is widely used in DNA polymerase reactions. There is an obvious temperature gradient from the center of the hot spring to its periphery and the designation of these zones reflects the presence of specific cyanobacteria in these areas. At temperatures greater that 73 °C cyanobacteria are absent, at 60–73 °C *Synechococcus* and *Oscillatoria* are present, at 30–60 °C *Phormidium* are abundant, and below 30 °C *Calothrix* are abundant. These cyanobacteria form biofilms and biomats of sizeable dimensions floating on the hot spring water. These cyanobacterial structures entrap minerals, algae, fungi, and bacteria. Interestingly, each cyanobacterial group has a unique coloration, with *Synechococcus* producing green growth, *Oscillatoria* orange growth, and *Calothrix* brown growth. Pigmentation in cyanobacteria and *Thermus* is attributed to carotenoids, which function as sunscreen compounds to reduce damage to UV exposure.

Cyanobacteria, heterotrophic and chemolithotrophic bacteria, and archaea have been studied in hot springs located in over a dozen countries. For example, sulfur-oxidizing *Proteobacteria* were recorded at the thermal springs in High Arctic springs on Svalbard (Reigstad et al. 2011) and the hot spring water on Kamchatka Peninsula, Russia contained bacteria actively oxidizing iron or metabolizing organic compounds (Belkova et al. 2007).

These observations have been summarized in numerous publications including the book by Reysenbach et al. (2001). While cyanobacteria, algae, and anoxygenic phototrophic bacteria are primary producers at or near the surfaces of hot springs, microorganisms growing in anaerobic zones are also important in providing nutrients to the spring water. An archaean of importance is a member of the class *Hadesarchaea* that appears to be of marine origin. This organism has been identified in the YNP, Great Basin (Miller-Coleman et al. 2012), numerous deep terrestrial environments, and is a member of the South-African Gold Mine Miscellaneous Euryarchaeal Group (SAGMEG) (see discussion in Chapter 3). Metabolism by archaea of the class *Hadesarchaea* was inferred from genome analysis and indicates that it can oxidize CO and H_2, fix CO_2 by the Calvin – Benson – Bassham cycle, and reduce nitrite to ammonia (Baker et al. 2016).

While there is a rich array of acidic thermophilic chemolithotrophic microorganisms present in the YNP environment, unusual microorganisms and viruses have also been isolated from the water from YNP. Bacteria and archaea isolated from these hot springs were resistant to mercury (Geesey et al. 2016), capable of oxidizing arsenite (Jackson et al. 2001), oxidizing iron (Kozubal et al. 2008), and using sulfur compounds for respiration (Boyd et al. 2007). *Mycobacterium parascrofulaceum*, an organism associated with health problems in humans, was isolated from the Norris Geyser Basin in the YNP (Santos et al. 2007). Additionally, a thermotolerant free-living amoeba *Naegleria,* which contains the pathogenic *Legionella* bacterium, was isolated from YNP and Grand Teton National Park (Sheehan et al. 2003). More recently, *Naegleria fowleri* and *Legionella pneumophila* were isolated in bath houses where the water was supplied from deep subsurface springs (Żbikowska et al. 2013).

Bacteria and archaea grow abundantly in alkaline thermal springs where the pH is 8.5–9.8. In Heart Lake Geyser Basin of YNP, the microbial community is influenced by temperature. There is a nonphotosynthetic bacterial community at 75 °C, phototrophic and heterotrophic bacteria of the *Deinococcus-Thermus* group present at 63 °C, and photosynthetic members of *Chloroflexi* and cyanobacteria growing at 44 °C (Bowen De León et al. 2013). Outside of YNP, numerous bacterial and archaeal species were detected in Lake Magadi and Little Magadi in Kenya (Kambura et al. 2016).

4.4.5 Microbial Mats

Microbial mats and biofilms are composed of heterogenous prokaryotic organisms and each of these communities has its own unique features (see Table 4.10). Microbial mats are diverse bacterial and archaeal communities which occur as layered environments where the microbial cells and minerals are imbedded in an extracellular matrix material consisting of extracellular polysaccharides and extracellular DNA (eDNA). Cyanbobacteria are present in the surface layer and filaments of these phototrophs intertwine to form a stable, dense mat. Historically, cyanobacterial mats were studied because the layered growth of cyanobacteria produce stromatolites that contain a fossilized record of early oxygenic organisms (see Chapter 1). Currently, cyanobacterial mats that are costal or intertidal mats, saline mats, and hot spring mats are of great interest. The cellular features of some common cyanobacteria are presented in Figure 4.15. Although desert mats have been reported (Antibus et al. 2012), they are discussed in this book in Chapter 3 as examples of desert or soil crusts.

Intertidal microbial mats cover many areas around the world and the ones most frequently studied include the beaches of: (i) the North Sea barrier islands in the Netherlands, Germany,

Table 4.10 Distinctions between biofilms and mats.

Biofilm	Mats
Covers solid surfaces	Covers sediments
1 μm or few mm thick	<1 mm to several cm thick. Often with several layers
Heterotrophs use nutrients supplied by surface or surrounding water	High density of photoautotrophs as primary producers
Found on most surfaces covered by a water film	Found in sediments of shallow water, hot springs and hyper saline waters

Source: Based on data from de Beer and Kühl (2001).

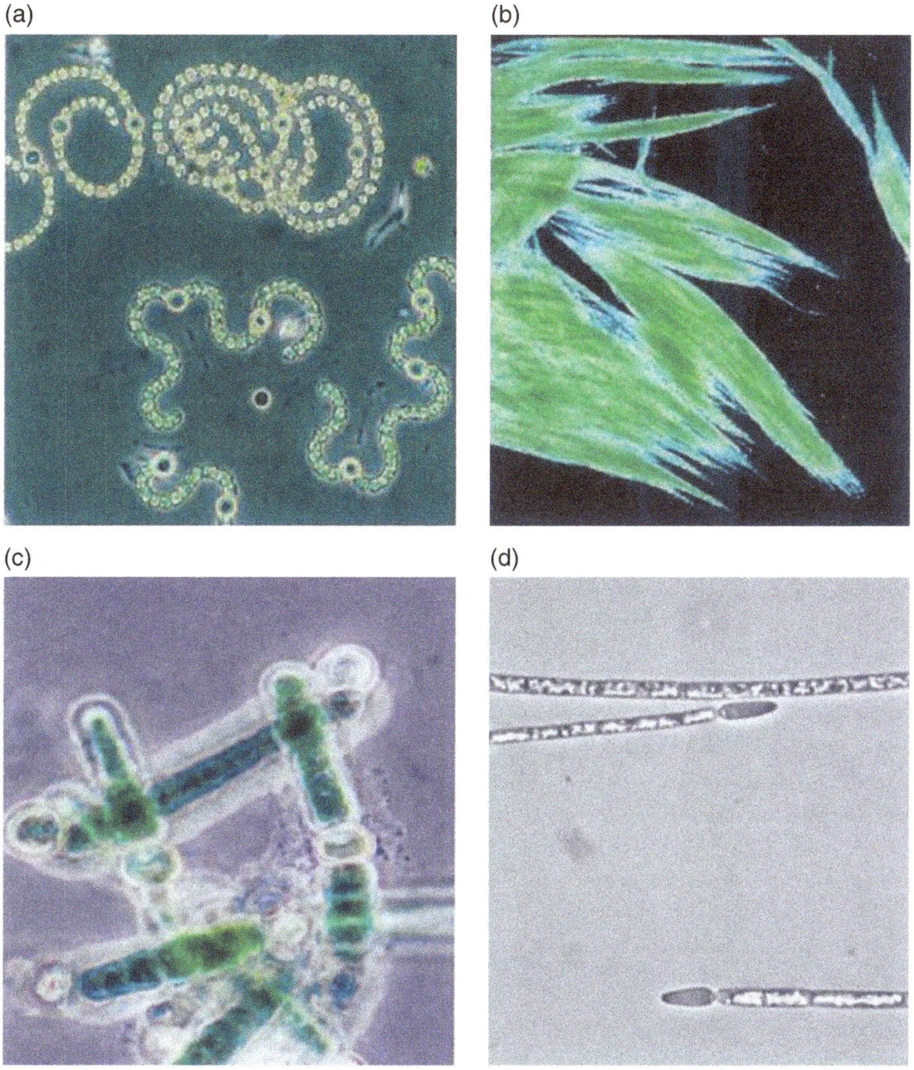

Figure 4.15 Important cyanobacteria producing fresh water blooms, showing members of heterocyst-containing N_2-fixing genera: (a) *Anabaena flos-aquae* (top of frame) and *A. spiroides*; (b) flake-like colonies of *Aphanizomenon flos-aquae*; (c) *Nodularia* sp.; (d) *Cylindrospermopsis raciborskii*. *Source:* Paerl et al. (2001). Licensed under CC BY 4.0.

and Denmark, (ii) the Sippewissett marsh in Massachusetts, (iii) an Arctic photosynthetic mat, and (iv) the Elkhorn Slough in California (Bolhuis et al. 2014). Analysis of these intertidal mats, which range from fresh to brackish waters, indicate that the organisms of the *Cyanobacteria* phylum (i.e. *Lyngbya aestuarii* and *Microcoleus chthonoplastes*) were dominant in the upper few millimeters of the mat and accounted for 80–90% of the active community. The green non-sulfide phototrophic anaerobic bacteria of the *Chloroflexi* phylum were frequently present buried in the bottom of the cyanobacterial layer. Beneath the cyanobacterial layer, purple sulfur bacteria of the *Gammaproteobacteria* phylum grew using sulfide as the electron donor for anoxic photosynthesis. Also abundant in the sub-layers of the intertidal mat were bacteria of the *Proteobacteria* (especially *Alpha-*, *Gama-*, and *Deltaproteobacteria*) and *Bacteroidetes* phyla. Bottom layers of the mat contained sulfate-reducing bacteria which produced hydrogen sulfide as they used sulfate as the terminal electron acceptor.

Hypersaline microbial mats are found around the world in natural salt lakes and constructed salterns. The best studied hypersaline mats are Guerrero Negro in Baja California Sur, Mexico, and Cuatro Ciénegas Basin, Mexico. Bacteria, including cyanobacteria, are the most abundant microorganisms in the mats of Guerrero Negro, with a ratio of bacteria/archaea/cellular eukaryotes of 90%/9%/1% (Bolhuis et al. 2014). An array of cyanobacterial species grow in the upper 5 mm of the saline mats and bacteria that are members of the *Spirochetes*, *Chloroflexi*, *Planctomycetes*, and *Proteobacteria* phyla are highly abundant. Members of 40 phyla of bacteria are present in the mats of Guerrero Negro, which is eight times greater than the number of bacterial phyla present in the human intestine (Bolhuis et al. 2014).

Moderately hot to temperately warm acidic spring mats have been examined in YNP and their primary productivity was attributed to algae, including *Cyanidioschyzon merolae* at >49 °C and *Chlorella protothecoides* var. *acidicola* and *Paradoxia multisita*-like algae at <39 °C (Ferris et al. 2005). Microbial mats in hot springs are found around the world and the major contributors to the primary productivity of these thermophiles includes oxygenic photosynthetic cyanobacteria (oxygenic mats), anaerobic photosynthetic bacteria (*Chloroflexus* mats), or acidophilic anaerobic photosynthetic bacteria (*Chlorobium* mats) (Bolhuis et al. 2014). Alkaline, low-sulfide mats from the Mushroom and Octopus Springs in YNP have a high population of oxygenic and anoxygenic phototrophic bacteria, chemolithotrophic bacteria, heterotrophic bacteria, and a few archaea. Metagenomic analysis of 16S rRNA genes of several layers of mat at Mushroom and Octopus Springs were examined and the upper green layer contained members of the *Cyanobacteria* phylum, *Synechococcus* Type A and Type B′. Additional primary productivity was attributed to *Roseiflexus castenholzii*, *Chloroflexus aurantiacus*, and *Thermanaerothrix daxensis* of the *Chloroflexi* phylum. There is an abundance of heterotrophic and chemolithotrophic bacteria present in the hot spring mats and while many of these represent novel uncultured species, they are separable into 12 different phyla (Thiel et al. 2016, 2017).

4.5 Maintaining Populations in Low Nutrient Environments

In many aquatic environments, the nutrients required for microbial growth are limiting and microorganisms are frequently adapted to the nutrient-limiting environment. There are two mechanisms that account for microbial growth in oligotrophic environments. The first is the adjustments of individual cells to have (i) structural features that are beneficial (i.e. holdfasts and sheaths) and (ii) modified metabolism with enhanced nutrient transport or energetic systems with increased efficiency. The second is a group activity where microorganisms work collectively to transform substrates, as for nutrient cycling, including bacterial loop activities.

Bacteria such as *Caulobacter* attach themselves using a holdfast to objects in the environment and collect nutrients that are flowing past. Nutrients are acquired by the holdfast stalk as well as the cell. *Caulobacter* cells divide only when attached to a site where sufficient nutrients are available to support growth; these immobilized cells are resistant to phage attack because virus receptors are present only on swimming cells. *Gallionella ferruginea,* an iron-oxidizing bacterium, has polymeric material extending from the cell, which becomes encrusted with iron (de Vet et al. 2011). Because *Gallionella ferruginea* resides in freshwater aquifers (Sayanthan et al. 2016), it is considered a nuisance organism because it contributes to an unpleasant flavor in drinking water. Well drillers may transfer the bacterium *Gallionella* from site to site and so

Table 4.11 Examples of bacteria that exploit their holdfasts and sheaths to assist in their acquisition of nutrients.

Genera	Structural features
Caulobacter	Holdfast attaches to a solid surface and numerous cells form a rosette. Grows in nutrient limited waters and acquires nutrients from flowing waters.
Beggiatoa	Cells form long filaments and these are collective found as bundles. There are marine and fresh water species and they oxidize sulfide.
Leucothrix	Forms trichomes or filaments 0.4 mm long, attaches to a solid by a holdfast and forms rosettes. It is a sulfide oxidizer.
Thiopioca	Cells form a bundle which is enclosed in a sheath. Form mats on sea floor off the shore of Peru and Chile. These are sulfur-oxidizing bacteria.
Thiothrix	Cells form long filaments and at the end of the filament is a holdfast. Many cells use hold fast to collect around the same surface and produce a rosette formation. These are sulfur-oxidizing bacteria.
Sphaerotilus	Fresh water bacteria that are enclosed in long straw-like sheaths and sheath attaches to solid surfaces. Iron and manganese oxides bind to sheath and organic nutrients bind to the metal oxides.
Leptothrix	Grows in flowing water. Cells are inside a sheath and the sheath attaches to sand or other surfaces. Cells acquire nutrients from the flowing stream and sheaths are often encrusted with iron and manganese oxides.

to prevent its transfer to a new aquifer, drill bits are cleaned with strong oxidizing agents. Bacteria which reproduce by budding would also have cells with a surface to volume ratio greater than cells with binary fission.

Several genera of bacteria grow within a tube-like or straw-like sheath. The sheaths provide protection against viruses and also nonspecifically bind nutrients that are present in marine or freshwater environments. Bacteria with sheaths are commonly found in low nutrient environments and some of these organisms are listed in Table 4.11. Sheathed bacteria can be sufficiently numerous to form a visible mat at the source of a spring or nutrient source. *Sphaerotilus natans* can become a nuisance when it grows in the waste stream from paper pulping mills or outlets with high levels of nutrients because the sheaths float on the water and promote the growth of facultative bacteria which convert the water stream to the anoxic state. In this instance, *S. natans* collects mineral precipitates in the sheath and becomes an environmental problem. Examples of bacteria with holdfasts and sheaths are given in Table 4.11.

Microorganisms that grow in nutrient-limiting environments have enhanced nutrient transport systems and unique metabolic processes. For shallow corals and mats located on hot and cold springs, photosynthetic bacteria and chemoautotrophic bacteria fix carbon dioxide with the formation of organic cellular material. The fixation of carbon dioxide is such an important event that bacteria use several different metabolic systems to generate organic carbon (see Figure 4.16). In water, inorganic carbon occurs as CO_2 and HCO_3^- and many autotrophic microorganisms have overcome a slow utilization of CO_2 by developing a system of using HCO_3^- (bicarbonate). Carbon dioxide is converted by to bicarbonate by carbonic anhydrase which is localized in the periplasm of the cell. An active transporter moves bicarbonate across the plasma membrane into the cytoplasm and proteins in the shell of the carboxysome converts it back to carbon dioxide. The enzyme ribulose-1,5-bisphosphate carboxylase/oxygenase (Rubisco) uses carbon dioxide for the production of sugars. When CO_2 concentrations are low, bacteria (such as the marine *Thiomicrospira crunogena*) with carboxysomes have an advantage

- **Calvin–Benson–Bassham cycle**

  CO_2 + Ribulose bisphosphate → Phosphoglycerate

  Cyanobacteria

- **Reverse tricarboxylic acid cycle**

  CO_2 + Succinyl - CoA → α-Ketoglutarate

  CO_2 + α-Ketoglutarate → Isocitrate

  Thermocrinis

  Planctomycetes

- **Wood–Ljungdahl pathway (reductive Acetyl CoA pathway)**

  $2CO_2 + 4H_2$ → Acetic acid + $2H_2O$

  Thermodesulfovibrio

  Thermodesulfobaacteria

  Anaerolineae

  Planctomyces

- **Hydroxypropionate cycle and related cycles**

  HCO_3^- + Propionyl - CoA → Methylmalonyl - CoA

  Chloroflexus

  Roseufkexus

Figure 4.16 Proposed metabolic pathways of CO_2 fixation by bacteria in mats associated with hot springs in Yellowstone National Park. Results are from analysis of metagenomic gene clusters. *Source:* Based on data from Thiel et al. (2017).

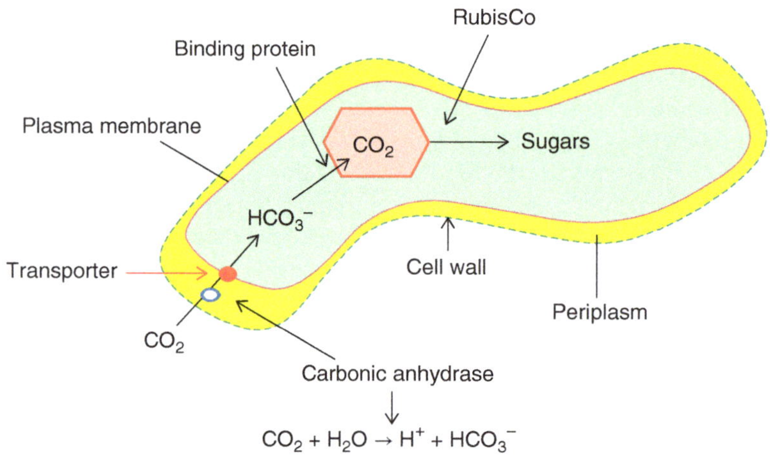

Figure 4.17 The active transport of HCO_3^- into cells is several hundred times faster than diffusion of CO_2 across cell membranes. Thus, bacteria with carboxysomes are capable of growing in low concentrations of CO_2. The formation of a bicarbonate ion from gaseous carbon dioxide is catalyzed by carbonic anhydrase from the periplasm of the gram-negative cells. The active transporter for the bicarbonate ion is energized by ATP.

because HCO_3^- is transported into the cell at a greater rate than CO_s can diffuse into the cell (Figure 4.17). Freshwater cyanobacteria have a high affinity bicarbonate transport system that is distinct form the high affinity bicarbonate transporter in marine cyanobacteria (Price et al. 2004). Many of the bacteria and archaea growing in oligotrophic waters have high affinity transport systems to acquire an array of nutrients in the water at extremely low concentrations. That is, even chemolithotrophic bacteria have systems to assimilate acquire amino acids and small organic molecules from the aquatic environment to enhance their metabolic status.

Reactions producing PP_i

Sulfate + ATP → APS + PP_i

Propionate + ATP + CoA → Propionyl – CoA + AMP + PP_i

Acetate + ATP + CoA → Acetyl – CoA + AMP + PP

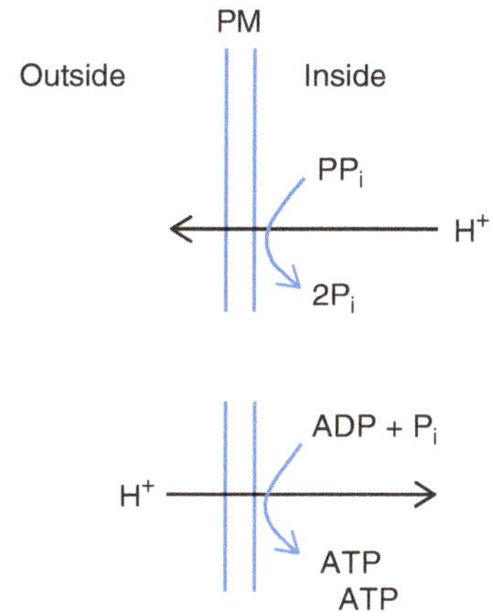

Figure 4.18 The conservation of energy using inorganic pyrophosphate (PP_i) to translocate protons across the plasma membrane. See text. *Source:* Based on data from Kleiner et al. (2012).

Some chemolithotrophic bacteria couple sulfide oxidation to the fixation of carbon dioxide and for deep water corals, sulfide is a product of abiotic conditions (see Section 10, Chapter 3). Heterotrophic and chemolithotrophic bacteria are involved in the metabolism of inorganic compounds of sulfur and nitrogen as part of their cellular energetics. Thus, with an appropriate group of bacteria, the cycling of carbon, nitrogen, sulfur, and hydrogen occurs at these nutrient-limited sites. Recently, it has been shown that some bacteria can use the energy of inorganic pyrophosphate (PP_i) to export protons and therefore energize the plasma membrane (Figure 4.18). Energy conserved by coupling pyrophosphate hydrolysis with proton translocation across plasma membranes has been proposed for *Desulfotomaculum* and genome analysis indicates this conservation of energy in PP_i may be important for *Desulfatibacillum alkenivorans*, *Desulfococcus oleovorans*, *Desulfotomaculum reducens*, and *Candidatus* Desulforudis audaxviator (Kleiner et al. 2012).

In open marine and fresh waters, food chains exist in which bacteria are consumed by grazing protists. The nutrients for this important growth of bacteria in the open water come from fecal material and other products of aquatic life. This reuse (or recycling) of animal wastes by bacteria has been called the "bacterial loop," see Figure 4.19. Due to the high activity of lytic viruses in aquatic environments, there is considerable release of DOM and POM. Heterotrophic bacteria rapidly convert the DOM into cellular mass and this provides more food for grazing organisms. The contribution of bacterial viruses to making nutrients available through bacterial cell lysis is referred to as the "viral loop" see Figure 4.19. Thus, bacteria and their viruses have a significant role in aquatic food webs.

In oligotrophic waters, the macronutrients of carbon (C), nitrogen (N), and sulfur (S) are limiting. Since C, N, and S are not only used for the synthesis of cellular building blocks but may also participate in respiratory activity as either electron donors or electron acceptors. Retaining compounds containing these elements in the immediate environment is critical for microbial populations. Thus, at the microbial community level, organisms with diverse metabolic capabilities transform the C, N, and S compounds into a variety of useful nutrients. While nutrient cycling is apparent in corals, gutless tube worms, and mat communities, at a larger scale the contribution of aquatic systems for the cycling of macronutrients (Figure 4.20) is also important (Canfield and Des Marais 1994; Dillon et al. 2007). Each site has its own specific bacteria that account for nutrient cycling and this becomes a distinctive ecological feature.

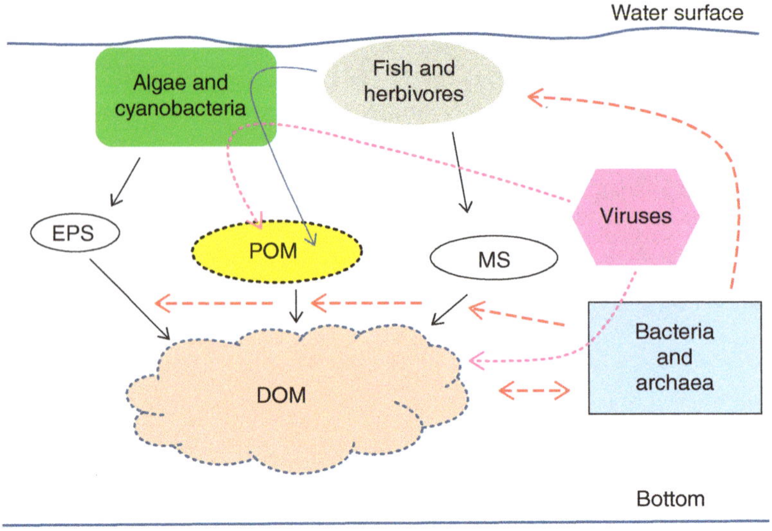

Figure 4.19 A model depicting the bacterial loop and the viral loop in aquatic environments. Bold dash lines = bacterial processes; dotted lines = role of the viruses. As the number of cells increases due to the breakdown of DOM, there is more food for the grazing populations. DOM = dissolved matter, POM = particulate organic matter, MS = marine (aquatic) snow consisting of fecal material, and EPS = Extracellular polysaccharide substances are products of algal/cyanobacterial secretions.

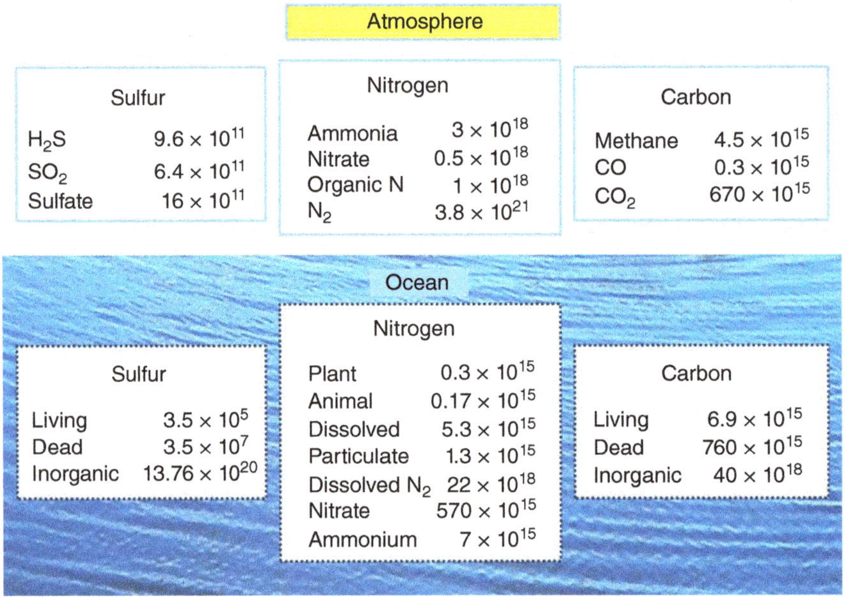

Figure 4.20 Global cycling of nutrients. Values represent total grams present.

4.6 Aquaculture Wastewater

Land-based aquaculture is an important industry to produce food for the increasing world population. Aquaculture is not a new science but was used by indigenous inhabitants of Australia in 6000 BCE to contain eels for human consumption (Turcios and Papenbrock 2014).

Information from 2010–2011 shows that 28.8 million metric tons (mt) of fish, 13.1 million mt of molluscs, and 5 million mt of crustaceans are produced annually from global aquaculture (Casline et al. 2013). About 90% of the seafood consumed in the USA is imported and 50% of this imported seafood (shrimp, catfish, salmon, carp, tilapia, bass, and sturgeon) is raised on farms in India, Indonesia, Ecuador, Vietnam, and Thailand.

Wastewater from land-based systems is treated for the removal of solids, phosphorus, and nitrogen using technologies similar to municipal wastewater systems (Casline et al. 2013). A primary treatment stage removes dense solids <100 μm that quickly settle out producing a sludge which is collected and discarded. Secondary treatment relies on the growth of microalgae and macroalgae, where management of algal biomass is required to optimally remove soluble nitrogen and phosphorus from wastewater. While the use of bacteria would be more efficient for nitrogen and phosphorus removal than using algae, bacterial treatment systems are more expensive and therefore less commonly used. The ideal system for nutrient removal uses bacterial reactors or digesters in the initial phase, cyanobacteria and algae to remove nutrients in the second phase, and a constructed wetland system to complete the wastewater treatment.

4.7 Hormone Degradation in Fresh Water

Humans and other mammals produce steroid sex hormones, with secretion primarily by the urinary tract. Both androgenic and estrogenic hormones (see Figure 4.21) are detected in many aquatic and terrestrial environments and there is a concern that estrogenic hormones in rivers and streams are detrimental to the reproductive systems of aquatic wildlife. In addition to the androgenic hormone, testosterone, natural estrogenic hormones such as estrone (E1), 17β-estradiol (E2), and estriol (E3) are excreted into the environment by humans, livestock, and wildlife (Fan et al. 2007). It is estimated that each pregnant woman secretes about 260 μg/day of E2. In an agricultural setting in the UK, the animal population excretes about four times more estrogenic hormones than humans (Johnson et al. 2006). Estrogenic hormones E1 and E2 are known to be degraded *Sphingomonas* strain KC8, *Rhodococcus zopfii*, *Rhodococcus equi*,

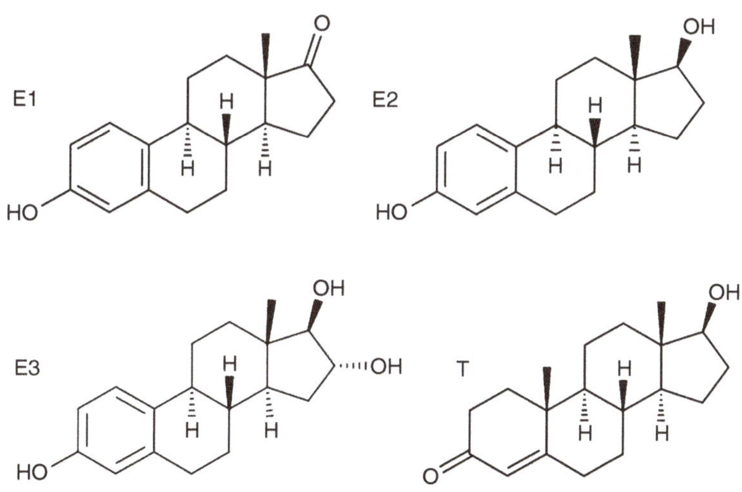

Figure 4.21 Structures of sex hormones found in rivers that are metabolized by bacteria. T = testosterone, E1 = estrone, E2 = 17β-estradiol, and E3 = estriol.

Ralstonia sp., *Nitrosomonas europaea*, *Achromobacter xylosoxidans*, and *Novosphinogobium tardaugens* (Roh and Chu 2010). Testosterone is degraded aerobically by *Comanonas testosteroni* and anaerobically by *Steridobacter denitrificans* (Leu et al. 2011). Steroid estrogens are degraded by mixed cultures of bacteria (Lee and Liu 2002; Bradley et al. 2009) but only incompletely degraded in urban wastewater treatment systems (Fan et al. 2007), so discharge waters containing these hormones contaminate rivers (Williams et al. 2003). Through optimizing management, it is anticipated that sex steroids can be decomposed in wastewater treatment facilities (Johnson et al. 2007).

4.8 Human Activities and Influence on Microbial Ecology

Fresh water represents an important need for humans and human activities including agriculture, industry, and transportation. Water used for drinking is removed from surface water, purified by filtration, and often disinfected by chlorination (Edzwald 2010). In some cases, rivers are dammed, which will certainly disrupt flow and mitigate flooding events. There are several instances where rivers have completely disappeared due to water being diverted for human use. As well, climate change has altered the flow and even existence of some rivers and lakes.

The introduction of wastewater into an environment has a major influence on its microbial ecology. A multi-interdisciplinary research field has developed over the past 20 years to address the impact of urban wastewater on lake and river water quality (Fisher et al. 2015).

There are several physical, chemical, and biological issues that relate to wastewater discharge. The presence of organic matter within wastewater stimulates the growth of heterotrophic bacteria and other microorganisms, which invariably leads to oxygen depletion within the water. In wastewater testing, two common tests are Biochemical Oxygen Demand (BOD), and Chemical Oxygen Demand (COD) which are performed to evaluate the potential oxygen loss by organic matter (Rice et al. 2012). BOD and COD tests are often used in water quality regulation so these tests are standardized. For example, a company that discharges wastewater to a municipal sewage treatment facility is charged on the basis of the sewage volume and the BOD content.

BOD is measured by taking a water sample and diluting it in a highly oxygenated buffer solution. The sample is then incubated for five days at 20 °C in the dark (to avoid photosynthesis). Dissolved oxygen measurements are taken before and after incubation and the BOD_5 is calculated as follows:

$$BOD_5 = (D_0 - D_5)/\text{dilution factor}$$

D_0 and D_5 represent the dissolved oxygen before and after five days' incubation and the dilution factor indicates how the original sample was diluted before testing.

Standard Methods describes variations to the traditional BOD test for samples that may have excessively high organic matter or other unusual characteristics (Rice et al. 2012). While some biological materials take longer than five days to decompose, various studies have shown that most oxygen consumption occurs during the initial five days, so this time and a temperature of 20 °C were chosen to standardize the procedure.

COD tests determine the quantity of oxygen required to oxidize the organic matter in wastewater sample using a chemical reagent. While the COD test can be performed more quickly than the standard five-day BOD test, some organic materials are not readily biodegradable but are amenable to chemical oxidation.

4.9 Drinking Water

Water is essential for human health and for various industrial and agricultural activities. With growing populations and climate change, accessibility to reliable and safe water resources is of increasing economic and political concern, as well as an environmental concern. The most common sources of water for human use are surface water (obtained from lakes and rivers) and ground water from aquifers. Other sources include rainwater collection (presently used at a small scale in localized houses and businesses), and seawater (which is increasingly being used in some countries in the Middle East). Seawater requires salt removal (desalination), which is typically done by reverse osmosis. Here, seawater is pressurized and passed through a selectively permeable membrane in the opposite direction to normal, which results in the dissolved ions (Na^+, Cl^-, and other ions) being blocked by the membrane and fresh water passing through. This process works quite well, although energy is required to generate the pressure. From a microbiology perspective, reverse osmosis causes biofilm growth and associated clogging (typically involving halophiles) on the membranes. Large-scale reverse osmosis facilities also generate large volumes of highly saline wastewater, referred to as brine (typically $>35\,g\,l^{-1}$, or 10× the salt content of seawater) that must be addressed.

The chemical and biological safety of potable water is a major concern. Gastrointestinal diseases are typically water-borne and are spread via fecal-contaminated water and food. They are very common and account for considerable morbidity and mortality in the developing world. As of May 2017, the World Health Organization estimates 1.7 billion cases of childhood diarrhea cases occur per year, resulting in the death of 525 000 children under the age of five. In the developed world, particularly in regions with good water-treatment facilities; water-borne diseases are rare. However, when natural or human-caused disasters (flooding, hurricanes, earthquakes, wars, etc.) disrupt infrastructure, water safety can readily emerge as a major public health problem. Examples of water-borne diseases include cholera, dysentery, diarrhea, and typhoid fever. Common pathogens associated with these diseases include a variety of enteric bacteria (diarrhea-causing bacteria including *E. coli* strains, *Salmonella* spp., *Vibrio* spp., *Campylobacter* spp., and *Shigella* spp.); protozoans including *Giardia* spp., and *Cryptosporidium* spp.; viruses including Enteroviruses and Noroviruses; and various parasites. Gastrointestinal pathogens enter water systems from improperly treated wastewater or alternatively from runoff from agricultural fields. Birds and wild animals can also contribute gastrointestinal pathogens. Two terms that are often used to describe the source of pollution are the phrases "point-source" and "non-point-source."

4.10 Municipal Water Treatment

Municipal water treatment involves several physical and chemical protocols to purify water and make it safe for human use. Physical barriers are used to prevent tree branches, leaves, trash, and other larger objects getting into the pumps and other machinery in the water treatment facility. Fine suspended materials are mixed with a chemical flocculant; aluminum sulfate (alum) is a common flocculant. When mixed with water aluminum sulfate reacts to form flocs of insoluble aluminum hydroxide, and this material adsorbs very small organic and inorganic particulates. The flocs are then allowed to settle in a sedimentation tank. The clear water is then passed through a sand filter to remove remaining flocs. Activated charcoal is sometimes used to remove soluble organic compounds that might affect taste.

The final stage in water purification is disinfection, which is a necessary to control pathogens. The most common disinfectants used are chlorine or chlorine-containing compounds, including sodium hypochlorite (bleach). Ozone (O_3) and ultraviolet light are used as alternative disinfectants. Organic compounds will react with and neutralize chlorine. The amount of chlorine added is calculated based on water chemistry; however, the practice is to ensure that a safe level of chlorine (defined by the World Health Organization as $\geq 0.5\,mg\,l^{-1} = 0.5\,ppm$) is present as the water leaves the treatment plant and goes into the water distribution system. The residual chlorine present in the distribution system is to control normal levels of microbial contamination that may enter the distribution pipes for drinking water.

4.11 Wastewater Treatment Systems

Household and municipal wastewater typically contains a mixture of human excrement (urine and feces) and other organic matter from cooking, laundry, personal hygiene, cleaning activities, and so on. Wastewater treatment is important for environmental protection (preventing eutrophication and associated oxygen depletion) and public health (inhibiting the spread of gastrointestinal pathogens into the environment). Microorganisms have an important role in wastewater treatment to mineralize organic soluble and particulate matter to carbon dioxide, inorganic nitrogen compounds, sulfate, and phosphate.

4.11.1 Septic Tanks

Septic tanks for rural households are essentially a large holding tank, often containing 4000–7500 l (~1000–2000 US gallons), coupled to a drainage field in which treated wastewater exits the septic tank and enters groundwater via porous pipes that are surrounded by gravel or less commonly sand. Larger septic tank systems are used for rural schools, commercial facilities, and other institutions that generate considerably more wastewater. In an average septic system, approximately 60% of the BOD reduction occurs via anaerobic microbial activity in the septic tank itself. The majority of the rest of the BOD reduction (almost 40% of the original BOD) occurs when the tank effluent enters the soil absorption field trenches in the drainage field, where some aerobic conditions are present (Lee et al. 2006). Septic tanks work well in most average household systems, although they do require maintenance, which involves periodic pumping of the tank to remove undigested sludge.

4.11.2 Municipal Wastewater Treatment

Some cities still permit raw sewage to be pumped directly into surface water but most communities treat wastewater to reduce BOD and to keep surface waters clean for public health as well as recreational and commercial uses. Raw sewage is pumped into a sewage treatment facility. In some cases, water from rainfall and snow melting (in storm sewers) is added. In other cases, storm sewer runoff is not treated prior to entering surface water. Although it can be advantageous to treat precipitation runoff to remove harmful materials that may enter storm sewers, the large volumes of water that may result from heavy precipitation, can overwhelm a municipal wastewater facility causeing the loss of the microbial inoculum in activated sludge (see later explanation). In some cases, this results in the temporary release of partially treated or untreated wastewater.

In most municipal wastewater treatment facilities, there are two broad categories of treatment designated as primary and secondary treatment (Figure 4.22). In some instances, tertiary treatment is also present. An estimated 40% of the original BOD in raw sewage is removed by primary treatment. Secondary treatment, which is microbiology based, removes an additional

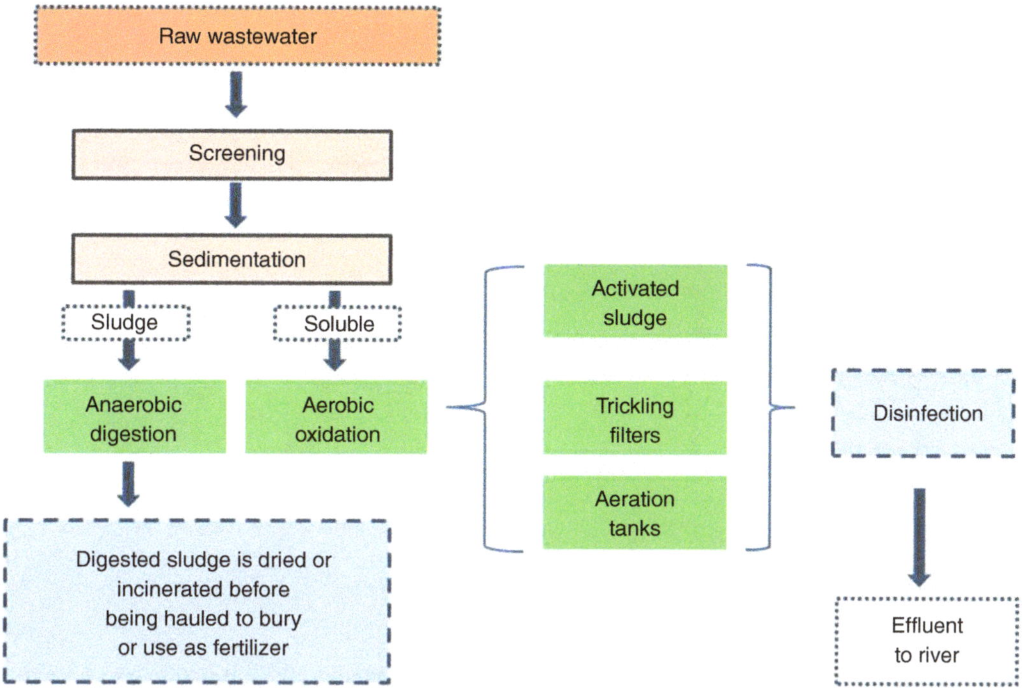

Figure 4.22 Systems generally used for processing municipal wastewater. Boxes: solid border = primary treatment; no border = secondary treatment; dashed border = tertiary treatment.

50% of the original BOD. One major concern with municipal wastewater treatment is the potential for harmful chemicals to enter the system. Such chemicals may inhibit bacteria, which are necessary for secondary treatment (explained below), or their toxicity may be unaffected by wastewater treatment and they may enter the environment. In cases where toxic chemicals have been identified, wastewater may be pretreated to remove or inactivate the toxic chemicals before conventional treatment.

4.11.2.1 Primary Treatment

Primary treatment involves mainly physical approaches to remove wastes. Incoming wastewater is passed through one or more screens that may contain metal bars spaced 1–3 in. (2–8 cm) apart to remove larger items (newspapers, bottles, and other garbage). These items are then removed and taken to landfill. The water then enters a large settling tank, where items will either sink to the bottom or rise to the surface. These items can then be removed and treated in digesters (see secondary treatment explanation).

4.11.2.2 Secondary Treatment

Secondary treatment uses enhanced microbial decomposition processes to remove the BOD. Activated sludge is the most commonly used aerobic approach. Here, wastewater continuously enters large tanks, which are continuously aerated, and remains there for an average of six hours. During this time, complex yet stable microbial communities metabolize the organic carbon and reduce the BOD. Recent metagenomic investigations of the microbial composition of several activated sludge communities (Ibarbalz et al. 2016) showed that 13 microbial phyla were commonly present in activated sludge. This included gram-positive *Actinobacteria* and

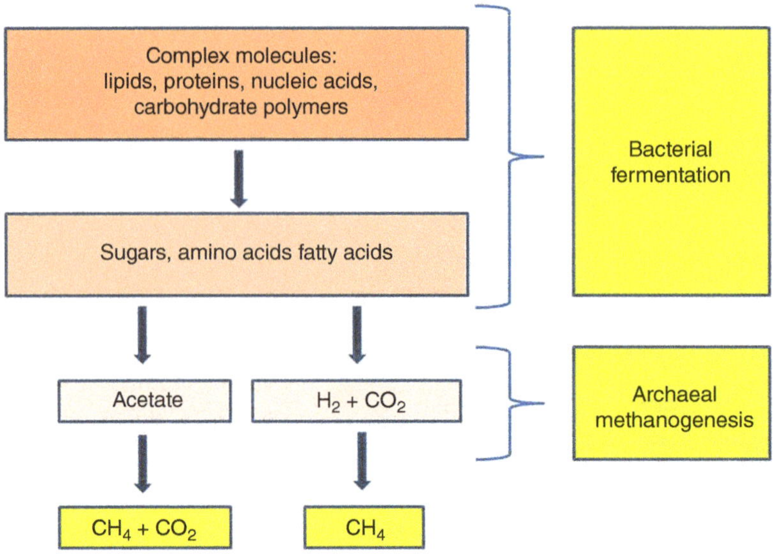

Figure 4.23 Metabolic activities of microorganisms in anaerobic digesters.

Firmicutes; gram-negative *Proteobacteria* and *Nitrospirae*; and the archaea phylum, *Euryarchaeota*. The microbes in activated sludge are thought to originate from wastewater microorganisms. Following activated sludge treatment, the treated water moves to a settling tank, where the sludge is removed by gravity. Some of the activated sludge is recycled back to the inlet region of the activated sludge tank to serve as an inoculum. The excess sludge, along with some solids from the primary treatment, is pumped to an anaerobic sludge digester. The purified water is then released into surface water or else treated further by tertiary treatment. Sometimes treated wastewater effluent may be chlorinated to reduce harmful bacteria.

Anaerobic sludge digestion occurs in closed tanks in which activated sludge and other particulates are degraded by anaerobic bacteria and archaea (Figure 4.23). The anaerobic microbial consortium associated with anaerobic sludge digestion will produce a "biogas" mixture containing methane (CH_4) along with other gases (primarily CO_2 and water vapor). Interestingly, anaerobic sludge digestion is much more efficient at moderately thermophilic temperatures, typically above 33 °C. The methane-containing biogas is often used to heat the sludge digesters and enhance the wastewater treatment process. Following anaerobic digestion, the digested sludge typically has very little odor and can be processed to remove water and then taken to a landfill, or can sometimes be used as an organic fertilizer in agricultural settings. Water removed from digested sludge is pumped into the activated sludge tanks and processed. Tertiary treatments are various specialized additional treatments for the removal of specific chemicals (such as excess nitrogen or phosphate) or other chemicals.

4.12 Alternative Approaches for Wastewater Treatment

Several other processes for treating wastewater have been proposed, which often involve microbial ecology. For example, in artificial wetlands or sewage lagoons, wastewater is pumped into a shallow pond and decomposed by natural microbial and biological processes. This can represent a low-cost option in remote communities where inexpensive land is readily available.

These ponds are sometimes designed to be wetlands (i.e. swamps) and may contain plants or fish. Another approach used in secondary treatment involves a trickling filter. Water is sprayed over a layer of rocks and biofilms that grow on the rocks remove organic matter (and reduce the BOD). This is sometimes used in lieu of activated sludge treatment.

4.13 Coliforms and Other Indicator Organisms

A major public health concern with wastewater is the presence of potential pathogenic bacteria from human and animal fecal wastes. Pathogenic bacteria in drinking water have been extensively reviewed and while the levels of pathogens are low in relation to other intestinal flora, indicator organisms are commonly measured (Cabral 2010). Details of these tests are in Standard Methods for the Examination of Water and Wastewater (Rice et al. 2012) and other public health references. Testing for indicator organisms is often mandated by the US Environmental Protection Agency, the US Food and Drug Administration, and other Public Health Agencies.

The most common group of indicator organisms are coliforms and their use in public health originated with the work of Theobald Smith (a public health microbiologist in New York State) in 1891, while lab detection was later further refined by a number of other microbiologists. Coliforms are defined as gram-negative, non-spore-forming bacilli (rod-shaped bacteria), capable of fermenting lactose with the production of acid and gas when incubated at 37 °C. These organisms are typically found in the intestinal tract of humans and warm-blooded animals. Common genera associated with coliforms include: *Escherichia* (i.e. *E. coli*), *Enterobacter*, *Citrobacter*, *Hafnia*, and *Klebsiella*. In common water testing protocols, coliforms are distinguished as fecal (mainly *E. coli*) or non-fecal. Other indicator organisms used are the fecal streptococci (e.g. *Enterococcus faecalis, E. faecium, Streptococcus bovis,* and *Streptococcus equinus*) (National Research Council 2004). Not only are these indicator organisms used to evaluate the safety of drinking and recreational waters, they are also used to test dairy products, fruit juices, and bottled water in the food industry.

The major criteria for using indicator organisms can be summarized as:

- Indicator organisms must be present only in polluted water.
- Indicator organisms must be present in water when pathogens are present.
- The quantity of indicator organisms correlates with the amount of pollution.
- Indicator organisms must survive longer than pathogens.
- Indicator organisms must be present in greater numbers than pathogens.
- Indicator organisms must be relatively easy to detect.

Coliform testing has been used since the early twentieth century and is still considered quite reliable. Lab and testing resources for coliform testing are relatively inexpensive, which allows the widespread use of this technology. However, some issues have arisen with the onset of molecular diagnostics and the improvement of testing protocols. In warmer environments, coliform survival tends to be quite prolonged, particularly if the organisms can grow and persist as a biofilm in sand or other porous media. One public health question related to climate change is the potential for the increased survival of intestinal flora (indicator organisms and pathogens) in the environment. Another concern relates to distinguishing between human and non-human sources of fecal pollution; conventional coliform testing does not allow sources to be distinguished.

Molecular sampling approaches are being used to test for the presence of gastrointestinal flora in wastewater. Realistically, these approaches are beyond the financial and technical capabilities of municipal public health labs, so are not widely used. In one study, McLellan et al. (2010) conducted a detailed metagenomics-based study on wastewater as it passed through a treatment plant in Milwaukee and then entered Lake Michigan. This study and subsequent studies found that the order *Bacteroidales* (which includes *Bacteroides* spp.) is commonly found in the intestinal tract of humans and many animals, has survival characteristics similar to many pathogens, and thus represents a candidate molecular indicator organism. There is also increasing data showing that some species and strains within *Bacteroidales* are host-specific, which enables the contribution of different fecal sources to be monitored. Members of *Bacteroidales* are anaerobic, so culture-based methods are challenging for many labs. Although there are challenges with conventional coliform testing, the testing protocol is considered to be reliable by public health authorities and so its use will continue.

4.14 Viruses in Aquatic Environments: Diversity and Activity

An important component of biodiversity in any environment is virus diversity and study of viruses has developed into a subset of the biosphere known as the viriosphere (Anderson et al. 2013). In marine ecosystems, estimates of viral diversity are up to 15-fold that of bacteria and archaea (Suttle 2007). The estimated ratio of virus abundance to bacteria abundance varies from <5 : 1 in some freshwater lakes to 10 : 1 in the Arctic Ocean, 40 : 1 in the Pacific, and about 100 : 1 in deeper ocean waters. Due to the size of the oceans and the numbers of bacteria and bacterial viruses in the oceans, it has been estimated that there are 10^{23} bacteria infected by viruses every second. Viruses require a living host to reproduce, so the diversity of the viruses reflects the diversity of other life forms. In that context, the marine viruses in greatest abundance are bacterial viruses which attack cyanobacteria, heterotrophs, and chemolithotrophs. Other microbe-targeting viruses are active against archaea, algae, and invertebrates (Jacquet et al. 2010).

Viruses have been detected in every aquatic environment sampled and range in size from 20 to 800 nm. While most viruses have 100 kilobases (kb), the small Sputnik virus has only 18 kb and the large mimivirus has 1185 kb. The mimivirus attacks amoeba and the host for the Sputnik virus is the mimivirus (Anderson et al. 2013). The double stranded DNA (dsDNA) viruses are the best studied but single stranded DNA (ssDNA) viruses are a part of the aquatic viral community. The virus community in the Arctic and Antarctic waters contains ssDNA and these viruses are remarkable because they are not related to any known ssRNA viruses (de Cárcer et al. 2015). RNA viruses are found in aquatic environments and are either ssRNA or dsRNA; most of the genomes are about 2 kb. Bacterial and archaeal viruses are visible using electron microscopy. Some viruses that attack archaeal cells are shaped like a lemon with a short tail at each end (Suttle 2007).

Bacterial viruses have various shapes but the most common are filamentous or with a head and a tail. In marine environments, there are three families of bacterioviruses: the *Podoviridae* have short non-contractile tails and infect a narrow range of hosts within the same species; the *Myoviridae* have contractile tails; and the *Siphoviridae* have long non-contractile tails. Viruses from the *Myoviridae and Siphoviridae* families have a broad host range and are responsible for interspecies gene exchanges (Anderson et al. 2013).

In general, virus distribution is restricted to the region where the bacteria or archaea are found but if there are currents mixing the waters, microbial viruses may be found in regions where the host has not been observed. It has been estimated that 20–40% of prokaryotic cells in the marine environment are lysed each day by viruses and this rapid conversion of bacterial biomass into DOM is available as nutrients for other bacteria. This process of controlling (killing) bacterial cells has been called a bottom up control and in terms of using organic material in the ocean it has been called the "viral shunt" (Suttle 2007). Bacterioviruses alter the population structure by lysing bacteria in a top down process, that is, in deep marine environments where grazing predators do not control bacterial numbers, the killing of bacterial cells is attributed to bacterioviruses. As bacterial cells become dominant in a specific environment, viruses that are capable of attacking the species of the most abundant bacteria will "kill the winner" (Anderson et al. 2013).

The viral contributions to microbial ecology are just beginning to be investigated. For example, red tide is associated with an overgrowth of red-pigmented phytoplankton. There is some evidence that marine viruses, which target the causative organisms, may be responsible for the sudden decline of these blooms. In other studies, the viral lysis of bacteria and the associated generation of particulate matter from viral host-cell debris accounts for the "viral shunt" which supplements the bacterial shunt in marine environments. Also, it has been proposed that viruses contribute to carbonate deposition in the open ocean. Viruses have been detected by microscopic procedures at depths of over 300 m below the seafloor and in subsurface sediments of the South Pacific Gyre, which is estimated to be ~14 Ma old (Engelhardt et al. 2014).

Aside from lytic infections, viruses have been associated with genetic exchange (transduction) as well as lysogeny, where the viral DNA incorporates into the host cell chromosome; viral DNA inserted into bacterial DNA is called a provirus. Archaea and intracellular pathogens generally do not harbor proviruses. About 40% of bacterial isolates from aquatic environments contain proviruses (Jacquet et al. 2010). As these bacteria become exposed to UV from sunlight, the provirus is released from the bacterial chromosome leading to the lytic phase. UV is also detrimental to viruses as it causes formation of thymine–thymine (T–T) dimers which damage DNA. It is likely that viral genomes with a high AT content are highly sensitive to UV exposure. Additionally, ssDNA and RNA viruses are less sensitive to UV damage than dsDNA viruses. DNA repair systems have long been known in bacteria such as *E. coli* and similar repair systems seem to be present in marine microorganisms (Mojica and Brussaard 2014).

Viruses associated with human and animal diseases have been studied for decades, but little is known about the roles and impact of environmentally important viruses, including those associated with aquatic environments. Freshwater ecosystems are known to harbor a variety of RNA viruses that produce serious illnesses in humans. Scientists have found RNA viruses from over 30 viral families in a Maryland lake (Djikeng et al. 2009). Since freshwater lakes contain the larval stages of numerous insects, these observations are extremely important and may explain emerging diseases and the persistence of certain viruses, including the virus thought to be responsible for the destruction of honey bee colonies.

In geothermal springs, control of prokaryotic cell populations is partly due to viruses and grazing protozoans; however, at temperatures of 70 °C grazing protozoans cannot grow while bacterial viruses maintain their antibacterial activity. Electron microscopy used to observe prokaryotic viruses in marine settings has shown that primary productivity in thermal springs is less than in oceans, so in most viral studies host cells are added to enrich the presence of viruses (Rice et al. 2001). Evaluations based on viral metagenomics indicated that archaeal viruses are more abundant than phages, Table 4.12 (Schoenfeld et al. 2008). Lytic viruses attacking prokaryotic cells release organic materials, which contributes to carbon cycling in hot springs. If the quantity of thermal waters around the world is 10^{19} l, the quantity of prokaryotic cells could approach 3.7×10^{29} and the number of viruses would be about 3.7×10^{30} (Breitbart et al. 2004). With abundance at these levels, the number of prokaryote cells and viruses in thermal springs and aquifers equals the number in the oceans.

Table 4.12 Classification of viruses in Bear Paw and Octopus thermal springs in Yellowstone National Park.

	Percentage of viral metagenomic contigs	
Viruses	**Bear Paw**	**Octopus**
Eukaryotic viruses	0.1%	4.2%
Archaeal viruses	4.7%	3.5%
Bacteriophages	0.4%	0.1%

Source: Based on data from Schoenfeld et al. (2008).

4.15 Summary

Microorganisms have a water-based metabolism and the presence of nutrients in water provides an environment for the growth of bacteria, archaea, and numerous eukaryotic microorganisms. Both marine and freshwater environments are discussed, with an emphasis on human activities associated with fresh water. In surface waters, cyanobacterial productivity is due to *Prochlorococcus* which uses high affinity transporters to acquire organic nitrogen compounds from water and *Synechococcus* which fixes nitrogen. Green algae and diatoms are frequently dominant in mesotrophic and oligotrophic environments. In summer, marine primary productivity is often due to *Roseobacter* and photoheterotrophs whereas later in the year, chemolithotrophs are present energized by oxidation of ammonia and sulfide. One of the most abundant heterotrophic bacteria in the ocean is *Candidatus* Pelagibacter ubique. Bacteria form endosymbiotic and exosymbiotic relationships with numerous marine invertebrates including gutless worms, clams, and corals. Core microbial populations of symbiotic microorganisms have sufficient diversity to recycle nutrients and maintain the host for the symbiont even when the environment is nutrient limited. Bacteria have an important role in efficient utilization of animal and plant products that make up DOM and particulate matter including marine snow; this process is referred to as the bacterial loop. Bacterial viruses cause significant mortality in bacteria, which releases organic compounds into the water that enable other bacteria to grow; this activity is called the viral loop. Thus, bacterial and viral loops have important roles in the marine food web.

Lakes, rivers, and wetlands waters contain considerable numbers of bacteria but a greater abundance of these microorganisms is present in the sediment. Lakes above the tree line are most likely pristine (oligotrophic) whereas lakes at lower altitudes may become mesotrophic or eutrophic due to influx of nitrogen and phosphorous. In spring, algal growth is dominant in lake waters and with the depletion of nitrogen, blooms of nitrogen-fixing cyanobacteria form. There are three zones of microbial activity in long rivers: diverse taxa of microorganism are carried into the river from points of origin; these are replaced downriver by the core river taxa that metabolize particulate organic matter in the river; and finally, river plumes extending into the oceans are enriched with microorganisms that are salt tolerant. Mats at the surface of hot and cold springs contain phototrophs at the surface which account for primary productivity and the anaerobic layer contains chemolithotrophic bacteria and archaea which promote nutrient cycling. Populations are controlled by bacterial and archaeal viruses because grazing protozoa are not present in hot springs.

Public health issues are of considerable importance for lake beaches and for wastewater management. Cities, companies, and individual family units generating wastewater use microorganisms to mineralize organic compounds to inorganic salts and destroy disease-producing bacteria and animal viruses before water streams enter the environment. Urban water from the streets and surrounding areas also needs to be treated.

Discussion Questions

1. Often suggestions are made in the literature that a "core microbial community" is present in a specific aquatic environment. What experimental procedures would you suggest to verify that a specific environment has a "core microbial community"?
2. The water from city streets is not treated prior to being drained into regional rivers. If *E. coli* is found in this urban water, how could it be determined if this bacteria was of human or mammal (dogs, raccoons, etc.) origin?
3. Consider the statement: "Lysogenic viruses are more prevalent than lytic viruses in oligotrophic waters." Do you agree or disagree with the statement? Defend your position.
4. What is the source of bacteria, archaea, and viruses in deep subsurface waters?
5. Discuss some environments where primary productivity in aquatic systems is *not* attributed to photosynthetic processes.
6. Acidic rivers occur in various parts of the world. Discuss what type of epibiotic bacteria would be on fish and other acid-tolerant animals in the river and how would you propose to make the necessary tests to confirm your hypothesis.

References

Abyzov, S.S., Philippova, S.N., and Kuznetsov, V.D. (1983). *Nocardiopsis antarcticus* – a new species of actinomyces, isolated from the ice sheet of the central Antartic glacier. *Izv. Akad. Nauk. USSR Ser. Biol.* 4: 559–568.

Anderson, R.E., Brazelton, W.J., and Baross, J.A. (2013). The deep viriosphere: assessing the viral impact on microbial community dynamics in the deep subsurface. *Rev. Mineral. Geochem.* 75: 649–675.

Antibus, D.E., Leff, L.G., Hall, B.L. et al. (2012). Cultivable bacteria from ancient algal mats from the McMurdo Dry Valleys, Antarctica. *Extremophiles* 16: 105–114.

Arnosti, C., Ziervogel, K., Yang, T. et al. (2015). Oil-derived marine aggregates – hot spots of polysaccharide degradation by specialized bacterial communities. *Deep Sea Res. II* 129: 179–186. https://doi.org/10.1016/j.dsr2.2014. 12.008.

Atlas, R.M. and Bartha, R. (1997). *Microbial Ecology: Fundamentals and Applications*, 4e. Menlo Park, CA: Benjamin Cummings Publishing.

Baker BJ, Saw JH, Lind AE, et al. (2016). Genomic inference of the metabolism of cosmopolitan subsurface archaea, *Hadesarchaea, Nat. Microbiol.* 1, 16002 (2016) https://doi.org/10.1038/nmicrobiol.2016.2

Barton, L.L. (2005). *Structural and Functional Relationships in Prokaryotes*. New York: Springer Science + Business Media. Inc.

Belkova, N.L., Tazaki, K., Zakharova, J.R. et al. (2007). Activity of bacteria in water of hot springs from southern and central Kamchatskaya geothermal provinces, Kamchatka Peninsula, Russia. *Microbiol. Res.* 162: 99–197.

Bertilsson, S., Burgin, A., Carey, C.C. et al. (2013). The under-ice microbiome of seasonally frozen lakes. *Limnol. Oceanogr.* 58: 1998–2012.

Bochdanskya, A.B., van Aken, H.M., and Herndl, G.J. (2010). Role of macroscopic particles in deep-sea oxygen consumption. *Proc. Nat. Acad. Sci. U.S.A.* 107: 8287–8291.

Bolhuis, H., Cretoiu, M.S., and Stal, L.J. (2014). Molecular ecology of microbial mats. *FEMS Microbiol. Ecol.* 90: 335–350.

Bowen De León, K., Gerlach, R., Peyton, B.M. et al. (2013). Archaeal and bacterial communities in three alkaline hot springs in Heart Lake Geyser Basin, Yellowstone National Park. *Front. Microbiol.* 4: 330. https://doi.org/10.3389/fmicb.2013.00330.

Boyd, E.S., Jackson, R.A., Encarnacion, G. et al. (2007). Isolation, characterization and ecology of sulfur-respiring Crenarchaea inhabiting acid-sulfate-chloride-containing geothermal springs in Yellowstone National Park. *Appl. Environ. Microbiol.* 73: 6669–6677.

Bradley, P.M., Barber, L.B., Chapelle, F.H. et al. (2009). Biodegradation of 17β-estradiol, estrone and testosterone in stream sediments. *Environ. Sci. Technol.* 43: 1902–1910.

Breitbart, M., Wegley, L., Leeds, S. et al. (2004). Phage community dynamics in hot springs. *Appl. Environ. Microbiol.* 70: 1633–1640.

Brown, B.E. and Bythell, J.C. (2005). Perspectives on mucous secretions in reef corals. *Marine Ecol. Prog. Series* 296: 291–309.

Busch, K., Endres, S., Iversen, M.H. et al. (2017). Bacterial colonization and vertical distribution of marine gel particles (TEP and CSP) in the Artic Fram Strait. *Front. Mar. Sci.* 4: 166. https://doi.org/10.3389/fmars.2017.00166.

Cabral, J.P.S. (2010). Water microbiology, bacterial pathogens and water. *Int. J. Environ. Res. Public Health* 7: 3657–3703.

Cai L, Zhou G, Tian R-M (2017). Metagenomic analysis reveals a green sulfur bacterium as a potential coral symbiont, *Sci. Rep.* 7: 9320(2017) https://doi.org/10.1038/s41598-017-09032-4

Canfield, D.E. and Des Marais, D.J. (1994). Cycling of carbon, sulfur, oxygen and nutrients in a microbial mat. In: *Microbial Mats*. NATO ASI Series (Series G: Ecological Sciences), vol. 35 (ed. L.J. Stal and P. Caumette), 255–263. Berlin: Springer.

Carini, P., Steindler, L., Beszteri, S. et al. (2011). Nutrient requirements for growth of the extreme oligotroph '*Candidatus* Pelagibacter ubique' HTCC1062 on a defined medium. *ISME J.* 7: 592–602.

Casline, S.A., McKinnon, A.D., Paul, N.A. et al. (2013). Wastewater treatment for land-based aquaculture: improvements and value-adding alternatives in model systems from Australia. *Aquacult. Environ. Interact.* 4: 285–300.

Chaturvedi, P., Reddy, G.S.N., and Shivaji, S. (2005). *Dyadobacter hamtensis* sp. nov. from Hamta glacier, located in the Himalayas, India. *Int. J. Syst. Evol. Microbiol.* 55: 2113–2117.

Chaturvedi, P. and Shivaji, S. (2006). *Exiguobacterium indicum* sp. nov. a psychrophilic bacterium from the Hamta glacier of the Himalayan mountain ranges of India. *Int. J. Syst. Evol. Microbiol.* 56: 2765–2770.

Chaudhary, A., Haack, S.K., Duris, J.W. et al. (2009). Bacterial and archaeal phylogenetic diversity of a cold sulfur-rich spring on the shoreline of Lake Erie, Michigan. *Appl. Environ. Microbiol.* 75: 5025–5036.

Christner, B.C., Moslely-Thompson, E., Thomson, L.G. et al. (2001). Isolation of bacteria and 16S rDNAs from Lake Vostok accretion ice. *Environ. Microbiol.* 3: 570–577.

Christner, B.C., Moslely-Thompson, E., Thomson, L.G. et al. (2003). Bacterial recovery from ancient glacial ice. *Environ. Microbiol.* 5: 433–436.

Cole, G.A. and Weihe, P.E. (2016). *Textbook of Limnology*, 5e. Long Grove, Ilinois: Waveland Press.

Dattagupta, S., Schaperdoth, I., Montanari, A. et al. (2009). A novel symbiosis between chemoautotrophic bacteria and a freshwater cave amphipod. *ISME J.* 3: 935–943.

de Beer, D. and Kühl, M. (2001). Interfacial microbial mats and biofilms. In: *The Benthic Boundary Layer* (ed. B.D. Boudreau and B.B. Jørgensen), 374–394. New York: Oxford Press.

de Cárcer, D.A., López-Bueno, A., Pearce, D.A. et al. (2015). Biodiversity and distribution of polar freshwater DNA viruses. *Sci. Adv.* 1: e1400127.

Decho, A.W. and Gutierrez, T. (2017). Microbial extracellular polymeric substances (ESPs) in ocean systems. *Front. Microbiol.* 8: 922. https://doi.org/10.3389/fmicb.2017.00922.

Defore, A.L., Weinke, A.D., Lindback, M.M. et al. (2016). Year-round measures of planktonic metabolism reveal net autotrophy in surface waters of a Great Lakes estuary. *Aquatic Microb. Ecol.* 77: 139–153.

DeLong, E.F., Franks, D.G., and Alldredge, A.L. (1993). Phylogenetic diversity of aggregate-attached vs. free-living marine bacterial assemblages. *Limnol. Oceanogr.* 38: 924–934.

DeLong, E.F., Preston, C.M., Mincer, T. et al. (2006). Community genomics among stratified microbial assemblages in the ocean's interior. *Science* 311: 496–503.

de Vet, W.W., Dinkla, I.J., Rietveld, L.C. et al. (2011). Biological iron oxidation by *Gallionella* spp. in drinking water production under fully aerated conditions. *Water Res.* 45: 5389–5398.

Dillon, J.G., Fishbain, S., Miller, S.R. et al. (2007). High rates of sulfate reduction in a low-sulfate hot spring microbial mat are driven by a low level of diversity of sulfate-respiring microorganisms. *Appl. Environ. Microbiol.* 73: 5217–5226.

Djikeng, A., Kuzmickas, R., Anderson, N.G. et al. (2009). Metagenomic analysis of RNA viruses in a fresh water lake. *PLoS One* 4 (9): e7264:https://doi.org/10.1371/ journal.pone.0007264.

Doherty, M., Yager, P.L., Moran, M.A. et al. (2017). Bacterial biogeography across the Amazon River-ocean continuum. *Front. Microbiol.* https://doi.org/10.3389/fmicb.2017.00882.

Edzwald, J.K. (2010). *Water Quality and Treatment*, 6. American Water Works Associatione. New York: McGraw-Hill.

Engelhardt, T., Kallmeyer, J., Cypionka, H. et al. (2014). High virus-to-cell ratios indicate ongoing production of viruses in deep subsurface sediments. *ISME J.* 8: 1503–1509.

Estepp, L.R. and Reavie, E.D. (2015). The ecological history of Lake Ontario according to phytoplankton. *J. Great Lakes Res.* 41: 669–687.

Fan, Z., Casey, F.X.M., Hakk, H. et al. (2007). Persistence and fate of 17β-estradiol and testosterone in agricultural soils. *Chemosphere* 67: 886–895.

Ferrantini, F., Fokin, S.I., Modeo, L. et al. (2009). "*Candidatus* Cryptoprodotis polytropus," a novel Rickettsia-like organism in the ciliated protest Pseudomicrothorax dubius (Ciliophora, Nassophorea). *J. Eukaryot. Microbiol.* 56: 119–129.

Ferris, M.J., Sheeham, K.B., Kühl, M. et al. (2005). Algal species and light microenvironment in a low-pH, geothermal microbial mat community. *Appl. Environ. Microbiol.* 71: 7164–7171.

Fisher JC, Newton RJ, Dila DK, et al. (2015). Urban microbial ecology of a freshwater estuary of Lake Michigan, *Elementa (Wash D C)*.; 3:000064: https://doi.org/10.12952/journal.elementa.000064

Flombaum, P., Gallegos, J.L., Gordillo, R.A. et al. (2013). Present and future global distributions of the marine cyanobacteria *Prochlorococcus* and *Synecholoccus*. *Proc. Natl. Acad. Sci. USA*, vol. 110, 9824–9829.

Fontana Jr. AJ (2001). Water Activity's Role in Food Safety and Quality, Food Safety Magazine. February/March 2001.

Fontana, A.J. Jr. (2008). Minimum water activity limits for growth of microorganisms, Appendix D (p. 405). In: *Water Activity in Foods, Fundamentals and Applications* (ed. G.V. Barbosa-Cánovas, A.J. Fontana, S.J. Schmidt and T.P. Labuza). Ames, IA: Blackwell Publishing.

Geesey, G.G., Barkay, T., and King, S. (2016). Microbes in mercury-enriched geothermal springs in western North America. *Sci. Total Environ.* 569–570: 321–331.

Gillan, D.C. and Dubilier, N. (2004). Novel epibiotic *Thiothrix* bacterium on a mane amphipod. *Appl. Environ. Microbiol.* 70: 3772–3775.

Goffredi, S.K., Jones, W.J., Erhlich, H. et al. (2008). Epibiotic bacteria associated with the recently discovered Yeti crab, *Kiwa hirsute. Environ. Microbiol.* 10: 2623–2634.

Goffredi, S.K., Warén, A., Orphan, V.J. et al. (2004). Novel forms of structural integration between microbes and a hydrothermal vent gastropod from the Indian Ocean. *Appl. Environ. Microbiol.* 70: 3082–3090.

Grzymski, J.J., Riesenfeld, C.S., Williams, T.J. et al. (2012). A metagenomic assessment of winter and summer bacterioplankton from Antarctica Peninsula coastal surface waters. *ISME J.* 6: 1901–1915.

Hallberg, K.B. and Johnson, D.B. (2004). Microbiology of a wetland ecosystem constructed to remediate mine drainage from a heavy metal mine. *Sci. Total Environ.* 338: 53–66.

Hofmann, A.F., Soetaert, K., and Middelburg, J.J. (2008). Present nitrogen and carbon dynamics in the Scheldt estuary using a novel 1-D model. *Biogeosci.* 5: 981–1006. http://www.biogeosciences.net/5/981/2008.

Ibarbalz, F.M., Orellana, E., Figuerola, E.L. et al. (2016). Shotgun metagenomic profiles have a high capacity to discriminate samples of activated sludge according to wastewater type. *Appl. Environ. Microbiol.* 82: 5186–5196.

Jackson, C.R., Langner, H.W., Donahoe-Christiansen, J. et al. (2001). Molecular analysis of microbial community structure in an arsenite-oxidizing acidic thermal spring. *Environ. Microbiol.* 3: 532–542.

Jacquet, S., Miki, T., Nobel, R. et al. (2010). Viruses in aquatic ecosystems: important advancements of the last 20 years and prospects for the future in the field of microbial oceanography and limnology. *Adv. Oceanogr. Linmol.* 1: 97–141.

Johnson, A.C., Williams, R.J., and Matthiesen, P. (2006). The potential steroid hormone contribution of farm animals to freshwaters, the United Kingdom as a case study. *Sci. Total Environ.* 362: 166–178.

Johnson, A.C., Williams, R.J., Simpson, P. et al. (2007). What difference might sewage treatment performance make to endocrine disruption in rivers? *Environ. Pollut.* 147: 194–202.

Kambura AK, Mwirichia RK, Kasili RW, et al., (2016). Bacteria and archaea diversity within the hot springs of Lake Magadi and Little Magadi in Kenya, *BMC Microbiol.* 16:136 DOI https://doi.org/10.1186/s12866-016-0748-x

Kellogg, C.A., Goldsmith, D.B., and Gray, M.A. (2017). Biogeographic comparison of *Lophelia*-associated bacterial communities in the Western Atlantic reveals conserved core microbiome. *Front. Microbiol.* 8: 796: https://doi.org/10.3389/fmicb.2017.00795.

Kellogg, C.A., Ross, S.W., and Brooke, S.D. (2016). Bacterial community diversity of the deep-sea octocoral *Paramuricea placomus. PeerJ* 4: e2529: https://doi.org/10.7717/peerj.2529.

Kitchell, J.F., Schindler, D.E., Herwig, B.R. et al. (1999). Nutrient cycling at the landscape scale. The role of diel foraging migrations by geese at the Bosque del Apache National Wildlife Refuge, New Mexico. *Limnol.Oceanogr.* 44: 828–836.

Kleiner, M., Wentrup, C., Lott, C. et al. (2012). Metaproteomics of a gutless marine worm and its symbiotic microbial community reveal unusual pathway for carbon and energy. *Proc. Natl. Acad. Sci. U.S.A.* 109: E1173–E1183.

Kozubal, M., Macur, R.E., Korf, S. et al. (2008). Isolation and description of a novel iron-oxidizing Crenarchaeon from acid geothermal springs in Yellowstone National Park. *Appl. Environ. Microbiol.* 74: 4942–4949.

Kwong, W.K., Engel, P., Koch, H., and Moran, N.A. (2014). Genomics and host specialization of honey bee and bumble bee gut symbionts. *Proc. Natl. Acad. Sci. U.S.A.* 111: 11509–11514.

Lee, H.B. and Liu, D. (2002). Degradation of 17β-estradiol and its metabolites by sewage bacteria. *Water Air Soil Pollut.* 134: 351–366.

Lee B, Jones D, Turco R (2006). Wastewater Biological Oxygen Demand in Septic Systems, Purdue University Agronomy Extension, HENV-14-W.

Leu, Y.-L., Wang, P.-H., Shiao, M.-S. et al. (2011). A novel testosterone catabolic pathway in bacteria. *J. Bacteriol.* 193: 4447–4455.

Li, G., Jiang, H., Hou, W. et al. (2011). Microbial diversity in two cols springs on the Qinghai-Tibetan Plateau. *Geosci. Front.* 3: 317–235.

Lopatina, A., Krylenkov, V., and Severinov, K. (2013). Activity and bacterial diversity of snow around Russian Antarctic stations. *Res. Microbiol.* 164: 949–958.

Malviya, S., Scalco, E., Audic, S. et al. (2016). Insights into global diatom distribution and diversity in the World's ocean. *Proc. Natl. Acad. Sci. U.S.A.* 113: E1516–E1525.

Margesin, R., Sproer, C., Schumann, P., and Schinner, F. (2003). *Pedobacter cryoconitis* sp. nov., a facultative psychrophile from alpine glacier cryoconite. *Int. J. Syst. Evol. Microbiol.* 53: 1291–1296.

Marsh, C.L. and Larsen, D.H. (1952). Characterization of some thermophilic bacteria from the hot springs of Yellowstone National Park. *J. Bacteriol.* 65: 193–197.

Mason OU, Canter EJ, Gillies LF, et al. (2016). Mississippi river plume enriches microbial diversity in the Northern Gulf of Mexico, *Front. Microbiol.*. doi. https://doi.org/10.3389/fmicb.2016.01048

McLellan, S.L., Huse, S.M., Mueller-Spitz, S.R. et al. (2010). Diversity and population structure of sewage-derived microorganisms in wastewater treatment plant influent. *Environ. Microbiol.* 12: 378–392.

Miller-Coleman, R.L., Dodsworth, J.A., Ross, C.A. et al. (2012). *Korarchaeota* diversity, biogeography, and abundance in Yellowstone and Great Basin Hot Springs and ecological niche modeling based on machine learning. *PLoS One* 7 (5): e35964: https://doi.org/10.1371/journal.pone.0035964.

Miteva, V. (2008). Bacteria in snow and glacier ice. In: *Psychrophiles: Principles from Biodiversity to Biotechnology, Heidelberg* (ed. R. Margesin et al.), 31–50. Springer-Verlag.

Mitsch, W.J. and Gosselink, J.G. (2007). *Wetlands*, 4e. New York: Wiley.

Mojica, K.D.A. and Brussaard, C.P.D. (2014). Factors effecting virus dynamics and microbial host-virus interactions in marine environments. *FEMS Microbiol. Ecol.* 89: 495–515.

Moore, C.M., Mills, M.M., Arrigo, K.R. et al. (2013). Processes and patterns of oceanic nutrient limitation. *Nat. Geosci.* 6: 701–710.

Neave, M.J., Rachmawati, R., Xun, L. et al. (2017). Differential specificity between closely related corals and abundant *Endozoicomonas* endosymbionts across global scales. *ISME J.* 11: 186–200.

Newton, R.J. and McLellan, S.L. (2015). A unique assemblage of cosmopolitan freshwater bacteria and higher community diversity differentiate an urbanized estuary from oligotrophic Lake Michigan. *Front. Microbiol.* 6: 1028: https://doi.org/10.3389/fmicb.2015.01028.

Ngai, J.T. and Jefferies, R.L. (2004). Nutrient limitation of plant growth and forage quality in Arctic coastal marshes. *J. Ecol.* 92: 1001–1010.

National Research Council (US) (2004). *Committee on Indicators for Waterborne Pathogens.* Washington, DC: National Academies Press.

Paerl, H., Fulton, R.S., Moisander, P.H. et al. (2001). Harmful freshwater algal blooms, with an emphasis on cyanobacteria. *Sci. World J.* 1: 76–113. https://doi.org/10.1100/tsw.2001.16.

Payne, J.T., Millar, J.J., Jackson, C.R. et al. (2017). Patterns of variation in diversity of the Mississippi river microbiome over 1,300 kilometers. *PLoS One* 12 (3): e0174890. https://doi.org/10.1371/journal.pone.0174890.

Perreault, N.N., Andersen, D.T., Pollard, W.H. et al. (2007). Characterization of the prokaryotes diversity in cold saline perennial springs of the Canadian High Arctic. *Appl. Environ. Microbiol.* 73: 1532–1543.

Petersen, M.J., Ramette, A., Lott, C. et al. (2010). Dual symbiosis of the vent shrimp *Rimicaris exocalata* with filamentous gamma- and epsilonproteobacteria at four Mid-Atlantic Ridge hydrothermal vent fields. *Environ. Microbiol.* 12: 2204–2218.

Price, G.D., Woodger, F.J., Badger, M.R. et al. (2004). Identification of a SulP-type bicarbonate transporter in marine cyanobacteria. *Proc. Natl. Acad. Sci. U.S.A.* 101: 18228–18233.

Raina, J.B., Tapiolas, D., Willis, R.L. et al. (2009). Coral-associated bacteria and their role in the biogeochemical cycling of sulfur. *Appl. Environ. Microbiol.* 75: 3492–3501.

Reed, A.J. and Hicks, R.E. (2011). Microbial ecology of Lake superior *Bacteria* and *Archaea*: an overview. *Aquat. Ecosys. Health Manage.* 14: 386–395.

Reigstad, L.J., Jorgensen, S.L., Lauritzen, S.-E. et al. (2011). Sulfur-oxidizing chemolithotrophic *Proteobacteria* dominate the microbiota in high Arctic thermal springs on Svalbard. *Astrobiology* 11: 665–678.

Reysenbach, A.L., Voytek, M., and Mancinelli, R. (eds.) (2001). *Thermophiles: Biodiversity, Ecology and Evolution*. Boston, MA: Springer.

Rice, E.W., Baird, R.B., Eaton, A.D. et al. (2012). *Standard Methods for the Examination of Water and Wastewater*, 22e. Washington, DC: American Public Health Association, American Water Works Association, Water Environment Federation.

Rice, G., Stedman, K., Snyder, J. et al. (2001). Viruses from extreme thermal environments. *Proc. Natl. Acad. Sci. U.S.A.* 98: 13341–13345.

Roh, H. and Chu, K.-H. (2010). A 17β-estradiol-utilizing bacterium, *Sphingomonas* strain KC8: part 1 – characterization and abundance in wastewater treatment plants. *Environ. Sci. Technol.* 44: 4943–4950.

Rudolph, C., Wanner, G., and Huber, R. (2001). Natural communities of novel archaea and bacteria growing in cold sulfurous springs with a string-of-pearls-like morphology. *Appl. Environ. Microbiol.* 67: 2336–2344.

Ruehland, C. and Dubilier, N. (2010). Gamma- and epsilonproteobacterial ectosymbionts of a shallow-water marine worm are related to deep-sea hydrothermal vent ectosymbionts. *Environ. Microbiol.* 12: 2312–2326.

Santos, R., Fernandes, J., Fernandes, N. et al. (2007). *Mycobacterium parascrofulaceum* in acidic hot springs in Yellowstone National Park. *Appl. Environ. Microbiol.* 73: 5071–5073.

Savio, D., Sinclair, L., Ijaz, U.Z. et al. (2015). Bacterial diversity along a 2600 km river continuum. *Environ. Microbiol.* 17: 4994–5007.

Sayanthan, A., Jashothan, P.T.J., Saravanan, S. et al. (2016). Characterization of iron bacterium *Gallionella ferruginea* isolated from the drinking water of the collector wells in Northern Sri Lanka. *Trop. Plant Res.* 3: 440–448.

Schoenfeld, T., Patterson, M., Richardson, P.M. et al. (2008). Assembly of viral metagenomes from Yellowstone Hot Springs. *Appl. Environ. Microbiol.* 74: 4164–4174.

Segawa, T., Miyamoto, K., Ushida, K. et al. (2005). Seasonal change in bacterial flora and biomass in mountain snow from the Tateyama Mountains, Japan, analyzed by 16S rRNA gene sequencing and real-time PCR. *Appl. Environ. Microbiol.* 71: 123–130.

Sheehan, K.B., Fagg, J.A., Ferris, M.J. et al. (2003). PCR detection and analysis of the free-living amoeba *Naegleria* in hot springs in Yellowstone and Grand Teton National Parks. *Appl. Environ. Microbiol.* 69: 5914–5918.

Sharrar, A.M., Flood, B.E., Bailey, J.V. et al. (2017). Novel large sulfur bacteria in the metagenomes of groundwater-fed chemosynthetic microbial mats in the Lake Huron Basin. *Front. Microbiol.* 8: 791: https://doi.org/10.3389/fmicb.2017.00791.

Shivaji, S., Chatervedi, P., Reddy, G.S.N., and Suresh, K. (2005). *Pedobacter himalayensis* sp. nov. from Hamta glacier located in the Himalayan mountain range of India. *Int. J. Syst. Evol. Microbiol.* 55: 1083–1088.

Sigman, D.M. and Hain, M.P. (2012). The biological productivity of the ocean. *Nat. Edu. Knowl.* 3: 1–16.

Simon, M., Grossart, H.-P., Schweitzer, B. et al. (2002). Microbial ecology of organic aggregates in aquatic ecosystems. *Aquat. Microb. Ecol.* 28: 175–211.

Slonczewski, J.L. and Foster, J.W. (2009). *Microbiology*. New York: Norton & Company, Inc.

Staley, C., Unno, T., Gould, T.J. et al. (2013). Application of Illumina next-generation sequencing to characterize the bacterial community of the upper Mississippi River. *J. Appl. Microbiol.* 115: 1147–1158.

Steindler, L., Schwalbach, M.S., Smith, D.P. et al. (2011). Energy starved *Candidatus* Pelagibacter Ubique substitutes light-mediated ATP production for endogenous carbon respiration. *PLoS One* 6 (5): e19725. https://doi.org/10.1371/journal.pone.0019725.

Sundareschwar, P.V., Morris, J., Koepfler, E. et al. (2003). P limitations of costal ecosystem processes. *Science* 299: 563–565.

Suttle, C.A. (2007). Marine viruses – major players in the global ecosystem. *Nat. Rev. Microbiol.* 5: 801–812.

Suzuki, Y., Suzuki, M., Tsuchida, S. et al. (2009). Molecular investigations of the stalked barnacle *Vulcanolepas osheal* and the epibiotic bacteria from the Brothers Caldera Kernadec Arc, New Zealand. *J. Marine Biol. Assoc. UK* 89: 727–733.

Terashima, M., Umezawa, K., Mori, S. et al. (2017). Microbial community analysis of colored snow from an alpine snowfield in northern Japan reveals the prevalence of *Betaproteovacteria* with snow algae. *Front. Microbiol.* 8: 1481: https://doi.org/10.3389/fmicb.2017.01481.

Thiel, V., Hügler, M., Ward, D.M. et al. (2017). The dark side of the Mushroom Spring microbial mat: life in the shadow of chlorophototrophs. II. metabolic functions of abundant community members predicted from metagenomic analyses. *Front. Microbiol.* 8: 943: https://doi.org/10.3389/fmicb.2017.00943.

Thiel V, Wood JM, Olsen MT, et al. (2016). The dark side of the mushroom spring microbial mat: life in the shadow of chlorophototrophs. I. Microbial diversity based on 16S rRNA gene amplicons and metagenomic sequencing. *Front. Microbiol.* 2016; 7: 919. https://doi.org/10.3389/fmicb.2016.00919

Trujillo, A.P. and Thurman, H.V. (2017). *Essentials of Oceanography*, 12e. Boston, MA: Pearson Education.

Turcios, A.E. and Papenbrock, J. (2014). Sustainable treatment of aquaculture effluents – what can we learn from the past for the future? *Sustainability* 6: 836–856.

Valentine, D.L., Fisher, G.B., Bagby, S.C. et al. (2014). Fallout plume of submerged oil from Deepwater Horizon. *Proc. Natl. Acad. Sci. U.S.A.* 111: 15906–15911.

Vannini, C., Ferrantini, F., and Schleifer, K.-H. (2010). "*Candidatus* Anadelfobacter veles" and *Candidatus* Cryptoprodotis polytropus two new *Rickettsiales* species hosted by the protest ciliate Euplotes harpa (Ciliophora, Spirotrichea). *Appl. Environ. Microbiol.* 76: 4047–4054.

Venter, J.C., Remington, K., Heidelberg, J.F. et al. (2004). Environmental genome shotgun sequencing of the Sargasso Sea. *Science* 304: 66–74.

Williams, R.J., Johnson, A.C., Smith, J.J. et al. (2003). Steroid estrogens profiles along river stretches arising from sewage treatment works discharges. *Environ. Sci. Technol.* 37: 1744–1750.

Yoshida, M., Yoshida, T., Kashima, A. et al. (2008). Ecological dynamics of the toxic bloom-forming cyanobacterium *Microcystis aeruginosa* and its cyanophages in freshwater, *Appl. Environ. Microbiol.* 74: 3269–3273.

Żbikowska, E., Walczak, M., and Krawiec, A. (2013). Distribution of *Legionella pneumophila* bacteria and *Naegleria* and *Hartmannella* amoebae in thermal saline baths used in balneotherapy. *Parasitol. Res.* 112: 77–83.

Zhang, D.C., Wang, H.X., Liu, H.C. et al. (2006). *Flavobacterium glaciei* sp. nov., a psychrophilic bacterium isolated from the China no.1 glacier. *Int. J. Syst. Evol. Microbiol.* 56: 2921–2925.

Zhu, F., Wang, S., and Zhou, P. (2003). *Flavobacterium xinjiangense* sp. nov. and *Flavobacterium omnivorum* sp. nov., novel psychrophiles from China no 1 glacier. *Int. J. Syst. Evol. Microbiol.* 53: 853–857.

Further Reading

Gerdes G (2007). Structures left by modern microbial mats in their host sediments, in Schieber J, Bose PK, Eriksson PG, et al., eds., *Atlas of Microbial Mat Features Preserved within the Classical Rock Record*. New York:Elsevier. pp. 5–38.

Orcutt BN, Sylvan JB, Knab NJ, et al. (2011). Microbial ecology of the dark ocean above, at, and below the seafloor, *Microbiol. Mol. Biol. Rev.* 75:361–422.

5

Life in Extreme Environments

5.1 Overview

The study of life in extreme environments instills a sense of wonder about the boundaries of life. The conditions where some extremophiles live (high temperatures, low pH, anaerobic conditions, low nutrients, etc.) resemble those that may have existed in the young Precambrian Earth, approximately 3–4 billion years ago (3–4 Ga). As such, they provide a link to the evolution and development of life. Extremophiles are also being studied as models for life beyond planet Earth and in astrobiology.

The biosphere is defined as the region or zone of Earth where life exists. As microorganisms can survive and grow in environments inhospitable to other organisms, microbial habitats effectively define the biosphere. Extreme environments may be related to pressure, temperature, water activity, pH, radiation, and nutrient levels. The organisms that live in such environments are called extremophiles (extreme condition-loving). There is certainly considerable scientific interest in the exploration of life at the extremes. Microbiologists and other scientists can investigate general biological concepts related to the limits of life, and possibly the origins of life given the high temperatures and other inhospitable conditions of the prebiotic young Earth. There is at least one scientific journal, *Extremophiles*, specifically devoted to this field of research. A second application is that extreme environments are used as model systems for exploring life outside Earth, in the developing field of astrobiology. Finally, there can be commercial applications of microorganisms in extreme environments. Thomas Brock, a notable microbial ecologist, studied thermophilic bacteria in a variety of environments including the hot springs of Yellowstone National Park (YNP) (USA). One of Brock's isolates from Yellowstone was the thermophile *Thermus aquaticus.* The DNA polymerase from this organism (referred to as Taq polymerase) is now a key component for the artificial DNA replication technique, the Polymerase Chain Reaction (PCR, Nobel Prize in 1993 to Kary Mullis).

Although some extremophiles have been cultured, most of our knowledge of their physiology and identity has been obtained through molecular approaches. Initially, molecular approaches to characterize extremophiles included cloning, expression of rare genes in another organism, and sequencing of genomic DNA; for example, the cloning and expression of a dehydrogenase complex gene from the thermophilic archaeon *Thermoplasma acidophilum* in *Escherichia coli* (Heath et al. 2007).

Here, we examine several extreme environments and representative organisms that live within them. For comparison purposes, Table 5.1 shows some representative organisms that live in various environments. It must be emphasized that an extremophile may have several different mechanisms that enable growth in a particular extreme environment. For example, deep sea bacteria may be psychrophilic and also resistant to high pressure (piezotolerant), and

Environmental Microbiology and Microbial Ecology, First Edition. Larry L. Barton and Robert J.C. McLean.

Table 5.1 Microorganism and temperature preference for optimal growth.

Term[a]	Temperature range	Representative organisms
Psychrophilic[b]	−30 °C to +10 °C	*Psychrobacter* spp., *Rhodoferax antarcticus*, other psychrophiles
Mesophilic	+11 °C to +45 °C	Many common bacteria including *E. coli*, *Pseudomonas aeruginosa*, *Bacillus* spp.
Thermophilic	+46 °C to +75 °C	*Geobacillus stearothermophilus*, *Clostridium thermobutyricum*
Hyperthermophilic	> + 75 °C	*Pyrococcus furiosus*, *Sulfolobus acidocaldarius*

[a] Adjective; the corresponding suffix for a noun is -phile, as in psychrophile, mesophile, etc.
[b] Extreme psychrophiles (temperature range from −30 °C to +1 °C) are sometimes referred to as cryophiles. We use the more common term, psychrophiles.
Source: Reproduced with permission from Zheng and Wu (2010).

grow in high salt concentrations (halotolerant). Therefore, it may be difficult to precisely state the reason for resistance of a microbial cell to one environment when it survives in several environments with different physical or chemical stresses.

5.2 Sampling in Extreme Environments

There are a number of challenges to studying extremophiles. While some extremophiles may be readily accessible to scientists, such as those in various hot springs, others may live in more remote locations (such as deep ocean and subsurface environments), so specialized equipment is needed to reach them; for example: (i) deep submersibles including the Alvin (Figure 5.1) owned by the United States Navy, used to study hydrothermal vent organisms and other deep ocean environments; (ii) balloons, used to collect and study atmospheric microorganisms, and (iii) the unmanned Mars rover (Figure 5.2), used to look for microorganisms or evidence of life on Mars. Studies like these are expensive and require the support of international organizations and national budgets.

Another environment where it is difficult to examine microorganisms is the terrestrial deep subsurface. A case study approach is used to describe the process for obtaining samples from the deep-subsurface regions. The deep gold, platinum, and diamond mines of South Africa provide a unique window into the deep-subsurface environment with tunnels extending over 4 km below the surface. As seen from a review of the authors' publications of these evaluations (Borgonie et al. 2011; Borgonie et al. 2015a; Borgonie et al. 2015b; Lau et al. 2014; Lau et al. 2016), an international team of scientists is required to establish aseptic procedures for collection and analytical assessment of water samples from these deep mines. The Driefontein Gold Mine (Figure 5.3) in the Witwaterstrand Basin of South Africa was one of the sites where fluid was collected from the fault zone, which was 1.4 km below ground. Scientists carrying sampling gear walked a considerable distance to the sampling site (Figure 5.4). The underground area for sample collection is pictured in Figure 5.5. A series of pipes were inserted into the fault zone (Figure 5.6) and regulators were attached in the sampling program to collect gases and water for analysis. About 12.8 million l of borehole water was collected over a two-year period of continuous operation and filtered through membrane with a pore diameter of 0.1 μm to collect material for molecular analysis of DNA, RNA, and protein (Figure 5.7). Other sites in

Figure 5.1 Alvin – the manned deep-ocean research submersible. *Source:* NOAA Photo Library.

Figure 5.2 The Mars Exploration Rover-2 (MER-2) is being tested in a controlled environment. *Source:* KSC-03PD-0786.jpg. NASA.

Figure 5.3 One of the mine shafts of the Driefontein Gold Mine in South Africa. This was the site used for deep subsurface fluid sampling where 12 845 647 l of borehole water was collected and filtered over a two year period. *Source:* Courtesy of Gaetan Borgonie.

Figure 5.4 A typical view of a mine corridor in the Beatrix Gold Mine. The mine corridor was at −1.4 km and scientists walked some distance to the sampling site. *Source:* Courtesy of Gaetan Borgonie.

South Africa where deep-subsurface samples were obtained included the Kopanang Gold mine where seismic activity was simulated (Figure 5.8) and the Zondereinde Platinum Mine where the water was 73 °C (Figure 5.9).

Analysis of the biological material collected from the deep-subsurface mines revealed a complex ecosystem containing *Archaea, Bacteria,* and *Eukarya*. Nematodes were isolated from the subsurface fluid where the dissolved oxygen level was 13–72 μM and a new species of

Figure 5.5 This borehole at Beatrix Gold mine (depth −1.4 km) was frequently sampled by Thomas Kieft, Gaetan Borgonie, and T.C. Onstott over several years. The borehole was also extensively sampled by Cara Magnabosco and Maggie Don Lau with some of the results in the paper of Lau et al. (2016). It is also the site where the first nematode *Halicephalobus mephisto* or the "devil worm" was found (Borgonie et al. 2011). This picture reveals the mine atmosphere, which is dark, dusty and sometimes damp with pools of water on the floor of the mine. The sampling in the picture was undertaken by Cara Magnabosco and lasted slightly less than four hours. A typical mine sampling crew consisted of four people and one mine geologist. *Source:* Courtesy of Gaetan Borgonie.

Figure 5.6 A detail of an area in the Beatrix Gold Mine sampling site (depth−1.4 km) shown in Figure 5.5. The borehole has a "packer" attached to it with multiple outlets for sampling. The white tube to the left was for gas sampling and the filter with water coming out was for bacterial analysis. *Source:* Courtesy of Gaetan Borgonie.

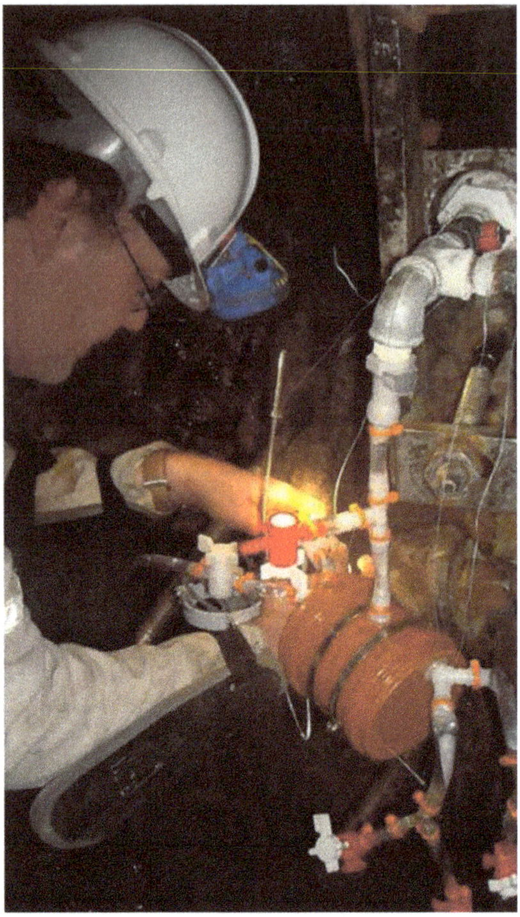

Figure 5.7 The DIY sampling device attached to the water source to look for Metazoan creatures in the Driefontein Gold Mine (depth −1.0 km). This is the site where 12 845 647 l of borehole water was collected and filtered over two years. Gaetan Borgonie is shown setting up the filter. *Source:* Courtesy of Gaetan Borgonie.

Figure 5.8 A borehole, at−1.3 km and 250 m deep inside the rock, frequently sampled in the Kopanang Gold Mine. The mine is now closed to the public. The borehole pipe was capped with a plug (gray) for several months before it was quickly reopened it to mimic seismic activity. This picture was taken shortly before the plug was removed. *Source:* Courtesy of Gaetan Borgonie.

Figure 5.9 A borehole sampled at a depth of −1.7 km in Zondereinde (previously Northam) Platinum Mine, South Africa. The Zondereinde Mine is very hot even at relatively shallow depths and is never easy to sample because the water is 73 °C. *Source:* Courtesy of Gaetan Borgonie.

nematode *Halicephalobus mephisto* was cultivated (Borgonie et al. 2011). A report of the nematode *Monhystrella parvella,* which requires a salt environment comparable to brackish or marine water, was isolated 1.3–3.1 km below ground level. *Monhystrella parvella* was found inside calcium carbonate stalactites attached to the ceiling of mine tunnels (Borgonie et al. 2015a). Bacterial biofilm was found on the surfaces of fissure rocks in mine boreholes and numerous eukaryotic organisms, which included Rotifera, Annelida, Platyhelminthes, and Arthropoda, were present in fissure fluid. The source of these eukaryotic organisms is thought to be replacement water from surface waters (Borgonie et al. 2015b). Metagenomic analysis of fracture water samples from five South African mines revealed the presence of several genes (*Nar*V, *NPD*, PAPS reductase, *Nif*D, *Nif*E, *Nif*H, *Nif*K, and *Nif*N) that could function in the cycling of nitrogen (Lau et al. 2014). Using metatranscriptomics and metaproteomic studies to evaluate the metabolic landscape of deep mine ecosystems, the major biomass producers were syntrophic autotrophic microorganisms energized by coupling sulfur oxidation reactions to denitrification (Lau et al. 2016). Assessment of the biosphere of deep surface terrestrial environments is challenging because of the difficulty in obtaining samples.

5.3 Extreme Temperature Environments

In 1808, the naturalist Alexander von Humboldt observed that plant and animal diversity was generally highest in tropical regions and decreased at higher latitudes. While several studies in plant and animal ecology support this concept, the role of extreme conditions on microbial diversity is now being re-examined using culture-independent (i.e. metagenomic) approaches. Soil and other environmental samples from Canada and New Zealand had the highest microbial diversity at 24 °C and this declined at higher or lower temperatures (Sharp et al. 2014). While temperature did influence microbial diversity, latitude had no effect, and some but not all microbial communities adjusted to changes of temperatures and pH.

Microorganisms have a range of temperatures in which they grow. The maximum growth rate typically occurs near the maximum growth temperature. Above the maximum growth temperature, denaturation of key proteins (enzymes and transport proteins) inhibits growth. At temperatures below the growth range, microorganisms are typically dormant due to the inhibited activity of key enzymes. With a few exceptions, refrigeration temperature (4 °C) is considered bacteriostatic for mesophiles and thermophiles. Freezing can cause cell death due to the formation of membrane-damaging ice crystals. In the laboratory, freezing damage can be minimized by adding a cryoprotectant, such as glycerol or dimethylsulfoxide (DMSO), to liquid-grown cultures, prior to freezing. National culture collections such as the American Type Culture Collection (ATCC) and the German Collection of Microorganisms and Cell Cultures (Deutsche Sammlung von Mikroorganismen und Zellkulturen – DSMZ), routinely use cryopreservation in which cultures are freeze-dried (lyophilized), and then stored at cool temperatures. Many research labs will maintain their own culture collections by adding sterile glycerol to an overnight broth culture (typical concentration is 12.5% (v/v)) and freezing cells at −80 °C. The frozen bacteria can be quickly revived by streaking frozen cells onto an agar plate and incubating at a temperature appropriate for that bacterial species for 24–72 hours.

5.3.1 Psychrophiles

About 80% of the Earth's biosphere is cold throughout the year and some regions are permanently frozen. While some microorganisms are obligate psychrophiles, which grow at <20 °C, bacteria such as *Listeria monocytogenes* grows optimally at 30–37 °C but it can also grow at 4 °C. Since the optimal temperature of growth for *L. monocytogenes* is greater than 15 °C, it would be classified as being psychrotolerant. Bacteria and archaea which are psychrophiles or psychrotolerant are of interest because of their potential use in biotechnology and in global nutrient cycles (De Maayer et al. 2014). Additionally, sampling cold desert soils such as those in the Antarctic is useful preparation for soil testing on Mars and other extraterrestrial environments (Goordial et al. 2016). Metagenomic studies indicate the presence of numerous bacteria in Antarctic soils and with prolonged incubation, bacteria have been cultivated from the permafrost soils and identified as *Lapillicoccus, Flavitalea, Quadrisphaera, Motilibacter,* and *Polymorphobacter* (Pulschen et al. 2017).

Many microorganisms are found in very cold environments such as glaciers, snow, and permafrost in polar and alpine regions (Miteva 2008). A partial list of psychrophilic and psychrotolerant microorganisms is presented in Table 5.2. In one recent study, 14 bacterial strains were isolated from glacial ice about 300 m below the surface in the Qinghan-Tibetan plateau of Western China; this ice field is over 750 000 years old. Sequences of 16 S rDNA were used to identify the bacteria present, which included members of the *Alphaproteobacteria, Betaproteobacteria, Actinobacteria,* and low G + C gram-positive groups (Christner et al. 2003). There are numerous bacteria in the Antarctic and Greenland ice sheets, with the number of microbial cells calculated as 9.61×10^{25} (Priscu and Christner (2004). Except for areas adjacent to geothermal activity, temperatures in the deep regions of oceans are around 3 °C. The microorganisms in frozen environments are highly diverse and they have an important environmental affect associated with thawing due to climate change (Jansson and Taş 2014). To be tolerant to cold environments, psychrophiles, like halophiles and alkaliphiles, may have high concentrations of intracellular small cryoprotectant molecules or osmolytes (Wilson et al. 2012). Some psychrophiles use antifreeze proteins to depress the freezing point and inhibit the formation of large ice crystals, which are detrimental to cells. The ice-nucleating protein is unique to the cell surface of some psychrophiles. As its name implies, this protein can promote the formation of ice crystals on the cell exterior; for example, the plant pathogen *Pseudomonas syringae*, in

Table 5.2 Examples of psychrophilic and psychrotolerant microorganism and selected characteristics.

Organism	Site of isolation	Optimum temperature or growth (°C)	References
Psychrophilic			
Cenarchaeum symbiosum[a]	sponges collected offshore of Santa Barbara, CA.	10	Hallam et al. 2006
Clostridium psychrophilum	a microbial mat near Lake Fryxel, Antartica	4	Guan et al. 2013
Colwellia psychrerythraea	Arctic marine sediments	8.5	Methé et al. 2005
Desulfotalea psychrophila	marine sediments off the coast of Svalbard, Norway	10	Knoblauch et al. 1999
Desulfofrigus fragile	marine sediments off the coast of Svalbard, Norway	7	Knoblauch et al. 1999
Idiomarina loihiensis	hydrothermal vent on Lōihi submarine volcano, Hawaii	4	Hou et al. 2004
Methanogenium frigidum[a]	Ace Lake in Antarctica	15	Franzmann et al. 1997
Moritella abyssi	sediment off West Africa coast	4	Xu et al. 2003
Moritella profunda	sediment *off the* West Africa coast	2	Xu et al. 2003
Pseudoalteromonas	Antartic seawater *haloplanktis*	4	Médigue et al. 2005
Psychromonas ingrahamii	sea ice from the Arctic	5	Riley et al. 2008
Psychrotolerant			
Desulfosporosinus hippel	permafrost Siberia, Russia	28	Vatsuina et al. 2008
Clostridium tagluense	PF gas-hydrate field Mackenzie Delta, Northwest Territories, Canada	15–20	Suetin et al. 2009
Methanococcoides burtonii	bottom of Ace Lake Antarctica	23	Goodchild et al. 2004
Psychrobacter arcticus	*PF*, Kolyma lowland Sediments, Siberia	22	Bergholz et al. 2009
Exiguobacterium sibericum	Kolyma Lowland northeast of Siberia	18	Rodrigues et al. 2008
Planococcus halocryophilus	Permafrost active-layer soil at Eureka, Ellesmere Island, Canadian high Arctic	25	Mykytczuk et al. 2013
Pseudoalteromonas	Antartic seawater *haloplanktis* TAC125	22	Médigue et al. 2005
Psychrobacter cryohalolentis	A cryopeg in the Kolyma lowland, Siberia, Russia	22	Bakermans et al. 2006
Sphingobacterium antarcticus	Soil from Schirmacher Oasis, Antarctica	25	Shivaji et al. 1992

[a] Archaea.

which ice nucleation is encoded by the gene *ina* (Lindow et al. 1982). The Ina protein has the same three-dimensional structure as ice crystals and so promotes ice-crystal-induced frost damage to plant cells. Frost damage releases nutrients for the bacteria. *Pseudomonas syringae* and other *ina*-containing, plant pathogenic bacteria have been found in the atmosphere and it is thought that ice nucleation by these organisms is a key contributor to cloud formation and precipitation (Pandey et al. 2016). Recently, some fungi have also been associated with ice nucleation ability (Fröhlich-Nowoisky et al. 2015). Liquid water is required for metabolic activity and so most psychrophiles are dormant when frozen; however, there is at least one report of essential enzymatic activity, DNA repair, that can occur at −15 °C (Dieser et al. 2013), and ongoing studies are investigating the structural properties of enzymes and other proteins in psychrophiles.

5.3.2 Thermophiles

Thermophiles have been defined as organisms living at elevated temperatures, typically with an optimal growth temperature >50 °C. Naturally occurring thermophilic environments include the surfaces of soils and rocks in hot desert environments, and areas with geothermal activity. While elevated temperatures increase enzymatic activity and metabolism, proteins may also become denatured. DNA strands may also separate (DNA melting), and so the stability of nucleic acids needs to be maintained. Table 5.5 shows some adaptations of thermophiles to elevated temperatures. Thermophiles often live in environments where there are other extremes including pH, salinity, available water, and pressure. For example, *Haloarcula quadrata, Haloferax elongans, Haloferax mediterranei,* and *Natronolimnobius aegyptiacus* are extremophiles that grow optimally above 50 °C but also require 20% NaCl and a pH of about 8.5 for optimal growth (Bowers and Wiegel 2011). Here, we examine thermophiles in three unique environments.

5.3.2.1 Alpine Environment – Yellowstone National Park

Yellowstone National Park (YNP) is in the northwestern corner of Wyoming state in the USA, and extends into the neighboring states of Montana and Idaho. It is readily accessible by road and as a national park, the wilderness has been preserved. As a result, YNP is a popular region to conduct work on extremophiles. Considerable early work on thermophiles was carried out by T.D. Brock in the geothermal environments of YNP (Brock 1978). The area throughout Yellowstone is underlain by magma (molten rock; the temperature in one type of molten rock (basalt) is 1100–1250 °C) and in the past this area has been the site of volcanic activity. The most recent eruption was 70 000 years ago. The heat produced by this molten rock heats groundwater, which can then rise through cracks in the rock to the surface as liquid water in hot springs or as steam in geysers (Figure 5.10).

The boiling point of water in the alpine environment (average elevation of 2400 m or ~8000 ft) of YNP is 93 °C. Some hot springs are quite colorful, with dissolved minerals responsible for the color in some cases, and the pigmentation of thermophilic microbial biofilms responsible in others.

Aside from *Thermus aquaticus,* other fascinating thermophiles found in YNP including various *Archaea,* notably *Sulfolobus* spp., which is the most common genus cultured. Archaea in YNP grow at pH ranges from 0.9 to 9.8, with a maximum temperature of 92 °C. Several metagenomic studies have been carried out to identify virus communities (Schoenfeld et al. 2008). Most viruses identified to date appear to be phages that target the thermophilic microorganisms present in the environment. Some eukaryotic microorganisms have been identified. One interesting organism is the fungus *Curvularia proturberata,* which is part of a

Figure 5.10 Daisy Geyser in Yellowstone National Park. *Source:* Brocken Inaglory, https://commons.wikimedia.org/wiki/File:Daisy_Geyser_erupting_in_Yellowstone_National_Park_edit.jpg. Licensed under CC BY-SA 3.0.

three-way symbiotic relationship. When this fungus and its viral partner *Curvularia* thermal tolerance virus (CThTV), associate with the roots of the hot springs panic grass (*Dichanthelium lanuginosum*), the grass can survive at soil temperatures of 55 °C; without the microbial symbionts, the survival maximum is <50 °C (Roossinick 2011).

Numerous thermophilic bacteria have been described in the literature and some are listed in Table 5.3.

One major hot spring in YNP is Mammoth Hot Springs. Photosynthetic bacteria, including the genera *Synechococcus, Chloroflexus,* and *Chlorobium,* have been associated with travertine formation, which is a prominent feature of Mammoth Hot Springs. Travertine is a sedimentary rock composed of two crystalline forms of calcium carbonate ($CaCO_3$) – calcite and aragonite. Travertine formation has been associated with several hot springs around the world. Travertine is used as a building material, notably in Italy, with one famous structure being the Colonnade of St. Peter's Square at the Vatican. Close inspection of travertine (Figure 5.11) reveals striations and holes in the rock. *Synechococcus* and other photosynthetic organisms associated with travertine formation typically grow as biofilms. Oxygen is produced as a byproduct of photosynthesis and oxygen bubbles can become entrapped in the biofilm matrix. Depending upon pH, dissolved carbon dioxide will take one of three forms:

$$\underset{(\text{pH} < 6.1)}{H_2CO_3(\text{carbonic acid})} \leftrightarrow \underset{(\text{pH } 6.1-10.3)}{HCO_{3^-} - (\text{bicarbonate})} \leftrightarrow \underset{(\text{pH} > 10.3)}{CO_3^{2-}(\text{carbonate})}$$

Table 5.3 Thermophilic bacteria from Yellowstone National Park, USA.

Name (genus)	pH and temperature	Description	Location
Cyanobacteria *Calothrix*	pH6–9 30–45 °C (86–113 °F)	Color: dark brown matsMetabolism: photosynthesis by day; fermentation by night	• Mammoth Hot Springs • Upper, Midway, and Lower geyser basins
Phormidium	pH6–8 35–57 °C (95–135 °F)	Color: orange mats Metabolism: photosynthesis	• Mammoth Hot Springs • Upper, Midway, and Lower geyser basins
Oscillatoria	pH6–8 36–45 °C (96–113 °F)	Color: orange mats Metabolism: photosynthesis; oscillating moves it closer to light sources.	• Mammoth Hot Springs • Chocolate Pots
Synechococcus	pH7–9 52–74 °C (126–165 °F)	Color: green mats Metabolism: photosynthesis by day; fermentation and nitrogen fixation by night	• Mammoth Hot Springs • Upper, Midway, and Lower geyser basins
Green Sulfur *Chlorobium*	pH6–9 32–52 °C (90–126 °F)	Color: dense, dark green mats Metabolism: anaerobic photosynthesis—producessulfate and sulfur, not oxygen.	• Mammoth Hot Springs • Calcite Springs
Green non-sulfur *Chloroflexus*	pH7–9 35–85 °C (95–185 °F)	Color: green mats Metabolism: anaerobic photosynthesis	• Mammoth Hot Springs • Upper, Midway, and Lower geyser basins
Aquifex *Hydrogenobaculum*	pH3–5.5 55–72 °C (131–162 °F)	Color: yellow and white streamers Metabolism: uses hydrogen, hydrogen sulfide and carbon dioxide as energy sources; can use arsenic in place of hydrogen sulfide.	• Norris Geyser Basin • Amphitheater Springs
Deinococcus-Thermus	pH5–9 40–79 °C (104–174 °F)	Color: bright red or orange streamers; contains carotenoid pigments that act as sunscreen.	• Lower geyser Basin

Source: Reproduced with permission from Yellowstone National Park website: https://www.nps.gov/yell/learn/nature/thermobacteria.htm.

During photosynthesis around neutral pH, bicarbonate is preferentially used during carbon fixation. Outgassing of dissolved CO_2 also occurs, leading to an increase in the pH of the microenvironment of the photosynthetic biofilm. Trace amounts of carbonate precipitate with dissolved Ca^{2+} ions forming calcium carbonate ($CaCO_3$) as calcite or aragonite. The precipitated calcite and aragonite are bound by the biofilm matrix, forming travertine (Figure 5.11). Travertine formation can be quite rapid in some instances, with a growth of up to 5 mm/day reported at Mammoth Hot Springs in YNP. Geochemical conditions related to the water chemistry (levels of dissolved Ca, various forms of dissolved CO_2, pH, temperature, other ions, etc.) will also influence travertine formation. Anionic microbial cell surfaces can bind and concentrate dissolved Ca^{2+} ions, which can promote calcite and aragonite mineral formation and thus enhance travertine formation (Fouke 2011).

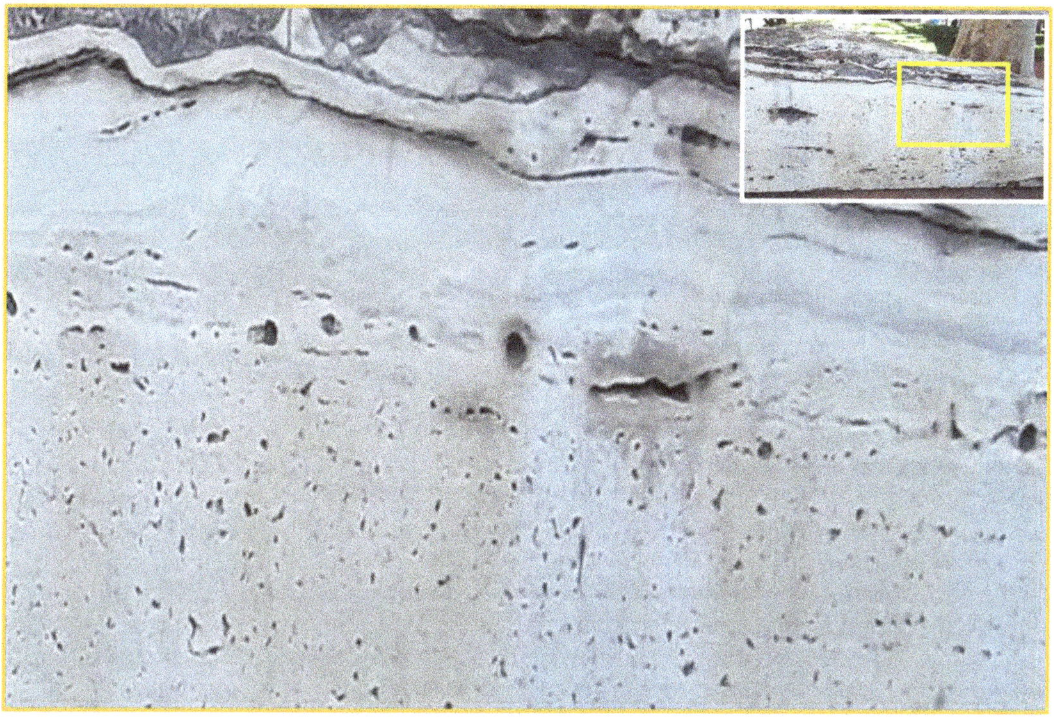

Figure 5.11 Travertine formation attributed to deposition of calcium minerals. The inset shows a large travertine stone and the outlined area in the inset is of the polished surface which is magnified to reveal the holes in the travertine. *Source:* Reproduced with permission from Larry Barton.

5.3.2.2 Hydrothermal Vent Communities

One of the most remarkable biological discoveries in the past 50 years, was the discovery of thriving biological communities by hydrothermal vents in the deep (abyssal) regions of the oceans. Prior to this discovery, biologists considered most lifeforms and ecosystems to rely on photosynthesis. In oceans, photosynthesis and associated primary production only occurs in the uppermost regions of the ocean (typically the top 10–80 m) and is limited by the depth of light penetration (see Chapter 4). Oceanographers and marine biologists had long considered the deepest regions of the oceans (depths > 1000 m) to be relatively barren, with the limited available food sinking from the surface regions. In 1977, Corliss et al. used a deep submersible submarine (Alvin) and explored the deep seafloor near the Galapagos Islands (Corliss et al. 1979). At depths of 2500 m, they observed thriving biological communities, consisting of large clams (*Calyptogena*), tube worms (*Riftia*) and other life forms. Work carried out by Holger Jannasch, Colleen Cavanaugh, and their colleagues, showed that primary production in hydrothermal vent communities arose from chemoautotrophic bacteria and that hydrothermal vent communities were totally independent of photosynthesis. Since the original discovery of hydrothermal vent communities at the Galapagos Rift and East Pacific Rise (latitude 21°N, longitude 109 °W), hydrothermal vent communities have been described throughout the world (Dick et al. 2013). While the abyssal ocean temperatures are cold (approximately 2 °C), water-exiting hydrothermal vents are considerably warmer and can range from 20 °C up to an estimated 380 °C (Spiess et al. 1980). At the extreme pressure (2500 m depth), the boiling point of water is estimated to be 400 °C. Upon mixing with the surrounding seawater, the temperature cools rapidly. Water flowing from hydrothermal vents contains a variety of reduced chemicals including (H_2S, H_2, Fe(II), and CH_4). The oxidation of these compounds by chemoautotrophic bacteria provides the energy for carbon fixation and chemosynthesis-based primary production. Hydrothermal vents are associated with plate tectonics (continental drift) and have been described throughout the oceans.

5.3.2.3 The Guaymas Basin

The Guaymas Basin in the Gulf of California (USA) contains hydrothermal mounds, chimneys, and enormous microbial mats (Teske et al. 2016). The floor of the basin is about 2 km below sea level and sampling at this site has been conducted by the *Nautile* and *Alvin* submersibles. The center of the basin is covered by several hundred meters of organic sediment that comes from the highly productive waters above the basin floor. Adding complexity to the basin environment is the presence of hydrothermal vents and cold seeps. Hydrothermal vent sites covered with 100–500 m sediment – hydrothermal mounds – support the growth of numerous bacterial and archaeal species. The quantity of carbon in these sediments may reach 2% and thermal action on the sediment carbon produces microbial nutrients such as methane, C1–C8 hydrocarbons, short chain fatty acids, alkanes, alkenes, polyaromatic compounds, and ammonia. The most commonly used electron acceptors to support microbial growth include nitrate, sulfate, and molecular oxygen.

The surface of the hydrothermal sediment is covered with a microbial mat rich with bacteria collectively referred to as filamentous sulfur-oxidizing *Beggiatoa*; however, based on 16S rRNA analysis, these mat bacteria are most closely related to *Thiomargarita* and the candidate taxa of *Maribeggiatoa* and *Marithioploca* (Teske et al. 2016). Other organisms isolated include *Persephonella* (microaerophilic chemolithoautrophic bacterium), *Thermodesulfobacterium* (thermophilic-sulfate reducing bacterium), *Archaeoglobus* (hyperthermophilic sulfate-reducing archaeon), *Thermococcus* and *Pyrococcus* (sulfur-reducing archaeon), *Methanocaldococcus* (hyperthermophilic methanogenic archaeon), and *Geoglobus ahangari* (hyperthermophilic iron-reducing archaeon) (Amend and Teske 2005).

The food web in hydrothermal vents and cold seeps in Guaymas Basin were examined and were found to differ even though seeps and vents had similar nutrients and environments for growth (Portail et al. 2016). Thiotrophs, microorganisms which couple oxidation of sulfide and other reduced sulfur compounds with nitrate as the electron acceptor, are dominant in both seep and vent environments. Microorganisms fix carbon dioxide by the ribulose-1,5 bisphosphate carboxylase/oxygenase (RuBisCO) pathway in both seeps and vents, while the fixation of carbon dioxide in the vent community is only by the reductive tricarboxylic acid cycle. Microorganisms in the vent environment rely on organic matter produced from petroleum matter and microorganisms in both seep and vent environments use ammonia/ammonium from the mat and not nitrate from the water column. Additional groups of microorganisms must be isolated to obtain a more complete understanding of food webs in the Guaymas Basin.

5.4 Xerophiles

Organisms that can grow in exceedingly dry or desiccating environments are called xerophiles. While the average annual rainfall for deserts is less than 400 mm, the Atacama Desert (Figure 5.12) in northern Chile is hyper-arid and receives less than 1 mm of rain a year. The barren Atacama Desert is important as a Mars model and it is used by NASA scientists and astrobiologists to test microbial isolation techniques (Azua-Bustos et al. 2012). The average temperature of the Atacama Desert is 16.5 °C, organic matter is present in trace amounts, the ultraviolet (UV) radiation is intense, and the soil contains halite, gypsum, and anhydrite.

Figure 5.12 The Atacama Desert between Antofagasta and Taltal, Chile. Inset: A map of South America with the arrow indicating the location of the Atacama Desert. *Source:* cobaltcigs, https://commons.wikimedia.org/wiki/File:Atacama_map.svg. Licensed under CC BY-SA 3.0.

Research describing the microbial communities in soils of the Atacama Desert have been reviewed recently (Crits-Christoph et al. 2013) and the diversity of species is low. Using high throughput DNA sequence data of microorganisms isolated from soil at depths of 0–10 cm, about 80% of the microorganisms were *Actinobacteria*, about 5.7% were *Acidobacteria*, and about 5.6% were *Proteobacteria*. About 4.7% of the bacterial sequences were unassigned and while archaea have not been detected in soil samples, colonization of halite by archaea (i.e. *Halococcus* and *Natrobacterium*) has been reported. Since very few species of *Firmicutes* were detected, it was assumed that the organisms present were vegetative cells and not spores. Saprophytic fungi were detected in the soil samples but their presence is likely from air dispersion of asexual spores. Members of the filamentous green non-sulfur bacteria of the *Chloroflexi* phylum were found in all soil samples. Bacteria in the hyper-arid Atacama Desert are in a "state of stasis" where the cells will become metabolically active if water is supplied. This preservation of vegetative bacteria in dry sand is scientifically accepted because it was used by early microbiologists to maintain bacterial culture collections.

Variations in the microbial communities in the Atacama Desert are due to selection imposed by salts in the environment and the availability of water. The deliquescence of salts in the soil could be important as a source of water and it is known that bacteria are found inside halite (hygroscopic) nodules present in ancient salt pans (Crits-Christoph et al. 2013). In some regions of the Atacama Desert, heavy fogs supply needed moisture. In regions of this desert where sufficient moisture is present to support some forms of life, cyanobacteria have been found growing in dew drops formed on spider webs, where the dew drops are attributed to frequent fogs. While damage to microbial cells occurs at a low rate in a dry environment, damage to desert microorganisms from ultraviolet light is an important event. Some desert microorganisms produce scytonemin, a UV protecting molecule, and most microorganisms grow on the underside of translucent quartz rocks where they are shielded from ultraviolet light but receive sufficient light for photosynthesis (Wierzchos et al. 2011). While colonization of halite in the Atacama Desert was attributed to a cyanobacterium closely related to *Halothece,* various types of bacteria, cyanobacteria, and algae colonize the bottom layer calcium sulfate crusts. Microorganisms that grow in the undermost layer of these crusts are called "hypoendoliths" (Wierzchos et al. 2011).

There are numerous fungi that can tolerate dry environments and some of these are also osmophiles or halophiles. Filamentous fungi that are xerophiles include *Aspergillus restrictus, Xerochrysium dermatitidis, Xerochrysium xerophilum, Aspergillus niger, Eurotium amstelodami,* and *Xerochrysium xerophilum* while xerophilic yeasts include *Debaryomyces hansenii* and *Hortaea werneckii* (Leong et al. 2015).

5.5 Piezophiles

Microorganisms that grow in environments of high pressure are designated as piezophiles and this term replaces barophiles. Piezophile microorganisms are found deep underground (see Section 3.5) or at ocean depths where the pressure may reach 100 megapascals (MPa). Piezophiles are difficult to isolate and grow because special pressure-inducing instrumentation is required, the organisms must be maintained at the temperature of the isolation environment, and exposure to light must be avoided because these organisms lack systems for DNA repair following UV damage. Because of these cultivation difficulties, relatively few bacteria and archaea are known to be obligate piezophiles. Piezophilic microorganisms are currently organized into two categories: (i) psychropiezophiles that grow at about 2 °C and pressures exceeding 10 MPa, and (ii) thermopiezophiles that grow at 98 °C and at a pressure of 50 MPa.

Mesophilic piezosensitive *Escherichia coli* and *Saccharomyces cerevisiae* have been used as model organisms to study the effects of high hydrostatic pressure (HHP) on microorganisms and these results have been summarized by Abe (2007). At 40–50 MPa, the FitZ ring proteins do not polymerize and cell division does not occur in growing cells of *E. coli,* with the cells becoming multinucleated and filamentous. Protein synthesis is inhibited at elevated HHP, which is attributed to dissociation of free 30S ribosomal subunits that are not protected by mRNA or tRNA. While HHP has been known to damage membranes, studies with *E. coli* and *S. cerevisiae* have described physiological changes to membranes of these microorganisms.

Many of the psychrophilic piezophilic bacteria isolated are members of *Shewanella, Photobacterium, Moritella, Colwellia,* and *Psychromonas* genera, with *Photobacterium profundum* SS9 used as a useful model to examine physiological adaptions to HHP. From the research by Lauro et al. (2007), the following proposals have been made: (i) psychrophilic piezophilic bacteria have evolved from psychrophilic bacteria and (ii) a similarity between Antarctic and Arctic psychrophiles and piezophiles would result from bacteria being carried by deep sea currents. Both piezophiles and psychrophiles have unique membrane lipids and protein structure. Specific lipids in cell membranes that prevent membranes transitioning from the soil to the gel state under HHP are found in *Acidilobus sulfurireducens* (Boyd et al. 2011). Protein function is enhanced in HHP due to the presence of only a few glycine and proline residues, which would decrease protein flexibility. Enzymes that are stable under HHP include DNA polymerases and hydrogenases (Abe and Horikoshi 2001).

Several investigations into gene structure and the regulation of gene expression have provided information on bacterial adaption to HHP. The stability of the 30S ribosome subunit to

HHP was proposed to be due to the unique structure of 16S RNA (Lauro et al. 2007). The exposure of a bacterium to an environment where the hydrostatic pressure is greater accounts for the upregulation of gene expression of outer membrane proteins (porins) that do not collapse under high pressure (Simonato et al. 2006). A transcriptomic study of the piezophile *Desulfovibrio hydrothermalis* revealed that 65 genes were differentially expressed with exposure to 0.1, 10, and 26 MPa and the major categories of these regulated genes included aromatic amino acid and glutamine metabolism, energy metabolism, and signal transduction (Amrani et al. 2014).

Relatively few thermopizeophiles have been isolated in deep sea environments and these include microorganisms of the *Archaea* domain (*Thermococcus barophilus, Pyrococcus abyssi,* and *Pyrococcus yayanosii*) and the bacterium *Marinitoga piezophila* (Michoud and Jebbar 2016). *Pyrococcus yayanosii* is the only obligate thermopiezophile as it will not grow at pressures less than 15 MPa; this archaeon has lost the pathways for aromatic amino acid biosynthesis. Proteomic and transcriptomic analysis of *Pyrococcus yayanosii* exposed to different levels of hydrostatic pressure indicates that 53 genes are upregulated and 36 genes are downregulated (Michoud and Jebbar 2016). The pressure-influenced genes were responsible for the regulation of translation and accounted for the increased production of ribosomal subunits and tRNA synthesis. They also influence chemotaxis by enhancing tumbling during swimming to enable the microorganism to more effectively search for nutrients in a nutrient-poor environment. These pressure-influenced genes also regulate the genes for hydrogenase, formate dehydrogenase and the CRISPR-*cas* system. The survival of *T. barophilus* under conditions of high pressure was dependent on genes associated with energy production, transport of inorganic compounds and carbohydrases, and metabolism of inorganic ions and carbohydrates (Vannier et al. 2015). While HHP affects thermopiezophilic organisms, this is only the initial phase of understanding how bacteria and archaea survive exposure to high pressure environments.

5.6 Acidophiles

Microorganisms that grow at pH 3 or less are called acidophiles and the metabolic activities of these organisms are reviewed by Johnson and Hallberg (2009). Although acidophiles are found in diverse areas, the number of species of bacteria and archaea that grow in extremely acidic environments is relatively small. For cells to be physiologically active in an acidic environment, the acidophiles must maintain a pH gradient of about 3 pH units across the plasma membrane, with the cytoplasm having the higher pH value. These organisms have considerable diversity for cell structure, respiration, metabolism, tolerance to inhibitors in the environment, and distribution.

Several members of the archaea display dual tolerance to high temperature and acidity. Generally, extreme acidiophilic archaea are members of two orders: *Sulfolobales* and *Thermoplasmatales* of the phyla *Crenarchaeota* and *Euryarchaeota*, respectively. Moderate acidophiles are found in the orders of *Acidilobales* and *Thermoproteales* of *Crenarchaeota*.

Sulfolobus acidocaldarius and *Sulfolobus solfataricus* are found in hot sulfur springs and solfateras and can grow at 80 °C and pH 3. These thermoarchaeal organisms can grow either as an autotroph or a heterotroph, lack a cell wall, but have a dominant S-layer. *Picrophilus oshimae* is an archaeal organism that is found in acidic solfataras. It has a cell wall and grows at pH 0 but displays cell lysis if the pH is raised to 4. It can be isolated from acidic soils and volcanic rocks. Other organisms isolated from hot springs or saline environments include: *Metallosphaera sedula*, an extremely thermoacidophilic archaeon which grows optimally at 73 °C and pH 2.0; *Halarchaeum acidiphilum*, which requires NaCl and Mg^{2+} for optimal growth at pH 4.4 and 37 °C; and *Acidianus brierleyi,* which grows at 70 °C and pH 1.5 using either autotrophic or heterotrophic metabolism. *Thermoplasma acidophilum* and *Thermoplasma volcanium* grow at pH 2 and are moderate thermophiles with growth at 60 °C. These archaeal members are found in acid soils and self-heating coal piles. They lack cell walls and S-layers and display a metabolism that utilizes organic compounds and elemental sulfur. An additional acidophilic archaeal member that lacks a cell wall and S-layer is *Ferroplasma acidarmanus.* This organism grows at 42 °C and pH 2, with the oxidation of Fe^{2+} to Fe^{3+}, which remains stable at this acidic pH, and the production of sulfuric acid when ferric ions react with iron sulfide.

Many of the thermophilic, acidophilic strains generate their own acidic environment by the aerobic oxidation of elemental sulfur to sulfate according to the following reaction:

$$2S^0 + 3O_2 + 2H_2O \rightarrow 2SO_4^{2-} + 4H^+$$

There are numerous acidic, hot springs in YNP and these waters host several archaea. To better understand the role of acidophiles in establishing their acidic environment, Colman et al. (2018), examined the genomes of archaea from 72 hot, acidic springs in YNP (Figures 5.13 and 5.14). The comparative genomics of organisms in the YNP springs showed that the protein profiles of *Thermoplasmatales* and *Sulfolobales* members were distinct from the protein array of the *Euryarchaeota* present and, from an evolutionary perspective, these archaea became adapted at several different times to the extremely acidic environment. The distinguishing

Figure 5.13 Acidic sulfur water from a hot spring in Yellowtone National Park. Yellow S^0 flocs surround the main runoff channel with the water at pH ~3 and ~80^0 C (Colman et al. 2018). *Source:* Courtesy of Daniel R. Colman.

Figure 5.14 Cinder Pool in Yellowstone National Park: this has a temperature of ~90 °C and is a mildly acidic spring (pH ~4) that is predominantly inhabited by archaeal microorganisms that use sulfur-based metabolism. The black spherules on the surface of the water contain S^0 with <1% pyrite, which came from a molten sulfur pool at the bottom of the spring (Colman et al. 2018). *Source:* Courtesy of Daniel R. Colman.

proteins that enable these archaea to maintain cellular pH homeostasis in an acidic environment include solute and amino acid transporters and permeases. The hyperacidophiles require O_2 for the conversion of reduced sulfur compounds to an acidic sulfate environment. Thus, it is probable that the acidic hot springs did not occur until ~0.8 Ga when atmospheric oxygen levels became similar to present day levels. Colman et al. (2018) argued that thermoacidophiles have evolved to fill the thermoacidophilic niche by incorporating highly energetic metabolisms (oxidative respiration of sulfur and sulfides) that contribute to the production of acid in hot springs.

Microbial leaching is an important process where acidophiles release the valuable metals present in low grade ores (Schippers et al. 2014). The basis of the approach for biomining is for bacteria such as *Acidothiobacilus ferrooxidans* and *Leptospirillum ferroxidans* to oxidize Fe^{2+} to Fe^{3+}, which chemically reacts with the metal sulfide. Biomining has been effectively used to recover copper and, more recently, cobalt, gold, uranium, and zinc from ores, and to process waste at mine dumps.

Bacterial and archaeal acidophiles dominate in areas with acidic drainage from mines and contribute to the solubilization of toxic minerals (Baker and Banfield 2003). The acid production is due to the oxidation/reduction activity of iron with production of a waste mine stream that is red because of the presence of high concentrations of Fe^{3+} (Figure 5.15). In the initial

Figure 5.15 Drainage from the Iron Mountain Mine, California. *Source:* United States Geological Survey.

reaction, ferrous ion is oxidized by atmospheric oxygen and this is followed by a second exothermic reaction that prduces sulfuric acid, as in the following reactions:

$$14Fe^{2+} + 3.5O_2 + 14H^+ \rightarrow 14Fe^{3+} + 7H_2O$$

$$FeS_2 + 14Fe^{3+} + 8H_2 \rightarrow 15Fe^{2+} + 2SO_4{}^{2-} + 16H^+$$

Deposits of FeS_2 (pyrite) may also contain chalcopyrite ($CuFeS_2$), arsenopyrite ($FeAsS_2$), sphalerite (ZnS), and galena (PbS), which are dissolved in the acid mine water and release toxic metals (Figure 5.15). The enzymatic oxidation attributed to bacteria and archaea under acidic conditions occurs at a rate that is about five orders of magnitude greater than if the oxidation of ferrous ion was abiotic. Based on 16S rRNA genes, over 50 different bacterial species have been detected in acid mine drainage including *Acidiphilum* spp., *Acidimicrobium ferrooxidans, Acidimicrobium acidophilus, Sulfobacillus thermosulfidooxidans, Sulfobacillus acidophilus, Leptospirillum ferrooxidans, Acidobacterium capsulatum,* and *Ferromicrobium acidophilus* (Baker and Banfield 2003). Few archaea members have been detected in acid mine drainage and these include *Thermoplasma* spp. and *Metallosphaera prunae.*

Unique microorganisms have been reported from mines in extreme environments. Acid drainage from coal mines in the high Arctic, halfway between Norway and the North Pole, was examined using culture-independent methods (García-Moyano et al. 2015). The samples were from the Bjørndalen site and the Sverdrup site near Longyearbyen in Svalbard, Norway, in a region where the mean annual temperature is −7 °C and in three summer months the temperature is about 10 °C. In winter, the rock pile near the point of drainage supports microbial activity because the temperature is 5 °C due to active leaching activity. The microbial community at the Bjørndalen site was highly diverse with bacteria assigned to nine phyla: *Proteobacteria* (*Alphaproteobacteria*, 7%; *Betaproteobacteria*, 15%; *Gammaproteobacteria*, 19%); Saccharibacter, 23%; *Actinobacteria*, 8%; *Bacteroidetes*, 8%; *Actinobacteria*, 7%; *Verrucomicrobia*, 6%; *Chlofoflexi*, 4%; *Planctomycetes*, 2%; and *Nitrospirae*, 1%. At the Sverdrup site, the sequences were assigned to five phyla: *Acidobacter, Actinobacteria, Chloroflexi, Nitrospirae, Proteobacter* (*Alphaproteobacter* and *Gammaproteobacter*), and *Rhodophyta*. At the Bjørndalen site, an abundance of dissimilatory iron reducers were detected including *Acidithiobacillus ferrivorans, Albidiferax ferrireducens, Acidiphilium* sp. *Acidocella* sp. *Aciditerrimonas ferrireducens*, and *Frateuria* sp. Some operational taxonomic units (OTUs) were investigated. One group was with the *Gammaproteobacteria* and these were closely identified with *Acidithiobacillus ferrivorans*, a psychrotolerant iron-oxidizing bacterium. Some OTUs were affiliated with the phylum *Candidatus* Saccharibacteria whose members are found in soils with metals. Phylotypes resembled *Candidatus* "Fodinabacter communificiens," an organism associated with an arsenic-rich mine. Some OTUs were found to have high identity with *Gallionella ferruginea* and *Sideroxydans paludicola*, which are iron-oxidizing neutrophilic bacteria. Additional studies on acid mine drainage is needed to provide information about community structure and other ecological details.

Biofilms are common in most environments including mines, and the biofilms in Richmond Mine, California (USA) has interesting *Archaea* present. The research by Jillian Banfield's group has analyzed these biofilms and reported that in addition to the bacterium *Leptospirillum* sp. and members of the archaeal order *Thermoplasmatales*, uncultivated nanosize archaea were detected. These unique cells passed through a filter with a pore diameter of 0.45 μm and were designated ARMAN (archaeal Richmond Mine acidophilic nanoorganisms) (Baker et al. 2010). Electron microscopy revealed that these ARMAN were <500 nm in diameter and were near the smallest size for living cells. Although many details are unknown, ARMAN are *Euryarchaea* with genes similar to those found in *Crenarchaea* and *Bacteria*.

Different members of the bacteria are moderate acidophiles that grow at moderate to extreme acidity levels and some are associated with human activities. Species within the spore-forming genus of *Alicyclobacillus* grow at pH 2–6 and are especially important in food spoilage. The gram-negative *Acetobacter aceti* is broadly distributed and is most recognized for vinegar production resulting from acetic acid production, which depresses the pH to 5.4–6.3. *Helicobacter pylori* organisms contribute to stomach ulcers in humans and these bacteria survive exposure to stomach acidity of about pH 2.

5.7 Alkaliphiles

Microorganisms that are obligate alkaliphiles have optimal growth above pH 9 and some may grow at pH 12. Alkaline environments where alkaliphiles are found include alkaline soda lakes (Figure 5.16), the hindgut of insects, serpentinizing ecosystems (Box 5.1), and soils subjected to alkaline processes. Several enzymes from alkaliphiles are used in industrial processes because these enzymes have unique properties (Preiss et al. 2015). Numerous bacteria have been isolated and a partial list of these isolates includes *Desulfonatronum thiodismutans, Halomonas campisalis, Alcalilimnicola nalodurans, Bacillus haloalkaliophilus, Pseudomonas alcaliphila, Bacillus pseudofirmus, Bacillus cohnii*, and *Halorhodospira halophito*.

Interesting aspects of alkaliphilic life are the homeostatic activity of maintaining the intracellular pH and the mechanism driving the F_1F_0–ATP synthase. In *Bacillus pseudofirmus*, the cytoplasmic pH is 8.3 when the external pH is 10.8 and the internal pH is 9.5 when the extracellular pH is 11.2. The maintenance of the pH difference across the plasma membrane to about 2 pH units is similar to that found in neutrophilic bacteria. This control of cytoplasmic pH with respect to external pH is attributed to a series of cation/proton anteporters; however, neutrophilic bacteria have an external pH of ~6 and this slightly acidic character contributed to the proton motive force driving with H^+ energizing the F_1F_0–ATP synthase. While it may be

Figure 5.16 Mono Lake is an alkaline and hypersaline lake in Mono County, California. *Source:* https://www.loc.gov/pictures/collection/highsm/item/2010630780. Library of Congress.

Box 5.1 Serpentinization Reaction

The abiotic serpentinization reactions involve heat, water, and ultramafic rock found at the tectonic plate boundaries of the oceans to produce serpentinite and H_2. The serpentinite rock has a surface which resembles the skin of a snake and this is the source of its name. Sometimes the abiotic formation of CH_4 occurs and this depends on the metals present in the ultramafic rock. H_2 and CH_4 produced by these abiotic reactions may be used by microorganisms as substrates to energize their chemolithotrophic metabolism.

tempting to propose that the F_1F_0–ATP synthase in extreme alkaliphiles is driven by Na^+ entry across the plasma membrane, that is not the case because research has demonstrated that the ATP synthase is driven by H^+ re-entry in both *Bacillus pseudofirmus* and *Pseudomonas alcaliphila* (Preiss et al. 2015; Matsuno and Yumoto 2015). The unique processes that enable gram-positive and gram-negative halophilic bacteria to regulate H^+ transfer on the surface of the plasma membrane may include a special role for cytochrome *c* and the topology of the F_1F_0–ATP synthase unit. It is envisioned that as more halophilic microorganisms are evaluated, additional mechanisms for supporting proton driven F_1F_0–ATP will emerge.

Alkaline environments may result from serpentinization reactions, in which bacteria may be involved. The abiotic reaction of water with low silica ultramafic rocks produces H_2, Ca^{2+}, OH^-, and minerals including magnetite, serpentinite, and brucite. Several novel bacteria were isolated from the alkaline water (pH 11.6) of a serpentinizing spring at The Cedars, California. These bacteria were members of the *Betaproteobacteria* and a new genus, *Serpentinomonas raichei,* was suggested (Suzuki et al. 2014). Members of this genus grow autotrophically using H_2 as the electron donor and O_2 as the electron acceptor, with CO_2 as the carbon source fixed by the Calvin–Benson–Bassham cycle.

5.8 Halophiles and Chaophiles

Some microorganisms are capable of growing in elevated salt (NaCl) concentrations and these cells are called halophiles. There are three categories of halophiles and the concentration of salt each tolerates is: slight, 0.3–0.8 M; moderate, 0.8–3.4 M; and extreme, 3.4–5.1 M. As a point of reference, seawater contains 0.6 M salt. With an elevated salt concentration in the environment, cells must contend with an osmotic imbalance and the movement of water out of the cell. The loss of water from a cell by osmosis can result in cell death and available water in an environment may be described water activity (a_w). (See discussion of water activity in Section 4.2.1.) One defense used by most halophilic bacteria and some fungi against dehydration of the cell is to increase the concentration of organic compounds, called compatible solutes, in the cytoplasm. The compatible solute used by most fungi is glycerol (Zajc et al. 2013) and depending on the bacterial or archaeal species, the compatible solute is charged, uncharged, a carboxylate, or an anionic solute (see Table 5.4). Generally it has been found that bacteria use neutral compatible solutes and that archaea use charged molecules (Roberts 2005).

An adaptive strategy used by a few halophilic microorganisms is to transport K^+ into the cytoplasm using K^+ uptake/efflux systems and, thereby, equilibrate the osmotic pressure across the cell membrane. Since the cytoplasmic salt level in halophiles would inhibit enzymes in non-halophiles, the protein structure of halophiles is unique as they have a higher level of acidic amino acids and a lower level of hydrophobic amino acids than found in non-halophiles (Mongodin et al. 2005; Kastritis et al. 2007).

The physiological activities of halophiles are relatively broad and include both aerobes and anaerobes, which have been found in the Red Sea and soda lakes around the world. Few bacteria have been isolated as halophiles and a partial list would include *Salinisphaera shabanensis*, a gram-negative, monotrichous coccus; *Haloplasma contractile,* a cell wall-less bacterium with a contractile bacterium; *Flexistipes sinusarabici,* flexible rods that are anaerobic heterotrophs; *Natranaerobius thermophilus,* an alkalithermophilic anaerobe that uses both K^+ and glycine

Table 5.4 Examples of compatible solutes of halotolerant and halophilic bacteria and archaea.

Class of solutes	Specific compounds
Zwitterionic (charged)	betaine, ectoine, hydroxyectoine, *N*-acetyldiaminobutyrate, *N*-acetyl-β-lysine, β-glutamine
Uncharged	α-glucosylglycerol, α-mannoslyglyceramide, trehalose, sucrose, *N*-α-carbamoyl-L-glutamine amide, *N*-acetylglutaminylglutamine-amide
Anionic (carboxylates)	L-α-glutamine, β-glutamate, hydroxybutyrate, poly-β-hydroxybutyrate, α-glucosylglycerate, α-mannosylglycerate
Anionic compounds	α-diglycerol phosphate, di-*myo*-inositol-1,1-phosphate, cyclic-2,3-containing phosphate diphosphoglycerate, sulfotrehalose or sulfate

Source: Reproduced with permission from Roberts (2005).

Box 5.2 Salterns

Collection sites for the production of salt from saline waters are referred to as salterns. Salterns may be an area where seawater or brackish water is concentrated by evaporation. In regions where evaporation is too slow, instillations are constructed where natural saltwater is placed in a large container, often referred to as a pan, and water is evaporated by heating. Salt crystals are removed from the brine by raking and additional salt water is added. Orange to red colors are characteristic of the concentrating brine and this color reflects the presence of haloarchaea, halophilic algae, and brine shrimp.

betaine to balance osmotic pressures; *Natranaerobius jonseii,* requires 1.4 M Cl^- for growth; *Halomonas elongate,* a gram-negative denitrifier; and *Thioalkalivibrio sulfidiphilus,* a sulfur oxidizing chemolithotroph (Ma et al. 2010). Some of the members of the *Archaea* are also halophiles that have been isolated from salterns (Box 5.2), saline soil, salted fish and hides, soda lakes, and the Red Sea. There is considerable metabolic diversity in the archaeal halophiles with of *Haloferax mediterranei, Haloarcula marismortui,* and *Halococcus saccharolyticus* using hexoses and pentoses while *Halobacterium salinarum* and numerous other archaeal strains use amino acids but not sugars (Falb et al. 2008). Several of the haloarchaea are also alkaliphiles, including *Natronobacterium, Natrinema, Natrialba, Natromonas, Natronococcus,* and *Natrorubrum.*

Salinibacter ruber was isolated by Antón et al. (2002) from saltern crystallizer ponds in Spain and accounts for a significant quantity of the microorganisms present. This halophilic red bacterium grows optimally at 2.5–3.9 M salt and will not grow if the salt concentration is less than 1.7 M. The genome of *S. ruber* contains a hypersalinity island, which is a collection of 19 genes from both bacterial and archaeal sources suggesting that this halophilic bacterium has obtained salt tolerance by horizontal gene transfer from members of the halophilic *Archaea.* From genomic analysis, *S. ruber* has four rhodopsin genes including genes with high similarity to the H^+ exporter and another gene to the Cl^- uptake pump found in archaeal genomes (Mongodin et al. 2005). Proton pumping in *S. ruber* is attributed to a unique xanthorhodopsin, which has the retinal chromophore plus a carotenoid pigment to assist in light harvesting. Some of the red color of salt concentrating ponds may be caused by *S. ruber.*

The greatest abundance of halophilic prokaryotes is associated with members of the *Archaea.* The presence of rhodopsin was first identified in *Halobacterium salinarum* (formerly called *Halobacterium halobium*) and these unique light sensitive proteins in the membrane are important for haloresistance. One of the proteins was initially identified as bacterial rhodopsin but this term is being changed to archaeal rhodopsin to accurately reflect that *Halobacterium* is a member of the archaea and not a bacterium. There are four rhodopsins in *H. salinarum* and two of these (bacteriorhodopsin and halorhodopsin) are light-driven ion pumps (Ernst et al. 2014). Bacteriorhodopsin is a major protein in the archaeal cell membrane that may represent half of the membrane protein and accounts for the red color characteristic of cultures of *H. salinarum.* Bacteriorhodopsin uses light to export H^+ from the cytoplasm and this H^+ may be used by the cell for proton-driven ATP synthesis. Halorhodopsin pumps chloride ions (Cl^-) into the cytoplasm to help stabilize the osmotic pressures on the cell membrane. The remaining two rhodopsins participate in chemotaxic activity of the cell. Since the discovery of these rhodopsins in *H. salinarum,* rhodopsins have been found in numerous marine microorganisms (Ernst et al. 2014).

The isolation of halophilic archaea from salt deposits that are millions of years old has stimulated discussion about the longevity of these organisms (Stan-Lotter and Fendrihan 2015 and references therein). Halophiles were cultivated from Permian or Triassic salt and the resulting

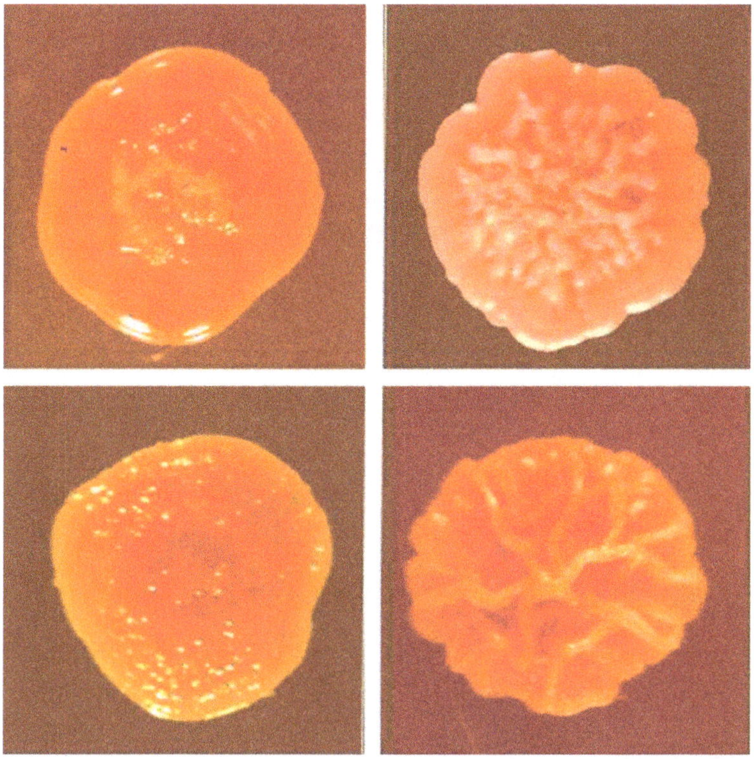

Figure 5.17 Haloarchaeal colonies on agar plates containing M2 medium and 20% salt. Permian or Triassic salt was dissolved under sterile conditions and the solutions were spread on the agar. Incubation was at 37–40 °C for three to four months (Stan-Lotter et al. 2000; Stan-Lotter and Fendrihan 2007). Pigmentation indicates the presence of carotenoids and bacterioruberin (a 50 carbon carotenoid pigment). Diameter of the colonies is about 7–8 mm. *Source:* Courtesy of Helga Stan-Lotter.

colonies were bright orange due to the presence of carotenoids and bacterioruberin (a 50 carbon carotenoid pigment (Figure 5.17) (Stan-Lotter et al. 2000; Stan-Lotter and Fendrihan 2007). *Halococcus salifodinae* was isolated from Permian and Permo-Triassic salt deposits and cultured in the laboratory (Figure 5.18) (Denner et al. 1994; Stan-Lotter et al. 1999). *Halobacterium salinarum* was isolated from fluid inclusions found inside 97 000-year-old salt crystal from Death Valley in California. It is unclear how these halophiles survive for very long periods because they do not produce spores or cysts. Under controlled conditions, it was observed that when *H. salinarum* is suspended in fluid inclusions from halite crystals, a rod-shaped cell is converted to four small spheres which are 0.4 μm in diameter. These haloarchaeal spheres are viable for years and become rod-shaped cells when placed in appropriate media. The change in surface to volume ratio with *H. salinarum* going from a rod to a small sphere may contribute to long term survival. Additionally, the viability of some strains of haloarchaea may be attributed to their polyploidy. Polyploidy in bacteria and archaea is dealt with in Chapter 11.

An intriguing halophile is *Haloquadratum walsbyi* which is unique because it is square (Figure 5.19). While each cell is 2–5 μm in diameter it is only 0.1–0.5 μm thick and this gives the organism a surface to volume ratio that is the highest found in the microbial world (Stan-Lotter and Fendrihan 2015). Cells of *H. walsbyi* have numerous gas vacuoles that enable the flat cells to float on the water surface. These cells are covered with halomucin, which protects against desiccation and viruses. Based on genomic analysis, this haolophile has several genes for

Figure 5.18 High-resolution field emission scanning electron micrograph of *Halococcus salifodinae* Blp which was isolated from a Permian salt deposit (Denner et al. 1994; Stan-Lotter et al. 1999). G. Wanner at Ludwig Maximilian University, Munich, removed the halophile from the M2 liquid growth medium at the end of the exponential phase and visualized the cells at the accelerating voltage of 3 kV. *Source:* Courtesy of Helga Stan-Lotter.

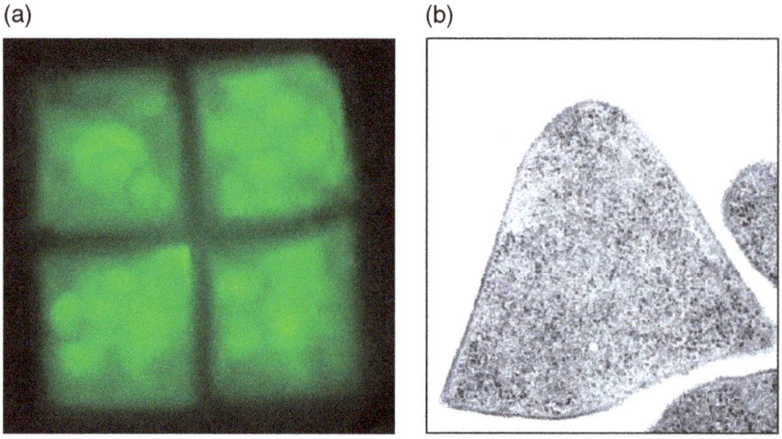

Figure 5.19 Unusual shapes for archaeal halophiles: (a) *Haloquadratum walsbyi*. Creative Commons: Attribution-ShareAlike License; Attribution: By Rotational (Own work) Public Domain, via Wikimedia Commons *Haloquadratum walsbyi*.; (b) triangular archaea, *Haloarcula japonica* strain TR-1. *Source:* Courtesy of Yayoi Nishiyama.

phosphate and phosphonate uptake and genes to encode for bacteriorhodopsin I and II (Dyall-Smith et al. 2011). While this square archaeal cell may be the dominant microorganism in a high salt environment, it is not the only uniquely shaped cell. Triangular halophilic archaea *Haloarcula japonica* strain TR-1 has also been reported (Nishiyama et al. 1992) (see Figure 5.19).

Environments may contain several different species of halophilic microorganisms and this is especially true if the environment contains a sufficient level of nutrients in addition to salt. In the evaporation ponds near San Francisco (Figure 5.20), those with low salinity are green where there is growth of green algae whereas those with moderate salinity are orange due to the presence of brine shrimp. In the ponds with high salt concentrations, the dark red color is due to *Dunaliella*, a haloresistant algae, and the pigmented haloarchaea. In addition to the halophilic microorganisms previously mentioned, those cultured from the environment include members of the *Halorubrum, Halobaculum, Haloferax, Haloarcula, Halococcus, Halogeometricum,* and *Haloterrigena* genera. Halophilic archaea are not only found in harsh environments, some are also associated with our brine-based foods. *Chromohalobacter beijerinckii* has been isolated from salted herring and salted beans and *Tetragenococcus halophilus* is commonly isolated from soy sauce.

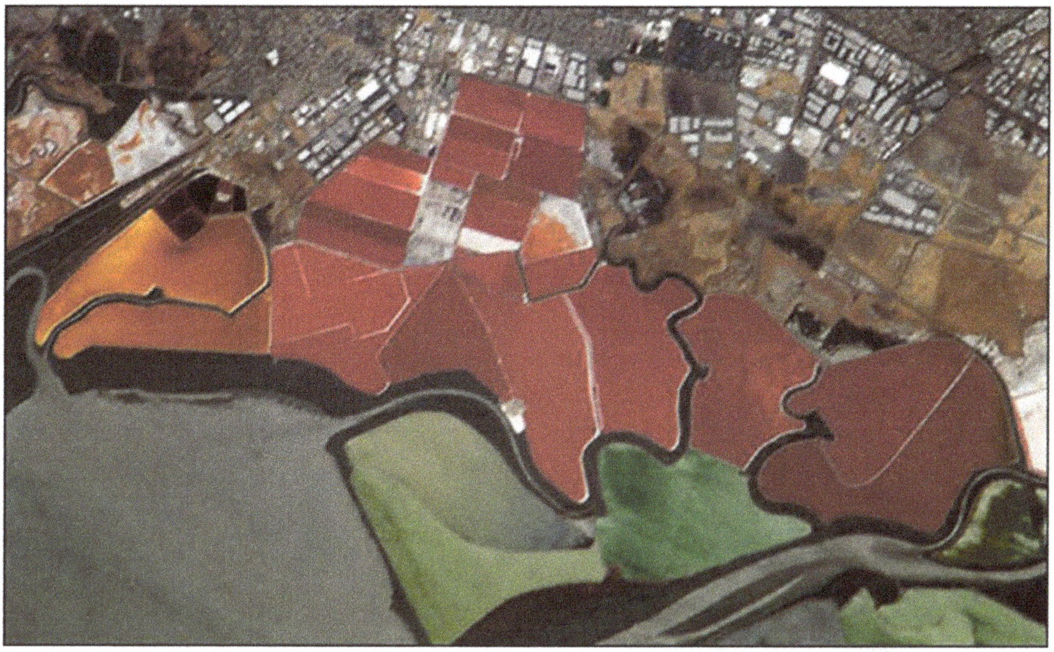

Figure 5.20 Satellite photograph of salt evaporation ponds at San Francisco Bay, California. *Source:* San Francisco Bay Salt ponds 2002.jpg. NASA.

Fungi can grow in many different environments and numerous fungi are halotolerant. Fungal cells respond to elevated NaCl concentrations by activating the high osmolytic glycerol signaling pathway, which causes the accumulation and export of glycerol to stabilize osmotic pressure in the cell. In high saline environments, ATP driven Na^+ export and other ion transport systems are used by yeast to maintain membrane potential. *Wallemia ichthyophaga* grows in solutions saturated with NaCl and this fungus is a true obligate halophile because it requires 10% NaCl for growth. In the past two decades, over 20 new species of *Emericella, Cladosporium, Aureobasidium,* and *Candida* were isolated from solar salterns. Additionally new isolates of *Trimmatostroma salinum* and *Eurotium halotolerans* were cultivated. The most abundant fungi in solar salterns are black yeasts that have been identified as *Hortaea werneckii, Phaeotheca triangularis, Aureobasidium pullulans,* and *Trimmatostroma salinum* (Gunde-Cimermann and Zalar 2014; Ma et al. 2010).

Halophilic microorganisms are found in salt marshes and solar salterns where seawater is collected in pools and allowed to evaporate. When the salt concentration of the evaporating solution reaches $300\,g\,l^{-1}$, NaCl forms crystals. The resulting brine is known as bittern and contains high concentrations of Mg^{2+}. Most of the halophilic fungi are inhibited by the presence of high concentrations of chaotropic salts which include $MgCl_2$ and $CaCl_2$. Fungi that grow in the presence of high concentrations of NaBr, $MgCl_2$, or $CaCl_2$ are called chaophilic or chaotolerant fungi (Zajc et al. 2014). There are relatively few chaotolerant fungi with most fungi growing on elevated concentrations of kosmotropic salts (NaCl, KCl, and $MgSO_4$).

Chaotolerant fungi have been isolated from the Dead Sea, solar salterns in Slovenia and Puerto Rico, salt marshes in Egypt, the rhizosphere of the Mediterranean, and the brine channels of sea ice (Zajc et al. 2014). Notable chaotolerant fungi include *Xeromyces bisporus,* a xerophilic filamentous ascomycete; *Hortaea werneckii,* a black yeast; *Aureobasidium subglaciale,*

Table 5.5 Tolerance of selected fungi to different salt concentrations.

	Highest molarity with growth				
Fungi	**NaCl**	**KCl**	**$MgSO_4$**	**$MgCl_2$**	**$CaCl_2$**
Aureobasidium subglaciale	2.0	2.0	2.0	0.75	1.2
Clasosporium cladosporoides	4.0	4.5	3.0	1.6	1.0
Eurotium chevalieri	4.0	4.6	3.0	2.1	1.9
Hortaea werneckii	5.0	4.5	3.0	2.1	1.7
Wallemia ichthyophaga	5.0	4.5	3.0	2.1	0.5
Xeromyces bisporus	2.5	3.5	2.5	1.6	1.0

Source: Reproduced with permission from Zajc et al. (2014).

a black yeast-like fungus; *Wallenia ichthyophaga*, a halophilic basidiomycete; *Cladosporium cladosporioides*, a saprophytic filamentous ascomycete; and *Eurotium chevalieri*, an ascospore producing fungus. The tolerance of several fungi to chaotropic salts and kosmotropic salts is given in Table 5.5. The mechanisms of resistance of fungi to salt concentrations that are toxic to most fungi remain to be established.

5.9 Radioresistant Microorganisms

One of the unique properties displayed by microorganisms is resistance to ionizing radiation and at this time there are only a few genera with high radiation resistance. The radiation resistant archaeon *Thermococcus gammatolerans* has been isolated from hydrothermal chimneys in the mid-Atlantic Ridge and Guyamas Basin (Zivanovic et al. 2009), but little research has been conducted on it to understand the mechanisms of resistance to ionizing radiation. Bacterial genera displaying considerable resistance to ionizing radiation include *Chroococcidiopsis* (a cyanobacterium), *Rubrobacter* (an actinobacteria), and *Deinococcus*. The bacterial genus *Deinococcus* was discovered in 1956, the genome was sequenced in 1999, and it has become the organism most studied for resistance to ionizing radiation (Munteanu et al. 2015). *Deinococcus radiodurans* is a polyextremophile because it can survive dehydration, nutrient starvation, cold exposure, and radiation exposure to account for its listing in the *Guinness Book of World Records* as the world's toughest bacterium. This bacterium is widely distributed in soils around the world and following nuclear reactor accidents that have released high levels of radiation, *Deinococcus* spp. are the dominant bacteria present. *Deinococcus radiodurans* survives an exposure of 3 million rads of gamma radiation, whereas spores of food spoilage bacteria are inactivated at this radiation level (Duggan et al. 1963). With an exposure of 10 kGy, *D. radiodurans* survives even though the cellular damage to its DNA results in over 100 double strand breaks (Daly et al. 2007). Since *Escherichia coli* and *Shewanella* spp. cannot survive after more than one or two double strand DNA breaks, the focus on *D. radiodurans* survival has been on DNA repair mechanisms. As reviewed by Daly et al. (2007), *D. radiodurans* uses DNA repair proteins found in radiation sensitive bacteria. Ionizing radiation damages cellular components creating oxidative stress in the cell, with the formation of oxygen and hydroxyl free radicals leading to double strand breaks in DNA. It is proposed by Daly et al. (2007) that the low level of iron in the cytoplasm may account for a low incidence of free radicals produced in the cytoplasm of *D. radiodurans* and without free radicals the DNA repair proteins are not damaged. Cells of *D. radiodurans* have iron localized outside the cell, with manganese replacing many of the iron-related metabolic requirements.

While *D. radiodurans* is the model organism for radiation resistance study, several radiation resistant bacteria have been isolated from Sonoran Desert soil (Rainey et al. 2005). When Sonoran Desert soil was subjected to 17 kGy, species of *Geodermatophilus* and *Hymenobacter* were isolated in addition to *Deinococcus*. Exposure of the desert soil to 5–9 kGy, resulted in isolation of *Deinococcus, Hymenobacter, Kineococcus, Kocuria,* and *Methylobacterium* species. The presence of radiation resistant bacteria in the desert raises the question: What is the connection between desiccation resistance and ionizing radiation resistance in bacteria?

5.10 Membrane Adaptations to Extreme Conditions

The plasma membrane functions in microorganism to retain cytoplasmic molecules and to control the passage of molecules into the cytosol. While O_2, N_2, H_2, and H_2O readily move across plasma membranes, the passage of ions and solute molecules across the plasma membrane is highly regulated. The basic structure of the plasma membrane is a lipid layer, which makes it impermeable to small charged ions, including protons. Hydrophobic peptide segments stabilize the proteins interspersed in the lipid layer of the membrane. A fundamental requirement of cell membranes is to maintain appropriate membrane fluidity or flexibility. The lipid of the membrane can become a liquid crystalline phase or a rigid gel phase dependent on temperature. In the rigid gel phase, the membrane is "frozen," which makes the membrane impermeable and the membrane proteins no longer function. The rigid gel phase occurs if the temperature is below the optimum for a given lipid membrane and if the temperature is above the optimum for that lipid, the matrix of the lipids become disoriented (melts) resulting in loss of membrane function. Therefore, the lipid molecules making up the membrane will determine the optimum temperature for growth of a specific microorganism. The core lipids are different for bacterial membranes and archaeal membranes (Siliakus et al. 2017). In bacterial membranes, fatty acids are joined to the glycerol backbone by ester bonds; however, ester bonds are susceptible to being broken by hydrolysis in hot acidic environments. In contrast, archaeal cells use ether bonds, which are not susceptible to hydrolysis and enable archaea to live in hot acidic environments. Extremophilic bacteria and archaea often grow in environments containing multiple stress factors (acidic, alkaline, hot, cold, salty, or high pressure) and this makes it difficult to identify a single component responsible for maintaining membrane motility. For example, the ability of microorganisms to grow under HHP and high osmotic pressure will reflect overlapping physiological adjustments to temperature and pH levels in the environment. While the characteristics of the membrane lipids of halophiles (Ma et al. 2010; Falb et al. 2008) and xerophiles (Leong et al. 2015) have been described, generalizations about lipid composition is based on relatively few microorganisms.

5.10.1 Low Temperatures

To enable bacteria to grow at a low temperature, specific lipids need to be present in the membrane to maintain the liquid crystalline phase (Siliakus et al. 2017). Psychrophiles are bacteria that grow optimally at a low temperature but many bacteria may grow at temperatures significantly lower than their optimum. Based on evaluations of psychrotolerant *Sphingobacterium antarcticus, Planococcus halocryophilus,* and *Micrococcus roseus* growing at low temperatures, membrane lipids have appreciable levels of trans-monounsaturated fatty acids, short chain fatty acids of <12 carbons, iso-branched chain fatty acids, and carotenoids.

Psychrophiles (i.e. *Clostridium psychrophilum, Colwellia psychrorerythraea, Psychromonas ingrahamii, Desulfotalea psychrophila*) grow in cold (~ 5 °C) environments and have lipids with appreciable levels of short chain fatty acids, unsaturated fatty acids, polar carotenoids, and glycolipids. The presence of omega-3 and omega-6 polyunsaturated fatty acids and cyclopropane fatty acids are common in lipids of psychrophilic bacteria. The lower limit of bacterial growth is likely between −20 and − 30 °C while few archaea grow at less than 10 °C (Clarke et al. 2013).

Of the relatively few psychrophilic archaea, the best studied are *Cenarchaeum symbiosum* and *Methanogenium frigidum,* which grow optimally at 10 °C and 15 °C, respectively (Siliakus et al. 2017). With this small sample, most generalizations concerning lipids in archaea growing at low temperatures are based on cold stress response and include the following: (i) high levels of unsaturated archaeols, (ii) the presence of hydroxyarchaeol with the hydroxyl at the C-3 position, (iii) the quantity of diether moieties is greater than the quantity of tetraether structures, and (iv) pentacyclization, which is the addition of several cyclopentane rings along the biphyanyl molecule.

5.10.2 High Temperatures

For growth at high temperatures, the lipid changes in the membrane are the opposite of those for a cold environment. Many bacteria can grow when the temperature of the environment increases by 10 °C or even 20 °C and these bacteria have adjusted their membrane lipids to maintain a liquid crystalline phase. To maintain membrane stability at high temperatures, bacteria have a high percentage of saturated fatty acids with a branched carbon chain (Siliakus et al. 2017). Thermophilic bacteria of the genera *Thermotogales, Thermodesulfotobacterium, Thermoanaerobacter, Thermosipho, Aquifex,* and *Fervidobacterium* grow above 70 °C and have ether bonded lipids as well as ester bonded lipids. Additionally, these thermophilic bacteria have lipids containing tetraethers, tertaesters, and diethers. Bacteria growing above 80 °C have only ether linked lipids suggesting that ester bonds in lipids are temperature labile. As reviewed by Siliakus et al. (2017), restricted lipid movements in the plasma membrane are needed for hyperthermophilic archaea that grow at temperatures approaching and exceeding 100 °C. To achieve restricted lipid movement in the membrane, archaea have tetraether lipids spanning the membrane and some have macrocyclic archaeols.

5.10.3 pH Extremes: Low and High

It is well accepted that neutrophilic bacteria quickly become nonviable when placed in environments that are extremely acidic or alkaline. The type of lipids in the membrane is considered to be one of the important features that enable bacteria to become specialized as acidophiles and alkaliphiles (Siliakus et al. 2017). Bacteria growing at <pH 2 use numerous membrane ion transporters and special lipids to maintain pH homeostasis (Baker-Austin and Dopson 2007). Using *Alicyclobacillus acidiphilus* (Matsubara et al. 2002) and *Acidithiobacillus ferrooxidans* (Mykytczuk et al. 2010) as model systems, membranes of acidithiophilic bacteria have appreciable concentrations of iso- and anteiso-branched chain fatty acids, cyclopropane fatty acids, saturated and unsaturated fatty acids, and the unusual β-hydroxy-, ω-cyclohexyl fatty acid. Note that archaea are capable of growing at pH <1 while *Picrophilus oshimae* (van de Vossenberg et al. 1998) with an optimum growth at pH 0.7 is the only bacterium to grow below pH 1. *Acidilobus sulfurireducens* (Shimada et al. 2008) and *Thermoplasma acidophilum*

(Boyd et al. 2011) are acidophilic archaea that have membrane spanning lipid molecules with methyl or cyclopentane rings attached giving closely packed lipids and membrane rigidity. Numerous acidophilic archaea have been isolated and their optimum pH for growth are: *Halarchaeum acidiphilum*, pH 4.5 (Minegishi et al. 2010); *Metallosphaera sedula*, pH 2 (Huber et al. 1989); *Acidianus copahuensis*, pH 3 (Urbieta et al. 2017); *Acidianus brierleyi*, pH 1.5–2 (Segerer et al. 1986); and *Metallosphaera sedula*, pH 2 (Huber et al. 1989). Because archaeal acidophiles are polyextremophiles, which are tolerant to several stresses in the environment, it is difficult to assess which of the lipids found in them is responsible for resistance to acids.

5.11 Astrobiology

Astrobiology is the search for life beyond planet Earth. Related questions could be about how life arose, and whether life-forms can move beyond their planet of origin. While it might be fun to imagine extraterrestrial life-forms based on science fiction books or movies, the most likely forms of extraterrestrial life are microorganisms, given their ability to survive in extreme environments. We now examine how the study of extremophiles relates to astrobiology.

If liquid water is a requirement for life, there are a small number of locations within our Solar System that are considered as potential candidates for extraterrestrial life or that may have evidence of prior life. These include Mars and some of the ice-covered moons of Jupiter and Saturn, notably Europa (moon of Jupiter) and Enceladus (moon of Saturn). Several unmanned spacecrafts have visited Mars. One of the more is the Martian Rover, *Curiosity*, which launched on 26 November 2011 and landed on 6 August 2012. Currently, the Martian Rover continues to actively take photographs and carry out spectrophotometric analyses of rock samples to look for chemical evidence of life and to investigate rock chemistry. Details of the equipment on board are on the mission website (https://mars.nasa.gov/msl/mission/instruments/spectrometers). One major biological concern during the planning for trips to other worlds (Mars, our moon, etc.) is that of planetary protection because we do not want to contaminate another planet (or moon) with microorganisms from Earth. Clean rooms are used for spacecraft assembly, and spacecraft disinfection protocols and tests are carried out as a precaution. Several organisms appear to be problematic, including oligotrophic, spore-forming gram-positive *Bacillus pumilus* and *Bacillus licheniformis* bacteria that resist the commonly used spacecraft disinfectant hydrogen peroxide (H_2O_2) and *Deinococcus radiodurans*, which is resistant to high levels of radiation (Munteanu et al. 2015). In future human expeditions, we do not want to jeopardize the human population with an extraterrestrial disease, and so precautions, including quarantine, are being explored. During the initial Apollo moon missions, astronauts were quarantined for 21 days upon return to Earth.

While there is a growing body of evidence that Mars may have had surface water in the distant past, the most likely location for water would be at the poles or within caves or other subsurface environments. Extremely arid environments are being studied as a terrestrial analog for a Martian environment. One such location is the dry valleys in Antarctica near McMurdo Station research centre has an average temperature range from −15 °C to −30 °C. The dry valleys are a polar desert with negligible precipitation (<100 mm/yr) and the average humidity ranges from 55% to 74%, although the humidity can range from a minimum of <10% to a maximum of 100% (Doran et al. 2002). The average temperature on Mars is −63 °C with a range from −140 °C to +30 °C. Although Mars is very dry, there have been reports of increased relative humidity and even transient frost during the Martian night on some Martian rovers (NASA Mars website). Work by E.I. Friedmann (1986) and others showed that microbial communities in deserts could survive in the surface layer of rocks (endolithic microbial communities). In the

McMurdo dry valleys, endolithic communities within porous rocks are most prominent in the first 1–2 mm depth, being limited by light penetration and associated primary production. The extreme weather conditions restrict life on the rock surface. These microbial communities obtain water from transient condensation on the rocks from the atmosphere or from melting snow.

Another terrestrial environment with similarities to Mars is the Atacama Desert in northern Chile. This is one of the driest environments on Earth with some weather stations having never recorded any precipitation. There are a number of lichens and other microorganisms that inhabit this environment. Although warmer than the dry valleys in Antarctica, it is more accessible and is being used as a terrestrial environment for testing prototypes of future Martian exploration vehicles. Using metagenomics, investigators have identified various microbiological communities containing Archaea, Eukarya, and Bacteria in the Atacama. Microbial environments tested included environments under rocks (hypolithic environments), soils at various depths, and stromatolites in saline pools. Current work in the Atacama and its potential application to Mars exploration are summarized in (Azua-Bustos et al. 2012).

Europa and Enceladus are considered as other potential sources of extraterrestrial life in the solar system. Europa is slightly smaller than Earth's moon, and is the second closest moon to Jupiter's surface. Enceladus is Saturn's sixth largest moon but is much smaller (500 km diameter) than Earth's moon (3400 km diameter). Although covered with ice, there is evidence from observations through the Hubble Space Telescope and data from several interplanetary NASA spacecraft missions that provide evidence for liquid water underneath the thick icy covering on both these moons. Although the surface temperatures of these moons are very cold (−160 °C Europa; −198 °C Enceladus), geothermal energy from the moons' interiors would provide sufficient heat for liquid water. If life does exist on these two moons, the likely primary production would arise from chemosynthesis (mediated by chemoautotrophs). Lake Vostok and Lake Whillans in Antarctica are considered to be terrestrial analogs to Europa and Enceladus. These two subglacial lakes in Antarctica have been buried under very thick ice (4 km thick ice in the case of Lake Vostok; 800 m in the case of Lake Whillans) and have been isolated from other environments for an estimated 15 million years. Recent metagenomic-based community analysis has been used to analyze communities in these two lakes and has shown the presence of genes associated with chemosynthesis, which implies that the primary production in these two remote lakes is facilitated by chemoautotrophs (Christner et al. 2014; Shtarkman et al. 2013). Hydrothermal vent communities with their chemosynthesis-based food chains (described earlier) represent another terrestrial model ecosystem in the search for life on Europa and Enceladus.

5.12 Nutrient Limited Environments

Microbiology students are no doubt familiar with the concept of growth in rich media, where organisms such as *Escherichia coli* may double in population every 30 min or less during exponential growth. In nature, organisms rarely encounter ideal nutrient levels and so starvation survival is a major aspect of microbial life. Limiting nutrients and Liebig's Law of the Minimum (described in Chapter 2) is certainly applicable in many ecosystems. There are several strategies used by many organisms to cope with nutrient limitation (listed in Table 5.6).

Dormancy is a widespread biological strategy for surviving conditions unsuitable for growth; for example, higher organisms may hibernate, which is typically used by some mammals and reptiles to survive winter. Microorganisms also frequently use a dormancy strategy; for example, endospores are formed by several genera, notably *Bacillus* and *Clostridium* spp. in

Table 5.6 Approaches used by microorganisms in response to low or limiting nutrients.

Strategy	Mechanism and concept	Example
Biofilm growth	Adhesion to surfaces Access to adsorbed or insoluble nutrients	Very common in most microorganisms
Use of alternate nutrients	Organisms adjust metabolism (gene regulation) when preferred nutrient unavailable (classic example is *E. coli* use of lactose when glucose unavailable and lactose is present)	*Pseudomonas* spp. can use a wide variety of nutrients. *Synechococcus* can fix nitrogen when nitrogen-limited
Increased affinity for essential nutrient	Expression of genes for high affinity nutrient uptake	Siderophore production (Fe-chelating molecules) and associated high-affinity transport for iron acquisition. Ferric uptake regulator (Fur) is a key component (Seo et al. 2014)
Stringent response and other starvation responses	Global shift in gene regulation in bacteria because of amino acid starvation. Results in drastic change in metabolism and gene expression, related to starvation-survival	Studied extensively in *E. coli* but is very widespread in both gram-negative and gram-positive bacteria (Gaca et al. 2015)
Dormancy	Organisms become dormant when conditions not suitable for growth, recover when conditions change	Endospore formation by *Bacillus, Clostridium* spp. Viable but non-culturable stage of some bacteria (e.g. *Vibrio* spp.)
Lysogeny	Bacteriophage (bacterial viruses) incorporate viral genes into host chromosome and are replicated along with host. Lysogenic phage may cause drastic alterations in host cells	Phage lambda and lysogeny in *E. coli* Genome analyses of most organisms shows incorporation of viral DNA in chromosomes
Association with other organisms (synergy)	Microbial synergy (allows organisms to jointly participate in metabolism of a compound)	Nitrifying bacteria – typically ammonia oxidizing bacteria (e.g. *Nitrosomonas* spp.) produce nitrite, that is then oxidized by nitrite-oxidizing bacteria (e.g. *Nitrobacter* spp.)
Competition (between organisms, A and B)	Differences in growth rate, nutrient acquisition, making environment unsuitable for competitor	*Lactobacillus reuteri* (resident in normal flora) produces lactic acid and lowers pH. It also produces an antimicrobial (reuterin) that inhibits some pathogens
	Interference; organism A interferes with signals in organism B	*Delisea pulchra* (red macroalga) produces furanones that disrupt signaling and biofilm formation in many bacteria
	Organism A kills organism B by antibiotic or bacteriocin production	*Streptomyces griseus* (soil isolate) produces the antibiotic streptomycin that can kill competing organisms in soil

response to nutrient limitation. Spores differ in structure from the vegetative (growing) cell in having a thicker wall, lower water content, and a unique compound (dipicolinic acid) th\t is involved in the protection of DNA and a few key proteins. Once conditions become favorable, the spore will germinate into a growing cell. The survival of endospores varies depending on other environmental conditions but is typically for years, decades, or longer. Gruinard Island near Scotland was the site of anthrax testing and release (the causative agent is *Bacillus anthracis*) in 1942 and 1943 during the Second World War and was quarantined for almost 50 years due to high concentrations of *B. anthracis* endospores and the associated risk of infection. The contaminated soil areas were successfully treated in the 1990s using formaldehyde and the island was eventually released from quarantine (Manchee et al. 1994).

There have been several reports of viable bacteria being isolated from very ancient rocks. In one instance, a bacterial spore was removed and cultured from an insect encased in amber that was estimated to be 25–40 million years old (Cano and Borucki 1995). This revived organism was shown by sequencing to be similar to *Bacillus sphaericus*. The oldest isolate found and revived was the halotolerant *Bacillus* strain 2-9-3 that was isolated from a salt crystal estimated to be 250 million years old (Vreeland et al. 2000). In both of these studies, extensive care was taken to select and sterilize samples to discount contamination. Other potentially extremely long-lived endospore-forming organisms have been found in deep subsurface investigations (Cano and Borucki 1995).

While microorganisms typically encounter a feast or famine experience, there are some environments with extremely oligotrophic conditions. Examples of microorganisms from natural environments include deep subsurface environments and the pelagic ocean (open water, far from land). Artificial environments such as clean rooms associated with computer manufacturing and spacecraft assembly, also have extremely oligotrophic bacteria. In contrast to other extreme environments, much less is known about the mechanisms that enable survival of extreme oligotrophs. One extreme oligotroph from the pelagic ocean (SAR-11, *Candidatus* Pelagibacter ubique), first identified by metagenomic analysis, has been grown and has been shown to play a key role in carbon cycling in marine ecosystems. However, for the most part, our knowledge of extreme oligotrophs and their ecological role is based on metagenomic and related molecular approaches (Itävaara et al. 2016; La Duc et al. 2007; Lever et al. 2015).

5.13 Volcanic Surfaces

Some of the most inhospitable environments on Earth are on newly formed volcanic cones because of the extreme heat and the release of toxic gases during a volcanic eruption. There are continental volcanoes which are characterized by eruption of Mount St. Helens and Eyjafjallajökull, Iceland as well as oceanic volcanoes which include submarine mud volcanoes and submarine volcanoes (Figure 5.21). The colonization of microorganisms following volcanic eruption at Eyjafjallajökull, Iceland was examined, and it was reported that bacteria were established three to five months after eruption (Kelly et al. 2014). After three to five months of eruption, specialist bacteria such as nitrogen-fixing organisms related to *Herbasprillum* and *Thiobacillus* sp. were isolated and following the appearance of the nitrogen fixers, the phototrophs, acidobacteria and actinobacteria become abundant in this harsh environment. The Sairecabur volcano in the Atacama Desert, Chile is considered to be a good Mars analog because of the extreme environment. As reviewed by Pulschen et al. (2015) and Schmidt et al. (2017), there is considerable diversity of bacteria, cyanobacteria, yeasts, and fungi in the Atacama Desert volcano soil but it is unknown if these are dormant cells brought

Figure 5.21 Eyjafjallajokull volcano plume. Overlooking the Eyjafjallajökull glacier and the ongoing volcano eruption from Hvolsvöllur on 17 April 2010. *Source:* Boaworm, https://commons.wikimedia.org/wiki/File:Eyjafjallajokull_volcano_plume_2010_04_17.jpg. Licensed under CC BY-SA 3.0.

there by air currents of if these represent growing populations. Microbial cells must be polytolerant to extreme environments because in addition to salt stress, desiccation, and temperature extremes, UV intensity at this high elevation can be detrimental to cell viability.

Following the eruption of an underwater volcano near El Hierro Island, Canary Archipelago, extensive mats of filamentous bacterium colonized near the top of the volcanic cone (Danovaro et al. 2017). The bacterial filaments called "Venus's hair" consisted of *Thiolava veneris,* a new order of the *Thiotrichales,* and this organism was the basis of a food web that supported a diverse ecosystem. Submarine mud volcanoes are an unusual phenomenon in which hydrocarbons including methane are released in large "bubbles" from the muddy sediments of the deep seafloor and methane becomes dispersed in the water without reaching the surface. It is generally considered that the microbial community in the Peru Margin, the Juan de Fuca Ridge near Oregon and near the Shimokita Peninsula of Japan is stratified with the organisms in the mud volcano sediments being different from those in the seawater over the mud volcanoes. Using 16S rRNA analysis, anaerobic heterotrophs associated with the *Atribacteria* taxon are present in the methane plumes but not in the seawater adjacent to the mud volcanoes (Hoshino et al. 2017). Obviously, the submarine mud volcanoes not only release methane from the subseafloor but the expulsion of water functions to disperse various bacteria, including *Atribacteria* from the sub-seafloor into the oceans. Since about half of the microbial cells in the ocean are associated with the sub-seafloor, the submarine mud volcanoes may have an important role in marine nutrient cycling.

5.14 Summary

Microorganisms are found in various extreme environments where bacteria and archaea frequently display resistance to multiple conditions. Using the deep subsurface mines of South Africa as an extreme environment, a diverse biological environment was detected that included Bacteria, Archaea and various examples of Eukarya, including small nematodes found inside stalactites extending from the ceiling of mine tunnels. Primary production in the extremely deep mines was attributed to sulfur-oxidizing bacteria with nitrate reduction. Unique strains of psychrophilic and psychrotolerant bacteria were isolated from glaciers, snow, and permafrost in polar and alpine regions. It has been estimated that the Antarctic and Greenland ice sheets are a considerable reservoir of bacteria and contain over 10^{25} cells. Thermophilic bacteria and hyperthermophilic archaea are found in the geothermal hot waters of hot spring and marine subsurface environments. Considerable research has addressed the types and distribution of microorganisms in the springs of the YNP, marine hydrothermal vents, and the Guaymas Basin. In the Atacama Desert of Chile, xerophilic microorganisms and their viruses have been reported and isolated using techniques similar to those used by instruments on Mars. Piezophilic microorganisms grow under HHPs and are distinguished by growing in environments that have a cold, hot, or moderate temperature. Acidophilic bacteria and archaea are present in the acidic hot springs of the YNP and in acidic waters from the coal fields of the high Arctic region. There are abundant nanosized archaea in the acidic stream from the Richmond Mine of California. Alkaliphilic bacteria are broadly distributed in alkaline soil, soda lakes, and insect guts. Halophilic microorganisms grow in high salt concentrations, including inside a salt crystal. Some cultures of halophilic microorganisms are bright red due to the presence of the red protein rhodopsin in their membranes. Ionizing radiation is lethal for most organisms; however, the bacterium *Deinococcus radiodurans* is an exception because it has a unique process for double stranded breaks in DNA induced by radiation. Extremophiles not only have structural modifications but also active transport mechanisms that enable them to grow in environments hostile to other life forms.

Discussion Questions

1. Make a list of some extreme environments where bacteria have not been reported. Discuss how the search for microorganisms could proceed in one of these environments.

2. Discuss the frequency of sampling and the number of times an environment needs to be sampled to provide a microbial description of an extreme environment.

3. Does the axiom "Everything is everywhere but the environment selects" adequately describe the presence of bacteria and archaea in extreme environments?

References

Abe, F. (2007). Exploration of the effects of high hydrostatic pressure on microbial growth, physiology and survival: perspectives from piezophysiology. *Biosci. Biotechnol. Biochem.* 71: 2347–2357.

Abe, F. and Horikoshi, K. (2001). The biotechnological potential of piezophiles. *Trends Biotechnol.* 19: 102–108.

Amend, J.P. and Teske, A. (2005). Expanding frontiers in deep subsurface microbiology. *Paleogeography, Paleoclimatology, Paleoecology* 219: 131–155.

Amrani, A., Bergon, A., Holota, H. et al. (2014). Transcriptomics reveal several gene expression patterns in the piezophile *Desulfovibrio hydrothermalis* in response to hydrostatic pressure. *PLoS One* 9 (9): e106831. https://doi.org/10.1371/journal.pone.0106831.

Antón, J., Oren, A., Benlloch, S. et al. (2002). *Salinibacter ruber* gen. nov., sp. nov., a novel, extremely halophilic member of the *Bacteria* from saltern crystallizer ponds. *Int. J. Syst. Evol. Microbiol.* 52: 485–491.

Azua-Bustos, A., Urrejola, C., and Vicuña, R. (2012). Life at the dry edge: microorganisms of the Atacama Desert. *FEBS Lett.* 586: 2939–2945.

Baker, B.J. and Banfield, J.F. (2003). Microbial communities in acid mine drainage. *FEMS Microbiol. Ecol.* 44: 139–152.

Baker, B.J., Comolli, L.R., Dick, G.J. et al. (2010). Enigmatic, ultrasmall, uncultivated archaea. *Proc. Natl. Acad. Sci. U.S.A.* 107: 8806–8811.

Baker-Austin, C. and Dopson, M. (2007). Life in acid: pH homeostasis in acidophiles. *Trends Microbiol.* 15: 165–171.

Bakermans, C., Ayala-del-Río, H.L., Ponder, M.A. et al. (2006). *Psychrobacter cryohalolentis* sp. nov. and *Psychrobacter arcticus* sp. nov., isolated from Siberian permafrost. *Int. J. Syst. Evol. Microbiol.* 56: 1285–1291.

Bergholz, P.W., Bakermans, C., and Tiedje, J.M. (2009). *Psychrobacter arcticus* 273-4 uses resource efficiency and molecular motion adaptations for subzero temperature growth. *J. Bacteriol.* 191: 2340–2352.

Borgonie, G., García-Moyano, A., Litthauer, D. et al. (2011). *Nematoda* from the terrestrial deep subsurface of South Africa. *Nature* 474: 79–82.

Borgonie, G., Linage-Alvarez, B., Ojo, A. et al. (2015a). Deep subsurface mine stalactites trap endemic fissure fluid *Archaea, Bacteria* and *Nematoda* possibly originating from ancient seas. *Front. Microbiol.* 6: 833. https://doi.org/10.3389/fmicb.2015.00833.

Borgonie, G., Linage-Alvarez, B., Ojo, A. et al. (2015b). Eukaryotic opportunists dominate the deep-subsurface biosphere in South Africa. *Nat. Commun.* (6): 8952. https://doi.org/10.1038/ncomms9952.

Bowers, K.J. and Wiegel, J. (2011). Temperature and pH optima of extremely halophilic archaea: a mini-review. *Extremophiles* 15: 119–128.

Boyd, E.S., Pearson, A., Pi, Y. et al. (2011). Temperature and pH controls on glycerol dibiphytanyl glycerol tetraether lipid composition in the hyperthermophilic crenarchaeon *Acidilobus sulfurireducens*. *Extremophiles* 15: 59–65.

Brock, T.D. (1978). *Thermophilic Microorganisms and Life at High Temperatures.* Springer Publishing.

Cano, R.J. and Borucki, M.K. (1995). Revival and identification of bacterial spores in 25- to 40-million-year-old Dominican amber. *Science* 268: 1060–1064.

Christner, B.C., Mosley-Thompson, E., Thompson, L.G. et al. (2003). Bacterial recovery from ancient glacial ice. *Environ. Microbiol.* 5: 433–436.

Christner, B.C., Priscu, J.C., Achberger, A.M. et al. (2014). A microbial ecosystem beneath the West Antarctic ice sheet. *Nature* 512: 310–313.

Clarke, A., Morris, G.J., Fonseca, F. et al. (2013). A low temperature limit for life on earth. *PLoS One* 8 (6): e66207. https://doi.org/10.1371/journal.pone.0066207.

Colman, D.R., Poudel, S., Hamilton, T.L. et al. (2018). Geobiological feedbacks and the evolution of thermoacidophiles. *ISME J.* 12: 225–236.

Corliss, J.B., Dymond, J., Gordon, L.I. et al. (1979). Submarine thermal springs on the Galapagos rift. *Science* 203: 1073–1083.

Crits-Christoph, A., Robinson, C.K., Barnum, T. et al. (2013). Colonization patterns of soil microbial communities in the Atacama Desert. *Microbiome* 2013 (1): 28. http://www.microbiomejournal.com/content/1/1/28.

Daly, M.J., Gaidamakova, E.K., Matrosova, V.Y. et al. (2007). Protein oxidation implicated as the primary determinant of bacterial radioresistance. *PLoS Biol.* 5 (4): e92. https://doi.org/10.1371/journal.pbio.0050092.

Danovaro, R., Canals, M., Tangherlini, M. et al. (2017). A submarine volcanic eruption leads to a novel microbial habitat. *Nat. Ecol. Evol.* 1 (6): 144. https://doi.org/10.1038/s41559-017-0144.

De Maayer, P., Anderson, D., Cary, C. et al. (2014). Some like it cold: understanding the survival strategies of psychrophiles. *EMBO Rep.* 15: 508–517.

van de Vossenberg, J.L., Driessen, A.J., Zillig, W. et al. (1998). Bioenergetics and cytoplasmic membrane stability of the extremely acidophilic, thermophilic archaeon *Picrophilus oshimae. Extremophiles* 2: 67–74.

Denner, E.B.M., McGenity, T.J., Busse, H.-J. et al. (1994). *Halococcus salifodinae* sp. nov., an Archaeal isolate from an Austrian salt mine. *Int. J. Syst. Bacteriol.* 44: 774–780.

Dick, G.J., Anantharaman, K., Baker, B.J. et al. (2013). The microbiology of deep-sea hydrothermal vent plumes: ecological and biogeographic linkages to seafloor and water column habitats. *Front. Microbiol.* 4: 124. https://doi.org/10.3389/fmicb.2013.00124.

Dieser, M., Battista, J.R., and Christner, B.C. (2013). DNA double-strand break repair at −15°C. *Appl. Environ. Microbiol.* 79: 7662–7668.

Doran, P.T., McKay, C.P., Clow, G.D. et al. (2002). Valley floor climate observations from the McMurdo dry valleys, Antarctica, 1986–2000. *J. Geophys. Res.* 107 (D24): 4772. https://doi.org/10.1029/2001JD002045,2002.

Duggan, D.E., Anderson, A.W., and Elliker, P.R. (1963). Inactivation of the radiation-resistant spoilage bacterium *Micrococcus radiodurans*. I. Radiation inactivation rates in three meat substrates and in buffer. *Appl. Microbiol.* 11: 398–403.

Dyall-Smith, M.L., Pfeiffer, F., Klee, K. et al. (2011). *Haloquadratum walsbyi*: limited diversity in a global pond. *PLoS One* 6 (6): e20968. https://doi.org/10.1371/journal.pone.0020968.

Ernst, O.P., Lodowski, D.T., Elstner, M. et al. (2014). Microbial and animal rhodopsins: structures, functions, and molecular mechanisms. *Chem. Rev.* 114: 126–163.

Falb, M., Müller, K., Königsmaier, L. et al. (2008). Metabolism of halophilic archaea. *Extremophiles* 12: 177–196.

Fouke, B.W. (2011). Hot-spring systems geobiology: abiotic and biotic influences on travertine formation at Mammoth Hot Springs, Yellowstone National Park, USA. *Sedimentol.* 58: 170–219.

Franzmann, P.D., Liu, Y., Balkwill, D.L. et al. (1997). *Methanogenium fngidum* sp. nov., a psychrophilic, H_2 -using methanogen from Ace Lake, Antarctica. *Int. J. Syst. Bacteriol.* 47: 1068–1072.

Friedmann, E.I. (1986). The Antarctic cold desert and the search for traces of life on Mars. *Adv. Space Res.* 6: 167–172.

Fröhlich-Nowoisky, J., Hill, T.C.J., Pummer, B.G. et al. (2015). Ice nucleation activity in the widespread soil fungus *Mortierella alpine. Biogeosci.* 12: 1057–1071.

Gaca, A.O., Colomer-Winter, C., and Lemos, J.A. (2015). Many means to a common end: the intricacies of (p)ppGpp metabolism and its control of bacterial homeostasis. *J. Bacteriol.* 197: 1146–1156.

García-Moyano, A., Austnes, A.E., Lanzén, A. et al. (2015). Novel and unexpected microbial diversity in acid mine drainage in Svalbard (78° N), revealed by culture-independent approaches. *Microorganisms* 3: 667–694.

Goodchild, A., Saunders, N.F., Ertan, H. et al. (2004). A proteomic determination of cold adaptation in the Antarctic archaeon, Methanococcoides burtonii. *Mol. Microbiol.* 53: 309–321.

Goordial, J., Davila, A., Lacelle, D. et al. (2016). Nearing the cold-arid limits of microbial life in permafrost of an upper dry valley, Antarctica. *ISME J.* 10: 1613–1624.

Guan, Z., Tian, B., Perfumo, A. et al. (2013). The polar lipids of *Clostridium psychrophilum*, an anaerobic psychrophile. *Biochim. Biophys. Acta.* 1831: 1108–1112.

Gunde-Cimermann, N. and Zalar, P. (2014). Extremely halotolerant and halophilic fungi inhabit brine in solar salterns around the globe. *Food Technol. Biotechnol.* 52: 170–179.

Hallam, S.J., Konstantinos, T., Konstantinidis, K.T. et al. (2006). Genomic analysis of the uncultivated marine crenarchaeote *Cenarchaeum symbiosum*. *Proc. Natl. Acad. Sci. U.S.A.* 103: 18296–18301.

Heath, C., Posner, M.G., Aass, H.C. et al. (2007). The 2-oxoacid dehydrogenase multi-enzyme complex of the archaeon *Thermoplasma acidophilum* recombinant expression, assembly and characterization. *FEBS J.* 274: 5406–5415.

Hoshino, T., Toki, T., Ijiri, A. et al. (2017). *Atribacteria* from the subseafloor sedimentary biosphere disperse to the hydrosphere through submarine mud volcanoes. *Front. Microbiol.* 8 (1135): 2017. https://doi.org/10.3389/fmicb.2017.01135.

Hou, S., Saw, J.H., Lee, K.S. et al. (2004). Genome sequence of the deep-sea gamma-proteobacterium *Idiomarina loihiensis* reveals amino acid fermentation as a source of carbon and energy. *Proc. Natl.Acad. Sci. U.S.A.* 101: 18036–18041.

Huber, G., Spinnler, C., Gambacorta, A. et al. (1989). *Metallosphaera sedula* gen. And sp. nov. represents a new genus of aerobic, metal-mobilizing, thermoacidophilic archaebacteria. *Syst. Appl Microbiol.* 12: 38–47.

Itävaara, M., Salavirta, H., Marjamaa, K. et al. (2016). Geomicrobiology and metagenomics of terrestrial deep subsurface microbiomes. *Adv. Appl. Microbiol.* 94: 1–77.

Jansson, J.K. and Taş, N. (2014). The microbial ecology of permafrost. *Nat. Rev. Microbiol.* 12: 414–425.

Johnson, D.B. and Hallberg, K.B. (2009). Carbon, iron and sulfur metabolism in acidophilic micro-organisms. *Adv. Microb. Physiol.* 54: 201–255.

Kastritis, P.L., Papandreou, N.C., and Hamodrakas, S.J. (2007). Haloadaptation: insights from comparative modeling studies of halophilic archaeal DHFRs. *Int. J. Biol. Macromol.* 41: 447–453.

Kelly, L.C., Cockell, C.S., Thorsteinsson, T. et al. (2014). Pioneer microbial communities of the Fimmvörðuháls lava flow, Eyjafjallajökull, Iceland. *Microbial Ecol.* 68: 504–518.

Knoblauch, C., Sahm, K., and Jørgensen, B.B. (1999). Psychrophilic sulfate-reducing bacteria isolated from permanently cold Arctic marine sediments: description of *Desulfofrigus oceanense* gen. nov., sp. nov., *Desulfofrigus fragile* sp. nov., *Desulfofaba gelida* gen. nov., sp. nov., *Desulfotalea psychrophila* gen. nov., sp. nov. and *Desulfotalea arctica* sp. nov. *Int. J. Syst. Evol. Microbiol.* 49: 1631–1643.

La Duc, M.T., Dekas, A., Osman, S. et al. (2007). Isolation and characterization of bacteria capable of tolerating the extreme conditions of clean room environments. *Appl. Environ. Microbiol.* 73: 2600–2611.

Lau, M.C.Y., Cameron, C., Magnabosco, C. et al. (2014). Phylogeny and phylogeography of functional genes shared among seven terrestrial subsurface metagenomes reveal N-cycling and microbial evolutionary relationships. *Front. Microbiol.* 5: 531. https://doi.org/10.3389/fmicb.2014.00531.

Lau, M.C.Y., Kieft, T.L., Kuloyo, O. et al. (2016). Oligotrophic deep subsurface community dependent on syntrophy is dominated by sulfur-driven autotrophic denitrifiers. *Proc. Nat. Acad. Sci. U.S.A.* 113: E7927–E7936. https://doi.org/10.1073/pnas.1612244113.

Lauro, F.M., Chastain, R.A., Blankenship, L.E. et al. (2007). The unique 16S rRNA genes of piezophiles reflect both phylogeny and adaption. *Appl. Environ. Microbiol.* 73: 838–845.

Leong, S.L., Lantz, H., Pettersson, O.V. et al. (2015). Genome and physiology of the ascomycete filamentous fungus *Xeromyces bisporus*, the most xerophilic organism isolated to date. *Environ. Microbiol.* 17: 496–513.

Lever, M.A., Rogers, K.L., Lloyd, K.G. et al. (2015). Life under extreme energy limitation: a synthesis of laboratory- and field-based investigations. *FEMS Microbiol. Rev.* 39: 688–728.

Lindow, S.E., Arny, D.C., and Upper, C.D. (1982). Bacterial ice nucleation: a factor in frost injury to plants. *Plant Physiol.* 70: 1084–1089.

Ma, Y., Galinski, E.A., Grant, W.D. et al. (2010). Halophiles 2010: life in saline environments. *Appl. Environ. Microbiol.* 76: 6971–6981.

Manchee, R.J., Broster, M.G., Stagg, A.J. et al. (1994). Formaldehyde solution effectively inactivates spores of *Bacillus anthracis* on the Scottish Island of Gruinard. *Appl. Environ. Microbiol.* 60: 4167–4171.

Matsubara, H., Goto, K., Matsumura, T. et al. (2002). *Alicyclobacillus acidiphilus* sp. nov., a novel thermo-acidophilic, omega-alicyclic fatty acid-containing bacterium isolated from acidic beverages. *Int. J. Syst. Evol. Microbiol.* 52: 1681–1685.

Matsuno, T. and Yumoto, I. (2015). Bioenergetics and the role of soluble cytochromes c for alkaline adaptation in gram-negative alkaliphilic pseudomonas. *BioMed Res. Int.* 2015 (847945): 14. https://doi.org/10.1155/2015/847945.

Médigue, C., Krin, E., Pascal, G. et al. (2005). Coping with cold: the genome of the versatile marine Antarctica bacterium *Pseudoalteromonas haloplanktis* TAC125. *Genome Res.* 15: 1325–1335.

Methé, B.A., Nelson, K.E., Deming, J.W. et al. (2005). The psychrophilic lifestyle as revealed by the genome sequence of *Colwellia psychrerythraea* 34H through genomic and proteomic analyses. *Proc. Natl. Acad. Sci. U.S.A.* 102: 10913–10918.

Michoud, G. and Jebbar, M. (2016, 2016). High hydrostatic pressure adaptive strategies in an obligate piezophile *Pyrococcus yayanasii. Sci. Rep.* 6: 27289. https://doi.org/10.1038/srep27289.

Minegishi, H., Echigo, A., Nagaoka, S. et al. (2010). *Halarchaeum acidiphilum* gen. nov., sp. nov., a moderately acidophilic haloarchaeon isolated from commercial solar salt. *Int. J. Syst. Evol. Microbiol.* 60: 2513–2516.

Miteva, V. (2008). Bacteria in snow and glacier ice. In: *Psychrophiles: From Biodiversity to Biotechnology* (ed. R. Margesin, F. Schinner, J.C. Marx and C. Gerday), 31–50. Berlin, Heidelberg: Springer.

Mongodin, E.F., Nelson, K.E., Daugherty, S. et al. (2005). The genome of *Salinibacter ruber*: convergence and gene exchange among hyperhalophilic bacteria and archaea. *Proc. Natl. Acad. Sci. U.S.A.* 102: 18147–18152.

Munteanu, A.-C., Uivarosi, V., and Adries, A. (2015). Recent progress in understanding the molecular mechanisms of radioresistance in *Deinococcus* bacteria. *Extremophiles* 19: 707–719.

Mykytczuk, N.C., Trevors, J.T., Ferroni, G.D. et al. (2010). Cytoplasmic membrane fluidity and fatty acid composition of *Acidithiobacillus ferrooxidans* in response to pH stress. *Extremophiles* 14: 427–441.

Mykytczuk, N.C.S., Foote, S.J., Omelon, C.R. et al. (2013). Bacterial growth at -15 °C; molecular insights form the permafrost bacterium *Planococcus halocryophilus* Or1. *ISME J.* 7: 1211–1226.

Nishiyama, Y., Takashina, D.W., Grant, W.D. et al. (1992). Ultrastructure of the cell wall of the triangular halophilic archaebacterium *Haloarcula japonica* strain TR-1. *FEMS Microbiol. Lett.* 99: 43–47.

Pandey, R., Usui, K., Livingstone, R.A. et al. (2016). Ice-nucleating bacteria control the order and dynamics of interfacial water. *Sci. Adv.* 2 (4): e1501630. https://doi.org/10.1126/sciadv.1501630.

Portail, M., Olu, K., Dubois, S.F. et al. (2016). Food-web complexity in Guaymas Basin hydrothermal vents and cold seeps. *PLoS One* 11 (9): e0162263. https://doi.org/10.1371/journal.pone.0162263.

Preiss, L., Hicks, D.B., Suzuki, S. et al. (2015). Alkaliphilic bacteria with impact on industrial applications, concepts of early life forms, and bioenergetics of ATP synthesis. *Front. Bioeng. Biotechnol.* 3: 75. https://doi.org/10.3389/fbioe.2015.00075.

Priscu, J.C. and Christner, B.C. (2004). Earth's icy biosphere. In: *Microbial Diversity and Bioprospecting* (ed. A.T. Bull), 130–145. Washington, DC: ASM Press.

Pulschen, A.A., Rodriguex, F., Duarte, R.T.D. et al. (2015). UV-resistant yeasts from a high altitude volcanic area on the Atacama Desert as eukaryotic models for astrobiology. *Microbiol. Open* 4: 574–588.

Pulschen, A.A., Brendia, A.G., Fricker, A.D. et al. (2017). Isolation of uncultured bacteria from Antarctica using long incubation periods and low nutritional media. *Front. Microbiol.* 8: 1346. https://doi.org/10.3389/fmicb.201701346.

Rainey, F.A., Ray, K., Ferreira, M. et al. (2005). Extensive diversity of ionizing-radiation-resistant bacteria recovered from Sonoran Desert soil and description of nine new species of the genus *Deinococcus* obtained from a single soil sample. *Appl. Environ. Microbiol.* 71: 5225–5235.

Riley, M., Staley, J.T., Danchin, A. et al. (2008). Genomics of an extreme psychrophile, *Psychromonas ingrahamii*. *BMC Genomics* 9: 210.

Roberts, M.F. (2005). Organic compatible solutes of halotolerant and halophilic microorganisms. *Saline Systems* 1: 5. https://doi.org/10.1186/1746-1448-1-5.

Rodrigues, D.F., Ivanova, N., He, Z. et al. (2008). Architecture of thermal adaptation in an *Exiguobacterium sibiricum* strain isolated from 3 million year old permafrost: a genome and transcriptome approach. *BMC Genomics* 9: 547.

Roossinick, M. (2011). The good viruses: viral mutualistic symbiosis. *Nat. Rev. Microbiol.* 9: 99–108.

Schippers, A., Hedrich, S., Vasters, J. et al. (2014). Biomining: metal recovery from ores with microorganisms. *Adv. Biochem. Eng. Biotechnol.* 141: 1–47.

Schmidt, S.K., Vimercati, L., Darcy, J.L. et al. (2017). A Naganishi in high places: functioning populations or dormant cells from the atmosphere? *Mycology* https://doi.org/10.1080/21501203.2017.1344154.

Schoenfeld, T., Patterson, M., Richardson, P.M. et al. (2008). Assembly of viral metagenomes from Yellowstone Hot Springs. *Appl. Environ. Microbiol.* 74 (13): 4164–4174.

Segerer, A., Neuner, A., Kristjansson, J.K. et al. (1986). *Acidianus infernus* gen. nov.? sp. nov. and *Acidianus brierleyi* comb. nov. Facultatively aerobic, extremely acidophilic thermophilic sulfur-metabolizing archaebacteria. *Internat. J. System. Bacteriol.* 36: 559–564.

Seo, S.W., Kim, D., Latif, H. et al. (2014). Deciphering Fur transcriptional regulatory network highlights its complex role beyond iron metabolism in *Escherichia coli*. *Nat. Commun.* 5: 4910.

Sharp, C.E., Brady, A.L., Sharp, G.H. et al. (2014). Humboldt's spa: microbial diversity is controlled by temperature in geothermal environments. *ISME J.* 8: 1166–1174.

Shimada, H., Nemoto, N., Shida, Y. et al. (2008). Effects of pH and temperature on the composition of polar lipids in *Thermoplasma acidophylum* HO-62. *J. Bacteriol.* 190: 5404–5411.

Shivaji, S., Ray, M.K., Rao, N.S. et al. (1992). *Sphingobacterium antarcticus* sp. nov. a psychrotrophic bacterium from the soils of Schirmacher Oasis, Antarctica. *Int. J. Syst. Bacteriol.* 42: 102–106.

Shtarkman, Y.M., Koçer, Z.A., Edgar, R. et al. (2013). Subglacial Lake Vostok (Antarctica) accretion ice contains a diverse set of sequences from aquatic, marine and sediment-inhabiting bacteria and *Eukarya*. *PLoS One* 8 (7): e67221.

Siliakus, M.F., van der Oost, J., and Kengen, S.W.M. (2017). Adaptations of archaeal and bacterial membranes to variations in temperature, pH and pressure. *Extremophiles* 21: 651–670.

Simonato, F., Campanaro, S., Lauro, F.M. et al. (2006). Piezophilic adaption: a genomic point of view. *J. Biotechnol.* 126: 11–25.

Spiess, F.N., Macdonald, K.C., Atwater, T. et al. (1980). East Pacific Rise: hot springs and geophysical experiments. *Science* 207: 1421–1433.

Stan-Lotter, H. and Fendrihan, S. (2007). Viable haloarchaea from ancient salt sediments and their response to simulated Mars condition. In: *ROME: Response of Organisms to the Martian*

Environment (ed. C. Cockell and G. Horneck), ESA SP-1299, 117–127. Noordwijk, The Netherlands: ESA Communications, ESTEC.

Stan-Lotter, H. and Fendrihan, S. (2015). Halophilic archaea: life with desiccation, radiation and oligotrophy over geological times. *Life (Basel)* 5: 1487–1496.

Stan-Lotter, H., McGenity, T.J., Legat, A. et al. (1999). Very similar strains of *Halococcus salifodinae* are found in geographically separated Permo-Triassic salt deposits. *Microbiology* 145: 3565–3574.

Stan-Lotter, H., Radax, C., Gruber, C. et al. (2000). The distribution of viable microorganisms in Permo-Triassic rock salt. In: *SALYT 2000, 8th World Salt Symposium*, vol. 2 (ed. R.M. Geertman), 921–926. Amsterdam, The Netherlands: Elsevier Science B.V.

Suetin, S.V., Shcherbakova, V.A., Chuvilskaya, N.A. et al. (2009). *Clostridium tagluense* sp. nov., a psychrotolerant, anaerobic, spore-forming bacterium from permafrost. *Int. J. Syst. Evol. Microbiol.* 59: 1421–1426.

Suzuki, S., Kuenen, J.G., Schipper, K. et al. (2014). Physiological and genomic features of highly alkaliphilic hydrogen-utilizing *Betaproteobacteria* from a continental serpentinizing site. *Nat. Commun.* 5: 3900. https://doi.org/10.1038/ncomms4900.

Teske, A., de Beer, D., McKay, L.J. et al. (2016). The Guaymas Basin hiking guide to hydrothermal mounds, chimneys, and microbial mats: complex seafloor expressions of subsurface hydrothermal circulation. *Front. Microbiol.* https://doi.org/10.3389/fmicb.2016.00075.

Urbieta, M.S., Rascovan, N., Vázquez, M.P. et al. (2017). Genome analysis of the thermoacidophilic archaeon Acidianus copahuensis focusing on the metabolisms associated to biomining activities. *BMC Genomics* 18: 445. https://doi.org/10.1186/s12864-017-3828-x.

Vannier, P., Michoud, G., Oger, P. et al. (2015). Genome expression of Thermococcus barophilus and Thermococcus kodakarensis in response to different hydrostatic pressure conditions. *Res. Microbiol* https://doi.org/10.1016/j.resmic.2015.07.006.

Vatsurina, A., Badrutdinova, D., Schumann, P. et al. (2008). *Desulfosporosinus hippei* sp. nov., a mesophilic sulfate-reducing bacterium isolated from permafrost. *Int. J. Syst. Evol. Microbiol.* 58: 1228–1232.

Vreeland, R.H., Rosenzweig, W.D., and Powers, D.W. (2000). Isolation of a 250 million-year-old halotolerant bacterium from a primary salt crystal. *Nature* 407: 897–900.

Wierzchos, J., Cámara, B., de Los Ríos, A. et al. (2011). Microbial colonization of Ca-sulfate crusts in the hyperarid core of the Atacama Desert: implications for the search for life on Mars. *Geobiology* 9: 44–60.

Wilson, S.L., Frazer, C., Cumming, B.F. et al. (2012). Cross-tolerance between osmotic and freeze-thaw stress in microbial assemblages from temperate lakes. *FEMS Microbiol. Ecol.* 82: 405–415.

Xu, Y., Nogi, Y., Kato, C. et al. (2003). *Moritella profunda* sp. nov. and *Moritella abyssi* sp. nov., two psychropiezophilic organisms isolated from deep Atlantic sediments. *Int. J. Syst. Evol. Microbiol.* 53: 533–538.

Zajc, J., Kogej, T., Galinski, E.A. et al. (2013). Osmoadaptation strategy of the most halophilic fungus, *Wallemia ichthyophaga*, growing optimally at salinities above 15% NaCl. *Appl. Environ. Microbiol.* 80: 247–256.

Zajc, J., Džeroski, S., Kocev, S. et al. (2014). Chaophilic or chaltolerant fungi: a new category of extremophiles. *Front. Microbiol.* https://doi.org/10.3389/fmicb.2014.00708.

Zheng, H. and Wu, H. (2010). Gene-centric association analysis for the correlation between the guanine-cytosine content levels and temperature range conditions of prokaryotic species. *BMC Bioinformatics* 11 (Suppl 11): S7. https://doi.org/10.1186/1471-2105-11-S11-S7.

Zivanovic, Y., Armengaud, J., Lagorce, A. et al. (2009). Genome analysis and genome-wide proteomics of *Thermococcus gammatolerans*, the most radioresistant organism known amongst the Archaea. *Genome Biol.* 200910: R70. https://doi.org/10.1186/gb-2009-10-6-r70.

Further Reading

Johnson, I.A. (1990). Cold adaptation in marine organisms. *Philos. Trans. R Soc. Lond. B Biol. Sci.* 326: 655–666.

Méndez-García, C., Peláez, A.I., Mesa, V. et al. (2015). Microbial diversity and metabolic networks in acid mine drainage habitats. *Front Microbiol.* https://doi.org/10.3389/fmicb.2015.00475.

Zeglin, L.H., Sinsibaugh, R.L., Barrett, J.E. et al. (2009). Landscape distribution of microbial activity in the McMurdo Dry Valleys: linked biotic processes, hydrology and geochemistry in a cold desert ecosystem. *Ecosystem* 12: 562–573.

Web Sites

http://metagenealyse.mpimp-golm.mpg.de/ (Analysis of Metagenome Systems)
http://www.genome.jp/kegg/kegg2.html (Kyoto Encyclopedia of Genes and Genomes)
https://volcanoes.usgs.gov/volcanoes/yellowstone/yellowstone_geo_hist_88.html (USGS website for volcanoes)
https://www.nps.gov/yell/learn/nature/thermophilic-bacteria.htm (Yellowstone National Park web page)
https://mars.nasa.gov/msl/mission/instruments/spectrometers (NASA Mars web site)

6

Mutualism: Microorganisms and Terrestrial Plants

6.1 Overview and Focus

As the microbial density increased and the diversity of plant species increased, a myriad of microbe–plant interactions ensued. While it is relatively easy to identify those interactions that are pathogenic, there are some associations where the activity is not obvious and careful examination is required. The interaction of bacteria and fungi with plants can be traced to the development of land based biosystems. A key feature of early evolution was the establishment of an oxygen environment on Earth and this is attributed to the photosynthetic activity of cyanobacteria. The successful adaptation of cyanobacteria as chloroplasts in plants is one of the earliest and most obvious of symbiotic development. As discussed in this chapter, cyanobacteria also engaged in mutualisms involving numerous organisms of different genera. Plants benefited greatly from symbiotic interactions with microbes that can use atmospheric nitrogen as the nitrogen source. This concept of symbiosis is not new but was initially introduced by Anton de Berry (1878) as he described cyanobacteria coexisting with *Azolla*. The taxonomic distribution of bacteria with symbiotic nitrogen-fixing (Box 6.1) capabilities involving plants is indicated in Figure 6.1 and represents a relatively limited number of bacteria. Phototrophs of the cyanobacteria group are greatly involved in mutualism with eukaryotic organisms and with the origin of the chloroplast. Legume nodules are attributed primarily to gram-negative rhizobia from the class *Alphaproteobacteria* and in a few cases to *Burkholderia* sp. from the class *Betaproteobacteria*. The actinorhizal symbiosis involves non-legume plants and is attributed to the gram-positive *Frankia* sp. of the *Actinobacteria* group. Another microbe–land plant interaction that was highly successful was the establishment of mycorrhizal symbiosis. Several examples of the symbiotic growth of microorganisms and plants are listed in Table 6.1. This chapter focuses on interactions between the host and symbiont and especially chemical signaling to understand selectivity in this process of co-evolution.

6.2 Cyanobacteria and the Chloroplast Ancestor

Cyanobacteria are unique prokaryotic cells that can photosynthesise but can also support growth by fermentation under anaerobic conditions (Sal and Moezelaar 1997). There are over 2000 known species of cyanobacteria, with estimates of the number of cyanobacterial species in the environment closer to 8000 (Nabout et al. 2013). The cyanobacteria are a highly diverse group and some of their characteristics are given in Table 6.2. Cyanobacteria significantly influence the global environment because they fix CO_2 and N_2, and produce O_2. The role of cyanobacteria in geochemical nutrient cycles is discussed in Chapter 8. Cyanobacteria are

Environmental Microbiology and Microbial Ecology, First Edition. Larry L. Barton and Robert J.C. McLean.

Box 6.1 Nitrogen Fixation

Nitrogen fixation is the conversion of N_2 to NH_3. In biology, only bacteria and archaea have enzymes for the production of ammonia; however, only bacteria can fix nitrogen when grown symbiotically with a plant host. Nitrogen fixation is accomplished by metalloenzymes containing metal ions of molybdenum, iron and in some instances iron is replaced by ionic vanadium. The nitrogenase enzyme is sensitive to oxygen and this necessitates an anaerobic environment in the cyanobacteria or on the plant root for nitrogen fixation to occur. The addition of six electrons to each atom of nitrogen is an energy demanding reaction, with 16 ATP required for conversion of one N_2 molecule to two molecules of NH_3. Because of this great demand for energy, cyanobacteria using photosynthetic respiration, use sunlight to produce ATP for nitrogen fixers while nitrogen- fixing legumes provide abundant quantities of sugars or organic acids to drive ATP synthesis. Numerous bacteria and archaea fix nitrogen; however, the quantity of NH_3 produced is markedly less than with cyanobacteria or rhizobia-containing plants.

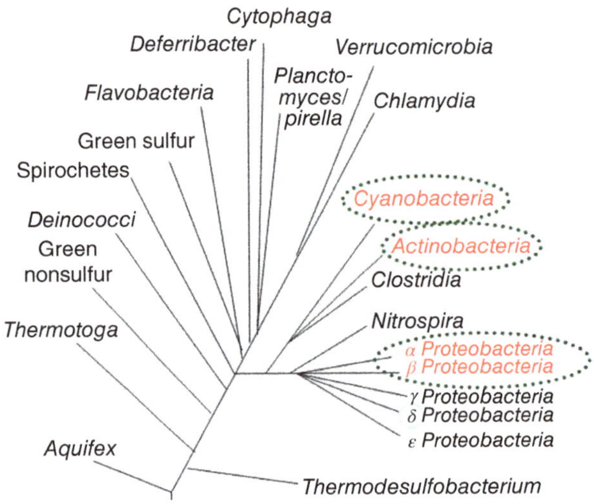

Figure 6.1 Distribution of bacteria capable of symbiotic nitrogen fixation. Taxonomic groups containing bacteria that fix nitrogen under symbiotic growth are listed in red.

nutritionally independent microorganisms that tolerate extreme environments, which enables cyanobacteria to be ubiquitous in aquatic and terrestrial habitats.

A hallmark feature of the cyanobacteria is the thylakoid membrane that contains the oxygenic photosynthetic apparatus. The photosynthetic lamellae are in the cytoplasm (Figure 6.2) and a model of the cyanobacterial cell with the photosynthetic apparatus is given in Figure 6.3. Associated with photosystem II (PS II) (Campbell et al. 1998) are phycobilisomes that function as accessory light receptor proteins to direct light energy to the reactive centers. These phycobilisomes contain protein complexes of allophycocyanin (AL), phycocyanin PC), and phycoerythrin (PE) that extend into the cytosol (Figure 6.4). Allophycocyanin is adjacent to PS II, which is imbedded in the thylakoid membrane. Phycocyanin proteins are stacked in columns on the allophycocyanin, with phycoerythrin at the surface of the stack. It is well accepted that phycobilisomes occur on PS II but recently there has been evidence that an antenna system specific for PS I is present in the cyanobacterium, *Anabaena* sp. (Watanabe et al. 2014).

Highly conserved throughout organisms that carry out aerobic photosynthesis is the presence of chlorophyll *a*, which absorbs light at 665–668 nm. However, the absorption of light by AL, PC, and PE at 650, 620, and 550 nm, respectively, enables cyanobacteria to harvest light over a broad spectrum (550–660 nm). The phycobiliproteins (AL, PC, and PE) contain open

Table 6.1 Examples of mutualism in microorganisms and plants.

Microorganisms	Plants as host
Cyanobacteria	
Nostoc sp. *Anabaena* sp.	*Azolla, Geosiphon, Gunnera,* cycads, diatoms, dinoflagellates, hornworts, lichens, liverworts, marine sponges, and mosses
Rhizobia	
Formation of nodules on aquatic plants	
Allorhizobium sp.	*Neptunia prostrata* (legume in tropics)
Azorhizobium sp.	*Sesbania rostrata* (woody legume native to Africa)
Formation of nodules on roots	
Bradyrhizobium japonicum	Soybeans
Mesorhizobium ciceri	Chickpea
Rhizobium leguminosarum biovar *phaseoli*	Bean
Rhizobium leguminosarum biovar *trifoli*	Clover
Rhizobium leguminosarum biovar *viciae*	Pea
Sinorhizobium[a] *meliloti*	Alfalfa
Formation of nodules on stems	
Azorhizobium caulinodans	Aquatic tropical legume
Methylobacterium nodulans	Clover
Frankia	
Formation of nodules on roots of woody plants	
Frankia spp.	<25 genera of dicotyledonous plants
Fungi	
Ectomycorrhizae	
Boletus elegans	*Larix* sp. (deciduous conifer native to central Europe. Common name larch.)
Boletus scaber	Birch, deciduous alder, hazels and hornbeam trees
Lactarius deliciosus	*Pinus pinea* (Italian stone pine)
Pisolithus tinctorius	*Pinus taeda* (loblolly pine)
Russula emetica	*Pinus pinaster* (maritime or cluster pine in the Mediterranean region)
Endomycorrhizae	
Gigaspora margarita	Cotton
Ericoid mycorrhizae	
Pezizella ericae	Heather
Orchid mycorrhizae	
Armillaria sp.	Orchids
Rhizoctonia sp.	Orchids

[a] Genus also classified as *Ensifer*. Although *Ensifer* has historical preference, *Sinorhizobium* is commonly used by investigators.

Table 6.2 Common cyanobacteria and their characteristics.

Organism	Characteristics
Anabaena	Spherical cells, 2–3 μm diameter, form an unbranched filament. Cellular differentiation to produce specialized resistance cells (akinetes) and nitrogen-fixing units (heterocysts). The *Anabaena* genome about 7.2 Mbp. Some strains have gas vacuoles to move cells vertically in the water column. Strains of *Anabaena* grow quickly in fresh water and may be responsible for algal blooms. Approximately 150 accepted species.
Prochloron	Unicellular cyanobacterium that grows as an extracellular symbiont on tissues of sea squirts; not successfully cultivated in the laboratory. These phototrophs are unique cyanobacteria because they produce chlorophyll *a* and chlorophyll *b* but not phycobilisomes or phycobiliproteins.
Nostoc	A cyanobacterium commonly found in soil and fresh water but rarely in marine settings. Spherical cells grow in chains or trichomes and form conspicuous gelatinous masses due to the extracellular polysaccharides it produces. Nitrogen fixation is due to the presence of heterocysts.
Synechococcus	A cyanobacterium found in marine and fresh water; cells 0.8–1.5 μm. The genome of *Synechococcus elongates* strain PCC7002 is 2.4 Mbp. This organism may account for 25% of CO_2 fixed in open seas.
Synechocystis	Bacterium capable of aerobic photosynthesis or heterotrophic growth in the dark; used as a model for cyanobacterial photosynthesis and stress response. Genome size of the mesophilic coccus *Synechocystis* PCC6803 is 3.947 Mbp.
Gleocapsa	Cells 40–60 μm diameter, surrounded by a thick gelatinous sheath, with some species halophilic. *Gleocapsa magma* is green, becoming dark brown to black if exposed to bright sunlight. It is a nuisance in Midwestern and Southeastern United States discoloring shingles on roofs of houses (Brook 1968).
Oscillatoria	Cells 1.5–3 μm long, 4–6 μm wide, with trichomes commonly formed. Green due to chlorophyll; moves by gliding. Often forms cyanobacterial mats, which are black "soil crusts" in dry environments. Some species can use H_2S instead of H_2O as the electron source in photosynthesis.
Microcystis	Aquatic cyanobacterium; individual cells aggregate into clusters. Capable of fixing nitrogen and responsible for cyanobacterial blooms in warm lake waters. Releases toxins that attack the nerves and are detrimental to small animals.
Prochlorococcus	Bacterium characterized as a photosynthetic picoplankton because the size (0.5–0.8 μm) qualifies it as a nanobacterium. It contains 1.81 Mbp of DNA with only about 2000 genes. Chlorophyll *a* in these organisms is replaced with chlorophyll a_2, chlorophyll b_2; some strains have phycoerythrin. It may be the most abundant organism with estimates of $2.8–3.0 \times 10^{27}$ cells present each year in the oceans (Partensky et al. 1999).

chain tetrapyrroles as the chromophore and these molecules account for the fluorescent character of PS II (Figure 6.5).

Cyanobacterial symbiosis reveals an ancient mutualism between eukaryotic organisms found in aquatic or terrestrial environments (Usher et al. 2007). The successful endosymbiosis of a cyanobacterium accounting for the origin of the chloroplast may have occurred only once and there is some evidence that a member of the order *Chroococcales* may have been the chloroplast ancestor (Falcón et al. 2010). Following endosymbiosis of cyanobacteria by the eukaryotic cell, the plastid (chloroplastid) retained only a fraction of the cyanobacterial genes. It is generally accepted that the gene content of plastids is ~200 while that of cyanobacteria is ~1500 genes. In a comparison of *Arabidopsis* genes with genes from *Synechocystis* sp., *Protococcus marinus*, and *Nostoc punctiforme*, ~18% of protein encoding genes in the *Arabidopsis* nuclear

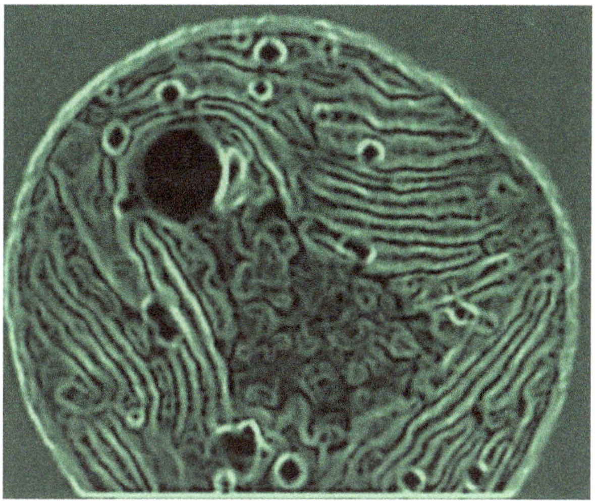

Figure 6.2 Intracellular photosynthetic membranes are readily observed in this cyanobacterial cell. *Source:* Reproduced with permission from Larry Barton.

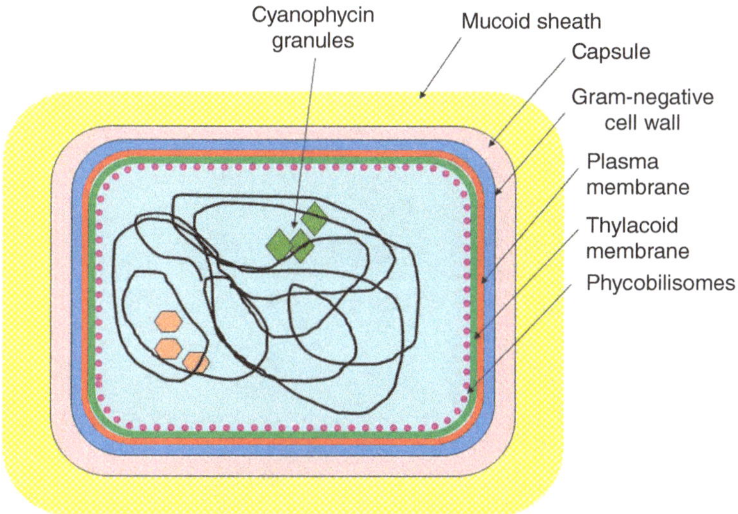

Figure 6.3 Model of a cyanobacterial cell. DNA is not localized in a true nucleus and intracellular membranes for photosynthesis are present.

genome are of cyanobacterial origin (Martin et al. 2002). Genes of cyanobacterial origin that are contained in the nucleus and expressed by the host, produce proteins that make up the chloroplast. Protein synthesis in chloroplasts is similar to bacteria because both have 50 S ribosomes and *N*-formyl methionine is the first amino acid in the newly synthesized peptide (Harris et al. 1994).

Although it is not clear how the endosymbiosis process occurred, it is apparent that there were several events between the uptake of the cyanobacterium and the stabilization of a chloroplast unit (Martin et al. 2002). In the primary endosymbiosis phase, the cyanobacterial cell avoids degradation or expulsion by the heterotrophic host. Replication of the endosymbiont occurs, with cell division of the cyanobacteria coordinated with the division of host cells. Resistance to oxidative stress must be accomplished by the host cell to avoid degradation of the photobiont or even destruction of the host cell. This oxidative stress would be attributed to

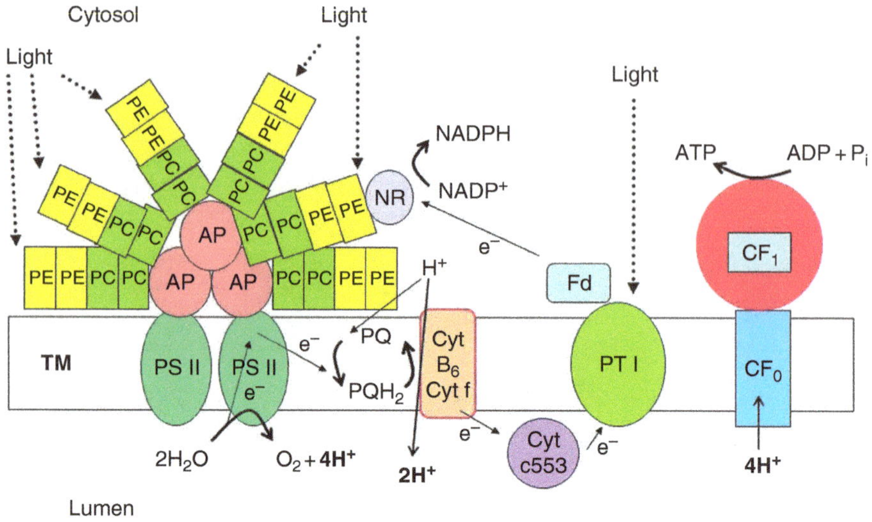

Figure 6.4 A simplified model of thylakoid electron transport based on characteristics of *Synechocystis* sp. strain PCC 6701. TM = Thylakoid membrane; PT I = photosystem I; PT II = photosystem II; PQ = plastoquinone; AL = allophycocyanin; PC = phycocyanin; PE = phycoerythrin; NR = NADPH oxidoreductase; Fd = ferredoxin; CF_0/CF_1 = ATP synthase. For a more detailed model see Campbell et al. 1998.

Figure 6.5 Images of *Anabaena* sp.: (a) colonies growing on an agar slant; (b) cells in a filament which fluoresce red when observed with a fluorescence microscope; (c) growth of *Anabaena* in a liquid medium collected at the bottom of the container, with resuspended cells clumped due to the mucoid sheath. *Source:* Reproduced with permission from Larry Barton.

reactive oxygen species (free radicals) released from the photoreceptor system of the newly acquired cyanobacterium. Early in the primary endosymbiosis process there is transfer of genes from the photobiont and nucleus of the host cell. This transfer could be attributed to DNA released from degradation of the endosymbiont or to RNA intermediates migrating from the endosymbiont to the nucleus of the host. Gene transfer would occur more quickly if the host acquired multiple cyanobacteria and if cyanobacteria were in the nucleus of the host as well as in the cytoplasm. In a few cases, most notably red algae, there were secondary and tertiary endosymbiotic events. The establishment of photosymbionts in a heterotrophic host may have occurred 60–200 million years ago (Martin et al. 2002).

6.3 Lichens: Cyanobacteria/Algae–Fungi Mutualism

6.3.1 Distribution and Organization

While lichens were initially considered as plants, Simon Schwendener in 1867 challenged the scientific community by proposing that lichens were a partnership consisting of algae and fungi (Honegger 2000). This concept of two species living as a single unit was highly controversial because at that time it was assumed that all living organisms were composed of a single life form. Subsequently, examinations have revealed that over 98% of lichens have fungi (mycobionts) belonging to the ascomycetes with the remaining 2% of mycobionts belong to *Basidiomycetes* or *Fungi Imperfecti* (Nash 2008). There are about 13 500 species of *Ascomycetes* that are found worldwide as lichens while basidiomycete fungi (basidiolichens) are principally found in tropical or subtropical environments. Eukaryotic algae from about 100 different species are the photosynthetic partners (photobionts) and 90% of the lichens have *Treboux* or *Trentepohlia* as the most dominant algal species. It has been estimated that 8–10% of the lichens have cyanobacteria as the photobiont and *Nostoc* is the principal species of cyanobacteria found in the cyanolichen (DePriest 2004). A few lichens contain both cyanobacteria and eukaryotic algae and in these cases, the cyanobacteria exist as an "island" surrounded by algae. From molecular phylogeny studies, it has been proposed that the symbiotic partnership formed multiple times throughout evolution (Gargas et al. 1995). Fossil records indicate that lichens were present 400 million years ago (Taylor et al. 1995) and that some lichen fossils may be over 600 million years old (Yuan et al. 2005). These fossil records suggest that the fungal–algal–cyanobacterial relationship existed before the presence of vascular plants.

Lichens are found on 6–8% of the Earth's terrestrial surface including rock surfaces, tree trunks, soil, and sand. Lichens grow in many different environments including tropical forests, deserts, and Arctic regions. The lichen body, thallus, has a unique form that is produced by the fungus but influenced by the specific alga/cyanobacteria. The three most common lichen morphologies are crustose (flat crusts present on rocks or tree bark), foliose (flat and leaf-like forms), and fruticose (branched and appears as a miniature bush). Examples of these lichens are given in Figures 6.6 and 6.7. To understand the mutualism between the fungus and algae, the foliose lichen thallus is often used as a model, see Figure 6.8. The top layer, called a cortex, consists of densely packed fungal hyphae that protect the algal cells in the second layer. In the third layer, the medulla, fungal hyphae form a loose connection between the algal cells and the fungal mat,which is also known as the lower cortex. The fungal mat is attached to the supporting surface by small "root-like" structures called rhizines. Fungal cells have appendages called haustoria that penetrate the algal cell to obtain sugars produced by the algae and this intimate contact between cells is carefully regulated so the algae are not killed. The fungal

Figure 6.6 Examples of lichens: (a) crustose lichens growing on tree bark; (b) crustose lichens growing on a head stone in a cemetery; (c) fruticose lichen at the tip of the green arrow and foliose lichen with leaf-like structure at the tip of the red arrow. Scale bar = 1 in (2.54 cm). *Source:* Reproduced with permission from Larry Barton.

Figure 6.7 Examples of fruticose lichens growing on a tree branch: (a) mini-shrub appearance; (b) extremely small tree-like appearance. For size comparison, note the size of the pine needles in the photograph. *Source:* Reproduced with permission from Larry Barton.

cortex protects algal cells from ultraviolet light and enables algal cells to remain viable in dry environments. When cyanobacteria are the photobiont, an additional benefit is the fixation of nitrogen, which supports the growth of the mycobiont.

Lichens grow slowly and colonies may be hundreds of years old. Lichen crusts are found on the sand in some deserts where the crusts remain viable but inactive even in extended dry

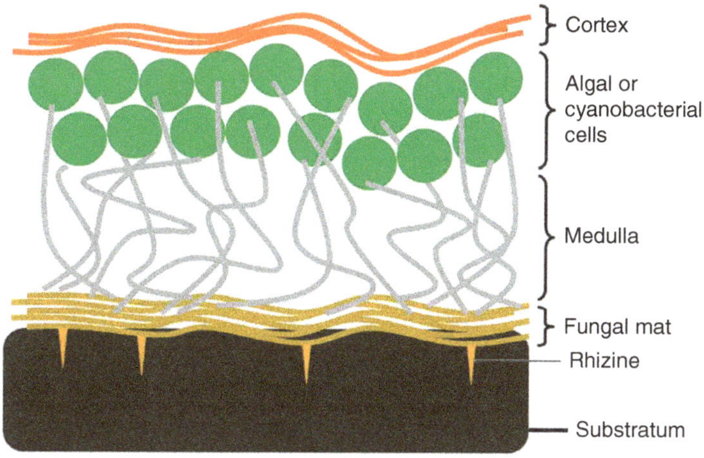

Figure 6.8 Model of lichen symbiosis showing fungal and algal (or cyanobacterial) cells.

periods and can quickly hydrate as a result of fog, dew, or rain. Desert lichen crusts are important because they stabilize the sand, help retain water and, if the photobiont is a cyanobacterium, the environment is enriched through nitrogen fixation (Box 6.1). Resistance of lichens to extreme environments was demonstrated when two species of lichen survived unprotected exposure to space for 15 days on an experiment conducted by the Russian Soyuz rocket in 2005 (Sancho et al. 2007).

6.3.2 Natural Products of Lichens

Unique secondary metabolites are produced by the fungus–alga mutualism but not when the symbionts are cultured individually. These compounds are called lichen acids and are present on the surface of the mycobiont hyphae in the cortex and medulla. There are over 1000 chemicals produced by lichens and many of these are phenolic compounds, anthraquinones, depsidones, depsones, γ-lactones, or pulvinic acid derivatives (Muggia et al. 2009). These chemicals are produced via polymalonate, shikimate, or mevalonate metabolic pathways encoded by genes on the genome of the fungal partner, with structural modifications from methylation, decarboxylation, or esterification reactions (Miao et al. 2001). The pulvinic acid compounds are products of the shikimic acid pathway, the steroids and terpenes are from the mevalonic acid pathway, but the acetate–polymalonate pathway produces most of the lichen compounds. Algal influence on biosynthesis of these secondary metabolites remains unresolved. As seen in Figure 6.9, the structures of the natural products of lichens are complex and unique. The secondary metabolites are brightly colored: usnic acid is yellow-green, anthraquinones are red-purple, and pulvinic acid derivatives are yellow. The secondary metabolites are identified using chemical spot tests or chromatographic separation (Culberson 1970, 1972). Lichen natural products may account for 20% of the dry weight of the thallus and may be distributed throughout the lichen tissue. Pigmented chemicals accumulate near the upper layers of the lichen and act as filters protecting the photobiont against ultraviolet light. Additional activities of these secondary metabolites include: preventing invertebrates from grazing on the lichen, enhancing mutualism, and binding of metal ions. The weathering of rock surfaces by lichens is covered in Chapter 10. Lichens are important commercially because their chemicals have been used as textile dyes, cold remedies, animal food, antibiotics, and additives in cosmetics. Lichen cultivation has become important because of the potential pharmaceutical use of lichen chemicals to inhibit growth of bacteria, fungi, viruses, and tumors. An industry has

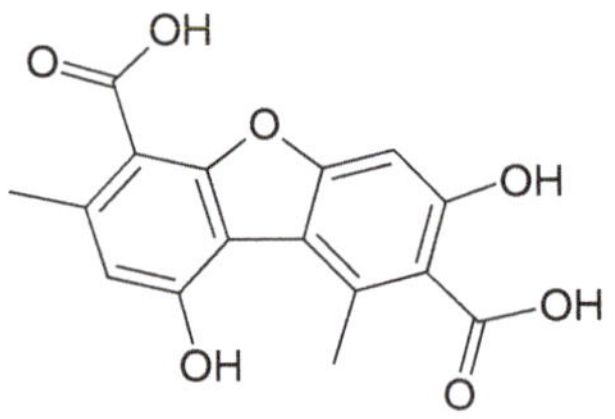

Pannaric acid

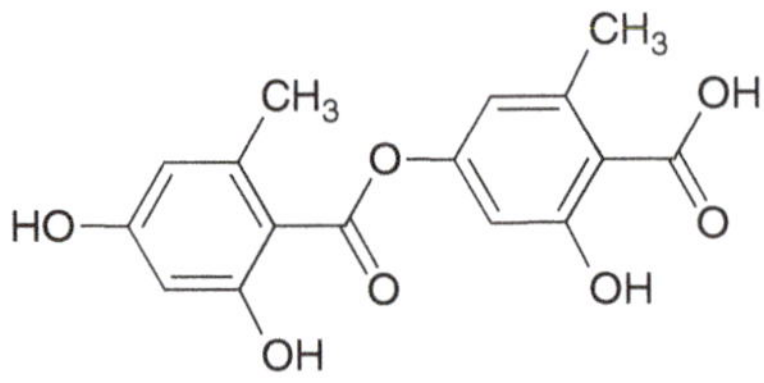

Lecanoric acid

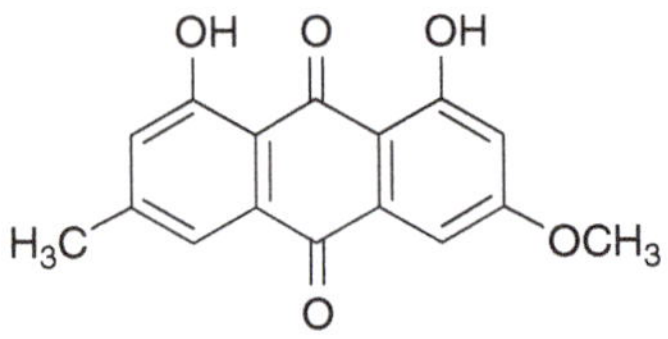

Parietin

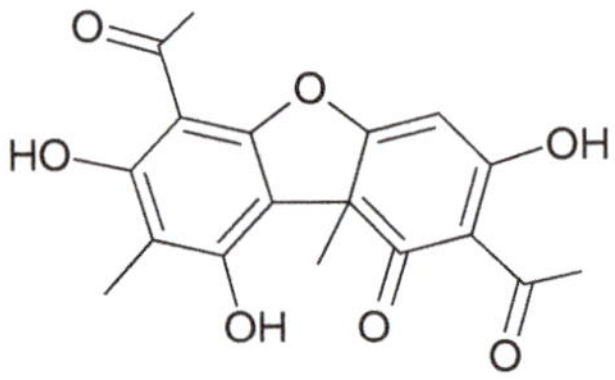

Usnic acid

Figure 6.9 Secondary metabolic products of lichens often called lichen acids.

developed that specializes in the production of unique chemicals following re-establishment of lichens from different fungal and algal species or from cultivation of lichen under laboratory conditions in tissue culture (DePriest 2004; Lobakova and Smirnov 2012). The production of the secondary metabolites is highly influenced by chemicals in the media used to cultivate suspension lichen cultures.

6.4 Mutualisms with Cyanobacteria as Intracellular or Epiphytic Organisms

Cyanobacteria are versatile organisms that have engaged in mutualism with numerous eukaryotes including bryophytes, the aquatic-fern *Azolla*, cycads, sponges, mosses, *Gunnera*, diatoms, *Geosiphon*, and protists (Rai et al. 2008; Barton and Northup 2011). The symbiotic interaction by cyanobacteria in the formation of lichens has been discussed and the following narrative addresses their symbiosis with bryophytes, cycads, *Azolla*, and *Gunnera*.

6.4.1 Bryophytes

Bryophytes are primitive nonvascular land plants only a few inches long. The taxonomic groups of bryophytes with cyanobacterial symbiosis include the hornworts and liverworts. *Anthoceros punctatus*, a hornwort, and *Blasia pusilla*, a liverwort, form an epiphytic or endophytic symbiosis with cyanobacteria (Adams and Duggan 2008). Hornworts and liverworts are only a few centimeters long, with dark spots (~0.5 mm diameter) on the flat thallus that indicate the presence of cyanobacteria. The major photobiont associated with hornworts or liverworts collected from the environment belong to the genus *Nostoc*; however, under laboratory conditions *Calothrix* and *Chlorogleopsis* may associate with liverworts. As reviewed by Adams and Duggan (2008), the establishment of the cyanobacteria–bryophyte symbiosis requires that the cyanobacteria can form motile hormogonia and heterocysts. For successful symbiosis to occur, the following specific interactions are observed:

- The cyanobacteria develop short motile units, known as hormogonia, that are infectious. As a result of rapid vegetative growth, hormogonia are released by fragmentation from the cyanobacterial filaments. The hormogonia use extracellular polysaccharide and type IV pili to move across the surface of the bryophyte thallus. Microarray analysis indicates that hundreds of genes in the hormogonium are upregulated and these genes are specific for sensory response and biosynthesis of extracellular polysaccharide and pili (Campbell et al. 2007).
- The plant host produces hormogonia-inducing factors (HIFs), which appear to be peptides, stimulating hormogonia formation as well as attracting the hormogonia to symbiotic tissues. The production of HIFs occurs when the plant host is starved for nitrogen.
- Chemotaxis of the hormogonia is believed to involve the chemotaxis-related proteins (CheA, CheB, Chew, CheD, and CheR), which are present in flagellated heterotrophic bacteria. Hormogonia move by gliding along a slime layer consisting of extracellular polysaccharides. Type IV pili on the cyanobacteria enhance a twitching movement and promote attachment to a specific plant surface.
- Following infection, there several changes in the host and endosymbiont. The cyanobacteria produce auxin and other plant stimulants, which promote the development of extensions from the plant cavity where the photobiont is located. This extension of the cell enhances the exchange of nutrients between plant and cyanobacteria. Differentiation of the cyanobacterial cells results in the formation of heterocysts, the nitrogen-fixing unit. Additionally, there is a reduction in the rate of cyanobacterial growth that prevents the symbiont overgrowing the host.

Because of the requirements for establishing symbiosis, it is not surprising that only *Marchantia* and *Porella,* which represent a small fraction of liverwort genera, have endophytic cyanobacteria. Endophytic associations of cyanobacteria, however, are common in hornworts and due to the unique cavities of the hornworts where the cyanobacteria can collect.

6.4.2 Mosses

Mosses are non-vascular plants that are generally only a few inches tall and grow in damp shaded areas. Commercially, *Sphagnum cristatum* and *Sphagnum subnitens* are the mosses commonly used as soil conditioners in gardening. Cyanobacteria are found growing as epiphytes on many of the 12 000 moss species. Since moss has highly effective water absorption and retention, it provides cyanobacteria with a moist environment for growth and, in acid bogs, the leaves secrete alkaline substances that neutralize local acidity and enable cyanobacteria to flourish. In the boreal forest, the “feather moss” covers 80% of the ground and cyanobacteria fix most of the nitrogen (Rousk et al. 2013).

6.4.3 Azolla

Azolla with *Anabaena azollae* as the photobiont is found in temperate regions of the world. This aquatic free-floating fern has leaves that are a few millimeters long, with the cyanobacteria found in ovoid cavities on the upper lobe of the leaves (see Figure 6.10). Cyanobacterial heterocysts fix nitrogen and the influx of nitrogen enriches the nutrition of *Azolla*. Due to this nitrogen-fixing capability of *Azolla*, this plant has been used as a biofertilizer in many parts of southeastern Asia. When phosphorus is not limiting, blooms of *Azolla* occur; this massive

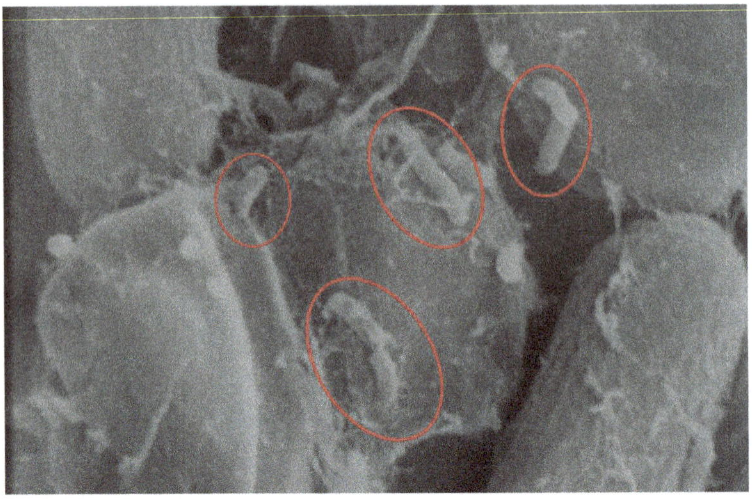

Figure 6.10 *Azolla* with bacteria on surface. Bacteria are present in the encircled area. *Source:* Courtesy of Gordon V. Johnson.

growth of *Azolla* prevents mosquitoes and other insects from laying eggs on the water, so it is commonly called the mosquito fern.

6.4.4 *Gunnera*

Gunnera are flowering plants known for their extremely large leaves. The largest plant is *Gunnera manicata,* which grows in Brazil and has leaves 1.52–1.83 m (5–6 ft) wide, 3.35 m (11 ft) long on 2.4 m (8 ft) stalks. Also in South America is *Gunnera albocarpa,* which has tiny 2 cm long leaves. *Nostoc punctiforme* is the nitrogen-fixing cyanobacterium associated with all the *Gunnera* spp. (Chiu et al. 2005). Although the details of the establishment of the symbiosis have not been extensively studied, it is known that *Nostoc* colonizes the glands at the base of the leaf stalks. These pockets of the photobiont are readily observed in a cross section of the stem of *Gunnera*.

6.4.5 Cycads

Cycads represent some of the oldest terrestrial plants with fossils dating to 280 million years ago. Depending on the genus, they are a few centimeters to several meters tall. Cycads are gymnosperms with a woody trunk and leaves at the top that resemble palm leaves. A cycad with a short stem is shown in Figure 6.11a. Unique to cycads is the formation of root nodules with nitrogen-fixing cyanobacteria. These cyanobacteria produce β-methylamino-L-alanine (BMAA), which is a neurotoxin found throughout the plant including in the cycad seeds; BMAA accumulates in the fat of animals that consume all parts (seeds, leaves, pulp) of the plant, with prolonged consumption causing neurological diseases similar to Parkinson's disease.

6.4.6 *Geosiphon*

Geosiphon is a coenocytic fungus found in moist soil that has *Nostoc* as an endosymbiont contained in a specialized unit called a bladder. The bladder is an extension of the fungal hyphae and extends into the atmosphere while the hyphae remains in the soil (see Figure 6.12).

Figure 6.11 Examples of plants that have bacteria symbiosis: (a) sago palm (*Cycas revoluta*) – a cycad from Java with symbiotic cyanobacteria; (b) alfalfa plant – foliage for cattle; (c) broad beans with rhizobial nodules; (d) soybeans – a major agricultural crop; (e) multicolored lupines – used in landscapes. *Source:* Reproduced with permission from Larry Barton.

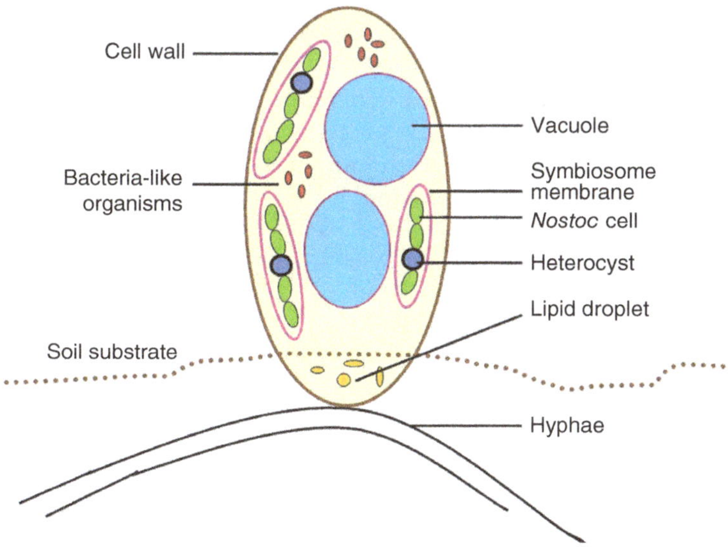

Figure 6.12 Symbiosis between *Nostoc* and *Geosiphon*. The bladder is 0.3 mm wide and 1–2 mm long. Cyanobacterial heterocysts are involved in nitrogen fixation.

The specificity of *Geosiphon* for *Nostoc* is due to the lectins produced by the fungus that are specific for the cyanobacterium. In the establishment of the symbiotic unit, trichomes of *Nostoc* are initially stimulated by red light to be motile. The trichomes become non-motile in the presence of green light and the cyanobacterium is taken up by the fungal bladder. Since *Nostoc* has nitrogen-fixing heterocysts, this cyanobacterium provides nitrogen for the fungus.

6.4.7 Diatoms

The diatom–cyanobacterium relationship has not been extensively studied and an evaluation of these cyanobacteria by 16S rRNA evaluation is needed. The marine diatoms *Epithemia turgida* and *Rhopalodia gibba* have two to five endosymbiotic cyanobacteria in each diatom cell (Janson 2002). The intracellular nitrogen-fixing cyanobacterium has been designated as *Richelia intracellular*. The intracellular cyanobacteria are transferred vertically from mother cell to daughter cell. Occasionally cyanobacteria have been reported (Janson 2002) as epiphytes on diatoms, where they may be an important support for diatoms living in nitrogen-limiting waters.

6.5 Rhizobia–Legume Symbiosis

Plants that are members of the family Leguminosae are commonly called legumes. The legumes relevant to the discussion about bacterial symbiosis are forage plants (alfalfa, clovers) and pod plants (soybeans, peanuts, peas, beans, lentils). Some of these legumes (Box 6.2) are shown in Figure 6.11. The presence of nodules on legume plants has been known for hundreds of years. Marcelo Malpighi, an Italian physician and biologist, in the latter part of the 1600s made numerous sketches of roots of legumes with nodules attached. In 1888, Martinus Beijerinck, a microbiologist from The Netherlands, provided the first scientific report of bacteria isolated from legume nodules and established that these legume bacteria were important for *in planta* nitrogen fixation.

6.5.1 Bacterial Species Involved

The research on nitrogen-fixing bacteria on the root and stems of legumes has been highly active over the past years. At one time all bacteria associated with the nitrogen-fixing nodules of legumes were designated as *Rhizobium*; however, there are now about 40 genera that have been reported with this legume-nodule activity. The taxonomy of these bacteria is evolving and based on the small subunit rRNA sequence analysis, these bacteria are divided into three groups that are members of the α-proteobacteria (Sadowsky and Graham 2006). One group includes the *Rhizobium, Agrobacterium, Allorhizobium, Sinorhizobium*, and *Mesorhizobium*;

Box 6.2 Legume Plants

The legume family *Fabaceae* is the third largest family of flowering plants, with over 18 000 species. In addition to the common legumes – including clovers, soybeans, alfalfa, lupines and peanuts – some shrubs, trees, and vines are also legumes. Commonly, legume plants enter into symbiotic relationships with bacteria that are collectively called rhizobia. As much as 100 lbs (45 kg) of nitrogen can be added to an acre of soil from legumes containing nitrogen-fixing bacteria.

a second group is the *Bradyrhizobium;* and a third group is the *Azorhizobium. Burkholderia caribensis* and *Ralstonia taiwanensis* of the β-proteobacteria are also capable of forming nitrogen-fixing nodules with legumes. In general, nodulation of a specific legume is attributed to one species of bacteria. A few bacterial species, however, can nodulate several legumes (Hadri et al. 1998); *Rhizobium* spp. strain NGR 234 is reported to nodulate over 125 different legume species (Pueppke and Broughton 1999) and due to the production of over 18 different Nod factors by this bacterial species.

6.5.2 Rhizospheric Rhizobia

The rhizobial cell undergoes numerous changes as it responds to the specific environment where it resides. In soil environments, rhizobia metabolize organic materials, produce siderophores if iron is limiting, and use general growth-related activities. Since iron is not readily available to biological systems in many soils, there has been much research into characterizing the various systems for iron uptake by rhizobia (see Table 6.3). In the rhizosphere, rhizobia have an active process for removal of hydrogen peroxide and other reactive oxygen species that are released from plants and microbes. As rhizobia interact with the root to initiate the infectious thread, there is considerable processing of signals to optimize the rhizobial cell to successfully initiate entry into the root tissue. This is followed by the symbiotic phase, in which

Table 6.3 Siderophores for utilization of iron by cyanobacteria and symbiotic nodules.

Molecules	Bacteria	Characteristics
Cyanobacteria[a]		
Anachelin Schizokinen Aerobactin	*Anabaena flos-aquae* *Anabaena cylindrical*	These are hydroxymate derivatives of citric acid
Legume symbiont[b]		
Rhizobactin 1021	*Sinorhizobium meliloti* strain 1021	A citrate–hydroxymate siderophore produced by many sinorhizobia
Rhizobactin	*Sinorhizobium meliloti* strain DM4	An amino polycarboxylic acid produced by few sinorhizobia
Vicibactin	*Rhizobium leguminosarum*	A cyclic trihydroxamate siderophore
Ferric citrate	Used by some strains of rhizobia	Most strains of rhizobia do not secrete citrate but can transport ferric citrate into the cell
Heme	Used by numerous rhizobia	Uptake of heme is due to hemophore proteins. Iron is released intracellularly following heme degradation
Ferrichrome Ferric rhodotorulate Ferrioxamine B	Used by some sinorhizobia	These siderophores are not produced by rhizobia but are present in the rhizosphere
Actinorhizal symbiont[c]		
Frankobactin	*Frankia* sp.	A hydroxamate siderophore with a unique phenyl oxazoline ring

[a] Nicolaisen and Schleiff (2010).
[b] Fabino and O'Brian (2012).
[c] Boyer and Aronson (1994).

rhizobia grow on nutrients provided by the plant and atmospheric nitrogen is fixed. As seen in the discussion of global regulation for the symbiosis of soybean with *Bradyrhizobium japonicum* (Loh and Stacey 2003), gene regulation leading to nodule formation involves several sensory response activities including a regulatory system with two components and quorum sensing. As rhizobia move from one environment to another in nodulation activities, it is important not only for new genes to be induced but also for some gene expression to be repressed to enable the rhizobia to display different physiological activities.

6.5.3 The Root Nodulation Process

The establishment of bacterial nodules on legumes is a result of the sensory response activity of both the host and the bacteria. This interaction is based on chemical signaling and accounts for the specificity of the bacterial species for the legume host. *Lotus japonicus* and *Medicago truncatula* are model legumes for the study of nodulation because they produce determinate (Figure 6.13) or indeterminate (Figure 6.14) nodule types, respectively.

The general process of legume root nodulation for *Rhizobium* and *Bradyrhizobium* is summarized below and a model of this activity is presented in Figure 6.15.

- As plants grow, several different organic acids (malic acid, citric acid, etc.) are released from the roots and rhizospheric bacteria migrate toward the roots to use the organic acids. In soils where nitrogen levels are adequate to support plant growth, nodules are not formed due to the production and release of ethylene compounds from the roots.
- In nitrogen-limited soils, legume roots release oxidative compounds that include $O_2^{\cdot-}$, $OH^{\cdot-}$, H_2O_2, and N_2O. Soil bacteria, including rhizobia, survive this oxidative stress due to the production of appreciable levels of detoxifying enzymes. As a result, considerable numbers of rhizobia cells may be found in the rhizosphere.
- Flavonoids are released from legume roots and each plant species has a specific flavonoid which it secretes. Examples of these complex aromatic compounds (see Figure 6.16) include flavone (luteolin) secreted by alfalfa roots, flavone glucoside (apigenin) from clover roots, and isoflavone (daidzein) from soybean roots (Maxwell et al. 1989; Hartwig et al. 1990; Dakora et al. 1993; Peters et al. 1986; Phillips et al. 1992). Rhizobia colonize the legume root surface because of the chemoattractive activity of the legume flavonoid.

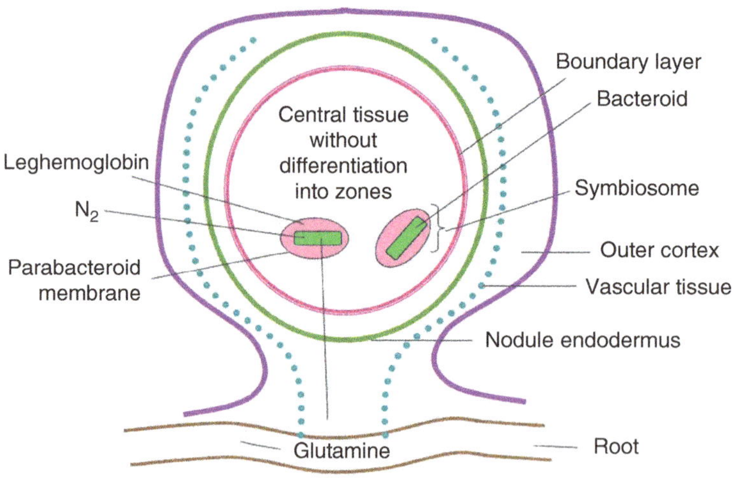

Figure 6.13 Model of determinate legume nodule.

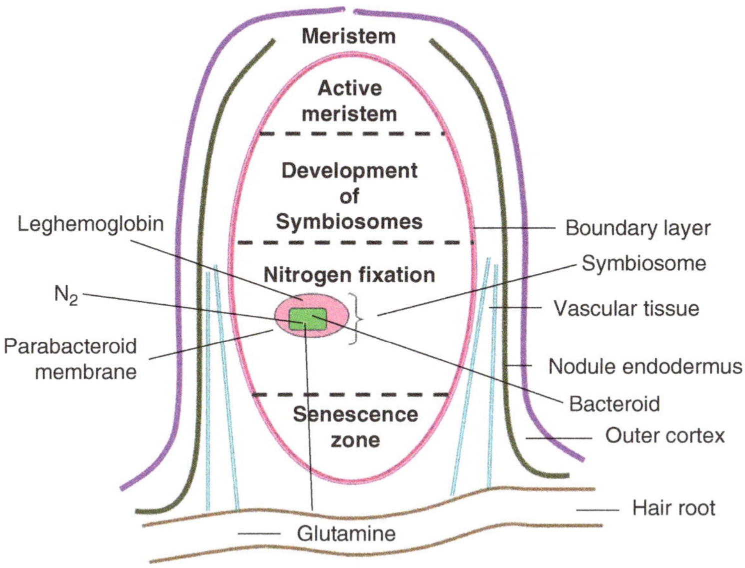

Figure 6.14 Model of indeterminate legume nodule. Four distinct zones of activity are observed in the intermediate nodule.

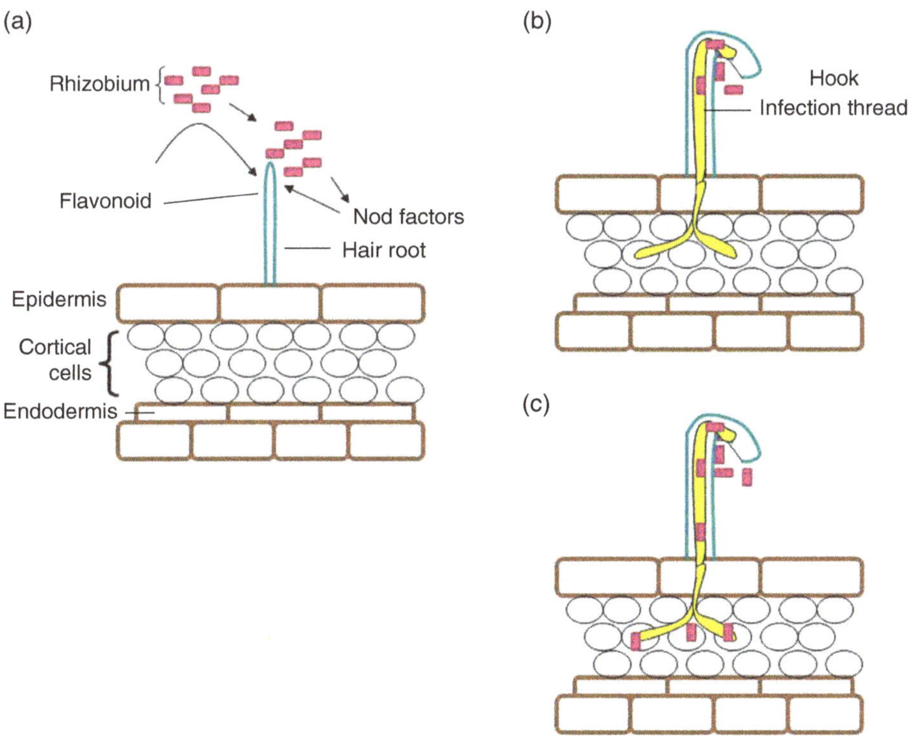

Figure 6.15 Development of the root nodule in legume plants: (a) the hair root releases flavonoids which attract rhizobia and stimulates these bacteria to secrete Nod factors; (b) in response to Nod factors, an infection thread is formed and rhizobia penetrate the root surface; (c) intracellular localization of rhizobia in the cortex cells occurs and after continued rhizobial multiplication, a nodule is formed.

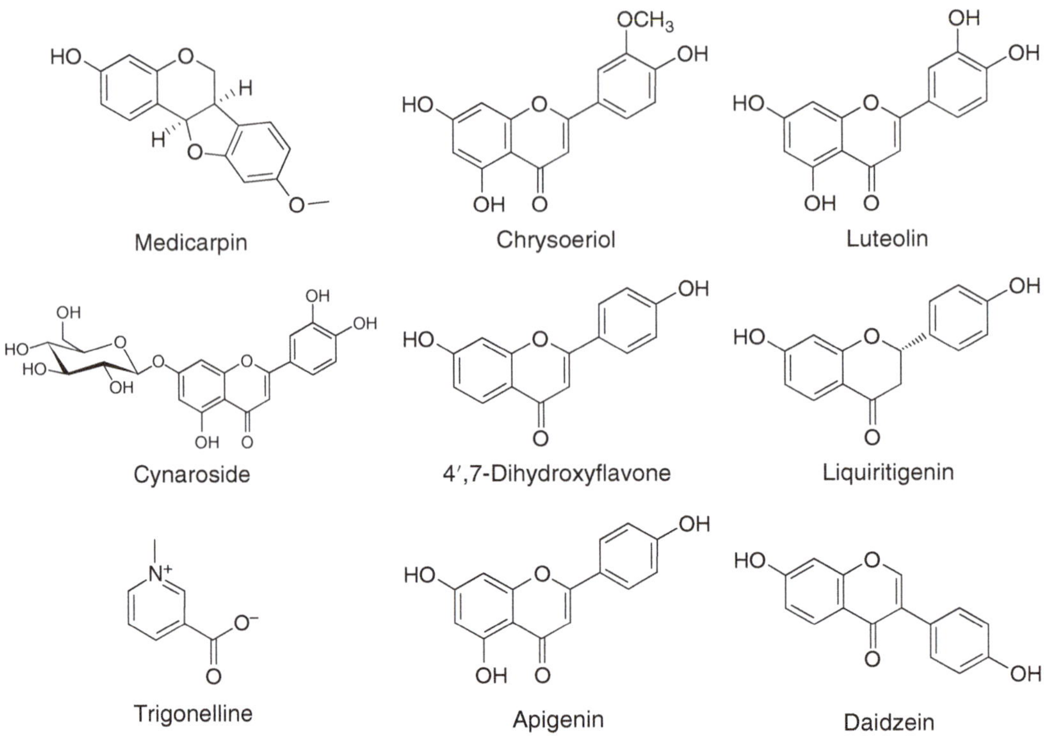

Figure 6.16 Examples of flavonoids produced by rhizobia that nodulate legumes.

- Attachment of the rhizobial cell to the surface of the root is due to rhicadhesin, an adhesion protein produced by the rhizobia and localized on the surface of the rhizobial cell. Lectins produced by the legume and on the root hairs may also have a role in attachment of rhizobia to root hairs.
- Flavonoids compatible for a specific species of rhizobia will bind to nodD, a rhizobial lipochitooligosaccharide signal molecule, and this activates the transcription of the *nod* genes in rhizobia to produce the Nod factor (Figure 6.17). Non-legume plants will also produce flavonoids and substantial levels of these incompatible flavonoids may interfere with the rhizobia nodulation process by inhibiting the production of Nod factors.
- The process of signal transduction across the plasma membrane of rhizobia appears to involve a sensory response mechanism with two components (Loh et al. 2002). Commonly present in nodulating bacteria are the *nodABC*, which encode for the synthesis of β-1,4-linked *N*-acetylglucosamine backbone structure. Other genes control host specificity and induced changes in the cells of the host root hairs.
- In addition to regulation by nodD, nodulation induction requires quorum sensing following an increased density of the rhizobial population (Loh et al. 2002). Bradyoxetin (Figure 6.18) is an extracellular signaling compound produced by *B. japonicum* at high population density. Bradyoxetin regulates the nodulation gene by repressing the production of the nodD component and inducing genes essential for nodular development.
- The legume responses to lipochitooligosaccharide (LCO) nod factors are complex and generally strain specific. The nod factors stimulate growth of roots, increased cell division, and expression of plant genes specific for nodule formation. Differential cell division of plant cells result in the formation of the hook formation, infection thread, and nodule (see Figure 6.15).

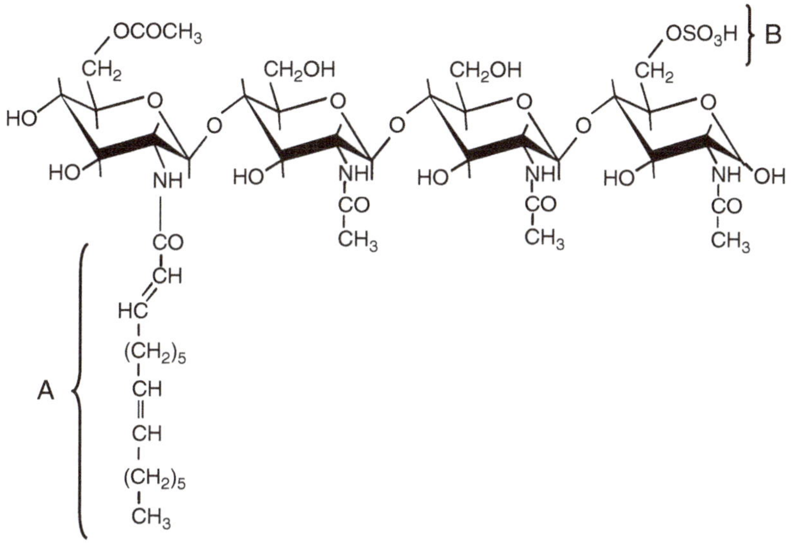

Figure 6.17 Structure of the nodulation factor produced by *Sinorhizobium meliloti*. Nodulation factors produced by other rhizobia differ at site A and site B. Lipids that have been reported at site A are 16C, 18C, or 20C and the unsaturation may vary with the species. The substitution of arabinose, fucose acetate, or methylfucose at site B is may also change.

- The rhizobial cells differentiate into bacteroides, which are found inside the symbiosomes. Leghemoglobin is synthesized by the plant and this reduces the concentration of O_2 around the bacteroid. Under reduced oxygen concentrations, nitrogenase activity is induced and bacteroides enzymatically convert N_2 to NH_3. This nitrogen is used to produce amino acids, which support bacterial growth, with excess amino acid released into the root to support plant nutrition.
- In this nodule–legume symbiosis, the plant provides the carbon source that enables the rhizobia to grow and to enhance plant growth by providing nitrogen.

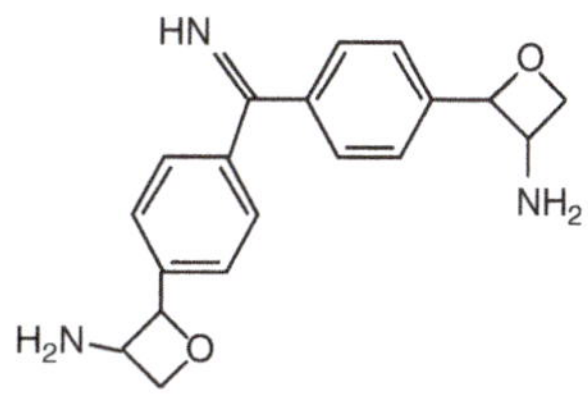

Figure 6.18 Structure of bradyoxetin which is a quorum sensing molecule produced by *Bradyrhizobium japonicum*. The role of bradyoxetin is to regulate expression of bacterial Nod factor genes.

There are two basic morphologies of legume nodules, with the type of nodule determined by the legume host (Ferguson et al. 2010). Intermediate nodules have a cylindrical or branched shape and are present in alfalfa, clover, pea, and *M. truncatula*. As indicated in Figure 6.14, the intermediate nodule has a gradient of developmental activities. There are four areas or zones of activity in the intermediate nodule. Zone I, The active meristem, is where new material is synthesized. Zone II is the region of development of symbiosomes, where bacteria from the infection thread have multiplied and cell differentiation occurs. Bacteroides in the intermediate area characteristically have low viability and occur as branched cellular forms inside the symbiosome. Zone III is the nitrogen fixation area, where leghemoglobin surrounds the bacteroid in the symbiosome and oxygen concentration is reduced to enable nitrogen fixation to occur. The accumulation of leghemoglobin gives this region a distinct red color. In Zone IV senescence occurs, with decay of plant cells and degradation of the symbiosomes. Determinate nodules, Figure 6.13, are spherical, without a developmental gradient and found with soybean, bean, and *Lotus* as the host legume. The determinate nodules are not associated with root hairs but are

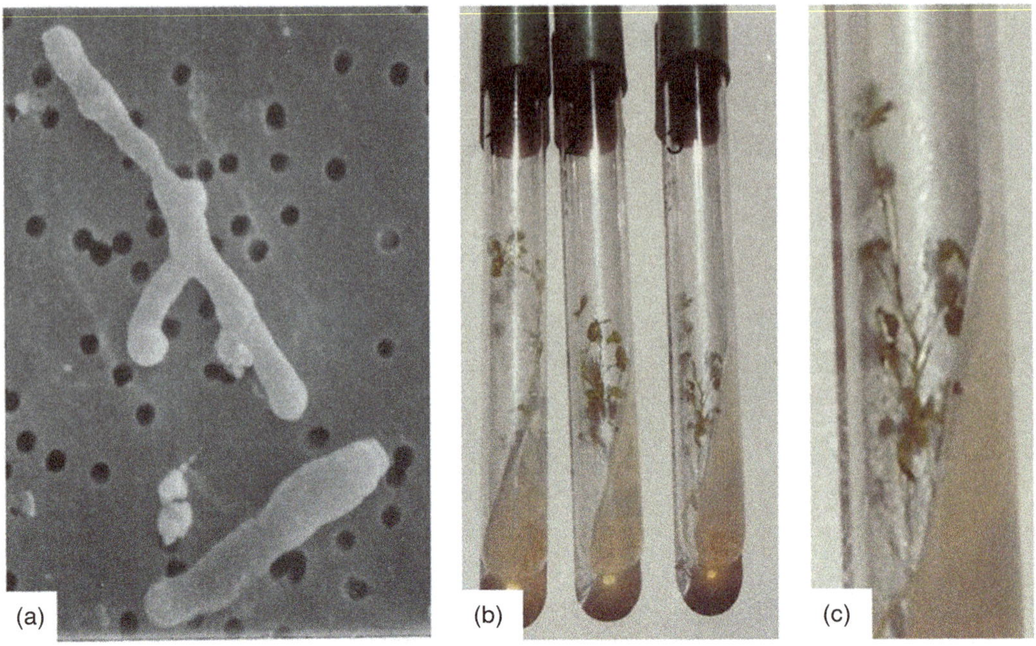

Figure 6.19 Alfalfa – rhizobia symbiosis. (a) Electron micrograph of *Sinorhizobium meliloti* bacterioid with an irregular shape. This is an example of bacteria present in the rhizobium-legume nodule. (b) and (c) Alfalfa plants growing in a test tube on agar solidified plant nutrients. This system allows one to inoculate the alfalfa plants with appropriate culture of bacteria and follow development of the nodules. These tubes can be sealed and the amount of nitrogen fixed can be assessed by using the acetylene reduction assay. *Source:* Reproduced with permission from Larry Barton.

found along the main root where the lateral roots are attached. Bacteroides in the intermediate nodule are of high viability.

With *Sinorhizobium meliloti*, several symbiosis-promoting factors have been recognized to participate in nodulation of alfalfa (Figure 6.19). In addition to quorum-sensing signals, exopolysaccharides (EPS) produced by *S. meliloti* are involved in biofilm formation and colonization on alfalfa roots. EPS I consists of repeating subunits of an octamer consisting of seven glucoses and one galactose unit plus succinate, pyruvate, and acetate residues. EPS II is composed of a polymer consisting of the galactose–glucose dimer with pyruvate linked to galactose and acetate attached to glucose. EPS I and EPS II participate in biofilm formation to protect planktonic or attached cells of *S. meliloti* from environmental stresses.

6.5.4 Nodules on Plant Stems

Nitrogen-fixing nodules on aerial stems of aquatic legumes contain bacteria of the genus *Azorhizobium* and studies are in progress to determine the mechanism of nodule formation. Legumes with stem nodules are restricted to tropical regions and have been reported for a few species of the genera *Neptunia, Aeschynomene, Sesbania*, and *Discolobium* (Dreyfus et al. 1988, Loureiro et al. 1994). Currently, *Azorhizobium caulinodans*, a non-photosynthetic bacterium, is the only species of stem-nodulating bacteria. Since legumes with stem nodules are used as green manure to fertilize rice and other crops, enhanced plant growth due to nitrogen fixation by stem nodulating bacteria is important and will receive attention in the future.

6.6 Frankia and the Non-legume Nitrogen-fixing Nodule

Frankia is a gram-positive soil actinomycete that forms nitrogen-fixing nodules on roots of specific trees and shrubs. Actinorhizal plants are dicotyledons that belong to 24 genera and over 200 species (Benabdoun et al. 2011). The symbiotic relationship between *Frankia* and plant roots is found in all continents except Antarctica and this broad distribution may account for as much as 15% of biological nitrogen converted to ammonia from atmospheric dinitrogen. The classification of *Frankia* is progressing slowly because isolates from actinorhizal roots are difficult to grow in the laboratory and at this time they comprise a single genus with four groups (Chaia et al. 2010). The genomes of three *Frankia* isolates have been sequence and scientists are examining the gene information to understand the basis for nodule development in actinorhizal plants. Since a genetic transformation system does not exist for *Frankia*, little is known about the molecular events leading to nodulation (Benabdoun et al. 2011).

Evaluation of nodule development and activity relies on descriptive information. A considerable history has been complied with respect to the environment where the actinonodules have been collected (Chaia et al. 2010). Since *Frankia* have not been isolated from the soil, all laboratory strains are derived from actinorhizal nodules. The presence of *Frankia* in soil can be demonstrated by planting actinorhizal plants and checking for nodule development or by hybridization and sequencing of DNA in soil samples (see review by Chaia et al. 2010 for specifics). An actinorhizal plant, Russian Olive, with root nodules is in Figure 6.20. In many instances the *Frankia* are unculturable from the nodule, lack the capability of fixing nitrogen,

Figure 6.20 Russian Olive tree with actinorhizal nodules. (a) The Russian Olive tree is often used as an ornamental in landscaping or as ground cover. (b) Actinorhizal nodule formed by *Frankia* sp. on roots of Russian Olive tree.

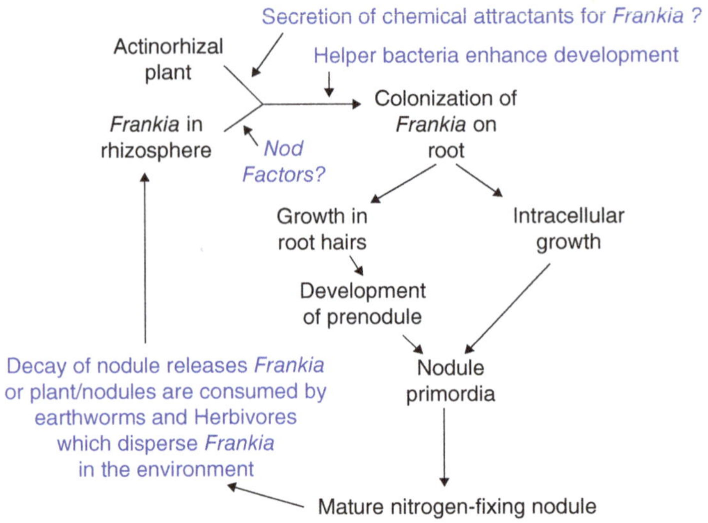

Figure 6.21 Model used to describe the free-living *Frankia* becoming an endophyte.

or unable to reinfect the original host plant (Mirza et al. 2009). *Frankia* cultured in the laboratory have three morphological types of cells: cells aligned in filaments, spores and vesicles. The thick wall vesicles account for nitrogen fixing *ex planta* but nitrogen fixation by *Frankia* in the soil has not been reported.

The model presented in Figure 6.21 highlights the major events in nodulation attributed to *Frankia* (Chaia et al. 2010). Vegetative growth of *Frankia* in the soil is the source of bacteria for nodulation and an appropriate host is selected. As with rhizobial nodule development, chemical signaling between the bacteria and plant root are involved in initiation of nodulation by *Frankia*. In roots of actinorhizal plants is the receptor-like kinase, *SymRK*, gene which is required for *Frankia* nodulation. Similarly, this *SymRK* gene is required for rhizobia-legume nodulation and symbiosis with arbuscular mycorrhiza (Gherbi et al. 2008). A root hair-forming factor has been reported for *Frankia* and it appears that this chemical produced by *Frankia* is structurally distinct from the nod factors produced by rhizobia (Gherbi et al. 2008). Soil bacteria other than *Frankia* are considered important to promote by unknown mechanisms the initial interactions between *Frankia* and the root. The infection process is either by the intracellular pathway involving root hairs or by an intercellular pathway. In the intracellular infection, root hairs become deformed and a structure analogous to the infection thread in legumes is formed in the hair root. Filaments of *Frankia* cells penetrate the root cortex and grow between the epidermal cells and the cortex cells. A fibrous nodule will develop from the apical meristem and following differentiation of bacterial cells, nitrogen fixation will occur. Stimulation of nodule development is also considered to result from quorum sensing activity by *Frankia*. Nitrogen fixation occurs after bacterial cells in the nodule develop vesicles and hemoglobin in the *Frankia* nodule is assumed to assist nitrogen fixation by reducing the level of oxygen (Schwintzer and Tjepkema 1990). To assist *Frankia* in obtaining iron from calcarious soil with low bioavailable iron, it may be anticipated that siderophores produced by *Frankia* would be beneficial (Table 6.4). A highly complex system of four-partner symbiosis has been reported for some plant roots that simultaneously harbor *Frankia* nodules, arbuscular mycorrhizae (AM) and ectomycorrhizal fungi (Chaia et al. 2010). Future discoveries are needed to elucidate the controlling factors for development of *Frankia* nodules.

Table 6.4 Characteristics of the different types of mycorrhiza.

Mycorrhizal associations	Characteristics
Ectomycorrhiza	Fungi found primarily on the root surface of woody plants. Thousands of fungal species occur. Some are generalists and form symbiotic associations with many different trees; some are specialists that form symbiosis with only one plant genus. Plants may have several fungal partners at the same time. Ectomycorrhizal fungi grow and sporulate in the rhizosphere and are not dependent on the symbiotic state for reproduction.
Arbuscular mycorrhiza	Formally known as vesicular arbuscular mycorrhizas, they have little fungal growth on the root surface and are found primarily in the cortex cells of the root. Fungal hyphae penetrate the host cells forming invaginations of the root cell membrane. This is the most widely occurring of the mycorrhizas and the oldest. The fungal hyphae produce glomalin, a unique glycoprotein that persists in the soil for many years and promotes desirable physical soil structure. Many cereal grains have arbuscular mycorrhizas making this association of economic importance.
Ericoid mycorrhiza	A form of arbuscular mycorrhiza where fungal hyphae are coiled inside the outer layer of root cells. These ericaceous fungi are ascomycetes or basidiomycetes that derive nutrients from decomposing materials and colonize evergreen shrubs.
Arbutoid mycorrhiza	The fungal group involved are similar to the ectomycorrhiza but, unlike the ectomycorrhizal fungi, these hyphae penetrate the root cells. This may also be known as ectendomycorrhiza. The fungi are ascomycetes that colonize conifers.
Orchid mycorrhiza	Various basidiomycete fungi form symbiotic relationships with orchids. The germinating orchid seed depends on the mycorrhizal fungal cells for carbon compounds because the seeds do not have nutrient reserves. Orchids are non-photosynthetic when mature and require mycorrhizal fungi to provide organic carbon and other nutrients.
Monotropoid mycorrhiza	The mycorrhizal fungi are either ascomycetes or basidiomycetes that colonize flowering plants that lack chlorophyll.

6.7 Mycorrhizae

The term mycorrhizae was used by Albert B. Frank in 1885 to describe the symbiotic association of fungi with plant roots. The fungal symbiosis can be divided into two major categories based on the root-associated growth of the fungi. There is the endomycorrhizae which has extensive fungal penetration into the root cortex and includes the AM, orchidaceous mycorrhizae and ericaceous mycorrhizae. The sheathed mycorrhizae have extensive fungal growth on the root surface and include the ectomycorrhizae (EcM), ectendomycorrhizae and monotropoid mycorrhizae. These six types of mycorrhizae are discussed in Table 6.5. Concerning the mycorrhizae, the AM and the EcM are the most common with over 80% of land plant species have mycorrhizal symbiosis. A highly prized plant for decoration is the orchid (Figure 6.22) and the mycorrhizae fungus (commonly a basidiomycete) is required for orchid growth. An interesting and unexpected finding is that some plants are able to exchange carbon to adjoining plants through mycorrhizae (Francis and Read, 1987).

The benefit of the fungus is most obvious if the plant is growing in nutrient-poor soils and in drought conditions. While neither the AM nor the EcM fungi have genes for nitrogen fixation, mycorrhizal fungi assist the host by providing phosphorus in the form of organophosphates and nitrogen as glutamine. It has been reported that AM and EcM influence the uptake of Cu and Zn under conditions of deficiency and toxicity (Nakmee et al. 2016). The acquisition of iron by plants with mycorrhizal fungi has been explored and under conditions of limiting iron, fungal siderophores enhance plant uptake of iron, see Table 6.5. From an ecological

Table 6.5 Activities of mycorrhizal fungi with respect to iron uptake by plants.

Mycobiont	Siderophore produced/ observed activity	References
Ectomycorrhizal fungi (EcM)		
Boletus edulis	Hydroxymate siderophore	Szaniszlo et al. (1981)
Laccaria laccata and *Laccaria bicolor*	Fusigen, ferricrocin, coprogen, and triacetylfusarinine C	Haselwander et al. (2013)
Cenococcum geophilum	Ferricrocin	Haselwander and Winkelmann (2002)
Suillus granulatus, Suillus luteus	Fusigen, ferricrocin, coprogen, and triacetylfusarinine C[a]	Haselwander et al. (2011)
Endomycorrhizal fungi (AM)		
Glomus albidum, Glomus mosseae, Glomus fasciculatum, Glomus macrocarpum	Enhanced iron uptake by mycorrhizal (AM) *Hilaria jamesii* (Galleta)	Cress et al. (1986)
Glomus etunicatum, Glomus intraradices	Siderophore production by mycorrhizal (AM) sorghum roots	Aliasgharzad et al. (2009)
Glomus mosseae	Hyphae from AM mobilize and transport iron to sorghum	Caris et al. (1998)
Ericoid mycorrhiza		
Hymenoscyphus ericae, Oidiodendron griseum, Rhodothamnus chamaecistus	Fusigen is produced by these fungi	Haselwandter et al. (1992)
Orchid mycorrhiza		
Ceratobasidium sp., *Rhizoctonia* sp.	Basidiochrome	Haselwander et al. (2006)

[a] Not produced by *Suillus luteus*.

Figure 6.22 Common ornamental orchids require mycorrhizal association. *Source:* Reproduced with permission from Larry Barton.

perspective, the benefit/cost ratios for mycorrhizas are greatest in slow growing plant species. The introduction of plants into barren soil results in limited plant growth unless rich soil containing mycorrhizal spores is introduced. In the ectomycorrhizal symbiosis with *Suillus tomentosus* and lodgepole pine, tuberculate structures are formed which host rhizospheric nitrogen-fixing bacteria thereby favoring the growth of lodgepole pine in nitrogen-limiting soils (Paul et al. 2007).

6.7.1 Arbuscular Mycorrhizae (AM)

The AM are the most ancient form of fungal-root association and are found in fossils of the Rhynie chert that date to over 400 million years ago (Wang and Qiu 2005). In the symbiotic state, plants provide sugars for fungal growth and the fungus enhances plant nutrition by providing water, nutrients and resistance to toxic metal ions. Plants with a developed AM are more resistant to diseases and insects. With AM, most of the fungal hyphae is associated with the interior of the root and a few hyphae will extend into the soil as shown in Figure 6.23. AM fungi do not grow independently in the soil but are dependent on symbiosis for growth. Asexual spores are produced by the AM fungi in sizeable aggregates and these spores can be collected by using screens for sieving soil. As depicted in the model (Figure 6.24), AM penetrates the cortex cells of roots producing vesicles or balloon-like arbuscules which facilitate nutrient exchange. The hyphae inside the cortex cells do not penetrate the membrane of the host cells but cause an invagination of the cortex cell membrane.

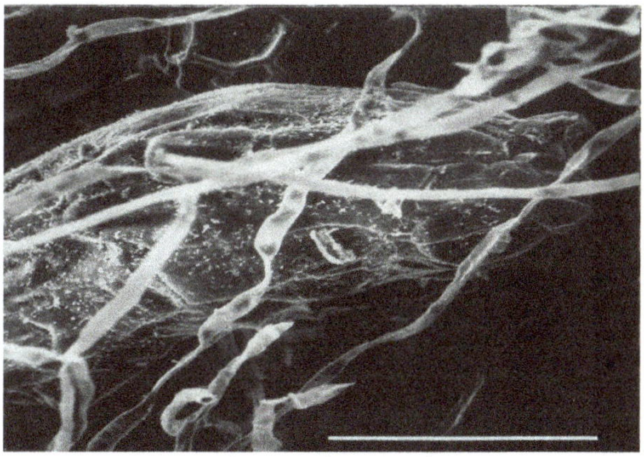

Figure 6.23 Example of fungal endomycorrhiza associated with a plant root. Hyphae of Glomus mosseae fungus extend from root tip of a grass. *Source:* Reproduced with permission from Larry Barton.

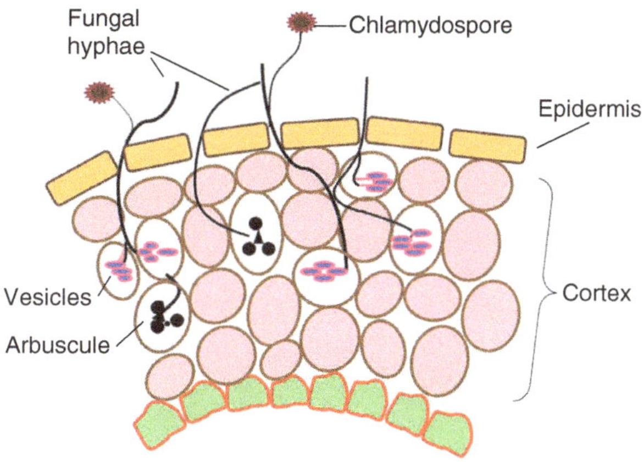

Figure 6.24 Model of a root infected with an endomycorrhizal fungus. Fungal hyphae penetrate cortex cells producing vesicles for nutrient exchange.

6.7.2 Ectomycorrhiza (EcM)

Ectomycorrhizal symbiosis occurs with about 2% of plant species and is found with roots of beeches, pines, birches, willows, oaks, Douglas firs, and eucalypts. Some of the mycobiont – tree partners are listed in Table 6.1. Ascomycetes or basidiomycetes are the mycorrhizal fungi commonly found as EcM and grow predominately on the surface of the roots (Figure 6.25) with hyphae extending into the rhizosphere. Fleshy forms of soil fungi often result as fruiting bodies of ectomycorrhizal fungi and these include truffles and mushrooms. Symbiosis by EcM is considered to have an ancestry that is more modern than AM with fossils showing the Harting net in plant roots that were > 50 million years old (LePage et al. 1997). The EcM partnerships are considered to have evolved numerous times (Tedersoo et al. 2010) and to have stimulated speciation as the land plants were evolving. In general, the mycorrhizal host plant has relatively little specificity for the fungus while the EcM fungi display a preference for host plants.

As indicted in the model (Figure 6.25), the growth of the EcM symbiont is associated with three areas of the root: extracellular hyphae growth, mantle formation by fungi and establishment of the Hartig net. One region is the exterior of the root where the fungal hyphae extend from the root into the rhizosphere. Fungal mycelium with a diameter many times smaller than the root provides a greater surface area than the root and enhances the uptake of water and nutrients from the environment. The EcM may form a connection between adjacent plants and this fungal network has been observed to share phosphorus, nitrogen, and organic carbon compounds between neighboring plants including trees of different species by a source-sink relationship (Simard et al. 1997). An extensive fungal network of *Armillaria ostoyae* in Oregon connects conifers in an area which is 3.8 km in diameter (Ferguson et al. 2003) and would provide a great site for study of nutrient translocations by fungi. Unlike endomycorrhizal fungi, ectomycorrhizal fungi grow freely in the rhizosphere releasing spores which germinate and initiate new symbiotic relationships.

At the soil root interface there is the growth of ectomycorrhizal fungi along the surface of the root and forms a structure designated as the mantle (Figure 6.25). The host root is totally enclosed by the mantle expanding the potential to adsorb nutrients form the environment and providing non-specific protection for the root against pathogenic fungi, bacteria, and soil nematodes. The mantel has an important role in providing protection to the plant salts, toxic

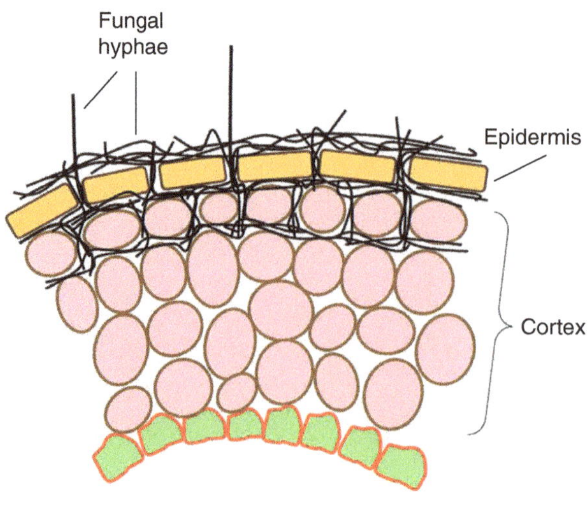

Figure 6.25 Model of a root infected with an ectomycorrhizal fungus. A mantel on the root surface is formed from the fungal hyphae and the Hartig net is the concentration of fungal cells at the periphery of the cortex.

heavy metals, organic pollutants and radionuclides. The fungal symbiont may increase root branching by altering the level of phytohormones (i.e. ethylene, cytokinin, jasmonic acid) that influence plant cell division.

The Hartig net is formed from intercellular fungal growth on the root surface and it functions to interface the mantel with the cells at the root interior. The thickness of the Hartig net varies with the various plant species with the fungal network limited to the surface cortical cells or in some species extending to the endodermis. As the fungal hyphae penetrate through the apoplastic space, cells in the plant root produce peroxidases and chitinases to limit further development of the Hartig net (Hilbert et al. 2001).

6.8 Patterns of Regulation for Plant–Microbe Mutualism

Except for a few examples, plants and bacteria or fungi are required to initiate the symbiotic state when the plant initiates growth. In the case of chloroplasts in plant cells, the mutualism has resulted in a high degree of complexity over the years and the symbiotic system is transferred vertically. In a few lichens, the establishment of lichens in a new environment results from fragments of a mature lichen with both the photobiont and mycobiont contained in a single fragment or the formation of a diaspore that contains algal cells surrounded by fungal cells. Cyanobacteria are highly receptive to establish new mutualistic states with different eukaryotic partners and this may reflect the surface or limited penetration of cyanobacteria into the host. In the case of nodulation of plants by nitrogen-fixing bacteria or the formation of mycorrhizas, each generation of plant requires interaction with the appropriate microbial culture. As a general theme, signaling between plants and symbiont is initiated by release of small molecules by both partners and this chemical communication results in genes expressed that assists the establishment of mutualism.

In the establishment of the fungal – root symbiosis, scientists have been searching for chemicals that may serve as signaling compounds. Recently there is information that the diffusible AM signals for the initiation of symbiosis are lipochitooligosaccharide and are designated as Myc LCO (Czaja et al. 2012; Maillet et al. 2011; Gough and Cullimore 2011). Although the mechanism for establishment of the AM symbiosis awaits clarification, there is an accumulating quantity of research to suggest that the genetic basis for establishment of *Frankia* nodules, rhizobial nodules and endosymbiosis are similar (Gherbi et al. 2008). In EcM there is the search for molecular changes that occur before the fungus and plant cells become associated. There is evidence that numerous genes are differentially expressed by the early phase of mycorrhizal symbiosis and that some of the plant related genes may be similar to those that encode for defenses employed by plants against challenges by pathogens (Sebastiania et al. 2009). Fungal genes induced during early symbiosis are associated with cytoskeleton formation, protein synthesis and mitochondrial respiration (Le Quéré et al. 2005). Clearly, there is considerable adjustment of gene expression to accommodate AM and EcM symbiosis.

Mutualism is a partnership where both partners benefit and there is a minimal amount of damage directed toward each partner. If the plant is too defensive, the bacterial or fungal symbiont is lost and if the microorganisms are too aggressive, a disease may occur with damage to the plant. Root nodule bacteria and perhaps mycorrhizal fungi may produce chitinase to enhance penetration of root cells. Plant cells may respond to chemical signals from the microorganisms and through a series of signal cascades produce various compounds including reactive oxygen species to inhibit microbial growth. The microbiont would produce symbiosis factors in a manner comparable to release of virulence factors by microbial pathogens. The neutralization of microbionts entry into the plant may be comparable to the defense system that plants employ in response to pathogenic microorganisms (Spoel and Dong 2012). In *Arabidopsis thaliana*, the induction of signaling activities to provide defense against pathogenic bacteria is attributed to oligosaccharides released from degradation of the plant wall by enzymes of released by the pathogenic bacterium (Norman-Setterblad et al. 2000).

The acquisition of iron is important for both the symbiont and the host. It is well established that microbial symbionts growing heterotrophically in the rhizosphere benefit for the production of siderophores but it is not clear what role iron may play in maintaining mutualism involving plants and microorganisms. The impact of mycorrhizal fungi on enhancing iron uptake from the environment is an area of research that requires further attention. While the role of quorum sensing is well established for many bacterial systems, the role for quorum signaling (i.e. bradyoxetin) in rhizobial nodules may be highly important. Bradyoxetin production by quorum sensing rhizobia is regulation by iron and there is a possible role of a similar quorum sensing molecule functioning in animal pathologies especially since *Brucella* and *Bartonella* occur in intracellular vacuoles which are similar to symbiosomes (Loh et al. 2002). As discussed by Fabino and O'Brian (2012), iron response regulator (Irr) protein and rhizobial iron regulator (RirA) protein are global regulators contributing to iron homeostasis in the *Alphaproteobacteria* rhizobia. Future studies are required to determine if global regulation of genes is associated with establishing and with maintaining symbiosis with plant-bacteria/fungus systems.

6.9 Bacterial-Fungal Interactions

For many years, there was a separation of research with mycologists studying fungi and bacteriologists studying bacteria but now scientists recognize that bacteria and fungi exist in the same environment and have substantial interactions. While some of these bacterial-fungal interactions (BFIs) may be relatively casual there are some situations where significantly new properties are displayed. These BFIs may promote plant growth, impact microbial communication, or directly impact the physiology and survival of the other microorganism (Frey-Klett et al. 2011). Examples of these processes are discussed below with an emphasis on the model systems used for research.

6.9.1 Direct Effects

Since bacteria and fungi may form communities within the same niche, several different effects have been reported where one microorganism has a direct effect on the other microorganism (Frey-Klett et al. 2011). In what may appear as a collection of fungi and bacteria occurring in the same area may be as a mixed biofilm producing new characteristics of one of the microorganisms involved. For example, the yeast *Candida albicans* promotes biofilm formation with *Staphylococcus aureus* and enhances the resistance of *S. aureus* to vancomycin. The homoserine lactone compound produced by *Pseudomonas aeruginosa* for quorum sensing inhibits filament development of *C. albicans*. *Amanita muscaria*, an ectomycorrhizal fungus, secretes organic acids which inhibit antibiotic synthesis by *Streptomyces* sp. By neutralization of acid in tomato juice, the yeast *Geotrichum candidum* promotes the growth of *Salmonella*. Ethanol production by *Saccharomyces cerevisiae* serves to stimulate the pathogenicity of *Acinetobacter baumannii* to the nematode *Caenorhabditis elegans*. In the cheese-ripening process desired flavors are imparted to the cheese and in one case *Brevibacterium linens* produces methanethiol which is an essential chemical to be metabolized by *Kluyveromyces lactis*.

Some examples involving species of *Burkholderia* reflect mutualism to symbiosis with soil fungi. Bacteria have a difficulty in migrating across air-filled spaces in soil but some bacteria form a mutualism with fungi that enables bacteria to "ride along" with hyphal networks in soil. *Burkholderia terrae* BS001 will produce a biofilm which enables it to migrate in the soil along with *Trichoderma asperellum, Rhizoctonia solani, Fusarium oxysporum,* and *F. oxysporum pv lini* (Nazir et al. 2014). *B. terrae* BS001 also enabled the soil bacterium *Dyella japonica* to be carried by fungal hyphae and it appears the bacteria with capability of degrading contaminants may also be comigrated throughout the soil. An interesting symbiosis has been reported for *Burkholderia rhizoxinica* and *Rhizopus microspores* (Partida-Martinez et al. 2007). *B. rhizoxinica* grows inside *R. microspores* and produces the phytotoxins which are released by this fungus. Additionally, *R. microspores* is dependent on *B. rhizoxinica* to support spore formation by the fungus.

Sometimes, antibiotic-producing bacteria are maintained by a host to defend against unwanted fungal invaders. The attine ants cultivate "gardens" of fungi which are food for the ant larvae. The fungus, *Escovopsis,* is a pathogen of the fungi in the gardens maintained by the ants and the ants employ antibiotic-producing soil bacteria (*Streptomyces* or *Pseudomonas*) to keep the population of *Escovopsis* under control. The attine ants have a special storage compartment where the antibiotic-producing bacteria are maintained and as the ants attend to the fungal gardens, *Escovopsis* is removed and placed in these compartments where the bacteria reside. Dentigerumycin, mycangimycin, and candicidin are antibiotics produced by bacteria that are effective *Escovopsis.* The southern pine beetle uses a similar strategy to maintain its symbiotic fungus (*Entomorticium* sp.) by carrying antibiotic-producing bacteria in a compartment to suppress an unwanted fungus (*Ophiostoma minus*).

6.9.2 Plant Growth-promoting Bacteria

Several strains of *Azospirillum, Agrobacterium, Azotobacter, Bacillus, Burkholderia, Delftia, Paenibacillus, Pantoea, Pseudomonas, Rhizobium,* and *Streptomyces* are available commercially as plant growth-promoting bacteria (PGPB) to supplement agricultural practices (Glick 2012). The activities of PGPB that support plant growth include fixing nitrogen, solubilizing phosphate, sequestering iron, and the production of cytokinins, gibberellins, indole acetic acid and ethylene. Additionally, the PGPB modulate the effect of environmental stress to facilitate near optimal growth under difficult conditions. Induced systemic resistance (ISR) in plants is attributed to PGPB and PGPB effects the signaling by jasmonate and ethylene to activate the plant's defense against pathogens. The ISR is a broader defense system than the systemic acquired resistance which targets a specific pathogen.

6.9.3 Systemic Induction of Plant Immunity

While plants do not have acquired immunity, they do have an innate immunity system which defends them against pathogens (Zhang and Zhou 2010). Pathogenic fungi and bacteria have effector proteins that activate sensory response systems which then induce plant disease resistance. There are hundreds of sensory-response systems in plants and pathogenic bacteria such as *Pseudomonas syringae* have over 30 different effector proteins that could potentially interact with the response proteins. Several of these molecular patterns have been conserved throughout evolutionary development to enable plants to recognize pathogen-associated molecular patterns (PAMPs). Lipopolysaccharides from gram-negative bacteria, bacterial flagellin (protein in flagella), bacterial elongation factor-TU, sulfated proteins and several other compounds from bacteria are important PAMPs. Receptor kinases on the plant cell surface recognize the PAMPs and activate other kinases that induce several changes within the cell, including the production of reactive oxygen species, depositing of callus at the cell wall, and activate mitogen regulated protein kinases. The response signal kinases in the cytoplasm lead to depolarization of membrane potential, disruption of Ca^{2+} homeostasis, and other changes in the plant cell. Because the transcription of over a thousand plant genes may be activated by PAMPs, there are a variety of phenotypic variations that may influence the pathogenicity of bacteria for plants.

Biocontrol of certain plant diseases may be achieved by PGPB (Compant et al. 2005). Plant growth-promoting rhizobacteria are usually found on the surface of the plant roots or in soil adjacent to the plant root but in sometimes endophytic communities are formed that enable the plant to be resistant to pathogens or abiotic stress. Another complex interaction is the insect–plant relationship in which uncultured microorganisms associated with the insect have an impact on the signaling systems that regulate the growth of insects in plants (Douglas 2013). Both interactions are based on the metabolic chemicals produced or used by rhizobacteria or resident microbiota.

6.10 Endophytic Microorganisms

Endophytes are microorganisms that grow in the interior of roots, leaves, stems, and branches without harming the plant. Mycorrhizae have hyphae extending from plant roots and, therefore, are not considered as endophytes. Additionally, the nitrogen-fixing rhizobia are not endophytes because they form nodules and are clearly distinguished from true endophytic root bacteria. Many endophytic microorganisms in plants are important for the production of phytohormones, the synthesis of insect repellents, the survival of plants under stress, and protecting the plant against pathogens or predators.

Various examples of fungal endophytes protecting plant hosts are described in the review by Azevedo et al. (2000). In Table 6.6, specific examples are provided to indicate the insects controlled by fungal endophytes. Note that the leafcutter ant selects endophyte-free plants over those with the fungus *Neotyphodium* sp. The plants with endophytic fungi are not always protected against insects and for appropriate anti-insect activity there are numerous variables

Table 6.6 Examples of endophytic fungi that control insects and pests.

Endophytic fungus	Plant host	Protection
Phomopsis oblonga	Elm tree	Prevents spread of Dutch elm disease attributed to the beetle *Physocnemum brevilineum*
Acremonium lolii	Rye grass	Decreases attack by Argentine stem weevil, *Listronotus bonariensis* Decreases attacks by aphids and leafhoppers
Acremonium alternatum	Brassicas	Destroys larvae of the moth *Plutella xylostella*
Rhabdocline parkeri	Oak tree	Active against the insect *Contarinia* sp.
Phyllosticta sp.	Balsam fir	Displays toxicity to *Paradiplosis tu*mifex, which produces galled needles

Source: Reproduced with permission from Azevedo et al. 2000.

involved, including plant nutrition, the age of plant parts, climate effects, and stress induced responses. The toxicity of the fungal endophytes on insects is due to ergot-like alkaloids of ergopeptine and clavinet and to lolitrems, which are neurotoxins. When the endophytic infection is especially dominant, herbivore mammals develop signs of illness. There has been interest in using genes for alkaloid synthesis to be genetically engineered into plants; however, this is still in the planning stage. It has been estimated that there are over a million different fungal endophytic species and molecular sequencing methods have been successfully used to identify fungi in this endophytic community (Sun and Guo 2012).

Bacteria are also identified as endophytes and many of these are non-nodulating nitrogen-fixing bacteria. Endophytic diazotrophic bacteria have been isolated from many different types of plants including cereal grains, palm trees, sugarcane, grasses, and tuber plants. The bacteria isolated from inside xylem vessels or inside cortical cells include members of *Azospirillum, Herbaspirillum, Acetobacter, Azoarcus,* and *Burkholderia*. The bacteria benefit because the plant provides them with nutrients, primarily sugars, and the bacteria fix nitrogen and provide excess ammonium to the plants. Many *Azospirillum* spp. also produce plant hormones, which enhance plant growth and development.

Non-diazotrophic endophytic bacteria isolated from *Zea mays* (maize also called corn) and other plants include *Bacillus, Pseudomonas, Clavibacter* (formerly *Corynebacterium*), *Erwinia,* and *Micrococcus*. These endophytic bacteria may be effective in reducing diseases in plants; however, more research is needed. It has been demonstrated that endophytic bacteria (*Alcaligenes piechaudii and Kluyvera ascorbata*) are effective in reducing the infection by the bacterium *Xanthomonas campestris* that causes cabbage black rot disease. The endophytic bacteria are passed from plant to plant by seeds, dead plant material, and sap-feeding insects. Medicinal plants have been found to contain many different species of the gram-positive filamentous members of the *Actinobacteria* phylum, which enable the endophytic plant to produce bioactive metabolites that are active against bacteria and cancerous cells (Golinska et al. 2015).

6.11 Microbiology of the Phyllosphere

The aerial portion of a plant is the phyllosphere and at this time, the bacteria, yeast, and filamentous fungi on the leaf surface have been the major focus of research. The microorganisms on the leaves are not infectious to the plant and easily escape detection unless one is specifically looking for them. The phyllosphere bacteria (also called bacterial epiphytes) are an important segment of our biosphere, with about a million bacteria on a square centimeter of leaf surface, which would suggest the presence of 10^{26} bacteria on the planet on plant surfaces (Lindow and Brandl 2003), with 2–13 million new bacterial species on the vascular plants in the Brazilian Atlantic forest (Whipps et al. 2008). Epiphytic yeast may form $10–10^{10}$ colony forming units (CFUs) per gram leaf and epiphytic filamentous fungi are estimated at $10^2–10^8$ CFU per gram leaf (Whipps et al. 2008). It is assumed that the phyllosphere bacteria promote plant growth and contribute to the suppression of plant pathogens whereas the fungi (yeast and filamentous fungi) enhance resistance to drought, discourage herbivores, and assist in preventing colonization of plant pathogens. Throughout the growing season there is a change in microbial populations, with bacteria dominant in young leaves, yeast most abundant near the end of the growing season, and filamentous fungi present at the end of leaf growth. The leaf surface is a nutrient limited environment for microorganisms, with carbon being the limiting factor for growth. Plant leaves release volatile methanol, long chain hydrocarbons, terpenoids, and various other carbon compounds that may be used to support microbial growth. The distribution of microorganism is not uniform along the leaf surface, with colonization at leaf epidermal cell wall junctions, at stomata, and at areas where the leaf surface is irregular.

While many different types of fungi are present in soil and air currents, the presence of relatively few different genera of yeast or filamentous fungi suggests that a selection process is involved in stabilizing the fungi on the leaf surface. Of the cultivatable microorganisms, the yeast usually isolated include members of the genera *Rhodotorula, Cryptococcus,* and *Sporobolomyces,* and the most commonly isolated filamentous fungi found on plant leaves include *Cladosporium, Alternaria, Penicillium, Acremonium, Mucor,* and *Aspergillus.* However, molecular techniques have revealed that there is an extremely diverse fungal community on the surface of oak leaves (Jumpponen and Jones 2009).

Bacteria are by far the most abundant microorganisms on plant leaves. In the tropical Brazilian forest, community analysis studies revealed that only 2% or 3% of all bacterial sequences are from described organisms. In a survey consisting of eight different plants, the phyllosphere contained mostly *Gammaproteobacteria* species, numerous species from *Alphaproteobacteria, Betaproteobacteria, Firmicutes,* and *Bacteroidetes,* and very few cyanobacteria, actinobacteria, or acidobacteria (Whipps et al. 2008). *Methylobacterium* spp. are found on the leaves of plants from almost every family and it has been proposed that these bacteria grow on methanol, which is released from plants when a new cell wall is synthesized. *P.* and *Erwinia* (*Pantoea*) spp. have been model systems to examine epiphytic bacteria. The formation of ice crystals on plant flowers and leaves is attributed to bacteria on the plant surface that serve as nucleation sites for ice formation and these bacteria are designated as ice$^+$ bacteria. To prevent frost damage, an ice$^-$ strain of *Pseudomonas fluorescence* can be sprayed on the plants to competitively exclude the ice$^+$ bacteria. Due to the concern of human pathogens being acquired from vegetables, numerous bacteria have been tested to access their capability of colonizing on plant leaf surfaces; however, the incidence of colonization of *Escherichia coli* and *Salmonella enterica* is quite low.

There are several characteristics that are commonly observed in phyllosphere bacteria. Over half of the *Pseudomonas* spp. isolated produces wetting agents which enable bacteria to move across the phylloplane (surface of the leaf) to new areas for nutrients. Many phyllosphere bacteria produce phytohormones (indole-3-acetic acid and cytokinin) which stimulate the release of polysaccharides from the plant and most produce extracellular polysaccharide to protect bacteria from environmental stress and promote bacterial aggregation. It may be that nonpathogenic phyllosphere bacteria have a type III secretion pathway similar to the ones in *P. fluorescens* and *Pseudomonas putida* and phyllosphere and use this secretion system to modify their environment. *Pseudomonas syringae* produces the toxin syringomycin that induces the plant to form ion channels, which release nutrients from a plant cell and this toxin could be introduced into plant cells using the type III secretion system.

6.12 Summary

The hallmark examples of mutualism in microbial-plant systems include lichens, root nodules with rhizobia, and root interactions with fungi. With the capability of photosynthesis, CO_2 fixation and N_2 fixation, the cyanobacteria are well prepared to grow independently; however, cyanobacteria form mutualistic associations with fungi to produce lichens, which enabled the cyanobacteria to grow in extreme environments. The cyanobacterial association with bryophytes, mosses, and other non-vascular plants provides cyanobacteria with protection from predators and the capability of becoming established in new environments. Highly developed signal–response systems in the rhizobia–root interaction resulting in the nodulation of legume roots, indicates the specificity of this association. The *Frankia*–actinorhizal root association appears to result from chemical signaling. For both the rhizobia–legume and *Frankia*–actinorhizal roots, the plant provides carbon compounds for bacterial growth and bacteria enhance plant nutrition by providing ammonia from the fixation of atmospheric nitrogen. The fungal-plant root association (i.e. mycorrhizae) has resulted in several different classes that range from growth of fungi on the root surface to growth inside the root cells. For the past 400 million years, the arbuscular mycorrhizal fungi have been mutualistic partners for plants.

Discussion Questions

1. What unique features of cyanobacteria enable it to establish so many symbiotic associations?
2. What limits the lateral gene transfer of nitrogenase genes from rhizobia or *Frankia* to plants? Would these limitations also preclude the transfer of nitrogenase from bacteria to fungi? Defend your answer.
3. What are some of the challenges that plant symbionts (i.e. legume-rhizobia and mycorrhiza-plants) have addressed through co-evolution?
4. What would be the situation where carbon dioxide fixation drives symbiosis and what would be the situation where dinitrogen fixation drives symbiosis? Give examples of each.
5. Would it be feasible to embark on a research program to construct mycorrhizal fungi to be able to fix nitrogen? What approaches would you suggest for such an effort?
6. If plants are infected with one or more endophytes, why has it taken so long to recognize their presence?

References

Adams, D.G. and Duggan, P.S. (2008). Cyanobacteria – bryophyte symbiosis. *J. Exp. Bot.* 59: 1047–1058.

Aliasgharzad, N., Shirmohamadi, E., and Oustan, S. (2009). Siderophore production by mycorrhizal sorghum roots under micronutrient deficient condition. *Soil Environ.* 28: 119–123.

Azevedo, J.L., Maccheroni, W., Pereira, J.O. et al. (2000). Endophytic microorganisms: a review on insect control and recent advances on tropical plants, *Electronic. J. Biotechnol.* 3 (1): 2000. https://doi.org/10.2225/vol3-isue1-fulltext-4.

Barton, L.L. and Northup, D.E. (2011). *Microbial Ecology*. Hoboken, NJ: Wiley.

Benabdoun, F.M., Nambiar-Veetil, M., Imanishi, L. et al. (2011). Composite actinorhizal plants with transgenic roots for the study of symbiotic associations with *Frankia*. *J. Bot.* 2011: 702947, 8 pages.

Boyer, G.L. and Aronson, D.B. (1994). Iron uptake and siderophore formation in the actinorhizal symbiont *Frankia*. In: *Biochemistry of Metal Micronutrients in the Rhizosphere* (ed. J.A. Manthey, D.E. Crowlley and D.G. Luster), 41–54. Boca Raton, FL: Lewis Publishers.

Brook, A.J. (1968). The discoloration of roofs in the United States and Canada by algae. *J. Phycol.* 4: 1529.

Campbell, E.L., Summers, M.L., Christman, H. et al. (2007). Global gene expression patterns of *Nostoc punctiforme* in steady-state dinitrogen-grown heterocyst-containing cultures at single time points during differentiation of akinetes and hormogonia. *J. Bacteriol.* 189: 5247–5256.

Campbell, D., Hurry, V., Clarke, A.K. et al. (1998). Chlorophyll fluorescence analysis of cyanobacterial photosynthesis and acclimation. *Microbiol. Mol. Biol. Rev.* 62: 667–683.

Caris, C., Hördt, W., Hawkins, H.-J. et al. (1998). Studies on iron transport by arbuscular mycorrhizal hyphae from soil to peanut and sorghum plants. *Mycorrhiza* 8: 35–39.

Chaia, E.E., Wall, L.G., and Huss-Danell, K. (2010). Life in soil by the actinorhizal root nodule endophyte *Frankia*. A review. *Symbiosis* 51: 201–206.

Chiu, W.-L., Peters, G.A., Levieille, G. et al. (2005). Nitrogen deprivation stimulates symbiotic gland development in *Gunnera manicata*. *Plant Physiol.* 139: 224–230.

Compant, S., Duffy, B., Nowak, J. et al. (2005). Use of plant growth-promoting bacteria for biocontrol of plant diseases: principles, mechanisms of action and future prospects. *Appl. Environ. Microbiol.* 71: 4951–4959.

Cress, W.A., Johnson, G.V., and Barton, L.L. (1986). The role of endomycorrhizal fungi in iron uptake by *Hilaria jamesii*. *J. Plant Nutr.* 9: 547–556.

Culberson, C.F. (1970). Supplement to chemical and botanical guide to lichen products. *Bryologist* 73: 177–377.

Culberson, C.F. (1972). Improved conditions and new data for the identification of lichen products by a standardized thin-layer chromatography method. *J. Chromatogr.* 72: 113–125.

Czaja, L.F., Hogekamp, C., Lamm, P. et al. (2012). Transcriptional responses toward diffusible signals from symbiotic microbes reveal *MtNFP-* and *MtDM13*-dependentant reprogramming of host gene expression by arbuscular mycorrhizal fungal lipochitooligosaccharides. *Plant Physiol.* 159: 1671–1685.

Dakora, F.D., Joseph, C.M., and Phillips, D.A. (1993). Alfalfa (*Medicago sativa* L.) root exudates contain isoflavonoids in the presence of *Rhizobium meliloti*. *Plant Physiol.* 101: 819–824.

de Berry, A. (1878). *Uber Symbiose – Tageblatt 51. Versammi. Deutscher Naturforscher u. Aerzte*, vol. 1878, 121–126. Cassel.

Dreyfus, B., Garcia, J.L., and Gillis, M. (1988). Characterization of *Azorhizobium caulinodans* gen. nov., sp. nov., a stem-nodulating nitrogen-fixing bacterium isolated from *Sesbania rostrata*. *Int. J. Syst. Bacteriol.* 38: 89–98.

DePriest, P. (2004). Early molecular investigations of lichen-forming symbionts: 1986–2001. *Annu. Rev. Microbiol.* 58: 273–301.

Douglas, A.E. (2013). Microbial brokers of insect-plant interactions revisited. *Chem. Ecol.* 39: 952–961.

Fabino, E. and O'Brian, M.R. (2012). Mechanisms and regulation of iron homeostasis in the rhizobia. In: *Molecular Aspects of Iron Metabolism in Pathogenic and Symbiotic Plant-Microbe Associations* (ed. D. Expert and M.R. O'Brian), 41–86. London, UK: Springer.

Falcón, L.I., Magallón, S., and Castillo, A. (2010). Dating the cyanobacterial ancestry of the chloroplast. *The ISME J.* 4: 777–783.

Ferguson, B.A., Dreisbach, T.A., Parks, C.G. et al. (2003). Coarse-scale population structure of pathogenic *Armillaria* species in a mixed-conifer forest in the Blue Mountains of Northeast Oregon. *Can. J. Forest Res.* 33: 612–623.

Ferguson, B.J., Indrasumunar, A., Hayashi, S. et al. (2010). Molecular analysis of legume nodule development and autoregulation. *J. Integr. Plant Biol.* 52: 61–76.

Francis, R. and Read, D.J. (1987). Floristic diversity in a model system using experimental microcosms. *Nature* 328: 420–422. https://10.1038/328420a0.

Frey-Klett, P., Burlinson, P., Deveau, A. et al. (2011). Bacterial-fungal interactions: hyphens between agricultural, clinical, environmental and food microbiologists. *Microbiol. Mol. Biol. Rev.* 75: 583–609.

Gargas, A., DePriest, P.T., Grube, M. et al. (1995). Multiple origins of lichen symbiosis in fungi suggested by SSU rDNA phylogeny. *Science* 268: 1492–1495.

Gherbi, H., Markmann, K., Svistppnoff, S. et al. (2008). SymRK defines a common genetic basis for plant root endosymbiosis with arbuscular mycorrhiza fungi, rhizobia, and *Frankia* bacteria. *Proc. Natl. Acad. Sci. U.S.A.* 105: 4928–4932.

Glick, B.R. (2012). Plant growth-promoting bacteria: mechanisms and applications. *Scientifica* 2012 (2012): https://doi.org/10.6064/2012/963401.

Golinska, P., Wypij, M., Agarkar, G. et al. (2015). Endophytic actinobacteria of medicinal plants: diversity and bioactivity. *Antoine van Leeuwenhoek* 108: 267–2898.

Gough, C. and Cullimore, J. (2011). Lipo-chitooligosaccharide signaling in endosymbiotic plant-microbe interactions. *Mol. Plant Microbe Interact.* 24: 867–878.

Hadri, A.-E., Spaink, H.P., Bisseling, T., and Brewin, N.J. (1998). Diversity of root nodulation and the rhizobial infection process. In: *The Rhizobiaceae* (ed. H.P. Spank, A. Kondorosi and P.J.J. Hooykaas), 347–360. Dorcdrecht, The Netherlands: Kluwer Academic Publishers.

Harris, E.H., Boydton, J.E., and Gillham, N.W. (1994). Chloroplast ribosomes and protein synthesis. *Microbiol. Rev.* 58: 700–754.

Hartwig, U.A., Maxwell, C.A., Joseph, C.M. et al. (1990). Chrysoeriol and luteolin released from alfalfa seeds induce *nod* genes in *Rhizobium meliloti*. *Plant Physiol.* 91: 842–847.

Haselwandter, K., Dobernigg, B., Beck, W. et al. (1992). Isolation and identification of hydroxymate siderophores of ericoid mycorrhizal fungi. *BioMetals* 5: 51–56.

Haselwander, K., Häninger, G., and Ganzera, M. (2011). Hydroxamate siderophores of the ectomycorrhizal fungi *Sillus granulatus* and *S. luteus*. *BioMetals* 24: 1153–1157.

Haselwander, K., Häninger, G., Ganzera, M. et al. (2013). Linear fusigen as the major hydroxamate siderophore of the ectomycorrhizal basidiomycota *Laccaria laccata* and *Laccaria bicolor*. *BioMetals* 26: 969–979.

Haselwander, K., Passler, V., Reiter, S. et al. (2006). Basidiochrome – a novel siderophore of the orchidaceous mycorrhizal fungi *Ceratobasidium* and *Rhizoctonia* spp. *BioMetals* 19: 335–343.

Haselwander, K. and Winkelmann, G. (2002). Ferricrocin – an ectomycorrhizal siderophore of *Cenococcum geophilum*. *BioMetals* 15: 73–77.

Hilbert, F., Duplessis, S., Ditengou, F. et al. (2001). Developmental cross talking in the ectomycorrhizal symbiosis: signals and communication genes. *New Phytol.* 151: 145–154.

Honegger, R. (2000). Simon Schwender (1829-1919) and the dual hypothesis in lichens. *Bryologist* 103: 307–313.

Janson, S. (2002). Cyanobacteria in symbiosis with diatoms. In: *Cyanobacteria in Symbiosis* (ed. A.N. Rai, B. Bergman and U. Rasmussen), 1–10. Dordrecht, The Netherlands: Kluwer Academic Publishers.

Jumpponen, A. and Jones, K.L. (2009). Massively parallel 454 sequencing indicates hyperdiverse fungal communities in temperate *Quercus macrocarpa* phyllosphere. *New Phytol.* 18: 438–444.

LePage, B.A., Currah, R.S., Stockey, R.A. et al. (1997). Fossil ectomycorrhizae from the middle Eocene. *Am. J. Bot.* 84: 410–412.

Le Quéré, A., Wright, D.P., Söderström, B. et al. (2005). Global patterns of gene regulation associated with the development of ectomycorrhiza between birch (*Betula pendula* Roth.) and *Paxillus involutus* (Batsch) Fr. *Mol. Plant Microbe Interact.* 18: 659–673.

Lindow, S.E. and Brandl, M.T. (2003). Microbiology of the phyllosphere. *Appl. Environ. Microbiol.* 69: 1875–1883.

Lobakova, E.S. and Smirnov, I.A. (2012). Experimental lichenology. In: *Advances in Applied Biotechnology* (ed. M. Petre). IntechOpen, https://doi.org/10.5772/30538, Available from:. ISBN: 978-953-307-820-5 https://www.intechopen.com/books/advances-in-applied-biotechnology/experimental-lichenology.

Loh, J., Carlson, R.W., York, W.S. et al. (2002). Bradyoxetin, a unique chemical signal involved in symbiotic gene regulation. *Proc. Nat. Acad. Sci. U.S.A.* 99: 14446–14451.

Loh, J. and Stacey, G. (2003). Nodulation gene regulation in *Bradyrhizobium japonicum*: a unique integration of global regulatory circuits. *Appl. Environ. Microbiol.* 69: 10–17.

Loureiro, M.F., de Faria, S.M., James, E.K. et al. (1994). Nitrogen-fixing stem nodules of the legume, *Discolobium pulchellum* Benth. *New Phytol.* 128: 283–295.

Maillet, F., Poinsot, V., André, O. et al. (2011). Fungal lipochitooligosaccharide symbiotic signals in arbuscular mycorrhiza. *Nature* 469: 58–63.

Martin, W., Rujan, T., Richly, E. et al. (2002). Evolutionary analysis of *Arabidopsis*, cyanobacterial, and chloroplast genomes reveals plastid phylogeny and thousands of cyanobacterial genes in the nucleus. *Proc. Natl. Acad. Sci. U.S.A.* 99: 12246–12251.

Maxwell, C.A., Hartwig, U.A., Joseph, C.M. et al. (1989). A chalcone and two related flavonoids released from alfalfa roots induce *nod* genes of *Rhizobium meliloti*. *Plant Physol.* 91: 842–847.

Miao, V.P.W., Coëffet-LeGal, M.-F., Brown, D. et al. (2001). Genetic approaches to harvest lichen products. *Trends Biotechnol.* 19: 349–355.

Mirza, B.S., Welsh, A., Rasul, G. et al. (2009). Variation in Frankia populations of the Elaeagnus host infection group in nodules of six host plant species after inoculation with soil. *Microb. Ecol.* 58: 384–393. https://doi.org/10.1007/s00248-009-9513-0.

Muggia, L., Schmitt, I., and Grube, M. (2009). *Lichens as Treasure Chests of Natural Products*, vol. 59, 85–97. SIM News.

Nakmee, P.S., Techapinyawat, S., and Ngamprasit, S. (2016). Comparative potentials of native arbuscular mycorrhizal fungi to improve nutrient uptake and biomass of *Sorghum bicolor* Linn. *Agric. Nat. Res.* 50: 173–178.

Nash, T.H. (2008). *Lichen Biology*, 2e. Cambridge, UK: Cambridge University Press.

Nabout, J.C., da Silva Rocha, B., Carneiro, F.M. et al. (2013). How many species of cyanobacteria are there? Using a discovery curve to predict the species number. *Biodivers. Conserv.* 22: 2907–2918.

Nazir, R., Tazetdinova, D.I., and van Elsas, J.D. (2014). *Burkholderia terrae* BS001 migrates proficiently with diverse fungal hosts through soil and provides protection from antifungal agents. *Front. Microbiol.* 5: 598. https://doi.org/10.3389/fmicb.2014.00598.

Nicolaisen, K. and Schleiff, E. (2010). Iron dependency of, and transport by, cyanobacteria. In: *Iron Uptake and Homeostasis in Microorganisms* (ed. P. Cornelis and S.C. Andrews), 200–227. Norfolk, UK: Caister Academic Press.

Norman-Setterblad, C., Vidal, S., and Palva, E.T. (2000). Interacting signal pathways control defense gene expression in *Arabidopsis* in response to cell wall degrading enzymes from *Erwinia carotovora*. *Mol. Plant Microbe Interact.* 13: 430–438.

Partensky, F., Hess, W.R., and Vaulot, D. (1999). *Prochlorococcus*, a marine photosynthetic prokaryote of global significance. *Microbiol. Mol. Biol. Rev.* 63: 106–127.

Partida-Martinez, L.P., Monajembashi, S., Greulich, K.O. et al. (2007). Endosymbiont-dependent host reproduction maintains bacterial-fungal mutualism. *Curr. Biol.* 17: 773–777.

Paul, L.R., Chapman, B.K., and Chanway, C.P. (2007). Nitrogen fixation associated with *Sillus tomentosus* tuberculate ectomycorrhizae on *Pinus contorta* var. *latifolia*. *Ann. Bot.* 99: 1101–1109.

Peters, N.K., Frost, J.W., and Long, S.R. (1986). A plant flavone, luteolin, induces expression of *Rhizobium meliloti* nodulation genes. *Science* 233: 977–980.

Phillips, D.A., Joseph, C.M., and Maxwell, C.A. (1992). Trigonelline, and stachydrine released from alfalfa seeds activate NodD2 proteins in *Rhizobium meliloti*. *Plant Physiol.* 99: 1526–1531.

Pueppke, S.G. and Broughton, W.J. (1999). *Rhizobium* sp. strain NGR234 and *R. fredii* USDA257 share exceptionally broad, nested host ranges. *Mol. Plant Microbe Interact.* 12: 293–318.

Rai, A.N., Söderback, E., and Bergman, B. (2008). Cyanobacterium-plant symbiosis. *New Phytol.* 147: 449–481.

Rousk, K., Jones, D.L., and DeLuca, T.H. (2013). Moss-cyanobacteria associations as biogenic sources of nitrogen in boreal forest ecosystems. *Front. Microbiol.* 4 (150): https://doi.org/10.3389/fmicb.2013.00150.

Sadowsky, M.J. and Graham, P.H. (2006). Root and stem nodule bacteria of legumes. In: *The Prokaryotes*, vol. 2 (ed. M. Dworkin, S. Falkow, E. Rosenberg, et al.), 818–841.

Sancho, L.G., De La Torre, R., Horneck, G. et al. (2007). Lichens survive in space: result from the 2005 LICHENS experiment. *Astrobiology* 7: 443–454.

Sal, L.J. and Moezelaar, R. (1997). Fermentation in cyanobacteria. *FEMS Microbiol. Rev.* 21: 179–211.

Schwintzer, C.R. and Tjepkema, J.D. (1990). *The Biology of Frankia and Actinorhizal Plants.* New York, NY: Academic Press.

Sebastiania, M., Figueiredo, A., Acioli, B. et al. (2009). Identification of plant genes envolved on the inicial contact between extomycorrhizal symbionts (*Castanea sativa* – European chestnut and *Pisolithus tinctorius*). *Eur. J. Soil Biol.* 45: 275–282.

Simard, S.W., Perry, D.A., Jones, M.D. et al. (1997). Net transfer of carbon between ectomycorrhizal tree species in the field. *Nature* 388: 579–582.

Spoel, S.H. and Dong, X. (2012). How do plants achieve immunity? Defense without specialized immune cells. *Nat. Re. Immunol.* 12: 89–100.

Sun, X. and Guo, L.-D. (2012). Endophytic fungal diversity: review of traditional and molecular techniques. *Mycology* 3: 65–76.

Szaniszlo, P.J., Powell, P.E., Reid, C.P.P. et al. (1981). Production of hydroxymate siderophore iron chelators by ectomycorrhizal fungi. *Mycologia* 73: 1158–1174.

Taylor, T.N., Hass, H., Remy, W. et al. (1995). The oldest fossil lichen. *Nature* 378: 244.

Tedersoo, L., May, T.W., and Smith, M.E. (2010). Ectomycorrhizal lifestyle in fungi: global diversity, distribution and evolution of phylogenic lineages. *Mycorrhiza* 20: 217–263.

Usher, K.M., Bergman, B., and Raven, J.A. (2007). Exploring cyanobacterial mutalisms. *Ann. Rev. Ecol. Evol. Syst.* 38: 255–273.

Wang, B. and Qiu, Y.L. (2005). Phylogenic distribution and evolution of mycorrhizas in land plants. *Mycorrhiza* 16: 299–363.

Watanabe, M., Semchonok, D.A., Webber-Birungi, M.T. et al. (2014). Attachment of phycobilisomes in an antenna-photosystem I supercomplex of cyanobacteria. *Proc. Natl. Acad. Sci. U.S.A.* 111: 2512–2517.

Whipps, J.M., Hand, P., Pink, D. et al. (2008). Phyllosphere microbiology with special reference to diversity and plant genotype. *J. Appl. Microbiol.* 105: 1744–1755.

Yuan, X., Xiao, S., and Taylor, T.N. (2005). Lichen-like symbiosis 600 million years ago. *Science* 308: 1017–1020.

Zhang, J. and Zhou, J.-M. (2010). Plant immunity triggered by microbial molecular signatures. *Mol. Plant* 3: 783–793.

Further Reading

Knief, C. (2014). Analysis of plant microbe interactions in the era of the next generation sequencing technologies. *Front. Plant Sci.* 5: 216. https://doi.org/10.3389/fpls.2014.00216. eCollection 2014.

7

Mutualism: Microorganisms and Animals

7.1 Overview and Focus

Microorganisms, like plants, have coevolved and formed extensive associations with animals and humans, often referred to as the microbiome. While the association of microorganisms with disease is well documented and will only be briefly addressed here, most microorganisms exhibit a commensal or mutualistic (Box 7.1) relationship with their host. In this discussion, "animals" includes vertebrates, invertebrates, and insects. Microbial-animal associations have been known for decades. Some notable advances in culture techniques, particularly for anaerobes, have aided our understanding of the role of bacteria in fiber digestion in ruminants and other herbivorous animals. As sequencing and bioinformatics technology has improved dramatically over the past 10–15 years, we are beginning to understand the microbiome.

7.2 Building a Microbial Community – The Role of the Host

Microbial colonization and development in animal and human hosts is controlled by many factors. Aside from environmental and nutritional factors (outlined below), the innate and adaptive immune responses of the host animal play a major role in shaping the microbiome. Although a full description of host immunity is beyond the scope of this text (see supplemental reading), some of the strategies involved in immunity are presented here.

7.2.1 Microbiology and Innate Immunity

Innate immunity and protection refer to the general and non-specific mechanisms that higher organisms use to protect themselves from microorganisms. Innate immunity is always present (constitutive) and has been documented in vertebrate and invertebrate animals as well as plants. In contrast, adaptive immunity is present only in vertebrates and is induced upon exposure to a specific antigen. A common example of innate immunity is mucus production in the mouth and digestive tract, whereas a common example of adaptive immunity would be the antibody response and host protection generated after vaccination.

General strategies involved with innate immunity (Table 7.1) include blocking microbial entry to the tissues through physical barriers, prevention of adhesion and removal of unattached organisms, and growth inhibition. There are also a variety of enzymes and other chemicals that can be toxic to microorganisms. Secretory immunoglobulin A (sIgA), although commonly associated with adaptive immunity, plays an important role in blocking microbial adhesion and toxin binding to tissue surfaces; this process is often called immune exclusion

Environmental Microbiology and Microbial Ecology, First Edition. Larry L. Barton and Robert J.C. McLean.

Box 7.1 Symbiosis vs Mutualism

While symbiosis and mutualism are often used interchangeably, each has a specific meaning. Symbiosis refers to the living together of two species for an extended time and is a larger category than mutualism. Symbiosis includes mutualism; commensalism, where the commensal organism cannot live without the host and the host is unaffected; and parasitism, where the host is harmed, Mutualism is the living together of two different species, where both species benefit from the relationship.

Table 7.1 Strategies involved in innate immunity.

Site	Characteristics	Protective strategy
Skin	Dry to moist Keratinized cells	Physical barrier Most microorganisms have limited ability to degrade keratin and penetrate unbroken skin
Epithelial cell turnover		Removal of adherent microorganisms
Body secretions in eyes and skin	Lysozyme	Attacks bacterial peptidoglycan
	High salt concentrations	Osmotic stress and growth inhibition
	Volatile fatty acids (VFAs)	C2–C4 organic acids are toxic to some bacteria
Mucus	Glycoproteins	Physical barrier that prevents microorganisms and harmful chemicals from underlying tissues
Tight junctions between cells		Watertight seal between animal or plant cells, provides physical boundary between cells
Liquid flow	Movement of gastrointestinal contents (peristalsis) and urine flow (urination)	Removal of unattached microorganisms
Transferrin and lactoferrin	Iron chelation	Inhibit microbial growth by restricting iron
sIgA	Secretory immunoglobulin A	Interferes with microorganisms and some toxin binding to tissue surfaces
TLRs	Toll-like receptors on leukocytes and other cells of immune system	Phagocytosis of microorganisms, inflammation, and other defenses activated

(Mantis et al. 2011) and is an important feature of mucosal defenses in vertebrates. Phagocytic cells and other cells associated with the immune system (lymphocytes) can recognize small molecular structures (referred to as microbial-associated molecular patterns, MAMPs) through structures called toll-like receptors (TLRs). TLR activation can trigger phagocytic cells (leukocytes) and signal the adaptive immune response through inflammation. Phagocytic cells often produce a number of highly reactive oxidizing compounds such as superoxide (O_2^-) and peroxide (O_2^{2-}) via a process called the oxidative burst.

7.2.2 Microbiology and Adaptive Immunity

Adaptive immunity (Table 7.2) is also an important defense strategy in vertebrate animals. Here, the host animal (or human) produces an immune response upon exposure to a specific organism, or more precisely molecular structures (antigens) associated with the foreign microorganism. The organisms can then be removed from the host due to aggregation with

Table 7.2 Components of the adaptive immune system.

Leukocytes	A variety of phagocytic cells	Engulf and destroy most microorganisms through enzymatic activity (defensins) or production of reactive chemicals (oxidative burst)
Antibodies	Highly specific recognition of antibodies, agglutination (clumping) of microorganisms	Microorganisms bound by antibodies are more likely to be engulfed and destroyed by leukocytes
Complement	Soluble proteins in the blood that upon activation by a microorganism can kill microorganisms	Membrane-attack complex formed upon complement activation lyses bacteria
Lymphoid tissue	Skin-associated lymphoid tissue (SALT), and mucosa-associated lymphoid tissue (MALT)	Contain resident leukocytes (macrophages) that phagocytose organisms that penetrate skin and mucosa

antigen-specific proteins (antibodies), attacked with soluble proteins present in the blood (complement), or engulfed and destroyed by leukocytes. Normally, the innate immune system provides the bulk of the protection to the host, whereas the adaptive immune system deals with the few organisms that get past the innate immune system.

7.3 Host Models to Study Parasite Relationships

7.3.1 Germ-free Animals

Prior to birth, the developing vertebrate fetus lives in a sterile world. During the 1950s, a few scientists developed procedures whereby small animals (often mice and other rodents) could be born by Caesarian section and raised in the absence of microorganisms (germ-free) or with a defined flora (gnotobiotic) (Reyniers 1959). When animals raised under germ-free conditions are colonized by normal flora (a process is often called conventionalization), tissue development and metabolism resemble that of conventionally reared animals. The contribution of specific microorganisms can be investigated in gnotobiotic animals. Based on this work, we now realize the important contributions of the normal flora to tissue development, nutrition, and protection from potential disease-causing organisms (pathogens).

7.3.2 *Caenorhabditis elegans*

Another experimental system used to explore host–microbe interactions involves the nematode *Caenorhabditis elegans* (Glavis-Bloom et al. 2012), which normally lives in temperate environments in association with rotting vegetation and ingests bacteria. This organism contains only 959 cells and was the first multicellular organism to be fully sequenced (The *C. elegans* Sequencing Consortium 1998). *Caenorhabditis elegans* is non-pathogenic, reproduces rapidly, and can be genetically manipulated. When grown in the laboratory, *C. elegans* is typically maintained on agar plates containing lawns of the non-pathogenic *Escherichia coli* OP50 strain. It has an innate immune system that can respond to different bacteria and is becoming widely used in screens for virulence or the function of various components of innate immunity. Although *C. elegans* does not have cellular immunity it does have systematic immunity, which is regulated by signaling pathways and consists of the release of antibacterial chemicals including lectins and lysozymes (Marsh and May 2012). While *C. elegans* does not have circulating immune lymphocytes, it has an epithelial or surface immunity that is distinct from intestinal immunity (Murfin et al. 2012). *Caenorhabditis elegans* is being used as a model organism to detect bioactive compounds effective in preventing infections (Kim et al. 2017).

7.3.3 *Drosophila melanogaster*

Drosophila melanogaster (fruit fly), which also has a well-characterized innate immune system and is fully sequenced (Adams et al. 2000), is also being used as a model for studying microbe–host interactions. Like *C. elegans*, *D. melanogaster* has extensive genetic tools available that include mutant collections as well as RNA interference (RNAi) technology. As reviewed by Kuraishi et al. (2013), the defense system of *Drosophila* that prevents microbial infections includes the physical barrier of the gut epithelium, enhanced production of reactive oxygen species by the peroxidase/dual oxidase (Duox) system, and the release of antimicrobial peptides following the recognition of the bacterial amino acid diaminopimelic acid. Using toll and immune deficiency (Imd) signaling pathways and related activities, *Drosophila* can discriminate between infectious bacteria and the approximately 30 different commensal bacteria in the fly's gut. A focus of this *Drosophila* research is to assess the role of metabolites produced by commensal gut flora to regulate the lipid and carbohydrate metabolism of the host by signaling pathways (Wong et al. 2016).

7.3.4 *Galleria mellonella*

Another multicellular non-vertebrate host used for virulence and innate immunity research is *Galleria mellonella*, the greater wax moth. Using larvae of the moth, and the fungal pathogens *Beauveria bassiana* and *Metarhizium anisopliae*, the process known as "transgenerational immune priming" in insects was examined (Vilcinskas 2016). Larvae fed a diet containing *E. coli*, *Micrococcus luteus*, *Pseudomonas entomophila*, and *Serratia entomophila* revealed changes in immunity-related gene expression in the midgut tissue. When fluorescent-labeled cells of *E. coli* were added to the diet of the larvae, the labeled cells not only crossed the midgut epithelium and induced cellular immunity in the hemocoel, but also became localized in the ovary and became closely associated with the eggs. Thus, this activity was used to explain maternal transgenerational immune priming in insects.

7.4 Digestive Tract Environment

In healthy animals, microorganisms are present in many anatomical regions, particularly those with constant exposure to microorganisms (notably the epidermis, mouth, and digestive tract); transiently present in some regions such as the circulatory system; and absent in other regions, notably the central nervous system. Animal-associated microorganisms are rarely found as monocultures but instead co-exist with other organisms as well as the host. Factors that influence microbial community colonization and development include the local environment (temperature, Eh, O_2, and other gases, pH, salt, moisture, volatile fatty acids (VFAs, sometimes referred to as short-chain fatty acids (SCFAs)), and other chemicals), available nutrients, colonization surfaces, host innate immunity, and interactions with other microorganisms. This chapter focuses on microbial ecology and function in the digestive tract. In humans and other mammals, this is the environment with the highest concentration of microorganisms. Based on the diet, vertebrates can be classified as omnivores, carnivorous and herbivores; examples of these animals are given in Table 7.3. Often the dietary habits of animals in the wild is not

Table 7.3 Classification of animals based on diet.

		Herbivores		
		Specialized digestive system		
Omnivores	**Carnivores**	**Foregut**	**Rumen**	**Hindgut**
Humans	Lions	Hoatzin	Cattle	Horses
Chimpanzees	Coyotes	Camel[a]	Sheep	Rhinoceros
Pigs	Wolves	Hippopotamus[a]	Goats	Rabbits
Bears	Foxes		Deer	Koalas
Opossums	Vultures		Antelope	Elephant
Skunks	Eagles		Giraffes	
Squirrels	Tigers		Buffalo	
Raccoons	Cheetahs			
Chipmunks	Hawks			
Mice				
Rats				

[a] Camel and hippopotamus have a pseudorumen because they have a stomach of three chambers and not four as in true rumen-containing animals.

obvious and it is known that caged animals have a diet different from those that range free, which impacts the bacteria present in the gut. Diet alone is not sufficient to select for the intestinal microbiota as it is a co-evolutionary process that also involves phylogeny (Ley et al. 2008).

7.4.1 Omnivores

There are notable differences when comparing the structure of the digestive tract of humans with other mammals. All organisms have an esophagus that transports food from the mouth into the stomach. Omnivores, including humans, have a one-chambered stomach in which food is initially acidified by HCl to pH 2 and then digested using pepsin, amylase from saliva, and other host produced enzymes. Depending upon the organism, stomach digestion in omnivores typically takes one to six hours. Mixing of the food in the stomach and transport through the gastrointestinal tract occurs as a result of smooth muscle contractions (peristalsis). The partially liquefied food (referred to as chyme) then passes into the small intestine. In the duodenum (first portion of the small intestine), additional protease and lipase enzymes are secreted from the pancreas to continue digestion. Bile salts – surfactant molecules that emulsify (suspend in water) fatty acids and aid in lipid digestion – are also secreted from the liver and gall bladder into the duodenum. The pancreas also secretes bicarbonate, which raises the pH of the duodenum contents to pH ~6. In humans, the pH of chyme continues to increase as the chyme progresses through the jejunum and ileum (middle and terminal sections of the small intestine), to the colon (large intestine). The majority of nutrients are absorbed by the small intestine, whereas water and some salts are absorbed by cells within the colon. Undigested food and a considerable number of microorganisms exit through the anus at the distal end of the colon as feces. At the beginning (proximal end of the colon), there is a blind sac referred to as the cecum.

While the cecum is relatively small in humans and carnivores, it can represent a significant component of the digestive tract in herbivores including ruminants (cattle, sheep, goats, deer) and non-ruminants (horses). In humans and some animals, there is a small organ called the appendix which is attached to the cecum. The appendix was thought to be a vestigial organ, but recent research shows that it may act as a reservoir for useful intestinal flora that may be lost due to disease (Bollinger et al. 2007); however, some studies dispute this assertion (Seretis et al. 2014).

7.4.2 Carnivores

Animals that have meat as a component of their diet are obligate carnivores if they only eat meat or facultative carnivores if they eat a few items other than meat. While terrestrial carnivores have a relatively low diversity of bacterial taxa, the diet of the carnivores greatly impacts their intestinal microbiome. Carnivores will acquire, as part of their gut community, bacteria that were in the guts of their omnivore or herbivore prey. Wild Iberian lynx are carnivores that prey only on wild European rabbits and receive about 1.2% of free sugars in this diet (Alcaide et al. 2012). From fecal samples, the greatest number of bacteria were associated with the phylum *Firmicutes*, which is similar to felines and canines, and the second largest number of bacteria were associated with the phylum *Bacteroidetes* with appreciable numbers of *Fusobacteria*. Not only is the microbiome of the lynx appropriate for digestion of structural protein but it also has bacteria that digest seed and plant material consumed by the rabbit. Having a dual metabiome with the characteristics of both herbivores and carnivores is found in other carnivores consuming prey that feeds on plants.

The microbiome of several obligate carnivores has been examined and as a group, they have few bacterial species dominating their intestine because they do not require bacteria to decompose herbs. Mosquitoes, leeches, sea lamprey, and other blood sucking animals have *Aeromonas* spp. as the dominant bacterium in the digestive tract and it is assumed that this bacterial species is needed for the digestion of blood (Hanning and Diaz-Sanchez 2015). Flesh-eating carnivores have appropriate bacterial taxa to produce enzymes that degrade sugar-containing biopolymeric materials, including sugars from glycoproteins, and convert protein from prey into useful metabolic intermediates. Polar bears (see Section 7.5.3) and minke whales live on a diet of seals and fish, respectively, and *Firmicutes* and *Bacteroidetes* are the dominant bacteria in their intestines. Members of the *Firmicutes* are well known for their ability to enhance fat storage in their hosts (Ogawa et al. 2010; Glad et al. 2010). The dominant bacterium in the gut of the American alligator is *Fusobacterium*, which is thought to reflect a relationship to prehistoric animals (Keenan et al. 2013).

Some animals have intestinal populations of bacteria that vary with fasting or hibernation. When animals fast, bacteria in the intestinal tract digests intestinal mucins to some extent but do not harm the gut lining so there is no leakage of bacteria from the intestine into the body cavity (Hanning and Diaz-Sanchez 2015). Snakes are obligate carnivores that have an unusual feeding pattern where prey are consumed periodically. In the cottonmouth snake, the dominant bacteria are members of the phyla *Firmicutes* and *Bacteroidetes*, with a significant number of *Proteobacteria* in the small intestine (Colston et al. 2015). *Firmicutes* and *Bacteroidetes* are dominant in the large intestine of Burmese pythons. When they fast, *Bacteroides*, *Rikenella*, *Synergistes*, and *Akkermansia* abundance is reduced (Costello et al. 2010), but a few days after feeding, *Clostridium* spp., *Lactobacillus* spp., and *Peptostreptococcacea* numbers increased. Hibernating ground squirrels have a reduction in the abundance of *Firmicutes* while the population of *Bacteroidetes* and *Verrucomicrobia* increases (Carey et al. 2013).

Carnivorous sponges from the Arctic Mid-Ocean Ridge were found to contain significant genes linkages that indicated that dominant bacteria were from *Gammaproteobacteria*, *Alphaproteobacteria*, *Flavobacteriaceae*, and *Thaumarchaeota* (Hestetun et al. 2016).

This dominance of *Alphaproteobacteria* and *Gammaproteobacteria* was also reported for noncarnivorous sponges; however, noncarnivorous sponges had considerable numbers of sulfur-oxidizing and ammonia-oxidizing species. Presumably, this difference in bacterial populations in sponges is attributable to diet.

Casual observation of baleen whales would conclude that as they eat fish, they are carnivores. However, the metabolic pathways of baleen whales for energy production, lipid synthesis, and pyruvate metabolism are similar to terrestrial herbivores, but their glutamate and amino acid metabolism is characteristic of terrestrial carnivores. These filter-feeding baleen whales consume small fish and crustaceans and use their multichambered foregut to convert chitin from the exoskeletons of the krill and crustaceans to metabolizable monomeric sugars (Sanders et al. 2015). The chitin-based fermentation by the baleen whale is comparable to the cellulose-based fermentation of land herbivores.

7.4.3 Herbivores

In contrast to carnivores, a considerable amount of herbivore digestion is carried out by microorganisms within a unique digestive tract. Ruminant animals have a four-chambered stomach, consisting of the reticulum, rumen, omasum, and abomasum (Figure 7.1). In the first two chambers (rumen and reticulum), insoluble plant fiber consisting of cellulose, starch, and related compounds are digested by microorganisms and converted to microbial protein, VFAs such as acetate (C2), propionate (C3), and butyrate (C4), B-type vitamins, and metabolic gases of CO_2 and CH_4. Over 70% of the energy needed by the animal is provided by VFAs, which are absorbed into the blood from the third chamber (omasum), transferred to the liver, and converted to glucose via gluconeogenesis (Mackie et al. 2000). Note that in rumen metabolism the goal is not the formation of a glucose pool for adsorbtion by the rumen membranes, but the production of glucose and other sugars to support microbial growth. The fourth chamber (abomasum) is a true stomach where the food and associated microorganisms are digested under acidic conditions by host enzymes.

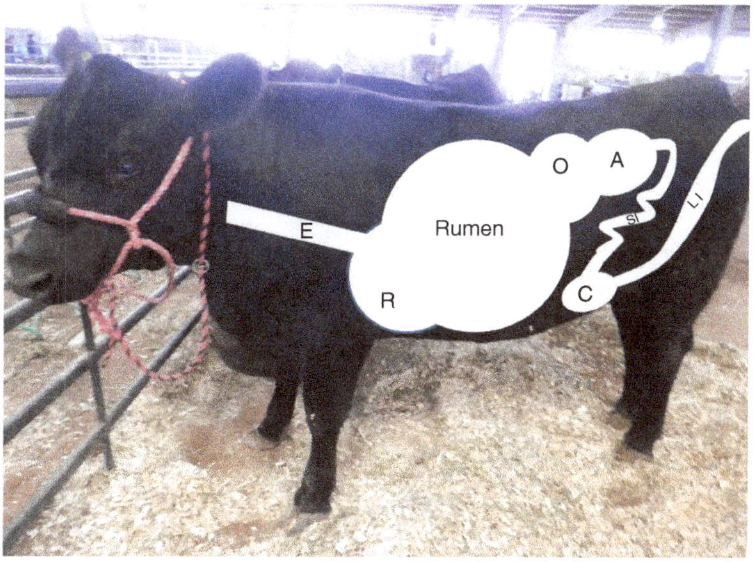

Figure 7.1 Outline of the bovine intestinal tract indicating rumen and additional stomachs. E = Esophagus, R = reticulum, O = omasum, A = abomasum. C = caecum, SI = small intestine, LI = large intestine.
Source: Reproduced with permission from Larry Barton.

Rumen microbiology and fermentation has been reviewed by Castillo-Gonzáleza et al. (2014) and the general characteristics of this microbial activity is found in general microbiology text books. While soluble components of plants (sugars, oils) are easily digested, insoluble plant fiber requires enzymes from microorganisms for its digestion. Adhesion of microbial cells and fiber digestion is enhanced when the food is chewed thoroughly. Ruminant animals will consume plant material and a few hours later regurgitate a bolus of food from the reticulorumen into their mouth, chew on it, and swallow it. This process of rumination or cud chewing lasts about eight hours daily for grass-fed animals. Rumination has the benefit of increasing the surface area of plant fiber for digestion, while the alkaline saliva neutralizes the organic acids, which are the end products of fermentation. Herbivores release fermentation gases (CO_2 and CH_4) into the atmosphere by eructation (belching). In the open environment, ruminants do not compete with non-ruminant herbivores for forage because ruminants can live on low quality plants with dry stems and twigs. Additionally, toxic substances produced by plants (i.e. *Astragalus* or locoweed, red clover, *Leucaena*, and *Heliotrophium*) are detoxified by rumen bacteria enabling ruminants to be less sensitive than nonruminants to these plants.

Over the years, the microbiological activity of the rumen has been extensively examined and a flow chart indicating the ruminant digestion of plant material is given in Figure 7.2. In cattle, the rumen contains 100–150 l of fluid whereas in sheep it is about 6 l. The rumen environment is anaerobic and 1 ml of the fluid characteristically contains 10^{10}–10^{11} cells of prokaryotes, ~10^4 protozoan cells, and a few fungi. Most introductory microbiology text books cover the activities of bacteria and archaea in the rumen but not the protozoa and the fungi. Perhaps, this is in part because these anaerobic protozoa and anaerobic fungi have not been grown in pure

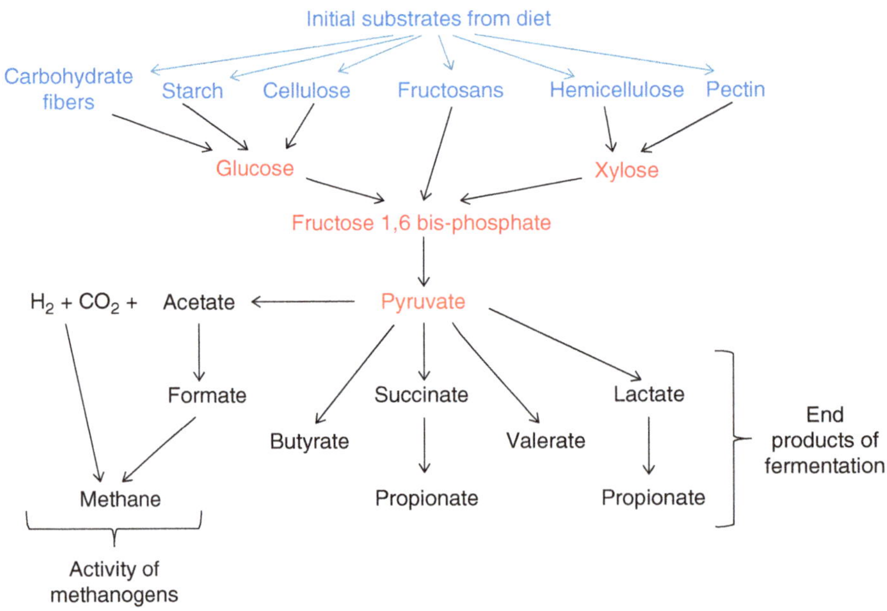

Figure 7.2 Metabolism of carbohydrates in the rumen. A mixture of complex carbohydrate polymeric substrates are consumed by the ruminant and microbial digestion results in production of glucose or xylose monomers. Using the glycolytic pathway, the sugars are converted to pyruvate, which is fermented to a mixture of end products that are characteristic of the diverse anaerobic bacteria present. Methane is produced by species of archaea present in the rumen. If the level of glucose is too high, the rate of mixed acid ferementation is also high and acid toxicity due to the organic acids may occur. If the gases are not released by belching, bloating occurs, which is also a dysfunction of rumen metabolism.

culture. Interest in the activities of fungi and protozoa in the rumen is increasing and the interaction of these three microbial types has prompted the use of "Rumenomics" and "RuminOmics" to describe the molecular studies of microorganisms in the rumen.

7.4.3.1 Bacteria and Archaea

As the rumen environment is anaerobic, special techniques are required for the cultivation of its bacteria and archaea. Robert Humgate pioneered the development of anaerobic procedures to grow anaerobes, and his early contributions using these anaerobic techniques are described in Hungate (1960). The plant material consumed by herbivores contains an array of biopolymers, which are hydrolyzed to produce substrates for bacterial fermentation. A general scheme for rumen fermentation is given in Figure 7.2. There are over 400 species of bacteria in the rumen with many of them in minor abundance and a central core of microorganisms are present with a grass diet or a protein supplemented diet. Some of the dominant bacteria in the rumen of cattle are listed in Table 7.4 along with their enzymatic capabilities. Enzymatic hydrolysis of cellulose and pectin provides a slow release of fermentable sugars, while glucose released from starch hydrolysis may lead to accumulation of organic acids. A neutral pH is required for optimal activity in the rumen and high levels of lactic acid produced by homofermentative activity of *Streptoccus bovus* could cause the pH of the rumen fluid to reach pH 6 or lower. To prevent acidosis, lactate degrading bacteria (*Megasphaera elsdenii* and *Selenomonas lactilytica*) ferment lactate to acetate, valerate, succinate, propionate, butyrate, CO_2, and H_2, with restoration to a neutral pH (Castillo-Gonzáleza et al. 2014).

The anaerobic environment of the rumen is an excellent site for growth of the methanogens and the species listed in Table 7.4 produce methane from H_2 and CO_2 (Hook et al. 2010). Some methanogens produce methane from acetate while others use H_2 and methanol. Due to the unique biochemical and physiological characteristics of methanogens (archaea), antibiotics that inhibit bacteria are not effective against methanogens. The methanogens are not involved in hydrolyzing polymeric materials supplied in the diet but rely on the end products of bacterial fermentation for their growth. A role for methanogens may be to act as a H_2 sink by consuming H_2 generated from different metabolic systems and, therefore, facilitate these reactions by preventing feedback inhibition. In terrestrial and aquatic environments, the production of H_2 would be attributed to the activities of sulfate-reducing bacteria; however, sulfate reducers are rarely present in the rumen.

7.4.3.2 Anaerobic Protozoa

Although the presence of protozoa in the rumen was first described in 1843, the role of protozoa in this microbial ecosystem of the rumen is largely unknown. While the number of protozoa is markedly less than the number of bacteria, the protozoa are significantly larger in size than bacteria and the protozoa account for about half of the biomass in the rumen. Due to cell density, however, the protozoans account for less than 20% of the microbial protein entering the intestine. Thus far, no axenic cultures of rumen protozoa have been obtained because of the presence of bacteria on the surface or inside the protozoa. It is known that the presence of protozoa in the rumen is not essential for the health of the animal and numerous studies have been conducted where the animal is free from protozoa (Newbold et al. 2015). From molecular biology studies, it appears that the protozoa have acquired the genes for hydrolysis of cellulose fibers from bacteria and that these genes are overexpressed in ruminal protozoa (Ricard et al. 2006). Protozoa graze on bacteria in the rumen and with the absence of protozoa in the rumen, the number of bacteria increase. It has been estimated that the amount of methane produced in the rumen decreases by 9–37% when protozoa are removed from the rumen. Methane production associated with protozoa in the rumen

Table 7.4 Rumen prokaryotes and their activity.

Group activity	Genus and species
Bacteria	
Cellulose digestion	*Fibrobacter succinogenes* *Butyrivibrio fibrisolvens* *Ruminococcus albus* *Clostridium lochheadii*
Starch digestion	*Bacteroides ruminicola* *Ruminobacter amylophilus* *Selenomonas ruminatium* *Succinomonas amylolitica* *Streptococcus bovis*
Lipase producing	*Anaerovibrio lipolytica*
Pectin digestion	*Lachnospira multiparus* *Prevotella ruminoicola* *Bacteroides ruminicola* *Butyrivibrio fibrisolvens*
Protein digestion	*Bacteroides amylophilus* *Bacteroides ruminicola* *Butyrivibrio fibrisolvens* *Streptococcus bovis* *Prevotella albensis*
Lactate degrading	*Selenomonas lactilytica* *Megasphaera elsdenii*
Archaea	
Methanogens	*Methanobrevibacter ruminantium* *Methanobrevibacter thaueri* *Methanobrevibacter gottschalkii* *Methanobrevibacter wolini* *Methanobrevibacter smithii* *Methanobacterium formicivum* *Methanomicrobium mobile* *Methanosarcina barkeri*

accounts for about 11% of total rumen methane and is due to H_2 production by hydrogenosomes (Figure 7.3) with episymbiotic or endosymbiotic methanogens. *Methanobrevibacter* is the dominant genera found as the endosymbiont of protozoa in the rumen and the presence of *Methanomicrobium* and *Methanomassiliiococcales* is variable depending on the protozoan genus. From the standpoint of energetics, the diversion of H_2 to methane production and not into bacteria for cellular growth, accounts for an energy loss. This loss of energy in dairy cattle, buffalo, and camels has been estimated to be 5.5–9.0%, 7.5–9.0%, and 7.0–9.0%, respectively (Hook et al. 2010).

7.4.3.3 Anaerobic Fungi

The presence of fungi in sheep rumen was demonstrated by C.G. Orpin in 1975 and since then fungi have been found in many different rumen (Orpin 1975). The five fungal genera listed in Table 7.5 are commonly found bound onto plant fibers in the rumen and may be the primary fiber decomposers in the rumen (Ho and Abdullah 1999). Most fungi produce zoospores,

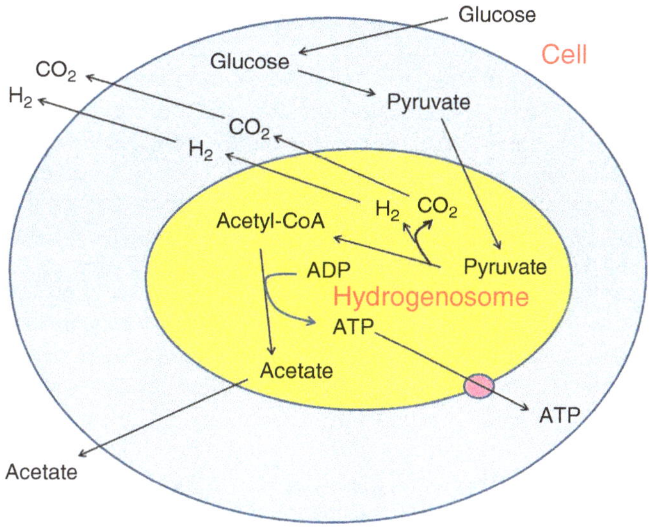

Figure 7.3 Some anaerobic ciliates, trichomonads, and fungi contain a membrane-containing organelle about the size of a bacterial cell that produces H_2. This organelle is called a hydrogenosome. In anaerobic eukaryotic microorganisms, the hydrogenosome provides ATP for the cell and this is important because the anaerobic eukaryotes do not have mitochondria for energy production.

Table 7.5 Eukaryotic microorganisms in rumen.

Microorganism	Genus and species
Protozoa	*Isotricha intestinalis* *Epidinium ecaudatum* *Isotricha prostoma* *Entodinium caudatum* *Dasytricha ruminantium* *Diploplastron affine* *Eudiplodinium maggi* *Polyplastron multivesiculatum*
Fungi	*Neocallimastix frontalis* *Piromyces communis* *Caecomyces equi* *Anaeromyces mucronatus* *Orpinomyces joyonii*

which have chemoreceptors for carbohydrates and phenolic acid structures characteristically located on plant fibers. The zoospores move across the plant material and use stomata or damaged segments of the fiber as attachment sites. Attachment takes about 30 min and is followed by the production of a germ tube that penetrates the plant structure. After two to three hours, fungal rhizoids penetrate the plant tissue and degrade it using extracellular enzymes. Higher plant fiber in the rumen is positively correlated with an increased quantity of fungal colonization. Fungi in the rumen produce endo-1,4-β-D-glucanase and β-d-glucosidase enzymes that hydrolyze cellulose faster than the terrestrial *Trichoderma reesei.* Also produced by rumen fungi are xylanases (β-xylanase, β-xylosidase, and xylobiase), endopectin lyase, α-amylase, and esterases that bond between phenolic acids (ferulic and coumaric acids). Most of these anaerobic fungi ferment the products of cellulose digestion to produce acetate, formate, lactate ethanol, CO_2, and H_2. The production of H_2 results from the metabolism of hydrogenosomes.

Box 7.2 FAO/WHO

The FAO is the Food and Agriculture Organization, which is a branch of the United Nations and the mission of FAO is to prevent starvation around the world. It was started in 1945 and its headquarters is in Rome, Italy. The WHO is the World Health Organization, it is a special segment of the United Nations and it is concerned with international world health problems.

7.4.3.4 Probiotics and Methane Mitigation Strategies

Ruminant animals are economically important and as a result, there is considerable interest in the agricultural industry to find mechanisms to improve feed efficiency (animal weight gain per unit feed). The production of methane does not benefit the host animal and the consumption of H_2 for methane formation is an energy loss of 6%. An additional concern is the impact of the methane produced by ruminants on greenhouse gases. About 50% of the methane released into the atmosphere is from the agricultural sector with beef cattle, dairy cattle, sheep, and goats as the primary sources (Hook et al. 2010). Therefore, the agricultural community is interested in reducing the production of methane by ruminants using additives to the diet other than antibiotics.

Prebiotics refers to the addition of chemicals that are not digested at specific amounts but would beneficially stimulate the growth of indigenous microorganisms. Most prebiotics are nondigestable oligosaccharides (NDOs) and include mannan-oligosaccharides, fructose-oligosaccharides such as inulin, galactose-oligosaccharides, or dietary fiber (Gaggìa et al. 2010). However, rumen systems contain microbial systems that can degrade most NDOs and prebiotics seem most effective for non-ruminant animals. Probiotics as defined by FAO/WHO (2002) (Box 7.2) refers to "live microorganisms which, when administered in adequate amounts, confer a health benefit on the host." In cattle, a mixture of *Saccharomyces cerivisae, Lactobacillus casei, Lactobacillus salivarius*, and *Pediococcus acidilactic* have been added to the diet and the health of the animals was assessed. With both prebiotics and probiotics, the results look promising in reducing methane production; however, more testing is needed to establish their benefit (Uyeno et al. 2015; Gaggìa et al. 2010). Approaches to diminish methane production by ruminants involve decreasing the presence of protozoa that have methanogens as symbionts. Reduction of ciliated protozoa has resulted from the use of vaccines (Hook et al. 2010), addition of mevastatin, lovastatin (Gottlieb et al. 2016) or moensin (Castillo-Gonzáleza et al. 2014), and defaunation, which is the physical removal of protozoa (Newbold et al. 2015). Many of these trials look positive and further study is needed to fully evaluate their benefit to the host and the cost of implementation.

7.5 The Human Microbiome

7.5.1 Skin

Skin covering the human body is not inert but is an organ with a broadly diverse group of microbial agents (i.e. bacteria, fungi, and viruses), which range from opportunistic pathogens to harmless microorganisms to bacteria that benefit the host. Using standard culture techniques, the dominant bacterium present on skin was found to be *Staphylococcus epidermidis*, with fewer numbers of the phylum *Actinobacteria* from the genera *Corynebacterium, Propionibacterium*, and *Brevibacterium*. Gram-negative bacteria are not part of the normal flora of the skin. From 16S rRNA analysis, the bacteria present on the skin represent four phyla: *Actinobacteria, Firmicutes, Bacteroidetes*, and *Proteobacteria*

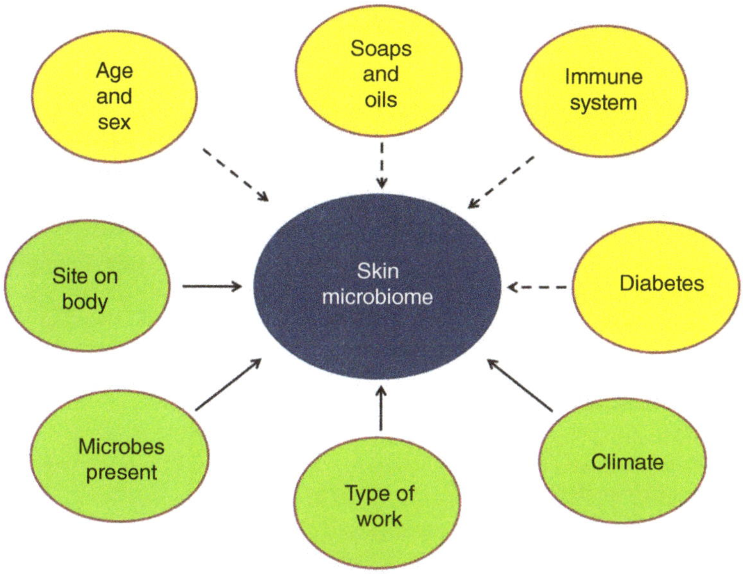

Figure 7.4 Factors influencing the microbiome of the human skin. Dashed lines – factors attributed to the individual; solid lines – environmental factors. For additional information see Grice et al. (2008).

(Grice and Segre 2011; Grice et al. 2008). The diversity of skin bacteria reflects the variations in the skin environment. Areas on the can be skin dry, oily, or moist and the number of bacteria in a square centimeter may range from a thousand for dry skin to a million for moist areas. Other variables contributing to the number and physiological types of bacteria present on the skin include the occupation, sex, and general health of the host and the type of clothing, soaps, and cosmetics used by the individual (Figure 7.4). Many scientists are interested in how the skin can discriminate between bacteria that may be detrimental to the individual and those bacteria that are beneficial and become part of the normal flora. While there is considerable interest in the human microbiome, skin residents have received less attention than oral or intestinal microorganisms.

The metabolic activities of gram-positive bacteria are important in selecting the bacteria present on the skin (Grice et al. 2008). Eccrine glands on the skin not only release water for temperature regulation but also supply electrolytic salts, which acidify the skin and prevent bacterial colonization. Secretions from apocrine glands in the armpit are metabolized by bacteria to produce volatile organic compounds and these compounds inhibit gram-negative bacteria. Unfortunately, these products are also associated with the undesirable sweaty smell so the application of deodorants targets the gram-positive bacteria that produce these offensive odors. There are sebaceous glands around the hair follicles that secrete a lipid which supports the growth of facultative anaerobes. *Propionibacterium acnes* grows on the lipids and colonizes the sebaceous glands causing acne. Due to the release of salts and organic acids, the skin maintains a slightly acidic pH that favors gram-positive bacteria.

7.5.2 Oral Microorganisms

Microorganisms in the mouth has fueled the curiosity of many including Antonie van Leeuwenhoek. He examined material from between his teeth through his microscope and in 1683 presented his observations to the Royal Society in London. van Leeuwenhoek became the first

Table 7.6 Bacteria associated with different sites in the human mouth.

Site	Dominant bacteria
Maxillary anterior vestibule	*Streptococcus mitis, Granulicatella* spp., *Gemella* spp.
Tongue dorsum	*Streptococcus mitis, S. australis, S. parasanguinis, S. salivarius, Granulicatella adiacens,* and *Veillonella* spp.
Lateral tongue	*Streptococcus mitis, S.australis, Granulicatella adiacens, G. hemolysans*, and *Veillonella* spp.
Hard palate	*Streptococcus mitis, S. infantis, Granulicatella elegans, G. hemolysans,* and *Neisseria subflava*
Soft palate	*Streptococcus mitis, G. adiacens* and *G. hemolysans*
Tonsils	*Prevotella* spp., *Porphyromonas* spp. *Firmicutes*
Tooth surface	*Streptococcus* sp. clone EK048, *S. sanguinis, S. gordonii, Rothia dentocariosa, Granulicatella hemolysans, G. adiacens, Actinomyces* sp. clone BL008, and *Abiotrophia defectiva*
Subgingival plaque	*Streptococcus* spp. and *Gemella* spp.

Source: Reproduced with permission from Aas et al. (2005).

person to study oral bacteria and to view individual bacterial cells (Ford 2008). To some extent, scientists are still trying to learn which bacteria are growing on our teeth. The human oral cavity is known to contain over 700 species of bacteria, only half of which have been cultivated. A culture-independent approach to identify bacteria in the mouth using 16S rRNA evaluations has revealed that there are about 140 species dominating the mouth, including 13 new phylotypes (Dewhirst et al. 2010; Aas et al. 2005). Bacteria from the different phyla and their dominant species are listed here: *Firmicutes* (*Streptococcus, Gemella, Eubacterium, Selenomonas, Veillonella*); *Actinobacteria* (*Actinomyces, Atopobium, Rothia*); *Proteobacteria* (*Neisseria, Eikenella, Campylobacter*); *Bacteroidetes* (*Porphyromonas, Prevotella, Capnocytophaga*); and *Fusobacteria* (*Fusobacterium, Leptotrichia*). The TM7 phylum has also been established; however, none of these species have been cultivated. It appears that different, specific bacteria have colonized designated sites of the mouth and these are summarized in Table 7.6. In a healthy individual, bacteria associated with periodontal disease are not readily detected.

Mixed species biofilms may be produced on several different surfaces in the mouth. There is a sequence of events that leads to the establishment of an oral biofilm, with the first being the coating of proteins from the host and bacteria onto the tooth or oral surface (often referred to as a conditioning film). A monolayer of gram-positive facultative anaerobes consisting of *Streptococcus* spp. are attached onto the coated surface by adhesins specifically produced by these bacteria (Avila et al. 2009; Kolenbrander et al. 2002). Colonizing onto the streptococci layer is a group of primarily gram-negative bacteria, including *Porphyromonas* spp., *Campylobacter* spp., *Tannerella forsythia, Treponema denticola*, and *Actinobacillus actinomycetemcomitans,* which if allowed to persist lead to periodontal inflammation. To avoid bleeding gums due to these pathogenic bacteria, some mouthwashes use antibacterial chemicals that specifically target gram-negative bacteria. To control the growth of bacterial biofilm, early colonizing *Streptococcus gordonii* produces hydrogen peroxide, which inhibits the growth of *Actinomyces naeslundii,* another early colonizer of the biofilm. An interesting observation is that *Streptococcus cristatus,* an early colonizer, can prevent the invasion of *Fusobacterium nucleatum* into host cells; however, the interaction between these two species may be either antagonistic or synergistic, depending on the other bacterial species in this mixed species

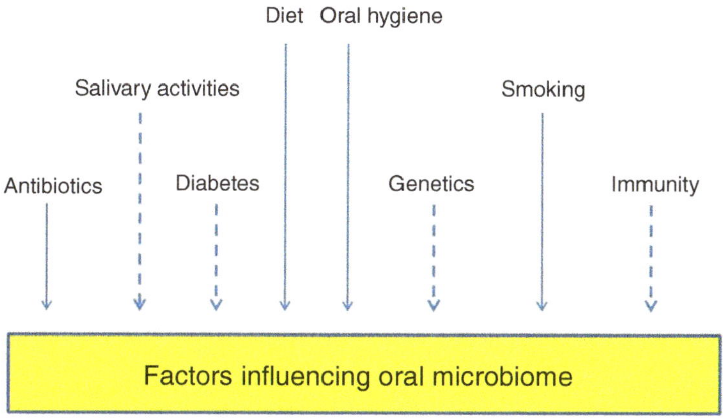

Figure 7.5 Factors influencing oral microbiome. Dashed lines – factors attributed to the individual; solid lines – environmental factors.

biofilm (Avila et al. 2009). Such observations underscore the role of communication between bacteria in a biofilm and the role that quorum sensing plays in oral biofilms.

Commensal organisms are thought to be effective in preventing pathogenic organisms from becoming established in the oral cavity. Competition for nutrients and space between commensal and potentially pathogenic bacteria is important. Oral microbiome homeostasis and health reflects a balance between factors promoting host-symbionts and factors opposed to a host–symbiont relationship (see Figure 7.5) (Kilian et al. 2016). The disruption of the commensal bacteria in the mouth precedes the release of antibacterial proteins (bacteriocins) and formation of VFAs as end products of metabolism. Facultative anaerobic bacteria or anaerobic bacteria grow in numerous anaerobic pockets – including crevices where the tissue meets the tooth surface – and produce VFA, which inhibits some bacteria. This may be why enteric bacteria, including *E. coli*, are not found in the mouth. The medical condition called candidiasis or thrush is attributed to an infection by *Candida albicans*. In most individuals, the commensal bacteria protect individuals against this fungal pathogen; however, depletion of the normal flora or weakening of the immune system can result in candidiasis. Thus, the oral flora reflects the presence of oral anaerobic pockets and individuals without teeth have a markedly different oral flora compared with those individuals with teeth.

7.5.3 Intestinal Microbiome

7.5.3.1 Establishment of Intestinal Flora

In humans and mammals, development of the digestive tract flora begins upon birth and continues throughout life. The fetus develops in a sterile environment and is introduced to microorganisms in the female reproductive tract during a normal (vaginal) birth, or to skin associated microorganisms if surgically removed from the womb (during a Caesarian section). At birth, the gastrointestinal environment is aerobic, but quickly becomes anaerobic within one to two days due to microbial colonization. The first organisms to colonize are facultative anaerobes, which deplete the oxygen due to their metabolic activity. In humans, the first bacteria are *Enterobacterium* sp., *Enterococcus* sp., and *Staphylococcus* sp. Obligate anaerobes are then able to colonize. In newborn mammals and humans, the first stools, called meconium, consist of amniotic fluid and are largely odorless. It was thought that the developing fetus is free from bacteria; however, Funkhouser and Bordenstein (2013) have reviewed the relevant

Table 7.7 Distribution of *Bacteroides* and *Bifidobacterium* in infants.

	Genus present as a percentage of total	
	***Bacteroides* (%)**	***Bifidobacterium* (%)**
Vaginal born	35	23
Cesarean	21	17
Breastfed	28	26
Formula fed	29	11

Source: Reproduced with permission from Pannaraj et al. (2017).

literature, which suggests that meconium from full term healthy infants often contains species from the *Enterobacteriaceae* and lactic acid bacteria. In preterm neonates, only a few bacterial species are found in the meconium and the microbiota of the meconium is different from that of the feces at one week after delivery (Moles et al. 2013).

Aside from the birth process, newborn animals are introduced to gastrointestinal flora during parental care, often from mothers licking the newborn and through nursing (Funkhouser and Bordenstein 2013). In humans, notable differences are observed in fecal flora between breastfed and formula-fed infants (Fan et al. 2013). Human milk contains a number of complex oligosaccharides that are largely indigestible by the infant. These sugar complexes function as a prebiotic as they promote colonization by *Staphylococcus* spp., *Streptococcus* spp., *Lactobacillus* spp., and *Bifdobacterium* spp. Furthermore, human milk oligosaccharides may also block adhesion of some intestinal pathogens (Chichlowski et al. 2011). Different oligosaccharides are present in infant formula and in other types of milk. As a result, infants drinking formula typically develop high levels of *Clostridium* sp., *Streptococcus* spp., *Enterobacteriaceae* spp., and low levels of *Bacteroides* spp. The relative presence of *Bacteroides* and *Bifdobacterium* in infants is presented in Table 7.7. Microbial development in the gastrointestinal tract is accelerated by major dietary changes. After birth, the first dietary change occurs when the newborn begins to nurse. When solid foods are introduced, the gastrointestinal flora transitions into the adult form, which stabilizes around age 3 years (Arrieta et al. 2014). With establishment of the bacteria in the intestine, the species remain with the individual despite antibiotic treatments or administration of salt water purges (laxative preparations).

7.5.3.2 The Healthy Gut

The human microbiome consists of over 10^{14} microorganisms which represent ~1000 species-level phylotypes (Neish 2009) and the health of an individual reflects the phylogenic diversity of only about 160 bacterial species. As reviewed by Gerritsen et al. (2011), the most studied gut microorganisms in humans are members of the phyla *Bacteroidetes* (also known as *Cytophaga-Flavobacterium-Bacteroides*) and *Firmicutes* as the dominant bacteria. The phylum *Firmicutes* is extremely large with over 200 genera and the dominant intestinal *Firmicutes* are taxonomically partitioned into two groups: the *Clostridium coccoides* group (*Clostridium* cluster XIVa) and the *Clostridium leptum* group (*Clostridium* cluster IV). In addition to bacteria of the genus *Clostridium, Eubacterium,* and *Ruminococcus,* members of the *Proteobacteria, Actinobacteria, Fusobacteria. Spirochetes, Verrucomicrobia,* and *Lentisphaerae* phyla are present in the human gut. The dominant methanogenic archaea in the human gut are *Methanobrevibacter smithii* and *Methanosphaera stadtmanae.* Members of fungi present in small numbers are from two fungal phyla: *Ascomycota* and *Basidiomycota.*

The microbiome of the gastrointestinal tract is not homogenous; each anatomical region has a characteristic microbial content. The number of bacterial cells in a gram of intestinal content is 10^1–10^4 in the duodenum, 10^4–10^8 in the jejunum and ileum, and 10^{10}–10^{12} in the colon (see review by Gerritsen et al. 2011). While host enzymes are most active in the duodenum, bacteria are not able to take advantage of the nutrients because bile acids and gastric acids are released in this the region of the small intestine. Bacterial numbers increase in the jejunum, with the greatest density of bacteria in the small intestine found in the ileum, which contains an abundance of nutrients. Bacteria that require amino acids and other factors include enterobacteria, *Lactobacillus, Turicibacter,* and other members of the *Proteobacteria, Lactobacillales,* and *Erysipelotrichales,* which grow in the rich nutrient environment of the ileum. Many of the indigenous bacteria in the gut help prevent invasion of pathogenic bacteria (Kamada et al. 2013). Zoetendal et al. (2012) have determined from metagenome and metatranscriptome data that *Streptococcus* spp. are functionally of greater importance than *E. coli* because they rapidly convert sugars to lactic acid and acetic acid, respectively. These bacteria transport sugars into the cell by phosphotransferase systems and use glycolysis and the pentose phosphate pathway to metabolize the sugars. Lactic acid is metabolized by *Veillonella* spp. with propionic acid and acetic acid as end products. Lactic acid plus acetic acid is metabolized by *Clostridium* cluster XIVa, with the production of butyric acid. The short chain fatty acids (SCFA) (acetic acid, butyric acid, propionate, and lactic acid) are used to energize the epithelial cells in the small intestine, with secretions from these cells providing a defense to invasion by bacteria.

The large intestine receives the complex carbohydrate polymers from the diet that are not digested in the small intestine, along with mucin and cellular material from the host. Since members of the *Lactobacillales* and enterobacteria like *E. coli* are unable to hydrolyze polymeric carbohydrates for energy, the bacteria prominent in the small intestine are in low abundance in the large intestine. Members of the *Clostridiales* and *Bacteroides* are abundant in the large intestine partly because of their ability to degrade mucins and plant fibrous materials. The bacterial species in the *Clostridiales* and *Bacteroides* phyla present in the large intestine varies with the type of polymeric plant material in the diet.

Intestinal gases are metabolic products from microorganisms in the colon and the quantity in a normal individual is 400–1200 ml per day. Competition between sulfate-reducing bacteria, acetogens, and methanogens determines if the gas released is hydrogen sulfide, carbon dioxide, or methane (Nakamura et al. 2010). Following hydrolysis of polymeric carbohydrate-containing material by extracellular enzymes of the *Clostridiales* and *Bacteroides* phyla, microbial fermentation produces H_2 and CO_2. The acetogenic bacteria synthesize acetic acid from CO_2; however, conversion of CO_2 to methane is not expected in the colon because the thermodynamics of acetate production from CO_2 is not as favorable as the reaction for methane production. It comes down to consumption of H_2 with competition between sulfate reducers and methanogens. Thermodynamically, the metabolic reaction with H_2 energizing sulfate reduction with production of H_2S is favored over the consumption of H_2 and CO_2 with the formation of methane. These metabolic activities are given in Figure 7.6. Thus, the competition between sulfate reducers and methanogens in the colon is dependent, in part, on the availability of sulfate in the colon. Sources and reactions by sulfate-reducing bacteria that contribute to production of H_2S have been reviewed recently (Barton et al. 2017).

7.5.3.3 Influence of the Intestine on Human Health

Metagenomics has been extremely useful in providing information about the healthy human microbiome. Initially, a healthy individual was considered to have a specific "bacterial core"; however, analysis of the bacterial microbiome of healthy individuals with different nationalities and diets has shown considerable variability in specific bacterial taxa (Lloyd-Price et al. 2016).

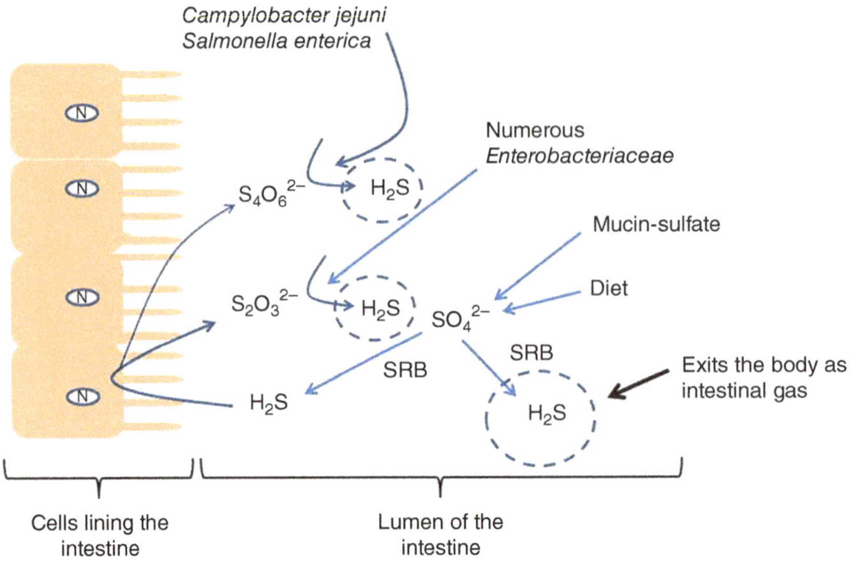

Figure 7.6 Transformations of sulfur in the large intestine. Diet provides sulfate from drinks, sulfate-containing organic molecules, and sulfur-containing amino acids. Mucin sulfate is from the intestinal lining of the host. SRB are sulfate-reducing bacteria in the intestine that reduce sulfate to sulfide. Some of the hydrogen sulfide gas is expelled from the body and some is oxidized by colonic cells. The intestinal cells oxidize sulfie to thiosulfate ($S_2O_3^{2-}$) and tetrathionate ($S_4O_6^{2-}$). Thiosulfate and tetrathionate are reduced by various bacteria commonly present in the colon while pathogenic *Campylobacter jejuni* and *Salmonella enterica* obtain energy from reduction of tetrathionate. Healthy colons have some levels of hydrogen sulfide but pathologies result if the level of hydrogen sulfide is too elevated.

A recent hypothesis proposes that there is a healthy "functional core", which contains a microbiome that provides the molecular and metabolic functions needed to provide the appropriate intestinal environment. Defining the "functional core" is extremely difficult but this has not deterred scientific investigators; over 10 000 publications have addressed intestinal microbiome during 2011–2016 (Lloyd-Price et al. 2016). As well as their roles in nutrition and in digestive processes, gastrointestinal flora serve several beneficial functions (listed in Table 7.8). One key activity is their role in protecting the host from potentially harmful microorganisms by maintaining a diversity of beneficial commensal microorganisms. The mechanisms of competition are shown in Table 7.9.

7.5.3.4 Obesity, Diabetes, and Health Issues

Adult humans show variation in the composition of their gastrointestinal flora. The role of the gastrointestinal microbiome in relation to obesity and many other medical conditions is an area of active research and points to the impact of microbial ecology on human health (de Moreno de LeBlanc and LeBlanc 2014). Of interest, investigations show a potential link between gut flora composition in human obesity, with obese individuals having a greater proportion of *Firmicutes* (typically gram-positive *Bacillus* and *Clostridium*) and lean individuals having a higher proportion of *Bacteroidetes* (anaerobic gram-negative *Bacteroides* spp.) (Turnbaugh et al. 2006). Many of the studies on the influence of gut microorganisms on human health have used mice because both have similar bacteria at the division level. Several studies concerning transplanting of gut microbiota from obese mice to germ-free mice have been conducted and reveal that bacteria in the intestine are the cause of weight gain (see Table 7.10). Some of the

Table 7.8 Interaction of commensal intestinal bacteria with host.

Activity	Linking microbiome with energy metabolism
Stimulating development of small intestine villi	Bacteria promote capillary development in small intestine epithelium. This enables more sugars to be extracted from the small intestine.
Production of short chain fatty (SCFA) in the colon	Fermentation of hydrolytic products in the colon results in SCFA, acids especially butyric acid which energizes colonocytes.
Reduces the release of AMP activated protein kinase (AMPK)	Gut bacteria prevent the release of AMPK which results in reduction of fatty acid oxidation, glucose uptake, and insulin secretion in the host tissues.
Activation of host immune system	Short chain fatty acids bind to G protein-coupled receptors (GPCRs) expressed in human adipocytes and colon epithelial cells.
Choline bioavailability	Gut bacteria convert choline to trimethylamine, which diminishes transport of very low density proteins which are needed for export of liver triglycerides.
Indole reduces inflammatory response	Tryptophanase produced by bacteria releases indole from tryptophane and indole stimulates intestinal epithelial cells to produce tight junctions, increase IL-10 secretions an inducing production of anti-inflammatory cytokines.

Source: Reproduced with permission from Boulangé et al. (2016); Bansal et al. (2010).

Table 7.9 Commensal intestinal bacteria preventing colonization by invaders.

Activity	Benefits of commensal intestinal microbiota
Production of antibacterial bacteriocins	Commensal *E. coli* produces colicins which inhibit growth of enterohemorrhagic *E. coli* (EHEC).
Production of short chain fatty acids	Fermenting bacteria produce acetic acid, propionic acid, and butyric acid, which inhibits growth of some bacteria.
Acidification of environment	*Bifidobacterium* spp. and other fermenting bacteria produce organic acids to prevent colonization of pathogenic bacteria.
Fermentative bacteria produce butyrate	Butyrate inhibits virulence genes in pathogenic *Salmonella* spp. including genes for expression of Type 3 secretion and toxin export.
Release of fucose from mucin	Fucosidase produced by *Bacteroides thetaiotaomicron* hydrolyzes mucin and released fucose regulates downregulates genes for virulence in EHEC.
Depletion of O_2	Members of *Enterobacteriaceae* and other facultative anaerobes consume O_2 in the intestine and this inhibits production of virulence factors by *Shigella flexneri,* which requires O_2 for virulence.
Bacterial production of indole	Reduces motility and colonization by *E. coli* O157:H7.

Source: Reproduced with permission from Kamada et al. (2013).

Table 7.10 Use of germ-free mice to study the role of intestinal microorganisms on obesity.

Donor mice	Recipient mice	Response by recipient mice
Fed western diet	Germ-free	Increase in body weight
Obese mouse	Germ-free	Increase in body weight
Lean mouse	Germ-free	Decrease in body weight

Transplanting gut microflora from animals fed a high calorie diet to germ-free mice results in an increase in body weight but transfer of intestinal bacteria from an animal fed a low calorie diet promotes a loss of body weight. For additional information see Million et al. (2013).

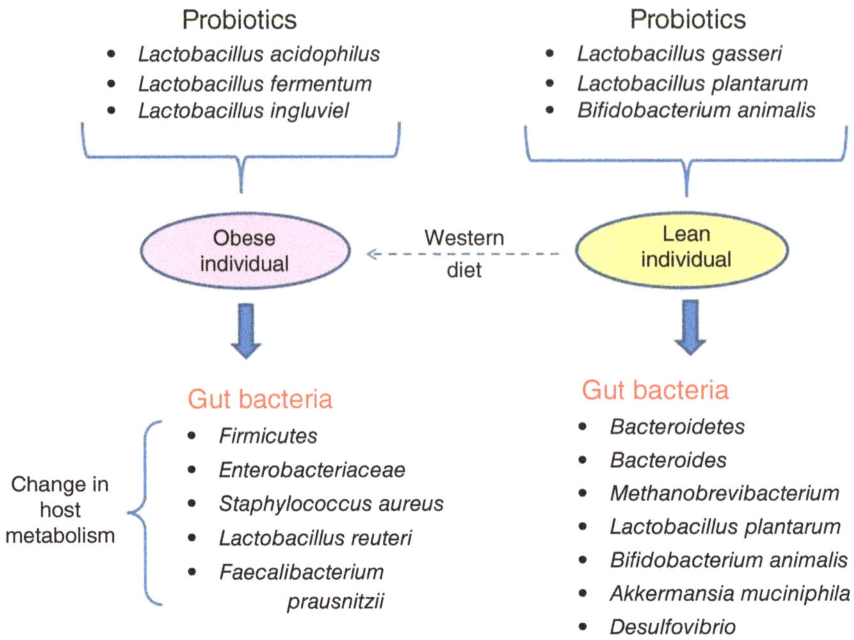

Figure 7.7 Probiotics used to modify intestinal flora and thereby reduce body weight. For additional information see Million et al. (2013).

bacterial interactions affecting obesity in humans are presented in Figure 7.7. The mechanisms used by gut microbiota in individuals predisposed to obesity and diabetes are numerous and have been extensively reviewed and provide an insight into how these bacteria affect the host (Musso et al. 2011).

There are numerous events that can cause dysbiosis (Box 7.3), an imbalance of gut microbiota, and one of these is the use of antibiotics (Hawrelak and Myers 2004). A disease that can result from prolonged use of antibiotics is chronic diarrhea attributed to *Clostridium difficile*. Although *C. difficile* is a minor component of normal flora, it can grow too much in some patients using antibiotics such as clindamycin. Due to the complexity of normal flora, scientists do not know precisely which factors trigger *C. difficile* increased growth or diarrhea. Normal treatment consists of discontinuing antibiotic use or switching treatment protocols. An experimental treatment showing promise is fecal transplants. Patients are administered fecal flora from a healthy donor, with the idea that these organisms will compete against *C. difficile* and return the gastrointestinal microbial ecology to normal (Staley et al. 2018).

Box 7.3 Dysbiosis

The term "dysbiosis" was introduced by the Russian Nobel prize laureate Elie Metchnikoff in the early twentieth century. He combined "dys" meaning "ill or diseases" and "symbiosis," which describes the relationship between two organisms. Dysbiosis can be used to describe an imbalance between microorganisms and organs or tissues (skin, sinuses, intestine, lungs, etc.) in the body. Most commonly it is used to describe dramatic changes of microbiota in the gut. There are various reasons for gut dysbiosis and the most common are: consumption of an excessive amount of fermentable sugars, or administering antibiotics or other chemicals. Dysbiosis of the gut may lead to bloat, diarrhea, headaches, and irritable bowel syndrome.

While intestinal homeostasis is desired, there are some situations such as antibiotic use, stress, and infection that influence the gut microbiota and result in serious health problems. Irritable bowel syndrome (IBS) is a chronic disorder of the bowel where a person has cramping followed by irregular diarrhea and constipation. Bacteria associated with IBS includes members of the *Enterobacteriaceae* and *Desulfovibrionaceae* and several illnesses are attributed to elevated hydrogen sulfide levels produced by sulfate-reducing bacteria in the gut (Singh and Lin 2015). Elevated levels of hydrogen sulfide may be one factor contributing to inflammatory bowel disease, which includes ulcerative colitis and Crohn's disease. As reviewed recently, there is a sulfur cycle in the human gut where hydrogen sulfide is oxidized by intestinal epithelial cells to produce thiosulfate and tetrathionate (Barton et al. 2017). The sulfoxy-anions in the intestine are reduced by sulfate-reducing bacteria but are also substrates for pathogenic bacteria (see Figure 7.7). Statins have been administered to individuals to manage cholesterol levels and it has been reported recently that statins inhibit methanogens in the intestine (Gottlieb et al. 2016). This is interesting because methane production by gut-associated archaea have been implicated in constipation.

7.5.3.5 Probiotics

Elie Metchnikoff (1908) recognized the importance of bacteria in yogurt fermentation (*Lactobacillus* and *Bifidobacterium*) as healthful additives to the human diet. Due to its importance in health and metabolism, there is renewed interest in manipulating the human gastrointestinal flora. In the context of nutrition, a prebiotic is a chemical (or chemical mixture) that will promote the growth or activity of normal flora. Prebiotics are short chains of hexose units that stimulate the growth of *Bifidobacterium* spp. and *Lactobacillus* spp. with the production of beneficial SCFAs. The specific compounds considered useful as prebiotics include fructose containing molecules such as inulin and oligofructose. These short chain polysaccharides are present in dairy products, vegetables, and whole grain cereals, which are often added to yogurt. Extracellular polysaccharides produced by *Pediococcus* spp., *Lactobacillus plantarum,* and *Lactobacillus sanfransiscensis* are metabolized by *Bifidobacteria* but not *Clostridium perfrengens* or *E. coli* and an industry is being built to market bacterial prebiotics (Rastall et al. 2005). Probiotics are viable bacteria that are consumed to promote a healthy gut microbiome and the most commonly used are *Bifidobacteria* and *Lactobacillus.* In a probiotic study with individuals consuming 200 g/day of yogurt with a starter culture of *Lactobacillus delbrueckii* ssp. *bulgaricus,* this bacterium was detected in the feces of 73% of the individuals, the level of *Enterobacteriaceae* was significantly decreased and the level of *Bifidobacterium* increased (Alvaro et al. 2007).

This probiotic study indicates that consumption of *Latobacillus* can affect the intestinal microbiome. The mechanisms proposed to promote a healthy microbiome and to prevent the growth of pathogenic bacteria include pH regulation, nutrient competition, and preventing bacterial colonization and the release of antibacterial compounds (Gerritsen et al. 2011; Kobyliak et al. 2016). To further enhance the health of an individual, both prebiotics and probiotics are administered at the same time and this activity is called synbiotics.

7.6 Gut Microbiota across the Animal World

With the advent of molecular technologies, over 60 different animal species have been examined with the intent of providing the intestinal metabiome. All of these examinations are extremely interesting; however, the number of individuals examined is often limited to make good generalizations. Additionally, colleagues in the gut microbiome community often question the reproducibility of the techniques used. The gut microbiome discussed in the following sections serve to illustrate some of the issues that complicate this research. Initially, there is the identification of the bacteria are present and this leads into the role of each species present. As for all reseach in ecology, additional information will provide a better understanding of gut microbiota.

7.6.1 Systems of Maternal Transmission

Bacterial symbiosis involving animals is well known and there is a considerable amount of information concerning transmission of microorganism from mother to offspring (see review by Funkhouser and Bordenstein 2013). The processes involved for human infants to acquire microorganisms, which ultimately become their normal flora is of great interest and has been discussed in the previous section. This process of providing appropriate bacteria to the new generation may involve external or internal maternal transmission and several examples are listed in Table 7.11. Because the animal host is dependent on the bacterial symbiont, it is not surprising that many different processes of bacterial transfer are used and these are all excellent examples of coevolution.

7.6.2 Microbiota of Ruminates and Hindgut Fermenters

The microbiome of bovine ruminants is focused on hydrolases that digest the cellulosic plant fibers (Brulc et al. 2009). The major phyla (~91% of all bacterial sequences) in the rumen of Hanwoo steers and Holstein-Friesian dairy cattle included *Bacteroidetes, Firmicutes, Fibrobacteres,* and *Proteobacteria,* with a very small quantity of bacteria aligned with the following phyla: *Lentisphaerae, Spirochaetes, Tenericutes, Actinobacteria,* TM7, *Chloroflexi, Synergistetes, Verrucomicrobia, Planctomycetes, Elusimicrobia,* SR1, and *Armatimonadetes.* About half of the sequences for species designation were unclassified but of those that could be matched, the most abundant species were *Prevotella, Fibrobacter, Anaerovorax, Succinivibrio, Ruminococcus,* and *Succiniclasticum.* In this study by Lee et al. (2012), the archaea were of the phylum *Euryarchaeota* and species of *Methanomicrobium* and *Natrialba.* However, in other studies with cattle on different diets, *Methanobrevibacter* sp. and *Methanobacterium* sp. have been reported.

Variations in microorganisms present in the rumen may vary from species to species and this may reflect differences in diet. Reindeer from the high Arctic archipelago of Svalbard (Figure 7.8) and Norwegian reindeer were examined and were found to have different bacteria in the rumen (Pope et al. 2012; Sundset et al. 2007). For over eight months of a year, the Svalbard reindeer forage for plants under the ice and snow because they must consume daily quantities as they do not store fat to endure the extreme winter. The diet of Svalbard reindeer is shrubs, evergreens, and mosses, which contain substantial levels of xyloglucans, mannans, and pectins that are in greater quantities than that found in most grasses. The Norwegian reindeer feeds on grasses in the summer and consumes lichens in winter on mainland Norway. Both reindeer groups were dominated by *Firmicutes,* with 76% from the order *Clostridiales* and 22% from the order *Bacteriodales.* The quantity of cellulolytic bacteria in the Svalbard reindeer is 15% in the summer and 35% in winter whereas in the Norwegian reindeer it is only 2.5%. From 16 S rRNA gene analysis, the abundance of bacteria in the rumen of Svalbard reindeer is as follows: *Firmicutes,* 27%; *Bacteroidetes* – unclassified *Prevotellaceae,* 23%; unclassified *Bacteroidetes,* 17%; *Bacteroidales*: SRM-1, 11%; unclassified *Bacteroidales,* 9%; *Chloroflexi,* 5%; other, 4%; TM7, 0.6%; unclassified – *Rikenellaceae,* 0.03%; *Planctomycetes,* 0.2%; *Proteobacteria,* 0.16%; *Spirochaetes,* 0.06%; and *Actinobacteria,* 0.02% (Pope et al. 2012). There was indication of

Table 7.11 Maternal transfer of bacteria to new generations.

Host	Transmission
Invertebrate marine	
Sponges	Extracellular microorganisms grow between the internal and external layers of the mesophyll. The symbionts consist of microorganisms from 10 different bacterial phyla and one phylum of archaea. These symbionts are not readily detected in the sea except associated with the sponge. Transfer to offspring is by symbionts on oocytes of egg laying sponges and in embryo of sponges with live birth.
Clams	The vesicomy clams use sulfur-oxidizing bacteria to supply nutrients and these symbionts grow in special host cells in the fleshy gills. Bacteria are transmitted to offspring through transovarian transmission.
Invertebrate terrestrial	
Insects	Transovarian transmission is used to provide symbionts including the intracellular *Wolbachia* to offspring.
Aphids	Pea aphids transfer the obligate endosymbiont *Buchnera aphidicola* by vertical transmission. Bacteriocytes located in the aphid body near the ovaries release the symbionts to the offspring.
Cockroach	Bacteriocytes with the symbiont *Blattabacterium* are transported from the fat cells to oocytes and release the bacteria, which enter the cytoplasm of the oocytes.
White flies	Transfer of the nutritional intracellular symbiont *Portiera aleyrodidarum* is by bacteriocytes to the eggs.
Wasp	The European beewolf is a digger wasp that deposits its offspring into the moist soil and supplies an antimicrobial agent, *Streptomyces philanthi,* to prevent environmental fungi and bacteria from decomposing the offspring this wasp. Females carry *S. philanthi* in special glands on the antennae and coats the brood cell with them. Developing larvae incorporate these bacteria into the cocoon and emerging female wasps cultivate these protective bacteria in their glands on the antennae.
Honey bee	Worker bees initially have no gut bacteria, but a core microbiota of *Snodgrassella alvi, Gilliamella apicola,* and *Frischella perrara* is established within four to six days by nurse workers. These gram-negative bacteria are fed to the workers through oral trophallaxis and gram-positive bacteria are acquired from components in the hive.
Vertebrates	
Birds	*Salmonella* spp. are commonly found in the intestine of birds. The domesticated chicken may have *Salmonella* sp. in the eggs as a result of surface contamination of the egg or from internal transmission before the eggs are laid. Direct transovarian transmission may result from *Salmonella* being deposited into the yolk, albumin, or egg membrane from an infected female. *Mycoplasma* spp. have been cultured from the yolk of eggs from chickens and from ducks.
Fish	*Renibacterium salmoninarum* is pathogenic for salmonids and rainbow trout while *Flavobacterium psychrophilum* is a pathogen of rainbow trout. Both of these pathogens are transovarian infections. Additionally, eggs infected with pathogenic bacteria may contribute to diseases of zebra fish and steelhead trout.
Turtles	Diseases of loggerhead sea turtles, South American river turtles, and green turtles may be attributed to a transovarial transfer process. Definitive information on this transfer awaits further study.

Source: Reproduced with permission from Funkhouser and Bordenstein (2013); Powell et al. (2014).

Figure 7.8 Examples of animals with distinctive intestinal flora: (a) turkey vulture eating a garter snake; (b) sea turtle; (c) Svalbard reindeer; (d) King penguin, Fortuna Bay, South Georgia. *Source:* (a) Andy Reago & Chrissy McClarren, https://commons.wikimedia.org/wiki/File:Turkey_Vulture_stalks,_catches_and_eats_live_garter_snake_(30570944315).jpg. Licensed under CC BY 2.0. (b) Courtesy of Larry Barton. (c) Per Harald Olsen, https://commons.wikimedia.org/wiki/File:Svalbardrein_pho.jpg. Licensed under CC BY-SA 3.0. (d) Godot13, https://commons.wikimedia.org/wiki/File:SGI-2016-South_Georgia_(Fortuna_Bay)%E2%80%93King_penguin_(Aptenodytes_patagonicus)_04.jpg. Licensed under CC BY-SA 4.0.

bacteria with cellulosomes (see Chapter 8, Box 8.2) in reindeer and evaluation of the operational taxonomic unit (OTU) suggests the many of the bacteria present are not closely related to cultured species. To some extent the reindeer rumen microbiota is unlike the bacteria in bovine rumen. When considering the gene fragments of glycoside hydrolase, genes encoding for over 20 enzyme families degrading pectins, cellulose, and hemicellulose are present in the reindeer. A functional core of bacteria are present in ruminants and the taxa harboring these enzymes may vary.

Several animals are foregut digesters and these are not true ruminants. Camels (Figure 7.9) are not true ruminants because they have three chambered stomach and lack the omasum compartment which is found in cattle and other ruminants. The hippopotamus has three chambers which are the parietal blind sac, the stomach and the glandular stomach which connects to the intestines. The hoatzin (Figure 7.10) is a unique South American bird because

Figure 7.9 Digestive systems of several mammals: (a) bison contain a rumen; (b) horse uses hind-gut fermentation; (c) Giant panda has a digestive system that inefficiently metabolizes bamboo; (d) camel uses foregut fermentation. *Source:* (a), (b) and (d) Courtesy of Larry L. Barton. (c) Aaron Logan, https://commons.wikimedia.org/wiki/File:Lightmatter_panda.jpg. Licensed under CC BY 1.0.

it is the only foregut digester of food. While many birds have a "crop" where seeds and plant material is stored prior to passage to the gizzard and on to the stomach, the crop in the hoatzin is greatly enlarged and accounts for about 10% of its body. The foregut of the hoatzin is somewhat comparable to the rumen in that it ferments leaves and plant fibers using members of the *Proteobacteria, Bacteroidetes* and *Firmicutes* (Waite and Taylor 2015). Numerous VFAs and other aromatic endproducts are produced by this fermentation in the crop of the hoatzin and the tremendous odor emanating from the bird has earned it the nickname "stinkbird."

The Tammar wallaby is a herbivore that has fermentation of grass in a forestomach that is about 15% of the size of the animal and they do not have a rumen or a pseudorumen. The forestomach is divided into an anterior and a posterior section where some digestion of complex plant polysaccharides occurs. While *Clostridiales* and *Prevotellas* are the most abundant bacteria in the bovine rumen and *Fibrobacteres* and *Spirochetes* are the predominant bacteria in the termite hindgut, the foregut of the wallaby has unique groups of bacteria belonging to the *Lachnospiraceae, Bacteroidales*, and *Gammaproteobacteria*. From metagenomic data sets, known cellulose enzymes have not been established for the wallaby but there are numerous genes for novel glycoside hydrolases (GH) including as β-1,4-endoglucanases and β-1,4- endoxylanases (Pope et al. 2010).

Figure 7.10 Digestive system of several birds and fish: (a) hoatzin near Lake Dandoval, Peru; (b) ostrich; (c) owl; (d) freshwater fish. *Source:* (a) Kate, https://www.flickr.com/photos/43555660@N00/5479759823. Licensed under CC BY-SA 2.0. (b), (c) and (d) Courtesy of Larry L. Barton.

The Asian elephant is considered to be a generalist rather than a specialist with respect to diet because it consumes grass, twigs, bark, roots, and even fruit from trees. The elephant is a hindgut fermenter with most of the decomposition of fibrous palnt material conducted in the cecum. Using fecal samples from a six year old elephant, it was established that the dominant bacteria were *Bacteroidetes* and *Firmicutes* at 47% and 36%, respectively (Ilmberger et al. 2014). Quantities of bacteria ranging from 2% to 5% belonged to *Lentisphaeria, Spirochaetes,* and *Fibrobacteres.* Archaea were represented by equal abundance of members of the *Methanomicrobia* and *Thermoplasma* classes. Genes for 84 glucoside hydrases were present and cellulolytic activity is proposed to be encoded by genes carried by *Bacteroidetes.*

Non-ruminant herbivores such as rabbits, hamsters, and horses (Figure 7.9) lack a foregut (a term used to describe the reticulum, rumen, and omasum). Instead, they have a very pronounced cecum (sometimes called a hindgut) where microbial fermentation occurs and VFA production is prominent. Horses are not highly efficient in their digestion of plant material. While sugars and soluble nutrients released by bacterial digestion of plants are absorbed from the cecum to the blood stream, bacterial cells produced in the cecum are not digested but are eliminated in the feces and this is a considerable loss of potential nutrition. Rabbits and hamsters are somewhat unique in that they exhibit coprophagy. Food is initially eaten and passes through the digestive tract. The rabbit then eats its own excrement (coprophagy, which typically is done at night) and digests it a second time. The second stools are not ingested. Thus, vitamin B-12 and microbial proteins, produced during hindgut fermentation, are absorbed in the rabbit's stomach and small intestine upon the second passage through the digestive tract.

7.6.3 Gut Microbiota of Bears

Brown bears actively consume food and increase their weight in the summer months and for about six months in the winter they fast while in hibernation. Evaluation of the intestinal microbiota using fecal samples indicates that the summer bacteria differ from those in the hibernating bear (Sommer et al. 2016). During active summer period, *Helicobacter* sp., *Streptococcus* sp. and *Helicobacter* sp., *Streptococcus* sp. and *Wolbachia* sp. were also most abundant sp. as well as members of *Firmicutes* and *Actinobacteria* were greater than during winter hibernation. Since *Wolbachia* are insect symbionts,the presence in the fecal material indicates an active consumption of insects in the summer. The abundance of *Bacteroides fragilis* increased markedly in hibernating bears and this coincides with the consumption of fats and gluconeogenesis in winter metabolism. An interesting experiment involved transferring summer and winter microbiota from the brown bear to germ-free mice (Sommer et al. 2016). The result was that germ-free mice receiving gut bacteria from a summer bear showed a significant increase in body weight while germ-free mice receiving gut bacteria from a hibernating bear had a marked loss of weight.

Bears are omnivorous and initially the giant panda bear was omnivorous but about 7 million years ago the giant panda (Figure 7.9) started eating bamboo. At about 2 million years ago, the giant panda switched to a diet of only bamboo; however, the gut microbiota did not switch to that of a herbivore but remained as that of a carnivore. From fecal analysis of the giant panda, *Streptococcus, Lactobacillus, Escherichia/Shigella, Clostridium* sensu stricto (cluster 1), and *Turicibacter* were the dominant bacteria (Xue et al. 2015). The clostridia residents were not present in herbivores nor were they known to be cellulolytic clostridia; however, genes carried by the clostridia in the giant pandas share characteristics with genes that encode cellulolytic enzymes. Truly, the giant panda is unique that it has carnivore-like intestinal bacteria even though it is a herbivore.

The carnivorous polar bears has a low diversity of bacteria in their intestinal tract as determined from fecal samples (Glad et al. 2010). Almost all of the molecular sequences obtained from the 16S rRNA gene library were of the phylum *Firmicutes* with 70% associated with the genus *Clostridium*. This intestinal flora reflects the diet of the polar bear which is essentially consumption of seals. Additionally, several phylobionts carried for genes encoding for ampicillin resistance. On the other hand, the dominant bacterial flora in wild and captive grizzly bears consists of *Enterobacteriaceae* and enterococci with the *Bacteroides-Prevotella-Porphyromonas* group being dominant. The *Clostridium coccoides* cluster was prominent in captive polar bears (Schwab et al. 2009). The abundance and type of SCFA in the feces of the grizzly bear was consistent with SCFA produced by bacteria in carnivores.

7.6.4 Microbiota of Birds

With over 10 000 species of birds living on Earth, it is not surprising that there is considerable diversity in types of birds and the foods they consume. While information concerning the avian gut microflora is less abundant than for vertebrates and insects, the metagenomic results of some species of birds indicate interesting coevolution trends (Waite and Taylor 2015). Bacterial densities of 10^{11} cells per g intestinal material is common and consists of about 140 genera and 640 species with only a small percentage of these can be grown in the laboratory (Dewar et al. 2013). The avian microbial intestinal community consists generally of *Firmicutes, Actinobacteria, Bacteroidetes,* and *Proteobacteria*; however, this is not to imply that all bird gut flora is the same. Cluster analysis of gut microbiota indicates that the avian group is markedly different in composition to that of mammals and insects. Additionally, there is a considerable range of phylotypes present in birds with over 40 bacterial phyla present in hoatzin (see Section 7.4.2) to a couple of phylotypes in the kakapo. The kakapo is a flightless parrot of New Zealand that is a herbivore. It does not consume plants but simply extracts the juices from the plant by striping leaves through their beak and does not require cellulolytic bacteria to facilitate digestion. Thus, the kakapo has a relatively simple fermentation with over 90% of the intestinal bacterial content attributed to *Escherichia* and *Streptococcus* species (Waite and Taylor 2015).

Some birds are carnivorous and their gut microbiots are distinct from birds that are herbivores. Turkey vultures (Figure 7.8), black vultures and other scavingers consuming decaying animals have members of *Clostridia* and *Fusobacteria* as their dominant microorganisms; however *Fusobacteria mortiferium* may constitute a third of the vulture's intestinal microbiota where it is an apparent commensal bacterium (Waite and Taylor 2015). In contrast, strains of *Fusobacteria* have been observed in human intestine but they were detrimental and pathogenic for the human host. Penguins (Figure 7.8), which are also carnivorous in that they feed on fish and squid, may have high percentages of their gut microbiota consisting of *Fusobacteria*. In a survey of gastrointestinal bacteria in king, gentoo, macaroni, little penguin and Adélie species, 13 different bacterial phyta were found; however, the dominant phyla were *Fusobacteria, Proteobacteria, Bacteroidetes* and *Firmicutes* (Dewar et al. 2013). The dominance of the *Bacteroidetes* and *Firmicutes* plyla in penguins is similar to that of other vertebrates but the presence of numerous members of the *Fusobacteria* and *Proteobacteria* distinguishes them from other vertebrates. As presented in Table 7.12, bacterial phyla varied and *Fusobacteria* were present in only the gentoo species. Examination of families of bacteria present indicated high distribution of *Firmicutes* and *Bacteroidetes* with *Proteobacteria* present in appreciable types in only the little penguin species (Table 7.13). Since only a few intestinal microbiota studies have been proposed, it is difficult to associate environmental and individual host activities with bacterial variables.

Large birds have an intestinal microbiota similar to other herbivorous birds. From the ceca of the ostrich, organisms from phyla *Firmicutes* (50.9%), *Bacteroidetes* (39.4%), *Fibrobacteres* (6.5%), *Euryarchaeota* (1.9%), *Spirochaetes* (1.0%), and *Verrucomicrobia* (0.3%) (Matsui et al. 2009). The emu is a flightless bird that is omnivorous feeding on plants and insects

Table 7.12 Dominant phyla of bacteria identified from penguin guano.

	Percentage of bacteria assigned to phyla				
Phyla	**Gentoo**	**King**	**Little**	**Macaroni**	**Adélie**
Fusobacteria	55	24			
Firmicutes	18	48	24	45	41
Proteobacteria	18		30	29	
Bacteroidetes	7	18	22	19	12
Actinobacteria			7		35
Planctomycetes			11		

Source: Reproduced with permission from Dewar et al. (2013).

Table 7.13 Dominant families of bacteria identified from penguin guano.

	Percentage of bacteria assigned to family			
Family	**Gentoo**	**King**	**Little**	**Macaroni**
Leuconostocaceae (F)[a]	6	19	3	31
Lachnospiraceae (F)	4		3	3
Streptococcaceae (F)	4	7	3	3
Lactobacillaceae (F)			7	
Porphyromonadaceae (B)	11	13	15	
Flavobacteriaceae (B)	3	8		
Bacteroidaceae (B)			4	
Moraxellaceae (P)	6	7	3	13
Campylobacteriaceae (P)		11		
Helicobacteriaceae (P)		8		
Neisseriaceae (P)			3	6
Enterobacteriaceae (P)			16	
Comamonadaceae (P)			3	
Phycisphaeraceae (Pl)			11	

Source: Reproduced with permission from Dewar et al. (2013).
[a] Indicates phyla: F = *Firmicutes*; B = *Bacteroidetes*: P = *Proteobacteria*; Pl = *Planctomycetes*.

with digestion by hindgut fermentation. Microbiota in the ceca contains nine phyla with the most abundant phyla are *Bacteroidetes*, 57%; *Proteobacteria*, 24%; *Fusobacteria*, 11.3%; and *Firmicutes*, 7%. They digest only 35–45% of food fiber which gives them about 50% of maintenance energy (Bennett et al. 2013).

With the examination of gut bacterial content in ceca of chickens and turkeys using 16S ribosomal DNA primers. Both bird species have bacteria from 13 phyla with 90% of the bacteria from the *Firmicutes, Bacteroidetes,* and *Proteobacteria* (Wei et al. 2013). Differences observed included the number of genera in each species with 117 genera in chickens (Figure 7.11), 69 genera in turkeys and only 16% similarity at the species level. However, there is an overriding similarity in the core bacteria with *Clostridium, Ruminococcus, Lactobacillus,* and *Bacteroides* as the most abundant bacteria in both chickens and turkeys. Due to economic importance of chickens, there have been many investigations into the bacteria associated with the chicken intestine and the population changes with different diets, age, use of antibiotics and environments. In general it has been found that the chicken cecum contains 10^{11} bacteria/g material and 38 species of anaerobic bacteria. As reviewed by Lu et al. (2003), 28% of the bacteria cultivated included 20% *Bacteroidaceae,* 16% *Eubacterium* spp., 9% *Bifidobacterium* spp., 6% budding cocci, 5% *Gemmiger formicilis,* and 5% *Clostridium* spp. Using 16S rRNA gene sequence analysis of bacteria in the ileum indicated 70% *Lactobacillus*, 11% *Clostridiaceae*, 6% *Streptococcus*, and 6% *Enterococcus*. Sequence analysis of bacteria from the cecum indicated 65% *Clostridiaceae*-related organisms, 14% *Fusobacterium*, 8% *Lactobacillus*, and 5% *Bacteroides*.

A concern is the contamination of eggs consumed by humans since the eggs may be contaminated with *Salmonella* which are pathogenic for humans. Birds commonly have low abundance of *Salmonella* as part of their normal flora in their intestinal tract. While contamination of eggs

Figure 7.11 Animals with different types of digestive systems: (a) pig; (b) red-tailed hawk; (c) chickens. *Source:* Courtesy of by Larry L. Barton.

may occur by fecal bacteria entering through tiny cracks in the egg shell, *Salmonella* may infect the chicken ovaries or oviduct before the shell is added and the egg is laid by the hen (De Buck et al. 2004). It is unresolved why *Salmonella enteritidis* and *S. typhimurium* are the most commonly occurring bacteria inside the chicken egg.

7.6.5 Intestinal Bacteria of Fish

Using culture independent and cultured bacteria, the intestinal bacteria of omnivorous, herbaceous and carnivorous fish cluster independently of the other. As reviewed by Wong and Rawls (2012), there is a major distinction of intestinal flora depending on whether the fish are from freshwater or saltwater. Greater representation by bacteria associated with the order *Aeromonadales* is found in freshwater fishes (Figure 7.10) while *Vibrionales* are increased in saltwater fishes. In salmon, the phyla with the greatest members represented in the intestine are *Proteobacteria, Firmicutes, Actinobacteria, Fusobacteria, Bacteroidetes, Verrucomicrobia* and *Spirochaetes* (Gajardo et al. 2016). While *Firmicutes* and *Bacteroidetes* are dominant in terrestrial mammals, in salmon the dominant phyla are *Proteobacteria* and *Firmicutes*. In fish and in herbivores, bacteria of the order *Clostridiales, Bacteroidales* and *Verrucomicrobiales* are dominant (Nielsen et al. 2017).

The mottled spinefoot rabbitfish consumes macroalgae in many regions including the Indo-Pacific coral reefs and has a gut microbiota typical for marine herbivores. Even though 25 phyla of bacteria were detected in the rabbitfish, the *Deltaproteobacteria, Firmicutes* and

Cyanobacteria were the dominant population of the gut (Nielsen et al. 2017). Methanogens and acetogens were not as prevalent as sulfate-reducing bacteria which suggests that metabolism of H_2 in the gut is attributed to sulfate reducers.

7.7 Insect–Fungus Symbiosis

Insects have three readily identifiable regions of their body: head, segmented thorax usually containing three pairs of legs, and a multi-segmented abdomen. While one may assume that insects serve to distribute microorganisms as they crawl around in the environment, some insects have developed a unique symbiotic relationship with fungi. While there is a continuing interest to describe and identify the partners involved in this insect-fungus symbiosis, many scientists are exploring the coevolution of the partners and examining the specificity of these partnerships.

7.7.1 Scale Insects and Septobasidium

Some of the members of the basidiomycetes are endophytes growing on leaves, branches and hardwood of trees and shrubs (Martin et al. 2015). A few of these endophytic fungi grow in a plant-fungus-insect relationship which is beneficial to the fungus and the insect. A specific group of scale insects are involved in this symbiotic event. There are over 8000 species of scale insects and these insects are relatively small (1–2 mm) and can be found growing on herbaceous plants throughout Asia, USA, Australia and New Zealand. *Septobasidium* sp., is one of the few fungal species commonly involved with scale insects and the mat of growth produced on the surface of the plant has often been described as a "felt-like" colony. With time, the *Septobasidium*–insect activity becomes detrimental and the Felt Disease will kill entire orchards (Choi et al. 2016).

The initial infection occurs with the scale insect infant established on the plant surface and inserting a sucking tube into to plant to obtain nutrients. Basidiospores of *Septobasidium* infect the scale insect, germinate with hyphae extending from the insect into the surrounding hyphal mat. A fungal mat grows over the scale insect providing protection against predators such as the parasitoid wasp. Fungal haustoria penetrate the scale insect to provide nutrients for the fungi and the scale insect continues to derive nutrients from the plant.

7.7.2 Attine Ant–Fungus Symbiosis

A relationship between ants and various species of basidiomycetes beneficial for both partners and many scientists believe that this relationship initially developed about 45–65 million years ago in South America. While some believe that ants exploited fungi for nutritional benefit, there is also the suggestion that the attine ant-fungus system evolved from a type of fungal myrmecochory where fungi relied on ants to disperse the fungi to new environments (Mueller et al. 2001). Some ants use fungal mycelium to stabilize the walls of their ant nest while ant species that farm fungi for nutritional benefit are referred to as attine ants. The attine ants comprise a group of about 230 different species and there are five major types of partners in this relationship (Mehdiabadi and Schultz 2009). There are five different types of fungi-ant relationships and they are characterized in Table 7.14. The best known symbiosis involves the fungus *Leucoagaricus gongylophorus* (formerly called *Attamyces bromatificus*) which is associated with the two genera of leaf cutting ants (*Acromyrmex* and *Atta*).

Table 7.14 Characteristics of attine–fungal mutualisms.

Type of agriculture	Characteristics
Lower	Fungal partners belong to the tribe *Leucocoprineae* also known as parasol mushrooms were initially cultivated by ants about 50 million years ago. The ants grew their fungus gardens on organic detritus.
Coral fungus	About 10–20 million years ago, the ant genus *Apterostigma* began growing a non-leucoprineaceous fungus that is closely related to fungi of the *Pterula* and *Deflexula* genus in the coral fungi clade.
Yeast	A primitive relationship existed between attine ants and 18 yeast species of the *Cyphomyrmex rimosus* group. This ant–yeast system evolved 5–25 million years ago. The yeast form was evident only when being cultivated by attine ants. The yeast were not parasitized by *Escovopsis,* which attacks the filamentous fungal forms.
Higher	The 63 species of higher attine ants selectively grew a clade of leucocoprineaceous fungi that was developed from the lower attine fungi. These fungi are unable to grow without the participation of attine ants and specialized nutrient-containing bulbs are at the tips of the fungal filaments.
Leaf-cutter	The leaf-cutting ants represent a group of ants that originated 8–12 million years ago. There are 40 species in two genera (*Atta* and *Acromyrmex*) of attine ants. Only the higher and leaf-cutting ants provide living plant material to the fungi whereas primitive fungi use decaying biomass.

Source: Reproduced with permission from Schultz and Brady (2008).

Leaf cutting ants have been known to destroy fruit trees and other crops and as such they have received the greatest research attention. *Atta secdens* is the most common leaf cutting ant in North America and a colony is started by a female that arrives at a site carrying the symbiotic fungus in a specialized pocket in her mouth. The female ant establishes a fungal colony and lays eggs on the fungal mat. Within a few years, a large nest results with many fungal gardens in subterranean chambers. The ants retrieve the cut leaves from outside the colony and deposit the pieces of leaves in the fungal gardens. Worker ants tend the gardens and remove foreign fungi to preserve a monoculture of desired fungi. The fungal symbiont, *Leucoagaricus gongylophorus,* produces a large number of lignocellulases, which are essential for the degradation of the fresh plant material used to support fungal growth (Aylward et al. 2013). The attine ants do not consume the fungal mat but eats the bulbus growths at mycelial tips to obtains nutrients. *Escovopsis kreiselii* is a microfungus that parasitizes fungi including those in the attine ant colony. While this parasitic fungus could decimate the ant colony, the attine ants use *Pseudonocardia sp.*, a bacterial actinomycete, to protect the fungal gardens against invasion by *Escovopsis*. This bacterium uses nutrients on the ant cuticle and working ants "feed" *Pseudonocardia* with glandular secretions from beneath their cuticle (Andersen et al. 2015).

7.7.3 Woodwasp–Fungus Symbiosis

The woodwasp of the species *Sirex noctilio* is native to Europe, Asia, and northern Africa. This is an important invasive species that is entering the USA and the fungi which this woodwasp carries is killing many newly planted pine trees. The damage to the tree is a result of the combined activity of enzyme secretion by this white-rot fungus and burrowing of larvae into the hard wood. The tree is weakened due to this initial infection but secondary infections of other insects and environmental fungi may contribute to destruction of the trees. While there are several native species of *Sirex* in the USA, the invasive species of concern is *Sirex noctilio*. The female *Sirex noctilio* drills through the bark into the xylem where she imbeds eggs and the symbiont fungus in a mucoid phytotoxic material (Long et al. 2009). When the eggs hatch, the larvae feeds on the fungus and when the female larvae matures, fungal oidia are placed in a special pouch near the ovipositors. Thus, horizontal transmission of the fungi to the new larvae is attained with each egg laying activity and this may be the only mechanism of dispersal for this symbiotic fungus.

This system of ectosymbiosis is of interest because it has long been considered that this interaction of the wood wasp–fungus mutualism was a species-specific association. However, it is now apparent that these interactions of wood wasp–fungus are flexible (Hajek et al. 2013). In some instances, *Sirex noctilio,*the invasive wood wasp, and *Sirex cyaneus*, the native species of wood wasp, grew symbiotically with only *Amylostereum areolatum* or *A. chailletii*, respectively, but in other samplings *S. noctilio* were detected growing with two different genotypes of *A. areolatum*. Additionally, horizontal transfer of strains of *Amylostereum* from *S. noctilio* to *S. cyaneus* were reported. It appears that *S. noctilio* is capable of altering the fungal symbiont to some extent.

7.7.4 Ambrosia Beetles–Fungus

While ambrosia in mythology refers to "food of the gods," here ambrosia refers to fungi used as food for beetles. There are several thousand different species of ambrosia beetles but only a few of these are associated with the serious economic loss of trees. These scolytid beetles create tunnels and chambers in tree trunks and branches often just below the bark layer. Ambrosia beetles consume the fungi growing on the woody materials in the tunnels and these fungi have been introduced into the tunnels by the ambrosia beetles. While the ambrosia beetles dig the tunnels in the wood, they do not consume the cellulosic materials but make it available for the lignolytic fungi (Klepzik and Six 2004). Yeast and nitrogen-fixing bacteria have also been isolated from the intestine of ambrosia beetles; however, their role in in the beetle– fungus symbiosis has not been pursued. Invasive species of the ambrosia beetle with its phytopathogenic fungi are responsible for lethal diseases in trees. For example, the invasive *Xyleborus glabratus* carries the phytopathogenic fungus, *Raffaelea lauricola* and it produces the lethal wilt disease in laurel. The Dutch elm disease which kills American elm trees is attributed to the elm bark beetle, *Scolytus scolytus*, (Figure 7.12) and this ambrosia beetle has *Ophiostoma ulmi* or *O. novo-ulmi*, as the fungal symbiont.

7.7.5 Termite–Fungus

While the ant-fungus symbiosis evolved from South America, the termite-fungus relationship originated in the tropics of Africa millions of years ago. It is proposed that there were at least four separate migrations of the fungus-growing termites from Africa to Asia and at least three migrations to Madagascar (Aanen and Eggleton 2005). There are about 750 species of termites present in Africa and of these about 11 genera with ~300 species of the subfamily Macrotermitinae are unable to digest cellulose or lignin. The fungus cultivated by the termites is a basidiomycete of the genus *Termitomyces* which comprises 30 species. The termites construct large mounds that may be 10–15 ft tall and this is honey-combed with fungal gardens. A large mound may contain a colony of a million termites which collect wood, dry grass and plant material to provide nutrients for the fungi. Studies addressing host specificity indicates that one termite species, *Macrotermes natalensis*, has a single fungal holotype while host species of the genus *Odontotermes* and *Microtermes* displayed low specificity for different strains of *Termitomyces* (Aanen et al. 2007). The patterns of interaction between the termite hosts and the fungal symbiont as well as the mode of fungal transmission continue to be of evolutionary interest.

Figure 7.12 Invertebrates relying on bacteria for help in digestion: (a) mealybugs on a flower (*Lantana* sp.) stem. Wikimedia Commons: This file is licensed under the Creative Commons Attribution-Share Alike 4.0 International license. (b) *Scolytus scolytus*, large elm bark beetle, grows to 3.5–6 mm and is an important vector in causing Dutch elm disease; (c) adult pea aphid on alfalfa. *Source:* (a) Crisco, https://commons.wikimedia.org/wiki/File:Mealybugs_of_flower_stem,_Yogyakarta,_2014-10-31.jpg. Licensed under CC BY-SA 4.0. (b) Fdcgoeul. https://commons.wikimedia.org/wiki/File:02_Scolytus_scolytus_Imago_rechte_Seite_besser.jpg. Licensed under CC BY 3.0. (c) Jpeccoud. https://commons.wikimedia.org/wiki/File:Peaaphidafalfa.jpg. Licensed under CC BY 3.0.

7.8 Mutualisms Involving Insects and Bacteria

To facilitate their lifestyle, commonly occurring insects such as aphids, whiteflies, mealybugs, cicadas, cockroaches and beetles accommodate bacteria in an intracellular symbiotic association (Baumann et al. 2000). Except for one instance when the endosymbiont is yeast, specialized bacteria are located in structures referred to as bacteriocytes and these symbionts establish an obligate relationship with the host.

7.8.1 Aphids–*Buchnera* and Endosymbionts

It has been known for some time that aphids harbor *Buchnera aphidicola* as an intracellular symbiont. Each aphid individual has 60–80 lobed structures called bacteriocytes, which contain the *Buchnera* cells. The pea aphid, *Acyrthosiphon pisum,* (Figure 7.12) has a codependency with the bacterium *Buchnera* sp. APS and their interdependency is so great that neither can grow without the other. Genome analysis reveals that *Buchnera* sp. APS has 640 681 base pairs (bp) and this DNA content is close to the smallest bacterial genome, *Mycoplasma genitalium*, with 580 070 bp (Shigenobu et al. 2000). *Buchnera* provides its host with essential amino acids and vitamins. The aphids do not excrete nitrogen compounds but recycle nitrogen to *Buchnera* for synthesis of amino acids. Glucose and mannitol are transported into the bacterial cell by phosphoenolpyruvate–carbohydrate phosphotransferase systems and the bacteria have few solute uptake transport genes. *Buchnera* sp. APS has complete gene sets responsible for glycolysis, the pentose phosphate cycle and aerobic respiration; however, it does not have genes for lipopolysaccharide synthesis or genes for the citric acid cycle. Due to the homeostatic condition as an endosymbiont, the *Buchnera* sp. APS has only rpoD and rpoH as sigma factors and is lacking regulatory genes.

The aphid, *Cinara cedri*, not only has *Buchnera aphidicola* as an endosymbiont but also has additional bacteria, *Wolbachia* sp., present in separate bacteriocytes. As reviewed by Gómez-Valero et al. (2004), secondary symbionts (S symbionts) occur in aphids and these include (i) the R-type, T-type and U-type which are linked to the *Gammaproteobacteria*; (ii) the pea aphid ricketssia (RPA) linked to the *Alphaproteobacteria*; and *Spiroplasma*. The S symbionts may be in various tissues of the aphid or associated with the primary bacteriocytes. The S symbionts may have been established by different infections and once established they are transferred maternally and horizontally to a second host.

7.8.2 *Wolbachia*–Insects

Wolbachia are gram-negative bacteria found in over 70% of insect species and nematodes. In some hosts, *Wolbachia* displays the characteristics of a parasite while in others it has a mutualistic relationship where insects cannot reproduce without it (Werren et al. 1997). *Wolbachia* are passed maternally through infected eggs and this infection can have several significant consequences: (i) some of the males may be killed in the larval development stage; (ii) males may become females; (iii) *Wolbachia*-infected males are unable to reproduce with uninfected females; (iv) *Wolbachia* can induce parthenogenesis where a female does not require a male for fertilization. A temperate bacteriophage is maintained in some *Wolbachia* (Masui et al. 2000) and this genetic component may be important in the horizontal transfer of this bacterium.

Wolbachia may have evolved from *Alphaproteobacteria,* possibly giving rise also to *Rickettsia, Ehrlichia, Anaplasma, Midichloria,* and the mitochondria (Correa and Ballard 2016). There are several groups within *Wolbachia* where specific groups infect insects or nematodes and these associations of *Wolbachia* may have occurred more than once. Some of the benefits that *Wolbachia* provide to the host include production of vitamin B and biotin for bedbugs, production of riboflavin and iron availability through supplying intermediates in heme synthesis to the fruit fly, *D. melanogaster*, and protection against RNA virus infection.

7.8.3 Mealybug–Bacteria

Mealybugs feed on low nutrient juices and are found in warm climates and in greenhouses. Mealybugs appear as fluffy white flakes on plants (Figure 7.12) and secrete material while they feed that earns them their name. There are several species of mealybugs and the ones that attack citrus plants are of great economic concern. the mealybug *Planococcus citri* has the endosymbiont *Tremblaya princeps* in its cytoplasm, and inside this bacterium resides another bacterium *Moranella endobia*. This is a highly dependent symbiotic structure where all three cellular types are required for growth and none can grow independently of the other two.

Table 7.15 Endosymbionts in different species of mealybugs.

Mealybug species	Host plants	Primary symbiont	*Gammaproteobacteria* symbiont
Phenacoccus avenae	Oat plant	*Tremblaya phenacola*	None present
Maconellicoccus hirsutus	Many trees	*Tremblaya princeps*	*Doolittlea endobia*
Ferrisia virgata	Cotton plants	*Tremblaya princeps*	*Gullanella endobia*
Planococcus citri	Citrus trees	*Tremblaya princeps*	*Moranella endobia*
Pseudococcus longispinus	~100 species	*Tremblaya princeps*	PLON1 and PLON2
Trionymus perrisii	Grasses	*Tremblaya princeps*	*Hoaglandella endobia*
Paracoccus marginatus	Papaya and tropical fruits	*Tremblaya princeps*	*Mikella endobi*

Source: Reproduced with permission from Husnika and McCutcheon (2016).

For example, the synthesis of phenylanaline requires nine enzymatic steps. *Tremblaya* synthesize six steps, *Moranella* make two more and *Planococcus citri* make the ninth, but none of the three individuals can synthesize phenylanine independently (Husnika and McCutcheon 2016). This illustrates the complex movement of genes and contraction of genomes in the three biological systems. While *Tremblaya* is a common primary endosymbiont for mealybugs, there is considerable variation in the second endosymbiont and this is illustrated in Table 7.15. Different species of *Gammaproteobacteria* occur as a result of either independent selection of a specific bacterial species or a replacement of an existing endosymbiont in *Tremblaya princeps*.

7.8.4 Termite Gut–Bacteria

There are two major types of termites that associate with microorganisms. Spirochetes of the genus *Treponema* are the dominant bacteria in the lower termite with digestion of wood attributed to the presence of cellulolytic flagellates. As reviewed by Peterson and Scharf (2016), lower termites have two groups of flagellated protists (the oxymonads and the parabasalids) in their intestine. Bacterial endosymbionts are present in the protists and they are members of the *Elusimicrobia, Bacteroidetes, Proteobacteria,* and *Actinobacteria* phyla. These bacteria ferment sugars to produce H_2, fix N_2, and produce numerous useful metabolites such as vitamins, cofactors and amino acids. Methane production by the archaea *Methanobrevibacter* spp., is present in the lumen and lining of tissue inside the intestine.

Higher termites belonging to the family *Termitidae* have a third proctodeal segment of the hindgut, called the paunch, which is filled with bacteria but lacks protozoans. As reviewed by He et al. (2013), the paunch of *Amitermes wheeleri*, which was collected from cow dung, displayed communities of *Firmicutes* and *Spirochaetes* while *Nasutitermes corniger* on a wood diet was dominated by *Spirochaetes* and *Fibrobacteres*. Archaeal representation was low and accounted for less than 0.2% of all sequences for both *A. wheeleri* and *N. corniger*. The degradation of lignin in the termite hindgut is perplexing because metagenomic analysis reveals that there are no genes for the degradation of lignin in these two higher termites but there is an abundance of glycoside hydrolases, which hydrolyze cellulose and hemicellulose. To consume H_2 produced from anaerobic metabolism, *A. wheeleri* uses acetogenic bacteria; these bacteria reduce carbon dioxide to acetic acid and have numerous genes for bacteriocin production.

7.9 Mutualisms Involving Invertebrates

7.9.1 Microbiome of Marine Worms

The giant tube worms, *Riftia pachyptila* (Figure 7.13), are several meters long and grow in high levels of sulfide several miles below the surface of the Pacific Ocean along the Galapagos Rift and East Pacific Rise. As reviewed by Stewart and Cavanaugh (2005), these invertebrates were first observed in 1977 by Lonsdale, but it was not until 1981 that the presence of endosymbiotic bacteria was determined to be essential for the existence of *R. pachyptila*. These tube worms do not have an intestinal system but rely on chemoautotrophic bacteria to obtain energy from the oxidation of sulfide to sulfate and the bacteria use this energy to fix carbon dioxide to organic matter.

Olavius algarvensis is a small oligochaete that is 12–25 mm long, has no digestive tract, and is dependent on symbiotic bacteria for growth (Figure 7.14). The bacteria are localized between the epidermis and cuticle, as well as inside epidermal cells (see Chapter 4 and Figure 4.3). In the review by Kleiner et al. (2012), the microbiome of *O. algarvensis* consists of five endosymbionts. Symbionts γ1 and γ3 are *Gammaproteobacteria* sulfur-oxidizing bacteria, δ1- and δ4-symbionts are *Deltaproteobacteria* sulfate-reducing bacteria and the fifth endosymbiont is a spirochete. There are segments of the sulfur cycle mediated by symbiotic bacteria with aerobic oxidation of sulfide to sulfate and the respiratory reduction of sulfate to sulfide. The process accounting for anaerobic respiration in an aerobic worm is unresolved. Based on metagenomic analysis, three bacteria (γ1, δ1- and δ4-symbionts) use CO as an energy source and sulfate-reducing symbionts are energized by H_2.

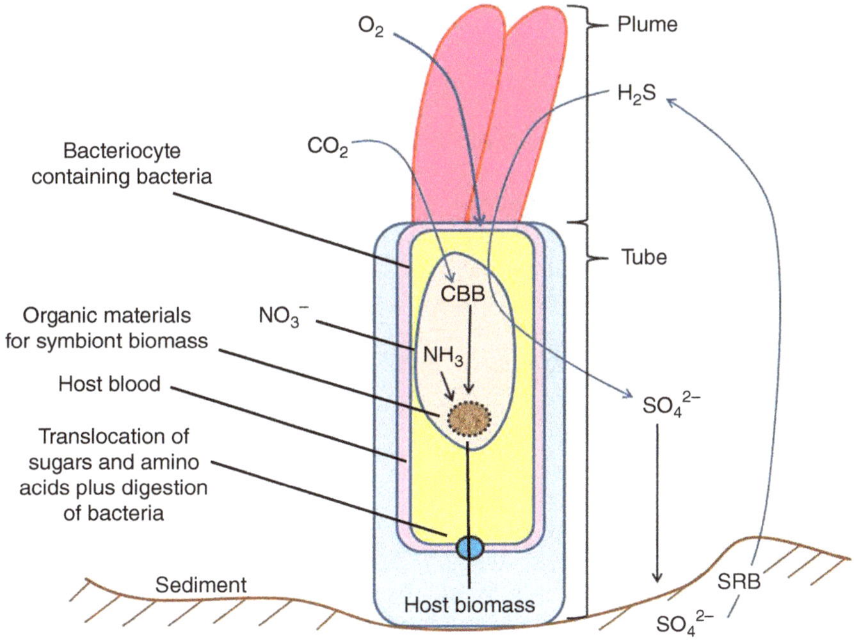

Figure 7.13 Model of *Riftia pachyptia* indicating symbiosis of sulfur-oxidizing bacterium. SRB = Sulfate-reducing bacteria. CBB = Calvin Benson Basaham cycle for CO_2 fixation. For additional information see Stewart and Cavanaugh (2005).

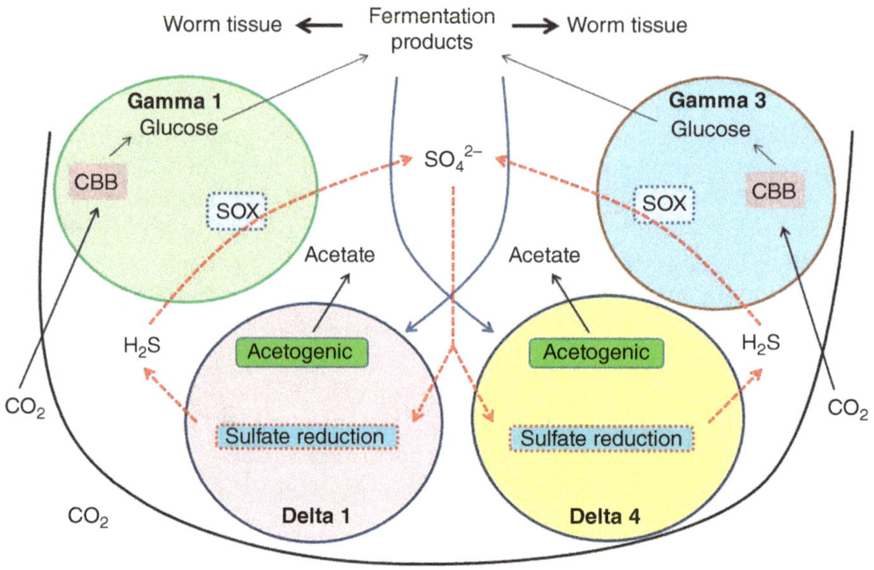

Figure 7.14 Model of multi-symbiont interactions in marine worms. Symbionts are members of *Proteobacteria* and are designated as members of *Gamma-* or *Deltaproteobacteria*. SOX = Sulfur-oxidizing reactions. CBB = Calvin-Benson-Basaham cycle for CO_2 fixation.

Olavius algarvensis grows in a nitrogen poor environment by maintaining high uptake systems for nitrogen from the environment and by recycling metabolic nitrogen products through the formation of urea or glycine betaine. Metagenomic analysis reveals that many of the symbionts can use pyrophosphate metabolism to conserve energy. Pyrophosphate could be produced from the following reactions:

$$\text{Sulfate} + \text{ATP} \rightarrow \text{Adenylylsulfate} + \text{PP}_i$$

$$\text{Propionate} + \text{ATP} + \text{CoA} \rightarrow \text{Propionyl-CoA} + \text{AMP} + \text{PP}_i \text{ and}$$

$$\text{Acetate} + \text{ATP} + \text{CoA} \rightarrow \text{Acetyl} - \text{CoA} + \text{AMP} + \text{PP}_i$$

These three reactions are catalyzed by sulfate adenylyltransferase, propionate-CoA ligase and acetate-CoA ligase, respectively. A proton translocating pyrophosphatase in the membrane exports protons that contribute to proton driven ATP synthase (Kleiner et al. 2012). Additional observations may provide more interesting insights into the cooperation between the symbionts and the gutless worm.

7.9.2 Squid (Euprymna)–Vibrio fischeri Symbiosis

The Hawaiian bobtail squid, *Euprymna scolopes* (Figure 7.15), has a symbiotic partner *Vibrio fischeri*. The squid is buried in the sand during the day and emerges near sundown. Details of the squid–vibrio symbiosis have been reviewed by McFall-Ngai (2014). After hatching, *V. fischeri* from the environment move along a series of ducts into deep crypts where they become established. If *V. fischeri* are not in the environment, other gram-negative but not gram-positive bacteria may make their way along the ducts; however, the squid can sense they are the wrong bacteria and has a system for excluding this unwanted population. *V. fischeri* in the crypts initiate a response with the host that stabilizes their residence and the symbionts

(a)

(b)

Figure 7.15 (a) Adult Hawaiian bobtail squid, *Euprymna scolopes*, with inset scale; (b) European medical leech, *Hirudo medicinalis* L., 1758, may reach 20 cm long. *Source:* (a) Margaret McFall-Ngai, http://journals.plos.org/plosbiology/article?id=10.1371/journal.pbio.1001783. Licensed under CC BY 4.0. (b) Karl Ragnar Gjertsen. https://commons.wikimedia.org/wiki/File:Sv%C3%B8mmende_blodigle.JPG. Licensed under CC BY-SA 3.0.

produce luminescence for communication with other squid. If mutant *V. fischeri* develop without the capability of producing luminescence, these mutant bacteria are eliminated from the crypt in about two days. The squid–bacterium system is on a daily rhythm where ~90% of *V. fischeri* are eliminated from the crypt by action of the squid. The circadian rhythm is controlled by two cryptochromes, blue-light receptive proteins, which are produced by the host.

7.9.3 Medicinal Leech–*Aeromonas* sp. and *Rikenella*-like Bacteria

The medicinal leech *Hirudo verbana* (Figure 7.15) has two symbiotic bacteria in the alimentary canal that can be grown in the laboratory: *Aeromonas* sp. and a *Rikenella*-like bacterium (Nelson and Graf 2012). Several different medicinal leeches are used, including North American medicinal leech, *Macrobdella decora*, and the European medicinal leech, *Hirudo medicinalis*. Because these leeches are similar in appearance, many leeches sold as *H. medicinalis* are *H. verbana* or *Hirudo orientalis*. The gram-negative β-hemolytic *Aeromonas* sp., the primary symbiont, was initially identified as *Aeromonas hydrophila* but recent results suggest that it may be *Aeromonas veronii*. The second symbiont isolated from leeches was an obligate anaerobe similar to *Rikenella*, which is a member of *Bacteroidetes*. To enable the anaerobic *Rikenella*-like bacteria to grow in the aerobic environment, *Aeromonas* sp. respiration depletes oxygen at a microsite. Some bacteria in low abundance in the leech have been cultivated and these include *Morganella morganii* and *Desulfovibrio* sp. (Nelson and Graf 2012). Using culture independent techniques, several bacterial symbionts have been identified in *H. verbana* and include an endosymbiont of the *Alphaproteobacteria* similar to the genus *Ochrobactrum*, two *Bacteroidetes* related to *Sphingobacterium* and *Niabella*, two *Betaproteobacteria* similar to *Comamonas* and *Sterolibacterium*, with a bacterium related to *Bdellovibrio* closely associated to the two *Betaproteobacteria* (Nelson and Graf 2012). Although leeches have had a medicinal use for over 2000 years, the story of the bacterial symbionts in the leech continues.

7.9.4 Nematode–Bacteria

There are over 25 000 species of nematodes distributed in every environment and while these would all be expected to have bacteria associated with them, attention here is focused on two examples of nematode–bacteria mutualism. One example is the bacteria in the gut of the free-living *Caenorhabditis elegans;* numerous studies have been used as a model to evaluate the bacteria–host interactions, including innate immune defense against pathogenic bacteria (Midha et al. 2017). The most abundant bacteria associated with the intestine of *C. elegans* are members of the *Proteobacteria* phylum with *Lactococcus, Lactobacillus, Acetobacter, Gluconobacter*, and *Gluconoacetobacter* present is smaller numbers. *Caenorhabditis elegans* display nonspecific bacterivory and therefore can be used as a model host of human infectious diseases (Tan et al. 1999).

Marine nematodes coated with bacteria are called stilbonematids and they establish an ectosymbiotic relationship with sulfur-oxidizing bacteria. *Laxus oneistus* is a stilbonematid and is interesting because of the nature of the bacteria attached to the cuticle of the nematode. The ectosymbiotic bacteria are rod shaped and attach to the surface of the nematode at the end of the cell, which means the bacteria are all perpendicular to the linear nematode. The molecular sequences from 16S rRNA analysis indicate that *L. oneistus* is related to the purple sulfur bacterium *Chromatiaceae* (Murfin et al. 2012). *Laxus oneistus* is an autotroph that fixes carbon dioxide using the RuBisCo system and is capable of nitrate and nitrite respiration. The benefit provided by the nematode beefits the sulfur oxidizing bacteria by carrying them along the aerobic zone for optimal sulfide oxidation, and the nematode benefits from periodically eating some of these ectotrophic bacteria. A hallmark feature of the ectosymbiosis is that the bacterial layer established contains only one species. The reason for this species specificity is unknown as is the mechanism of bacterial cell division. When the bacteria divide while attached at one pole to the nematode surface, longidutinal cell division would be required, which has never been established for cells with rod morphology. If cell division is followed by the release of the bacteria, special features to attract bacteria to the nematode surface would be implicated.

7.10 Summary

It is apparent that wherever there are animals, microorganisms are there also. The presence of microorganisms, and especially bacteria, on and in humans has been extensively studied and their mutualism has been explored. As an alternate to using primates to study infectious bacteria, animal models have been developed that use animal models including the fruit fly and

C. elegans. The microbiota of the digestive tracts of omnivores, carnivores, and herbivores is extremely complex and often contains hundreds of bacterial species. The intestinal tract of omnivores and carnivores is the region of bacterial activity for digestion and dominant intestinal bacteria include members of the phyla *Firmicutes* and *Bacteroidetes*. Carnivores that consume herbivorous prey have intestinal bacteria capable of digesting both the prey and the plant food present in the the prey's intestine. Carnivores with intermittent feeding patterns (i.e. snakes and hibernating animals) may display a changing microbial flora with initiation of feeding. The intestinal flora for blood sucking animals and carnivorous corals are specific for their diet. Digestion of plant materials by animals is distinguished by those animals with a rumen and those lacking a forestomach. Anaerobic digestion of foliage by ruminants requires the participation of bacteria, protozoa, and fungi to produce a variety of short chain organic acids and VFAs. Numerous bacterial fermentative reactions result in the production of H_2 and CO_2, which archaea (methanogens) use to produce CH_4.

Studies of the human microbiome have focused on bacteria present on skin, in the mouth, and in the intestine. The microbial activities associated with the gut are associated with good health and with health issues, including obesity, diabetes, and intestinal bowel disorders. Prebiotic and probiotic additives have been used to enhance the health of humans and although the specific additions are different, probiotic additions to diet are being used enhance the growth of agronomic animals. The intestinal bacteria of birds reflect whether the bird is carnivorous or herbivorous, while distinction in fish microbiota is correlated to the freshwater or marine habitat. A plant–fungus–insect relationship indicating the symbiotic relationship of specific fungi with scale insects, Attine ants, and wood wasps is discussed and includes the activities of the fungus to obtain nutrients from the plant for the insect. For the Ambrosia beetle and termites, the insect consumes the fungus growing on plant materials. *Buchnera* and *Wolbachia* occur as bacterial endosymbionts in aphids and some symbiotic interactions are so linked that neither the bacterium nor the plant can exist without the other. Lower termites have protists, which degrade cellulose in their intestine, and these protists have endosymbiotic bacteria that produce numerous fermentation products including hydrogen and carbon dioxide, which are converted to methane by methanogens. Mealybugs attack citrus plants, and some have a highly dependent symbiosis involving a specific bacterium inside which is another endosymbiotic bacterium. A symbiotic relationship occurs between marine nematodes and bacteria where sulfur-oxidizing bacteria are the primary producers in a dark non-photosynthetic environment.

Discussion Questions

1 Discuss some reasons why the digestive tract of the Great panda has not adjusted to the bear's diet.

2 Which microbe–animal relationships would you expect to be influenced the most during a climate change in which the Earth becomes significantly warmer? Which animals would be affected the least by a change in climate?

3 Which animals would you expect to NOT to have any microbes in them or on them?

4 To what extent are viruses important in gut microbiology?

References

Aanen, D.K. and Eggleton, P. (2005). Fungus-growing termites originated in African rain forest. *Cur. Biol.* 15: 851–855.

Aanen, D.K., Ros, V.I.D., de Fine Licht, H.H. et al. (2007). Patterns of interaction specificity of fungus-growing termites and *Termitomyces* symbionts in South Africa. *BMC Evol. Biol.* 7: 115. https://doi.org/10.1186/1471-2148-7-115.

Aas, J.A., Paster, B.J., Stokes, L.N. et al. (2005). Defining the normal bacterial flora of the oral cavity. *J. Clin. Microbiol.* 43: 5721–5732.

Adams, M.D., Susan, E., Celniker, S.E. et al. (2000). The genome sequence of *Drosophila melanogaster*. *Science* 287: 2185–2195.

Alcaide, M., Messina, E., Richter, M. et al. (2012). Gene sets for utilization of primary and secondary nutrition supplies in the distal gut of endangered Iberian lynx. *PLoS One* 7 (12): e51521. www.plosone.org.

Alvaro, E., Andrieux, C., Rochet, V. et al. (2007). Composition and metabolism of the intestinal microbiota in consumers and non-consumers of yogurt. *Br. J. Nutr.* 97: 126–133.

Andersen SB, Yek SH, Nash DR, et al. (2015). Interaction specificity between leaf-cutting ants and vertically transmitted Pseudonocardia bacteria, *BMC Evol. Biol.*; 15: 27. https://doi.org/10.1186/s12862-015-0308-2

Arrieta, M.-C., Stiemsma, L.T., Amenyogbe, N. et al. (2014). The intestinal microbiome in early life: health and disease. *Front. Immunol.* 5: 427. https://doi.org/10.3389/fimmu.2014.00427.

Avila, M., Ojcius, D.M., and Yilmaz, O. (2009). The oral microbiota: living with a permanent guest. *DNA Cell Biol.* 28: 405–411.

Aylward, F.O., Burnum-Johnson, K.E., Tringe, S.G. et al. (2013). *Leucoagaricus gongylophorus* produces diverse enzymes for the degradation of recalcitrant plant polymers in leaf-cutter ant fungus gardens. *Appl. Environ. Microbiol.* 79: 3770–3778.

Bansal, T., Alaniz, R.C., Wood, T.K. et al. (2010). The bacterial signal indole increases epithelial-cell tight-junction resistance and attenuates indicators of inflammation. *Proc. Natl. Acad. Sci. USA* 107: 228–233.

Barton, L.L., Ritz, N.L., Fauque, D.G. et al. (2017). Sulfur cycling and the intestinal microbiome. *Dig. Dis. Sci.* 62: 2241–2257.

Baumann, P., Moran, N.A., and Baumann, L. (2000). Bacteriocyte-associated endosymbionts of insects. In: *The Prokaryotes* (ed. M. Dworkin), 1–67. New York: Springer-Verlag.

Bennett, D.C., Tun, H.M., Kim, J.E. et al. (2013). Characterization of cecal microbiota of the emu (*Dromaius novaehollandiae*). *Veterinary Microbiol.* 166: 304–310.

Bollinger, B.R., Barbas, A.S., Bush, E.L. et al. (2007). Biofilms in the large bowel suggest an apparent function of the human vermiform appendix. *J. Theor. Biol.* 249: 826–831.

Boulangé, C.L., Neves, A.L., Chilloux, J. et al. (2016). Impact of the gut microbiota on inflammation, obesity, and metabolic disease. *Genome Med.* 20168: 42. https://doi.org/10.1186/s13073-016-0303-2.

Brulc, J.M., Antonopoulos, D.A., Miller, M.E. et al. (2009). Gene-centric metagenomics of the fiber-adherent bovine rumen microbiome reveals forage specific glycoside hydrolases. *Proc. Natl. Acad. Sci. USA* 106: 1948–1953.

Carey, H.V., Walters, W.A., and Knight, R. (2013). Seasonal restructuring of the ground squirrel gut microbiota over the annual hibernation cycle. *Am. J. Physiol. Regul. Integr. Comp. Physiol.* 304: 33–42.

Castillo-Gonzáleza, A.R., Burrola-Barrazab, M.E., Domínguez-Viveros, J. et al. (2014). Rumen microbiology and fermentation. *Arch. Med. Vet.* 46: 349–361.

Chichlowski, M., German, J.B., Lebrilla, C.B. et al. (2011). The influence of milk oligosaccharides on microbiota of infants: opportunities for formulas. *Annu. Rev. Food Sci. Technol.* 2: 331–351.

Choi, I.-Y., Lee, W.-H., Lee, J.-J. et al. (2016). Characterization of a *Septobasidium* sp. associated with felt disease of *Schisandra chinensis*. *Mycobiology* 44: 58–62.

Colston, T.J., Noonan, B.P., and Jackson, C.R. (2015). Phylogenetic analysis of bacterial communities in different regions of the gastrointestinal tract of *Agkistrodon piscivorus*, the cottonmouth snake. *PLoS One* 10 (6): e0128793. https://doi.org/10.1371/journal.pone.0128793.

Correa CC, Ballard JWO (2016). Wolbachia associations with insects: winning or losing against a master manipulator, *Front. Ecol. Evol.*, https://doi.org/10.3389/fevo.2015.00153

Costello, E.K., Gordon, J.I., Secor, S.M. et al. (2010). Postprandial remodeling of the gut microbiota in Burmese pythons. *ISME J.* 4: 1375–1385.

De Buck, J., Van Immerseel, F., Haesebrouck, F. et al. (2004). Colonization of the chicken reproductive tract and egg contamination by *Salmonella*. *J. Appl. Microbiol.* 97: 233–245.

Dewar, M.L., Arnould, J.P.Y., Dann, P. et al. (2013). Interspecific variations in the gastrointestinal microbiota in penguins. *Microbiol. Open* 2 (1): 195–204.

Dewhirst, F.E., Chen, T., Izard, J. et al. (2010). The human oral microbiome. *J. Bacteriol.* 192: 5002–5017.

Fan, W., Huo, G., Li, X. et al. (2013). Diversity of the intestinal microbiota in different patterns of feeding infants by Illumina high-throughput sequencing. *World J. Microbiol. Biotechnol.* 29: 2365–2372.

FAO/WHO (2002). Joint FAO/WHO (Food and Agriculture Organization/World Health Organization) working group report on drafting guidelines for the evaluation of probiotics in food. London, Ontario. Canada guidelines for the evaluation of probiotics in food. Joint working group report on drafting, London, Ontario. 2002:1–11.

Ford, B.J. (2008). Available at http://www.ucmp.berkeley.edu/history/leeuwenhoek.html. University of California Berkeley, Berkeley

Funkhouser, L.J. and Bordenstein, S.R. (2013, 2013). Mom knows best: the universality of maternal microbial transmission. *PLoS Biol.* 11 (8): e1001631. https://doi.org/10.1371/journal.pbio.1001631.

Gaggìa, F., Mattarelli, P., and Biavati, B. (2010). Probiotics and prebiotics in animal feeding for safe food production. *Internat. J. Food Microbiol.* 141: S15–S28.

Gajardo K, Rodiles A, Kortner TM, et al. (2016). A high-resolution map of the gut microbiota in Atlantic salmon (*Salmo salar*): a basis for comparative gut microbial research, *Sci. Rep.* 6, 30893 (2016) https://doi.org/10.1038/srep30893

Gerritsen, J., Smidt, H., Rijkers, G.T. et al. (2011). Intestinal microbiota in human health and disease: the impact of probiotics. *Genes Nutr.* 6: 209–240.

Glad, T., Bernhardsen, P., Nielsen, K.M. et al. (2010). Bacterial diversity in faeces from polar bear (*Ursus maritimus*) in Arctic Svalbard. *BMC Microbiol.* 201010: 10. https://doi.org/10.1186/1471-2180-10-10.

Glavis-Bloom, J., Muhammed, M., and Mylonakis, E. (2012). Of model hosts and man: using *Caenorhabditis elegans, Drosophila melanogaster* and *Galleria mellonella* as model hosts for infectious disease research. *Adv. Exp. Med. Biol.* 710: 11–17.

Gómez-Valero, L., Soriano-Navarro, M., Pérez-Brocal, V. et al. (2004). Coexistence of *Wolbachia* with *Buchnera aphidicola* and a scondary symbiont in the aphid *Cinara cedri*. *J. Bacteriol.* 186: 6626–6633.

Gottlieb, K., Wacher, V., Sliman, J. et al. (2016). Review article: inhibition of methanogenic archaea by statins as a targeted management strategy for constipation and related disorders. *Alimentary Pharmacol. Therapeutics* 43: 197–212.

Grice, E.A. and Segre, J.A. (2011). The skin microbiome. *Nat. Rev. Microbiol.* 9: 244–253.

Grice, E.A., Kong, H.H., Renaud, G. et al. (2008). A diversity profile of the human skin microbiota. *Genome Res.* 18: 1043–1050.

Hajek, A.E., Nielsen, C., Kepler, R.M. et al. (2013). Fidelity among *Sirex* woodwasps and their fungal symbionts. *Microbial Ecol.* 65: 753–762.

Hanning I, Diaz-Sanchez S (2015). The functionality of the gastrointestinal microbiome in non-human animals, *Microbiome* 3: 51. https://doi.org/10.1186/s40168-015-0113-6

Hawrelak, J.A. and Myers, S.P. (2004). The causes of intestinal dysbiosis: a review. *Alternative Med. Rev.* 9: 180–197.

He, S., Ivanoval, N., Kirton, E. et al. (2013). Comparative metagenomic and metatranscriptomic analysis of hindgut paunch microbiota in wood- and dung-feeding higher termites. *PLoS ONE* 8 (4): e61126. https://doi.org/10.1371/journal.pone.0061126.

Hestetun, J.T., Dahle, H., Jørgensen, S.L. et al. (2016). The microbiome and occurrence of methanotrophy in carnivorous sponges. *Front. Microbiol.* 7: 1781. https://doi.org/10.3389/fmicb.2016.01781.

Ho, Y.W. and Abdullah, N. (1999). The role of rumen fungi in fiber digestion – review. *AJAS* 12: 104–112.

Hook, S.E., Wright, A.-D.G., and BW, M.B. (2010). Methanogens: methane producers of the rumen and mitigation strategies. *Archaea* 2010 (2010): 945785. https://doi.org/10.1155/2010/945785.

Hungate, R.E. (1960). Microbial ecology of the rumen. *Bacteriol. Rev.* 24: 353–364.

Husnika, F. and McCutcheon, J.P. (2016). Repeated replacement of an intrabacterial symbiont in the tripartite nested mealybug symbiosis. *Proc. Natl. Acad. Sci. USA* 113: E5416–E5424.

Ilmberger, N., Guüllert, S., Dannenberg, J. et al. (2014). A comparative metagenome survey of the fecal microbiota of a breast- and a plant-fed Asian elephant reveals an unexpectedly high diversity of glycoside hydrolase family enzymes. *PLoS ONE* 9 (9): e106707. https://doi.org/10.1371/journal.pone. 0106707.

Kamada, N., Chen, G.Y., Inohara, N. et al. (2013). Control of pathogens and pathobionts by the gut microbiota. *Nat. Immunol.* 14: 685–690.

Keenan, S.W., Engel, A.S., and Elsey, R.M. (2013). The alligator gut microbiome and implications for archosaur symbioses. *Scientific Reports* 3: 2877. https://doi.org/10.1038/srep02877.

Kilian, M., Chapple, I.L.C., Hannig, M. et al. (2016). The oral microbiome – an update for oral healthcare professionals. *British Dental Journal* 221: 657–666.

Kim, W., Hendricks, G.L., Lee, K. et al. (2017). An update on the use of C. Elegans for preclinical drug discovery: screening and identifying anti-infective drugs. *Expert Opin. Drug Discov.* 12: 625–633. https://doi.org/10.1080/17460441.2017.1319358.

Kleiner, M., Wentrup, C., Lott, C. et al. (2012). Metaproteomics of a gutless marine worm and its symbiotic microbial community reveal unusual pathways for carbon and energy use. *Proc. Natl. Acad. Sci. U S A.* 109: E1173–E1182.

Klepzik, K.D. and Six, D.J. (2004). Bark beetle – fungal symbiosis: context dependency in complex associations. *Symbiosis* 37: 189–205.

Kobyliak, N., Conte, C., Cammarota, G. et al. (2016). Probiotics in prevention and treatment of obesity: a critical view. *Nutr. Metabol.* 13: 14. https://doi.org/10.1186/s12986-016-0067-0.

Kolenbrander, P.E., Andersen, R.N., Blehert, D.S. et al. (2002). Communication among oral bacteria. *Microbiol. Mol. Biol. Rev.* 66: 486–505.

Kuraishi T, Hori A, Kurata S (2013). Host-microbe interactions in the gut of *Drosophila melanogaster, Front. Physiol.*; 4: 375. https://doi.org/10.3389/fphys.2013.00375

Lee, H.J., Jung, J.Y., Oh, Y.K. et al. (2012). Comparative survey of rumen microbial communities and metabolites across one caprine and three bovine groups, using bar-coded pyrosequencing and 1H nuclear magnetic resonance spectroscopy. *Appl. Environ. Microbiol.* 78: 5983–5993.

Ley, R.E., Hamady, M., Lozupone, C. et al. (2008). Evolution of mammals and their gut microbes. *Science.* 320: 1647–1651.

Lloyd-Price, J., Abu-Ali, G., and Huttenhower, C. (2016). The healthy human microbiome. *Genome Med* 8: 51. https://doi.org/10.1186/s13073-016-0307-y.

Long, S.J., Williams, D.W., and Hajek, A.E. (2009). *Sirex* species (hymenoptera: Siricidae) and their Parasitoids in *Pinus sylvestris* in eastern North America. *The Canadian Entomologist* 141: 153–157.

Lu, J., Idris, U., Harmon, B. et al. (2003). Diversity and succession of the intestinal bacterial community of the maturing broiler chicken. *Appl. Environ. Microbiol.* 69: 6816–6824.

Mackie, R.I., Aminov, R.I., White, B.A. et al. (2000). Molecular ecology and diversity in gut microbial ecosystems. In: *Ruminant Physiology: Digestion, Metabolism, Growth and Reproduction* (ed. P.B. Cronje), 61–78. Oxon, United Kingdom: CABI Publishing.

Mantis, N.J., Rol, N., and Corthésy, B. (2011). Secretory IgA's complex roles in immunity and mucosal homeostasis in the gut. *Mucosal Immunol.* 4: 603–611.

Marsh, E.K. and May, R.C. (2012). *Caenorhabditis elegans,* a model organism for investigating immunity. *Appl. Environ. Microbiol.* 78: 2075–2081.

Martin, R., Gazis, R., Skaltsas, D. et al. (2015). Unexpected diversity of basidiomycetous endophytes in sapwood and leaves of *Hevea*. *Mycologia.* 107: 284–297.

Masui, S., Kamoda, S., Sasaki, T. et al. (2000). Distribution and evolution of bacteriophage WO in *Wolbachia*, the endosymbiont causing sexual alterations in arthropods. *J. Mol. Evol.* 51: 491–491.

Matsui, H., Kato, Y., Chikaraishi, T. et al. (2009). Microbial diversity in ostrich ceca as revealed by 16S ribosomal RNA gene clone library and detection of novel *Fibrobacter* species. *Anaerobe* 16: 83–93.

McFall-Ngai, M. (2014). Divining the essence of symbiosis: insights from the squid-vibrio model. *PLoS Biol.* 12: e1001783. https://doi.org/10.1371/journal.pbio.1001783.

Mehdiabadi, N.J. and Schultz, T.R. (2009). Natural history and phylogeny of the fungus-farming ants (hymenoptera: Formicidae: Myrmicinae: Attini). *Myrmecological News* 13: 37–55.

Metchnikoff, E. (1908). *Prolongation of Life. G. P.* New York: Putnam and Sons.

Midha, A., Schlosser, J., and Hartmann, S. (2017). Reciprocal interactions between nematodes and their microbial environments. *Front. Cell. Infect. Microbiol.* 7: 144. https://doi.org/10.3389/fcimb.2017.00144.

Million, M., Angelakis, E., Maraninchi, M. et al. (2013). Correlation between body mass index and gut concentrations of *Lactobacillus reuteri, Bifidobacterium animalis, Methanobrevibacter smithii* and *Escherichia coli*. *Int. J. Obes. (Lond)* 37: 1460–1466. https://doi.org/10.1038/ijo.2013.20. 22.

Moles, L., Gómez, M., Heilig, H. et al. (2013). Bacterial diversity in meconium of preterm neonates and evolution of their fecal microbiota during the first month of life. *PLoS ONE* 8 (6): e66986:https://doi.org/10.1371/journal.pone.0066986.

de, de Moreno, LeBlanc, A. and LeBlanc, J.G. (2014). Effect of probiotic administration on the intestinal microbiota, current knowledge and potential applications. *World J. Gastroenterol.* 20: 16518–16528.

Mueller, U.G., Schultz, T.R., Currie, C.R. et al. (2001). The origin of the attine ant-fungus mutualism. *Q. Rev. Biol.* 76: 169–197.

Murfin, K.E., Dillman, A.R., Foster, J.M. et al. (2012). Nematode-bacterium symbioses - cooperation and conflict revealed in the 'Omics' age. *Biol. Bull.* 223: 85–102.

Musso, G., Gambino, R., and Cassader, M. (2011). Interactions between gut microbiota and host metabolism predisposing to obesity and diabetes. *Ann. Rev. Med.* 62: 361–380.

Nakamura, N., Lin, H.C., McSweeney, C.S. et al. (2010). Mechanisms of microbial hydrogen disposal in the human colon and implications for health and disease. *Annu. Rev. Food Sci. Technol.* 1: 363–395.

Neish, A.S. (2009). Microbes in gastrointestinal health and disease. *Gasteroenterol.* 136: 65–80.

Nelson, M. and Graf, J. (2012). Bacterial symbioses of the medicinal leech *Hirudo verbena*. *Gut Microbes.* 3: 322–331.

Newbold, C.J., de la Fuente, G., Belanche, A. et al. (2015). The role of ciliate protozoa in the rumen. *Front. Microbiol.* 6: 1313. https://doi.org/10.3389/fmicb.2015.01313.

Nielsen, S., Walburn, J.W., Vergés, A. et al. (2017). Microbiome patterns across the gastrointestinal tract of the rabbitfish *Siganus fuscescens*. *Peer J.* 5: e3317. https://doi.org/10.7717/peerj.3317.

Ogawa, G., Ishida, M., Kato, H. et al. (2010). Identification of facultative anaerobic bacteria isolated from intestine of the minke whale *Balaenoptera acutorostrata* by 16S rRNA sequencing analysis. *Fish Sci.* 76: 177–181.

Orpin, C.G. (1975). Studies on the rumen flagellate *Neocallimastix frontalis*. *J. Gen. Microbiol* 91: 249–262.

Pannaraj, P.S., Li, F., Cerini, C. et al. (2017). Association between breast milk bacterial communities and establishment and development of the infant gut microbiome. *JAMA Pediatr.* 171: 647–654. https://doi.org/10.1001/jamapediatrics.2017.0378.

Peterson, B.F. and Scharf, M.E. (2016). Lower termite associations with microbes: synergy, protection, and interplay. *Front. Microbiol.* 7: 422. https://doi.org/10.3389/fmicb.2016.00422.

Pope, P.B., Denman, S.E., Jones, M. et al. (2010). Adaptation to herbivory by the Tammar wallaby includes bacterial and glycoside hydrolase profiles different from other herbivores. *Proc. Natl. Acad. Sci. USA* 107: 14793–14798.

Pope, P.B., Mackenzie, A.K., Gregor, I. et al. (2012). Metagenomics of the Svalbard reindeer rumen microbiome reveals abundance of polysaccharide utilization loci. *PLoS ONE* 7 (6): e38571. https://doi.org/10.1371/journal.pone.0038571.

Powell, J.E., Martinson, V.G., Urban-Mead, K. et al. (2014). Routes of acquisition of the gut microbiota of the honey bee *Apis mellifera*. *Appl. Environ. Microbiol.* 80: 7378–7387.

Rastall, R.A., Gibson, G.R., Gill, H.S. et al. (2005). Modulation of the microbial ecology of the human colon by probiotics, prebiotics and synbiotics to enhance human health: an overview of enabling science and potential applications. *FEMS Microbiol. Ecol.* 52: 145–152.

Reyniers, J.A. (1959). Germ-free vertebrates – current status. *Annal. N.Y. Acad.Sci.* 78: 1–400.

Ricard, G., McEwan, N.R., Dutilh, B.E. et al. (2006). Horizontal gene transfer from bacteria to rumen ciliates indicates adaptation to their anaerobic, carbohydrates-rich environment. *BMC Genomics* 7 (22): https://doi.org/10.1186/1471-2164-7-22.

Sanders, J.G., Beichman, A.C., Roman, J. et al. (2015). Baleen whales host a unique gut microbiome with similarities to both carnivores and herbivores. *Nature Communications* 6: 8285. https://doi.org/10.1038/ncomms9285.

Schultz, T.R. and Brady, S.G. (2008). Major evolutionary transitions in ant agriculture. *Proc. Natl. Acad. Sci. USA.* 105: 5435–5440.

Schwab, C., Cristescu, B., Boyce, M.S. et al. (2009). Bacterial populations and metabolites in the feces of free roaming and captive grizzly bears. *Can. J. Microbiol.* 12: 1335–1346.

Seretis, C., Seretis, F., and Goonetilleke, K. (2014, 2014). Appendicectomy and *Clostridium difficile* infection: is there a link? *J. Clin. Med. Res.* 6: 239–241.

Shigenobu, S., Watanabe, H., Hattori, M. et al. (2000). Genome sequence of the endocellular bacterial symbiont of aphids *Buchnera* sp. APS. *Nature* 407: 81–86.

Singh, S.B. and Lin, H.C. (2015). Hydrogen sulfide in physiology and diseases of the digestive tract. *Microorganisms* 3: 1-x manuscripts; https://doi.org/10.3390/microorganisms30x000x.

Sommer, F., Ståhlman, M., Ilkayeva, O. et al. (2016). The gut microbiota modulates energy metabolism in the hibernating brown bear *Ursus arctos*. *Cell Rep.* 14: 1655–1661.

Staley, C., Kaiser, T., Vaughn, B.P. et al. (2018). Predicting recurrence of Clostridium difficile infection following encapsulated fecal microbiota transplantation. *Microbiome.* 6: 166. https://doi/10.1186/s40168-018-0549-6.

Stewart, F.J. and Cavanaugh, C.M. (2005). Symbiosis of thioautotrophic bacteria with *Riftia pachyptila*. In: *Progress in Molecular and Subcellular Biology: Molecular Basis of Symbiosis* (ed. J. Overmann). Heidelberg: Springer-Verlag.

Sundset, M.A., Praesteng, K.E., Cann, I.K. et al. (2007). Novel rumen bacterial diversity in two geographically separated sub-species of reindeer. *Microb. Ecol.* 54: 424–438.

Tan, M.W., Mahajan-Miklos, S., and Ausubel, F.M. (1999). Killing of Caenorhabditis elegans by Pseudomonas aeruginosa used to model mammalian bacterial pathogenesis. *Proc. Natl. Acad. Sci. USA.* 96: 715–720. https://doi.org/10.1073/pnas.96.2.715.

The C. elegans Sequencing Consortium (1998). Genome sequence of the nematode *C. elegans:* a platform for investigating biology. *Science* 282: 2012–2018.

Turnbaugh, P.J., Ley, R.E., Mahowald, M.A. et al. (2006). An obesity-associated gut microbiome with increased capacity for energy harvest. *Nature* 444: 1037–1041.

Uyeno, Y., Shigemori, S., and Shimosato, T. (2015). Effect of probiotics/prebiotics on cattle health and productivity. *Microbes Environ.* 30: 126–132.

Vilcinskas, A. (2016). The role of epigenetics in host–parasite coevolution: lessons from the model host insects *Galleria mellonella* and *Tribolium castaneum. Zool.* 119: 273–280.

Waite, D.W. and Taylor, M.W. (2015). Exploring the avian gut microbiota: current trends and future directions. *Front. Microbiol.* 6: 673. https://doi.org/10.3389/fmicb.2015.00673.

Wei, S., Morrison, M., and Yu, Z. (2013). Bacterial census of poultry intestinal microbiome. *Poult. Sci.* 92: 671–683.

Werren, J.H. (1997). Biology of Wolbachia. *Ann. Rev. Entomol.* 42: 587–609.

Wong, S. and Rawls, J.F. (2012, 2012). Intestinal microbiota composition in fishes is influenced by host ecology and environment. *Mol. Ecol.* 21 (13): 3100–3102.

Wong, A.C.N., Vanhove, A.S., and Watnick, P.I. (2016). The interplay between intestinal bacteria and host metabolism in health and disease: lessions from *Drosophilia melanogaster. Dis. Model Mech.* 9: 271–281.

Xue Z, Zhang W, Wang L, et al. (2015). The bamboo-eating giant panda harbors a carnivore-like gut microbiota, with excessive seasonal variations, *mBio* 6 3, e00022–15, https://doi.org/10.1128/mBio.00022-1519

Zoetendal, E.G., Raes, J., van den Bogert, B. et al. (2012). The human small intestinal microbiota is driven by rapid uptake and conversion of simple carbohydrates. *ISME J.* 6: 1415–1426.

Further Reading

Brooks, S.P., McAllister, M., Sandoz, M. et al. (2003). Culture independent phylogenetic analysis of the fecal flora of the rat. *Can. J. Microbiol.* 49: 589–601.

Glavis-Bloom, J., Muhammed, M., and Mylonakis, E. (2012). Of model hosts and man: using *Caenorhabditis elegans, Drosophila melanogaster* and *Galleria mellonella* as model hosts for infectious disease research. *Adv. Exp. Med. Biol.* 710: 11–17.

Lyte, M. and Freestone, P.P.E. (2010). *Microbial Endocrinology: Inter-Kingdom Signalling in Health and Infectious Disease.* New York: Springer.

Websites

http://nccam.nih.gov/health/probiotics (NIH website for probiotics)
http://www.genome.jp/kegg/ (Kyoto Encyclopedia of Genes and Genomes website)
flybase.org (Curated database on *D. melanogaster* research)
www.wormbase.org (Curated database on *C. elegans* research)

8

Microbes Driving the Nutrient Cycles

8.1 Overview and Focus

Biology is dependent on specific chemicals for growth and, to sustain life through an indefinite number of generations, these compounds must be continuously available. The elements required to support life are separated into three general categories: essential major elements (C, O, H, N, P, and S), minor elements (Ca, Mg, and K) and trace elements (Co, Cu, Fe, Mn, Mo, Ni, and Zn). All microorganisms require C, O, H, N, P, and S because these elements are present in amino acids, nucleic acids sugars, and other molecules present in cells. The elements of C, N, and S are unique because not only are they used for biosynthesis but they also display a range of oxidation states that microorganisms may use to support cellular respiration. Minor elements serve to facilitate various cellular processes and trace elements are involved as metal cofactors to facilitate enzymes function. Metals are present in the environment and while some are of low bioavailability, metal ions are acquired by specific cellular transport systems. This chapter addresses the role of microorganisms in the transformation of nutrients and minerals through the biogeochemical cycles.

8.2 Nutrient Cycles and What Drives Them

Life on Earth is dependent on an appropriate supply of nutrients, which can be divided into macronutrients and trace-level nutrients. Microorganisms require a carbon source, a nitrogen source, bioavailable phosphorus, and a series of trace elements for biosynthesis of cellular biomass. Additionally, all cells require an electron donor and an electron acceptor to energize the biosynthetic reactions. For heterotrophs, cellular energetics may be integral with carbon or nitrogen metabolism, and for chemolithotrophs, energy yielding reactions involve the oxidation or reduction of minerals. An important feature influencing the growth capability of bacteria is their ability to mediate oxidation–reduction reactions. There are numerous chemical reactions that are energetically important and the half-reactions of several of these are listed in Figure 8.1.

The range of biological applications for conversion of electron flow to chemical energy (ATP production) is approximately −0.61 to +0.816 V. In microbiological metabolism, electron flow is from the electronegative to the electropositive. To predict if the coupled reactions in Figure 8.1 are sufficiently energetic to promote growth, the voltage of the half reaction of the electron donor is compared to the voltage of the electron acceptor. If the difference is approximately −200 mV, enough energy is released to produce ATP. Another

Environmental Microbiology and Microbial Ecology, First Edition. Larry L. Barton and Robert J.C. McLean.

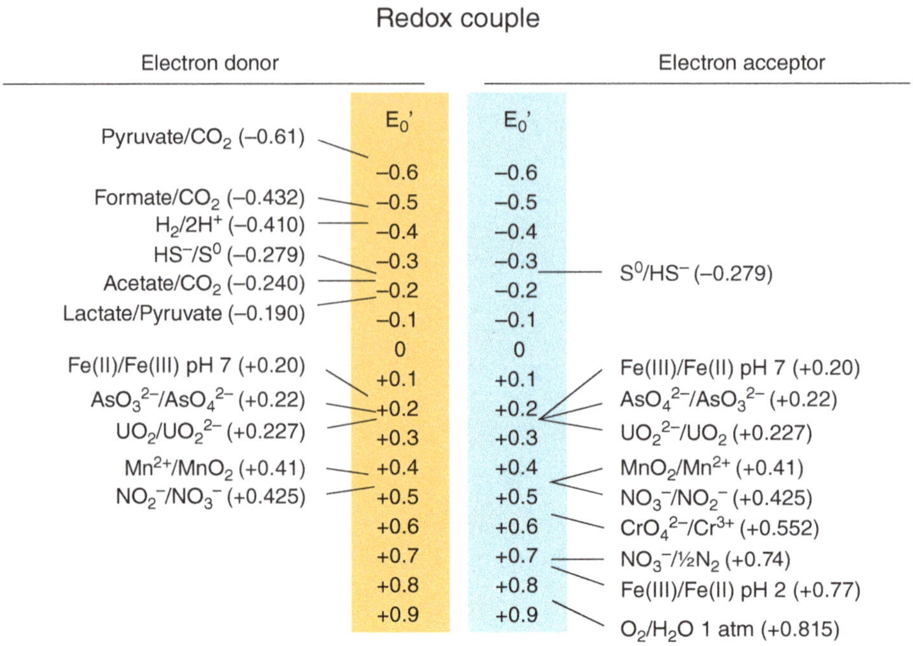

Figure 8.1 Half-cell reactions of common electron donors and electron acceptors. See text on how to predict which electron donor/electron acceptor reactions are sufficiently energetic to support growth.

Table 8.1 Energetics of reactions associated with the carbon, nitrogen, and sulfur cycles.

Energy-Yielding Reactions Involving C, N and S	$\Delta G^{0\prime}$ = kJ/Reaction
Carbon Cycle	
Glucose + $6O_2 \rightarrow 6H_2O + 6CO_2$	–2,870.0
Glucose → 2 ethanol + $2CO_2$	–238.8
Glucose → 2 lactate + H^+	–196.0
$4H_2 + CO_2 \rightarrow CH_4 + 2H_2O$	–136.0
$4H_2$ + 2 CO_2 → acetic acid + $2H_zO$	–104.8
H_2 + fumarate → succinate	–86.2
Acetic acid → $CH_4 + CO_2$	–37.0
Nitrogen Cycle	
Glucose + $8NO_2^- + 8H^+ \rightarrow 6CO_2 + 4N_2 + 10H_2O$	–3,144.0
$3H_2 + NO_2^- + 2H^+ \rightarrow NH_4^+ + 2H_2O$	–436.8
$NH_3 + 2O_2 \rightarrow NO_2^- + 2H^+ + 2H_2O$	–266.7
Sulfur Cycle	
$S^{2-} + 2O_2 \rightarrow SO_4^{2'}$	–790.5
$S^0 + 1.5O_2 + H_2O \rightarrow SO_4^{2-} + 2H^+$	–580.9
$4H_2 + SO_4^2 \rightarrow HS + 3H_2O + OH^-$	–152.0
Acetate + $SO_4^{2-} + 3H^+ \rightarrow 2CO_2 + HS^- + 2H_2O + H^+$	–57.5
$2H_2 + S^0 \rightarrow HS^- + H^+$	–29.0

Table 8.2 Energetics of reactions associated with iron, manganese, selenium, and arsenic metabolism.

Energy-Yielding Reactions Involving Fe, Mn, Se, and As	AG$^{0\prime}$ = kJ/Reaction
Iron reactions	
Acetate + $8Fe^{3+}$ + $4H_2O$ → $2HCO_3^-$ + $8Fe^{3+}$ + $9H^+$	−233.0
Fe^{2+} + $0.25O_2$ + H^+ → Fe^{3+} + $0.25H_2O$	−44.2
Manganese reactions	
MnO_2 + $4H^+$ $2e^-$ → Mn^{2+} + $2H_2O$	−233.0
Mn^{2+} + $0.5O_2$ +H_2O → MnO_2 + $2H^+$	−44.2
Selenium reactions	
Lactate + $2SeO_4^{2-}$ → acetate + $2SeO_3^{2-}$ + HCO_3^- + H^+	−343.1
Lactate + $2SeO_3^{2-}$ + H^+ → acetate + $2Se^0$+ HCO_3^- + H_2O	−529.5
Arsenic reactions	
Lactate + $2AsO_4^{2-}$ + H^+ → acetate + $2H_2AsO_3^-$ + HCO_3^-	−140.3
Lactate + $2H_2AsO_3^-$ → acetate + $2AsH_3$ + HCO_3 + H^+	−138.4

way of predicting the reactions that will support bacterial growth is to calculate the energy yield of a reaction, as given in Tables 8.1 and 8.2. The high-energy phosphate bond in ATP is −31.8 kJ or ~8 kcal and the reactions listed in Tables 8.1 and 8.2 can accommodate the formation of several ATP molecules. However, the bacterium must also have the enzymology to accomplish the steps of oxidation, reduction, and electron flow. The cycling of nutrients reflects the flow of energy between the individual components to the carbon, nitrogen, and sulfur cycles.

Phototrophic bacteria are energized by the appropriate wavelength of light. The unique chlorophylls found in aerobic and anaerobic photosynthetic bacteria absorb light at discrete wavelengths that are specific for each species and enable different phototrophs to be layered or stacked in water columns. With longer wavelengths of light penetrating greater depths in water than shorter wavelengths, anaerobic phototrophic bacteria can exploit multiple regions in a lake. While light is the energy source in phototrophs, electron donors, electron acceptors, and an electron flow that is coupled to ATP synthesis is a characteristic of all photosynthetic bacteria.

8.3 The Aerobic Environment

8.3.1 The "Great Oxidation Event"

For the first 2.3 billion years, planet Earth lacked an aerobic atmosphere and the development is outlined in Table 8.3. While photolysis of H_2O and N_2O produces a minute quantity of O_2, 99.9% of atmospheric O_2 is produced by the photosynthetic activity of cyanobacteria. Photosystem II of the cyanobacteria (Figure 6.4) uses the energy from sunlight and manganese as a cofactor to release oxygen from water by the following reaction: $H_2O \rightarrow O_2 + 2H^+ + 2e^-$.

Table 8.3 Stages in the development of life on Earth.

Time / Ga	Event
4.6	Earth was formed
4.6–2.5	(Archean) was sulfur limited
3.8–3.6	Biosphere not developed yet
3.5	Anoxic photosynthesis by bacteria
2.8	First evidence for oxygenic photosynthesis
2.3	Seawater was <2.5 mM sulfate, which is less than present day values
1.8	Banded iron formations

The photocatalytic production of oxygen by cyanobacteria is projected to have initiated about 2.7 billion years ago, giving rise to the Great Oxidation Event. Fossil evidence of microbial forms, presumably cyanobacteria, is contained in the stromatolites of Western Australia that are ~3.5 billion years old (Schopf et al. 2007). Stromatolites contain layers of bacterial mats that are bound tightly to fine grain material producing stratified or conical morphologies in the rock formation. Stromatolites can be found in many regions of the Earth and examples are given in Figure 8.2. Biofilms containing cyanobacteria and benthic microorganisms (i.e. bacteria and archaea), can be traced to the early Archean and are contained in microbial induced sedimentary structures (MISSs) as well as in stromatolites (Noffke and Awramik 2013). As O_2 was generated, it was initially sequestered in silicates and oxide minerals of the Earth's lithosphere. An important oxidation of Fe(II) by O_2 resulted in the formation of Fe_2O_3, which is observed in the banded iron formations (BIFs) of Minnesota and Western Australia. These BIFs were formed 2.7–2.5 billion years ago and are attributed to oxygen from cyanobacteria. However, BIFs older than 2.7 billion years are probably the product of anoxygenic photosynthetic bacteria (Koehler et al. 2010). With cyanobacterial activity, O_2 content in the Earth's atmosphere gradually increased and profound changes in the biosphere occurred about 2.3 billion years ago. With the presence of oxygen in the atmosphere, aerobic respiration in bacteria evolved, leading to the development of diverse aerobic organisms.

8.3.2 Oxygen Cycle

There are three reservoirs for oxygen: the atmosphere, biological material or the biosphere, and the crust of the Earth or lithosphere. As summarized in Figure 8.3, the greatest amount of oxygen is in the lithosphere while the least is in the biosphere. The biosphere has the shortest residence time for oxygen and this high flux of oxygen is attributed to the intense metabolic activity of biological systems. Approximately one-sixth of the oxygen removed from the atmosphere is attributed to microbial activity. Because of microbial photosynthetic activity in the oceans, the release of oxygen into the atmosphere is attributed to cyanobacteria and algae.

Aerobic forms of life are supported by the current level of O_2, which is ~21% of the Earth's atmosphere. Currently, ocean-based photosynthesis accounts for ~45% of O_2 generated and about half of this amount is produced by the nanobacterium *Prochlorococcus*. The *Prochlorococcus* cell is unique because it is only 0.6 μm in diameter (Chisholm et al. 1988)

Figure 8.2 Some of the oldest records of life are preserved in stromatolites: (a) a stromatolite from the north shore of Lake Superior; (b) conical stromatolite formation in red and black chert which is about 2000 million years old, from Gunflint Iron Formation in Ontario, Canada; (c) cross section of a 150-year-old stromatolite with layers built up by cyanobacteria from the Late Jurassic Morrison Formation, collected by Spencer Lucas in Union County, New Mexico USA. *Source:* (a) Courtesy of Roy L. Johnson, Jr. *Source:* (b), (c) Reproduced with permission of Larry L. Barton. Samples in (b) and (c) are on display at the Natural History Museum, Albuquerque, New Mexico.

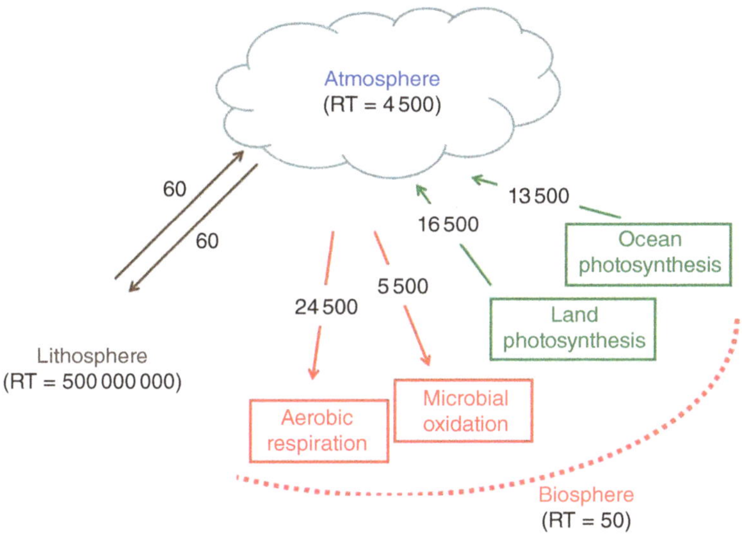

Figure 8.3 Oxygen flux and contributions of microorganisms to global oxygen change. RT = residence time in years. Values over and under arrows = 10^{10} kg O_2 per year. *Source:* Based on data from Walker (1980).

and contains chlorophyll *b*, which is unusual for a prokaryotic phototroph. Marine waters may contain 10^5 cells of *Prochlorococcus* per ml and on a global perspective, *Prochlorococcus* is the most abundant individual on Earth with $\sim 3 \times 10^{27}$ cells produced each year (Flombaum et al. 2013). Thus, it is apparent that throughout time, phototrophic bacteria (cyanobacteria and *Prochlorococcus*) have been important in the global enrichment of atmospheric oxygen.

Marine environments are examined here to better understand the complexities of carbon cycling. Biological, physical, and chemical processes in the oceans contribute to the "ocean carbon pump" (Sigman and Haug 2006). The concentration of CO_2 (including carbon dioxide, carbonate ions, and bicarbonate ions) in marine waters is several times greater than in the atmosphere. In the first phase of the biological pump, CO_2 is fixed by planktonic photosynthetic organisms into cellular components. With death of photosynthetic cells, the carbon may be recycled in this aerobic euphotic zone or the second phase of the biological pump is initiated where single cells aggregate with organic matter to form marine snow. Non-photosynthetic bacteria use marine snow as a nutrient and release CO_2 from the dead organic material via aerobic catabolism. Protozoa and related systems graze on these bacteria and this process of recycling organic material in the deep marine system is called the "microbial loop" (Fenchel 2008). In the final phase of the biological pump, the organic material is deposited on the floor of the ocean where it collects and serves as a reservoir for organic carbon.

There are numerous areas in the environment where a gradient of O_2 occurs between aerobic and anaerobic zones. Certain bacteria prefer to grow in these regions where the O_2 level is less than that found in the atmosphere. These microaerophilic environments are populated by bacteria that have unique cytochromes with a high affinity for O_2. Some bacteria, facultative anaerobes, migrate from aerobic to anaerobic zones and adjust their respiratory systems accordingly. Obligate anaerobic bacteria and archaea are killed by exposure to O_2 and, for methanogenic archaea to grow, the environment must have a low oxidation–reduction (redox) value. Often, the oxygen content of culture media or an environment may be estimated using dyes with known redox values.

While the major process for production of O_2 is attributed to photosynthesis, there are only three other systems that account for biological oxygen production (Figure 8.4). One of these bacterial processes is the detoxification of reactive oxygen species (ROS) and this activity is discussed in Section 8.3.3. Two other processes for O_2 production are associated with bacterial reduction of nitric oxide (NO) and chlorate (ClO_3^-). Under anaerobic conditions, bacterial isolates that are members of the alpha, beta, and gamma divisions of *Proteobacteria,* reduce chlorate to chlorite (ClO_2^-) and chlorite is enzymatically converted to Cl^- plus O_2 by a dismutase (Coats et al. 1999). Nitric oxide, an intermediate compound resulting bacterial

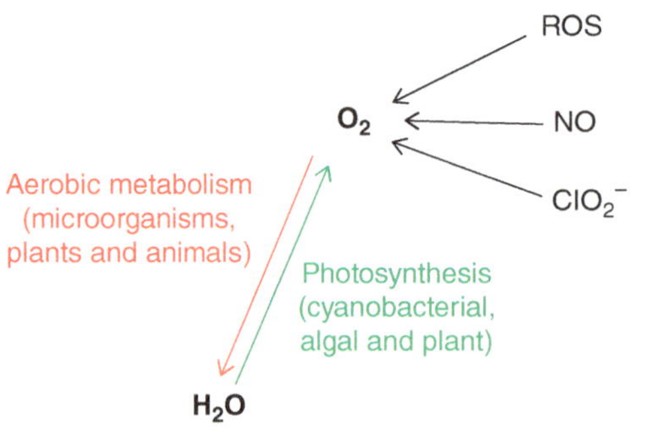

Figure 8.4 The major scale and microscale of oxygen production. The liberation of O_2 by aerobic photosynthesis is important for the global oxygen cycle while reactions with ClO_2^-, NO and ROS are of importance in a biofilm on a microscale.

dissimilatory reduction of nitrite, is dismutated by *Methylomirabilis oxyfera* to produce N_2 and O_2 (Ettwig et al. 2010). While nonphotosynthetic oxygen production is not a significant contributor to global oxygen levels, it may be highly relevant in dense bacterial communities such as in mat formations.

8.3.3 Hydrogen Peroxide and ROS

Microorganisms may produce ROS under specific aerobic conditions and these oxygen-containing molecules may cause considerable damage to cells. The ROS are byproducts of oxygen metabolism that include hydrogen peroxide (H_2O_2), superoxide anion radicals ($O_2^{\cdot -}$), and the reactive hydroxyl radicals ($OH^{\cdot -}$). Usually, ROS is produced by the movement of single electrons from the bacterial electron transport system to molecular oxygen. Reactions associated with single electron release from microorganisms interacting with molecular oxygen are given in Table 8.4.

By oxidation of membranes, DNA, RNA proteins and lipids, ROS may destroy bacteria adjacent to those producing hydrogen peroxide (Cabiscol et al. 2000). Several microorganisms produce hydrogen peroxide and a few of these are presented in Table 8.5. Several species of bacteria produce hydrogen peroxide as a virulence factor to enhance their pathogenicity or to prevent other bacteria from growing in their environment. Both white rotting and brown rotting fungi produce hydrogen peroxide to assist in the decomposition of wood. Microorganisms producing hydrogen peroxide are not destroyed because they produce sufficient levels of catalase, which detoxifies hydrogen peroxide.

The presence of hydrogen peroxide is important in producing oxygen stress in microorganisms and the response of bacteria is the synthesis of catalase, superoxide dismutase (SOD), and superoxide reductase (SOR). The reactions of these oxygen metabolizing enzymes are listed in Table 8.4. Catalase is a heme-containing enzyme found in aerobic bacteria, facultative anaerobes, and aerotolerant bacteria. A catalase isolated from thermophilic bacteria and thermophilic archaea is unique in that it contains manganese and not heme (Amo et al. 2002). Catalase production protects bacteria and archaea from transient exposure to hydrogen peroxide; in a mixed biofilm, catalase production by a few bacterial species protects all the bacteria present. The metal-containing enzyme SOD in microorganisms may have copper, zinc, iron, manganese, or nickel in the ionic form as the metal cofactor. Most aerobic bacteria and aerotolerant anaerobes produce SOD. Many bacteria and

Table 8.4 Reactions involving hydrogen peroxide, superoxide, or hydroxyl radical.

Reaction	Equation
Catalase reaction	$2H_2O_2 \rightarrow 2H_2O + O_2$
Superoxide formation	$O_2 + e^- \rightarrow O_2^{\cdot -}$
Detoxification of superoxide	
By superoxide reductase (SOR)	$O_2^{\cdot -} + e^- \rightarrow H_2O_2$
By superoxide dismutase (SOD)	$2O_2^{\cdot -} + 2H^+ \rightarrow H_2O + O_2$
Hydroxyl radical formation	$H_2O_2 + e^- + H^+ \rightarrow H_2O + HO^{\cdot -}$
Detoxification of hydroxyl radical	$HO^{\cdot -} + e^- + H^+ \rightarrow H_2O$

Table 8.5 Hydrogen peroxide production by microorganisms.

Microorganisms	Associated target or activity	Reference
Bacteria		
Anacystis nidulans	Production results from chemicals coupled to electron transport chain	(Roncel et al. 1989)
Bacteria in the human colon	May contribute to ulcerative colitis Crohn's disease	(Strus et al. 2009)
Beggiatoa leptomitiformis	Interacts with products of sulfur cycle	(Dubinina et al. 1990)
Enterococcus avium	May perpetuate inflammatory bowel disease	(Strus et al. 2009)
Enterococcus faecium	A virulence factor for nematodes	(Moy et al. 2004)
Escherichia coli	Intracellular concentration of H_2O_2 is 0.2 µm	(González-Flecha and Demple 1995)
Lactobacillus delbrueckii	Inhibitor for food-borne pathogens	(Baldori et al. 2007)
Lactobacillus johnsonii	Kills *Salmonella enterica*	(Pridmore et al. 2008)
Mycoplasma pulmonis	A virulence factor	(Brennan and Feinstein 1969)
Pseudomonas aeruginosa	Enhances release of DNA from bacteria	(Das and Manefield 2013)
Streptococcus oralis	A cytotoxin for macrophages	(Okahashi et al. 2013)
Streptococcus pyogenes	Production is linked to lactate oxidation	(Seki et al. 2004)
Fungi		
Brown rot fungi	Basidiomycetes commonly produce brown-rot of wood	(Koenigs 1974)
White rot fungi	Commonly produced and levels are influenced by media	(Zhao et al. 1998)

archaea contain SOR and its function is to detoxify the superoxide radical by reduction, with NADPH used as the reductant to reduce the oxidized SOR protein. Bacteria and archaea are capable of surviving exposure to ROS because of the function of catalase, SOR, and SOD (see Table 8.4).

8.4 Carbon – A Renewable Resource

Carbon forms many different types of organic molecules that serve as the basis of various biomolecules and, therefore, carbon is an essential element. Approximately 0.04% (400 ppm) of the Earth's atmosphere is CO_2, which is 750–800 GTC (1 GTC is 10^{15} metric tons of carbon), and is the smallest reservoir of the Earth's carbon. Land plants and ocean biota contain about 600 and 3 GTC, respectively. Most of the Earth's carbon resides in the sedimentary rocks, oceans, fossil fuels, and soil, with 1×10^7, 4×10^4, 6×10^3, and 2×10^3 GTC, respectively. The carbon cycle (Figure 8.5) is highly dynamic and the major microbial involvement includes CO_2 fixation (photosynthesis, lithotrophy, and heterotrophy), CO_2 released primarily from mineralization of organic compounds, and methane metabolism.

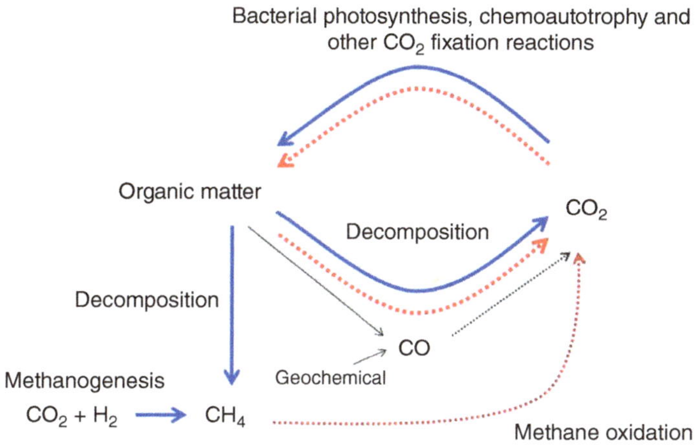

Figure 8.5 Microbial processes that contribute to the global carbon cycle. Solid lines, anaerobic activities; broken lines, aerobic reactions.

8.4.1 Carbon Dioxide Fixation and Carbonate Reduction

Microorganisms use several systems for CO_2 fixation. Photosynthetic aerobic bacteria and chemolithotrophic bacteria are autotrophs that obtain all their carbon for cellular synthesis from CO_2. The carbon dioxide fixation pathway is often referred to as the Calvin cycle or the Calvin–Benson–Bassham cycle and the key enzyme is ribulose 1,5-bis-phosphate carboxylase/oxygenase (RuBisCO). RuBisCO is broadly distributed in cyanobacteria, purple phototrophic bacteria (*Rhodospirillum* and *Rhodobacter*), and bacteria that obtain energy from inorganic forms of sulfur or nitrogen. In some bacteria, RuBisCO is organized in internal structures called carboxysomes and while the function of carboxysomes is not completely understood, their role in CO_2 fixation may be most important when the CO_2 concentration is low. For a recent review of prokaryotic carboxysomes consult Heinhorst et al. (2014). While there are some DNA homologies for RuBisCO in archaea, the function of these homologies is unclear. It should also be noted that chloroplasts of eukaryotes also fix CO_2 using RuBisCO.

In anaerobes and archaea, alternative CO_2 fixation systems may occur. The reductive tricarboxylic acid (TCA or Krebs) cycle is used to fix CO_2 in some thermophilic proteobacteria, green sulfur phototrophs such as *Chlorobium* and archaeal hyperthermophilic sulfur oxidizers (*Thermoproteus* and *Pyrobaculum*). The reductive acetyl-CoA pathway is used by the anaerobic acetogenic and sulfate-reducing bacteria as well as the methanogens. The 3-hydroxypropionate cycle is used by the green phototrophs (*Chloroflexus*) and *Sulfolobus* of the Archaea domain.

Carbonate reduction occurs by reactions that are coupled to reactions that are energetically favorable. *Methanococcus* and *Methanobacter* are archaea that couple carbonate reduction to hydrogen oxidation with the production of methane. Homoacetogenic clostridia use hydrogen oxidation to reduce two carbonate ions to produce acetate.

Some microorganisms take up small organic compounds as they fix CO_2 and this process is called myxotrophy (Box 8.1). The carbon compounds supplement the CO_2 fixed and enhance the growth rate of the microorganism by providing part of the carbon balance. It has long been suggested that members of sulfate-reducing bacteria and archaea are capable of myxotrophic growth (Offre et al. 2013). Myxotrophic organisms can grow either under autotrophic or heterotrophic conditions and can switch from heterotrophic growth when the environmental organic carbon level is insufficient for growth.

Box 8.1 Myxotrophic Growth

The metabolic and energetic capabilities of microorganisms are commonly considered as heterotrophs or autotrophs. Some bacteria, however, are capable of growing heterotrophically and autotrophically at the same time. If the spectrum of growth possibilities for microorganisms are heterotrophy and autotrophy at the two extremes, there are varying degrees of myxotrophy between them. Additionally, some microorganisms may be autotrophic only under aerobic conditions, heterotrophic under both aerobic and anaerobic states, and myxotrophic under anaerobic growth conditions.

8.4.2 Methanogenesis, Methanotrophy, and Methylotrophy

Only members of the domain Archaea have the appropriate enzymology for methane production (Hedderich and Whitman 2006). The biochemical transformation of carbon compounds to methane requires the presence of unique cofactors and coenzymes (i.e. F420, methanofuran, methanopterin, coenzyme B, and coenzyme M) and these are present in archaea. Under extreme anaerobic conditions, methanogenic organisms use carbon dioxide or acetate as final electron acceptors to produce methane as indicated here.

$$CO_2 + 4H_2 \rightarrow CH_4 + 2H_2O$$

$$CH_3COOH \rightarrow CH_4 + CO_2$$

Methanogenesis is characteristically associated with decomposition of organic materials (Figure 8.5) and is of concern because it is a greenhouse gas that contributes to global warming. Although there is considerable variation in the levels of methane produced from different sources, the annual production of methane is about 600 Tg (Tg = 10^{12} g). About 70% of methane released into the atmosphere comes from: the energy industry (coal, gas, and oil production), wetlands, rice cultivation, and enteric fermentation. The remaining 30% of methane can be attributed to landfills, sewage treatment, animal waste, invertebrates (termites, millipedes), and ocean environments. Enteric fermentation in cows, sheep, and other ruminates are major contributors of methane, which is released by belching or flatulence. Due to the social pressure to reduce global methane production, there is a focus on the management of feeding of cattle, which may reduce methane production by as much as 15–30% (Knapp et al. 2014). The recent isolation (Dridi et al. 2012) of a new methanogenic archaeon (*Methanomassiliicoccus luminyensis*) has indicated that methanogens may contribute to unexplained human intestinal disorders.

Bacteria that obtain all their cellular carbon from methane oxidation are designated as methanotrophs. There two pathways for assimilation of carbon from methane into cellular carbon: the ribulose monophosphate pathway (Type I) or the serine pathway (Type II). Bacteria (i.e. *Methylomonas*, *Methylobacter*, *Methylomicrobium*, and *Methylocaldum*) using the Type I pathway belong to the *Gammaproteobacteria* whereas bacteria (i.e. *Methylosinus*, *Methylocystis*) using the Type II pathway for methane oxidation are members of the *Alphaproteobacteria*. *Methylococcus* is a member of the *Gammaproteobacteria*, but it contains segments of both Type I and Type II methane monooxygenase (MMO). These methanotrophic bacteria characteristically have an extensive internal membrane system to facilitate methane oxidation and they have been isolated from soil, landfills and aquatic environments. Most of the methane-oxidizing bacteria isolated are obligate methanotrophs but a few are facultative and can use other carbon sources.

Disruption of the carbon–hydrogen bond in methane requires a special enzyme (MMO), which occurs in two forms. A soluble form (sMMO) is found in the cytoplasm of methanotrophs and has two iron atoms in the active site whereas the particulate form (pMMO) has a copper center. The oxidation of methane by sMMO produces methanol:

$$CH_4 + O_2 + NADH + H^+ \rightarrow CH_2OH + H_2O + NAD^+$$

If oxidation is by pMMO, reduced cytochrome *c* replaces NADH to catalyze the reduction of methane to methanol. Characteristically, MMO enzymes have low substrate specificity and *Methylococcus capsulatus* will oxidize numerous one-carbon compounds, *n*-alkanes, ethers, and aromatic compounds (Anthony 1988). For this reason, bacteria producing MMO are frequently considered for bioremediation of sites where diverse hydrocarbons occur.

Oxidation of methane by microorganisms in anaerobic environments has been a topic pursued by many scientists. With mixed cultures, methane is slowly degraded and recently a new bacterium, *Methylomirabilis oxyfera*, has been isolated that will couple methane oxidation to nitrite reduction (Ettwig et al. 2010). While this organism grows anaerobically with reduction of nitrite to NO, N_2 and O_2 are produced from dismutation of NO. This could explain the source of O_2 for the MMO reaction in an anaerobic environment. Genome analysis has revealed that *Methylomirabilis oxyfera* lacks genes for known N_2 production but contains genes for aerobic methane oxidation.

For many decades, the disappearance of methane in deep marine environments has been studied. The anaerobic oxidation of methane has been correlated with the reduction of sulfate, iron, and manganese (Beal et al. 2009; Knittel and Boetius 2009) and not by MMO reactions. The concomitant oxidation of methane and reduction of sulfate is possibly due to a marine consortium of methanogens with sulfate-reducing bacterial cells (Boetius et al. 2000). The anaerobic oxidation of methane is proposed to involve a reversal of methanogenesis.

Methylotrophic bacteria use one-carbon compounds such as methanol, formaldehyde, and formate but not methane as the carbon and energy source. The bacteria capable of growing on methanol are broadly distributed in the environment and may be either facultative or obligate in their requirement for methanol. For most bacteria, methanol oxidation is dependent on methanol dehydrogenase, which requires pyrroloquinoline quinone (PQQ) or NAD^+ as the cofactor. The oxidation of methanol occurs in the periplasm and produces formaldehyde as indicated here:

$$CH_3OH + \text{electron carrier} \rightarrow CHO + \text{reduced electron carrier}$$

For methylotrophs, formaldehyde is incorporated into cell carbon by either the serine pathway or the ribulose monophosphate pathway. Energy to support biosynthesis through formaldehyde assimilation comes from the oxidation of formaldehyde to formic acid. Depending on the bacteria, formate dehydrogenase may require either NAD^+ or cytochrome as a cofactor. Because formaldehyde is used both for building cell structure and for energy, the cell must control the fate of formaldehyde in the cell. The final step in methanol utilization is the oxidation of formate by formate dehydrogenase to produce CO_2 as the final metabolic product.

A continuous source of methanol is released from plant leaves as cell wall precursors are condensed to form plant cell walls. It has been estimated that there are about a million phyllospheric bacteria colonized on each square centimeter of plant surface; globally this would account for about 10^{26} epiphytic bacteria. While the three main genera present on aerial plant surfaces are *Pseudomonas, Sphingomonas,* and *Methylobacterium,* the principle epiphytic methylotroph *Methylobacterium mesophylicum* has been found with members of most plant families (Delmotte et al. 2009). As *Methylobacterium mesophylicum* grows by consuming methanol, it secrets phytohormones and this supports the suggestion that a symbiotic state has evolved between this methylotrophic bacterium and plants (Kutschera 2007).

8.4.3 Mineralization of Carbon Compounds

In terrestrial environments, the liberation of CO_2 focuses on decomposition of plant and animal materials. Plants and animals produce a myriad of organic compounds which serve as carbon and often energy sources for microorganisms (Table 8.6), and decomposition of organic material entering the aerobic environment occurs in stages (Barton and Northup 2011). Initially, bacteria account for a rapid consumption of soluble low molecular weight compounds with a rapid increase in the growth of microbial cells. Bacteria outcompete fungi in the consumption of sugars, amino acids, polysaccharides, and proteins because bacteria have robust active transport systems which rapidly sequester soluble nutrients from the environment. A second metabolic phase occurs where fungi and actinomycetes are dominant microorganisms because they secrete extracellular enzymes that slowly convert polymeric plant or animal material into small, soluble organic materials. The final stage is the digestion of highly complex organic material that may persist in the environment for a considerable time and it is markedly slow.

Complete oxidation of carbon compounds is most efficiently achieved by aerobic microorganisms using the TCA cycle, also known as the citric acid cycle or the Krebs cycle. End products from the catabolism of organic compounds by anaerobic bacteria may include CO_2; however, most of the carbon from anaerobic bacterial metabolism remains as diverse carbon compounds. In anaerobic environments (i.e. rumen, anaerobic sludge digester, or mud), methanogens convert low molecular weight carbon compounds into methane.

The complex carbon compound cellulose has received considerable attention over the years. Cellulose is composed of polymerized glucose monomers and because cellulose is the most abundant polymer on Earth, glucose is the most abundant sugar in our biosphere. This abundance of glucose in the environment may account for most metabolic processes using glucose as the preferred (or default) starting molecule. The linear glucose polymers in cellulose are covalently bonded by β-1,4 glucosidic linkages and hydrolysis of cellulose is attributed to various forms of 1,4-β-glucanase by fungi and bacteria. The interest in conversion of plant biomass into biofuel or bioplastics has given impetus to enhance cellulase or hemicellulase activities of microorganisms (Mazzoli et al. 2012).

The biological process associated with decomposition of fresh plant material has several steps. As the decaying leaves in Figure 8.6b indicate, enzymatic digestion of tree leaves reveals the areas of the leaf with different structural types of molecules. Initially, pectin is released from the cell walls by propectinase activity and this is followed by hydrolysis of pectin by pectinases, and degradation of hemicellulose and lignin. The last component of the plant cell to be degraded is cellulose. Cellulose is a difficult substrate for microbial degradation because cellulose is insoluble and does not easily interface with microbial cellulases. Cellulose released from plant cells may be digested by 1,4-β-glucanases that are secreted by fungi (i.e. *Phanerochaete chrysosporium, Trichoderma reesei,* and *Poria placenta*) and not associated with the fungal cell walls. While some bacteria (*Cellulomonas fimi, Thermobifidia fusca, Cellvibrio japonicus, Acetivibrio cellulolyticus,* and *Clostridium stearcorarium*), release the cellulolytic enzymes from the bacterial cell, several anaerobic bacteria assemble the cellulolytic enzymes in units called cellulosomes on the cell surface (Box 8.2). The cellulosome structure consists of 60–90 proteins, which range in function from attachment to the bacterial cell surface, serving as a scaffold for attachment of enzymes, and binding substrate. The review by Bayer and Lamed (2006), provides technical information on the structure of bacterial cellulosomes. Bacteria with cellulosomes include *A. cellulolyticus, Bacteroides cellulosovens,* and several species of

Table 8.6 Natural carbon materials that are slowly decomposed.

Complex compounds	Characteristics of decomposition
Pectin	Pectin is a fiber consisting of galacturonic acid and neutral sugars such as arabinose, xylose, and galactose. It forms a gel matrix which holds cells together and is in high concentration in citrus fruits. Propectinases release pectin from cell walls and pectinases release soluble sugars from pectin molecules. Many different soil bacteria and fungi produce propectinase and pectinase.
Cellulose	About 90% of cotton and 40–50% of wood consists of cellulose. Cellulose is a rigid molecule consisting of thousands of glucose units held together by a β-1,4 glucosidic bond and it may assume several crystal formations. Several different cellulose enzymes are required for complete conversion of cellulose to glucose.
Hemicellulose	About 23–32% of wood is hemicellulose which binds cellulose microfibrils. Hemicellulose is an amorphous structure consisting of hexose and pentose sugars. Various soil bacteria and fungi produce hemicellulases.
Lignin	Lignin is a polymeric substance consisting of aromatic alcohols (i.e. coniferyl, sinapyl and coumaryl alcohols) and it forms a covering over cellulose and hemicellulose. Lignin accounts for about 15–25% of the woody material and is degraded only under aerobic conditions by peroxidase and laccase.
Chitin	Subunits of N-acetylglucosamine are covalently bonded to produce a linear polymer and parallel chains of this linear polymer are held in place by hydrogen bonding. Chitin is present in exoskeletons of invertebrates, insects, and fungi. Glucosidases digest chitin and these enzymes are produced by bacteria (*Cytophaga, Streptomyces,* and *Vibrio*) and by soil fungi (*Trichoderma, Penicillium,* and *Mortierella*).
Keratin	Keratin consists of amino acids and these insoluble proteins are found in feathers, horns, hooves, wool, and hair. Microorganisms that produce proteases capable of degrading keratin are broadly distributed throughout the environment and keratin degradation occurs slowly in nature.
Fibroin	Silk produced by spiders and mulberry silkworms consists of protein fibers arranged in antiparallel β-sheets and these β-sheets account for resistance to enzyme attack. Several soil bacteria and fungi produce proteases that degrade fibroin.
Collagen	Collagen is a protein found in tendons and skin. It is extremely difficult to hydrolyze because about half of L-proline is hydroxyproline and a sizeable fraction of the lysine residues are hydroxylysine. The hydroxyl amino acids cross link the collagen fibers and collagenase is produced by numerous bacteria and fungi in the environment.

Clostridium and *Ruminococcus*. *Clostridium thermocellum* and *Ruminococcus albus* have been used as model organisms to study cellulose degradation. Cellulose digestion is recognized as important in the nutrition of ruminants, termites, marine borer (*Bankia setacea*) and cellulolytic beetle (*Stromatium barbatum*). Recently microbes have been used to produce biofuels by biomass conversion, which has carried cellulosome research to new directions where bacterial systems are genetically engineered (Ding et al. 2008).

The decomposition of woody plants not only results in the release of neutral sugars into the environment (especially glucose) but also removes the clutter of dead trees from the landscape. The degradation of branches, trunks, and roots of trees is attributed to white rot (e.g. *P. chrysosporium*) or brown rot (example *P. placenta*) fungi that are widely dispersed in the aerobic environment (Figure 8.6d). Lignin decomposition is essential for the mineralization of dead trees because the amorphous lignin structure permeates the hemicellulose and cellulose structures. However, due to the random formation of covalent bonds between the phenolic

Figure 8.6 Some biomaterial is decomposed by microorganisms slowly. When the feathers of a duck (a) and goose (c) are shed, mineralization of feathers occurs slowly; (b) structural material in plant leaves decomposes more slowly compared with cells located between the veins; (d) cellulose in logs from trees is decomposed slowly by white fungi.

Box 8.2 Cellulosomes

Some cellulolytic bacteria have enzymes clustered as a unit on the cell surface and this was initially reported by Raphael Lamed and Ed Bayer at Tel Aviv University, Israel (1983). The binding of cellulose to a site adjacent to the enzymes that degrade cellulose not only enhances the digestion rate but localizes the soluble sugar products adjacent to the cell producing the cellulose enzymes. The cellulosome may be attached to the cell surface through covalent linkage to the S-layer. The cellulosome has been described by some microbiologists as a nanomotor.

alcohols serving as monomers for lignin, a vast array of enzymes would be required for complete lignin digestion. The lignin peroxidases and laccases release free radicals that oxidize the phenolic lignin structure (Janusz et al. 2013). This free radical attack on lignin is relatively nonspecific and contributes to lignin destruction and exposure of cellulose and hemicellulose. Most of the lignin peroxidases contain Mn^{2+} as a cofactor and generate free radicals only in aerobic environments.

There are several notable examples of tree trunks in anaerobic environments persisting in a pristine form for hundreds to thousands of years. Several hundred years ago, it was common to transport cut trees by log drives on the rivers or lakes. These logs became water logged and settled to the bottom of waterways. Additionally, millions of trees were flooded when hydroelectric dams were constructed (Tenenbaum 2004). Due to the cold and anaerobic environment the logs and trees never decayed but can be processed in conventional lumber mills. Logs from Lake Superior and other cold waters have been raised and are being sold by numerous commercial groups. On the beach near Borth, Ceredigion, Mid Wales, Australia, the stumps of trees that were buried under peat for over 5000 years have been revealed following a severe storm (Wilkes 2014). In the peat bogs of the North Island, New Zealand, trees that are over 18 000 years old and not decomposed have been found (Newnham et al. 1988). In the past few years, logs have been retrieved from a peat bog in New Zealand and from a sandpit in South Carolina that have been carbon dated to be 44 000–50 000 years old (Gardner 2004).

Keratins are widely distributed in nature and based on their chemical structure are classified as either α-keratin or β-keratin. Hair, wool, horns, and hooves have long protein molecules with an α-helix and numerous disulfide cross bridges that provide a physically strong structure which requires specific enzymes for digestion. Feathers and scales of birds (Figure 8.6) and reptiles consist of proteins with a β-sheet structure and are designated as β-keratins. This β-sheet structure produces a highly compact protein that is difficult to degrade. The rate of decomposition of feathers is dependent on the pigment molecules added to the keratin protein. The presence of carotenoids produces red to yellow feathers, melanins produce a brown to black feather, porphyrins account for green feathers and blue is dependent on the arrangement of melanin on the keratin microstructure. Parrot feathers are often highly colored with the pigment attributed to psittacofulvins produced only by parrots. These pigmented feathers, which are resistant to bacterial decomposition, may have a role in communication or camouflage (Burtt et al. 2011). Feathers without pigments degrade faster than pigmented ones and melanin-containing feathers degrade before carotenoid-containing feathers. Dermatophyte fungi, *Bacillus licheniformis*, and several soil bacteria produce enzymes to slowly hydrolyze α- and β-keratins. The feather degrading microorganisms attach to bird feathers as the birds dust themselves in soil and, if the atmosphere is humid, feather decomposition occurs even when the bird is in flight. Thus, birds in dry climates are thought to have lightly pigmented feathers whereas birds in very humid regions have darkly colored feathers (Burtt and Ichida 2004).

8.4.4 Production and Utilization of CO

The early atmosphere of the Earth was anaerobic and contained appreciable levels of methane and carbon monoxide (CO). There is an annual release of about 2500 Tg (teragram) of CO into the atmosphere from biological and anthropogenic sources. Aerobic and anaerobic microbes consume about 20% of the Earth's atmospheric CO each year (Jeoung et al. 2014). Carbon monoxide dehydrogenase (COD) is reversible and produces either CO or CO_2:

$$CO + H_2O = CO_2 + 2H^+ + 2e^-$$

Aerobic CO-fixing bacteria from the *Alphaproteobacteria, Firmicutes*, and *Actinobacteria* are known as carboxydotrophs because they tolerate 10% CO. A diverse taxonomic group of aerobes in *Alphaproteobacteria, Betaproteobacteria*, and *Gammaproteobacteria* use CO only at concentrations of about 1000 ppm and are designated carboxydovores. In an examination of the bacteria oxidizing CO in the coastal waters of Maine, USA, 15% of the CO oxidizers were of the *Roseobacter*-associated clade of the alpha subclass of *Proteobacteria* (Tolli et al. 2006). The CO_2 produced is used as a carbon source via the Calvin–Benson–Bassham cycle or the Wood–Ljungdahl pathway. Some bacteria (i.e. *Stappia stellulata* and *Ruegeria pomeroyi*) are only capable of obtaining energy from CO oxidation and are unable to fix CO_2 to carbon material. Anaerobes such as acetogens, hydrogenogens, methanogens, and sulfate reducers use the reductive acetyl-CoA pathway to metabolize CO with the production of cell mass and cellular energy (Mörsdorf et al. 1992). Based on their metal content, there are two types of COD enzymes. The nickel-containing iron-sulfur COD is found in anaerobic microorganisms and is the most ancient while the copper-molybdenum COD is found in aerobes.

8.4.5 Production and Utilization of Hydrogen Cyanide

Hydrogen cyanide (HCN) is a highly toxic compound that inhibits metalloenzymes, including cytochrome oxidase. Although several species of bacteria are capable of producing cyanide, most research has focused on rhizospheric fluorescent pseudomonads that suppress the growth of plant roots (Kremer and Souissi 2001) or function as a biocontrol agents inhibiting the growth of pathogenic fungi on plant roots (Blumer and Haas 2000). Cyanide is a product of secondary metabolism in which bacteria in the stationary phase induce the synthesis of an enzyme that releases CO_2 and HCN from glycine. Rhizospheric cyanide may also result from bacterial and fungal hydrolysis of cyanide-containing glucosides produced by plants.

Bacterial cyanide metabolism is regulated and subject to various controls. Not only are cyanogenic bacteria found in the rhizosphere, but they are also associated with human pathologies. *Pseudomonas aeruginosa* and *Burkholderia cepacia* are important opportunistic pathogens in humans and these bacteria become cyanogenic when cultured in colonies or as biofilms (Ryall et al. 2008). Quorum sensing processes induce virulence in numerous pathogens and signal molecules for quorum sensing appear to influence cyanogenesis in *P. aeruginosa* and *B. cepacia*. Bacterial cyanide production in patients with cystic fibrosis and other pseudomonad infections may contribute to the pathogenicity of these bacteria.

Under appropriate conditions microorganisms will use and degrade cyanide (Knowles 1988). Cultures of bacteria and soil fungi can be induced under nitrogen limited conditions to convert cyanide to CO_2 and NH_3, with the assimilation of ammonia for amino acid synthesis. Some bacteria will add cyanide to cysteine to form β-cyanoalanine. Another detoxification process is demonstrated by fungi, with the hydration of cyanide by cyanide hydratase to produce formamide.

Because various industrial and mining processes generate considerable quantities of cyanide, several biological processes for bioremediation have resulted (Knowles 1988). Filters containing fungi known to produce cyanide hydratase have been marketed to detoxify waste streams containing low levels of cyanide. Another commercial process uses bacteria producing the enzyme cyanidase to convert cyanide to formic acid and ammonia.

8.5 Nitrogen for Biosynthesis and Energy

Microorganisms are dependent on nitrogen for the synthesis of nucleic acids, amino acids, and a vast array of other organic nitrogen compounds associated with cellular biosynthesis. About 15% of bacterial cell mass is nitrogen, which may be acquired by bacteria from ammonia, nitrate, nitrite, and even dinitrogen (N_2) from the atmosphere. The greatest reservoir of nitrogen is the Earth's atmosphere since it contains 78% nitrogen. As indicated in Figure 8.7, nitrogen conversion involves several inorganic forms. For the most part, animals are limited to using organic compounds produced by plants. Nitrogen compounds in plants come from the assimilation of nitrate or, to a lesser extent, nitrite. Estimates of the amount of organic nitrogen in land animals, soil, and oceans are 5×10^{15}, 6.5×10^{15}, and 8×10^{17} g N, respectively (Ward 2012; Schlesinger 1991).

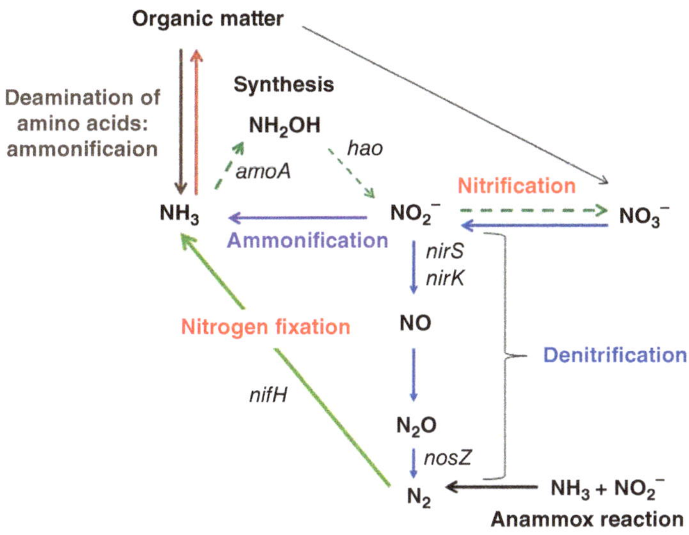

Figure 8.7 Components of the nitrogen cycle. Genes associated with steps in the cycle are indicated in italics. Solid lines, anaerobic activities; broken lines, aerobic reactions. Synthesis and decomposition of organic matter occurs under both aerobic or anaerobic conditions. *Source:* Courtesy of Larry L. Barton.

Table 8.7 Oxidation states of the various molecular forms of nitrogen in the nitrogen cycle.

Oxidation state	Compounds
+5	Nitrate (NO_3^-)
+3	Nitrite (NO_2^-)
+2	Nitric oxide (NO) – a gas
+1	Nitrous oxide (N_2O) – a gas
0	Dinitrogen (N_2) – a gas
−1	Hydroxylamine (NH_2OH)
−3	Ammonia (NH_3) – a gas
	Ammonium (NH_4^-)
	Amino acids, amino sugars (R-NH_2)

A significant feature of the nitrogen cycle is the oxidation state of the nitrogen atom. There is a range of −3 to +5 for the oxidation state of nitrogen in the biosphere (see Table 8.7) and microorganisms may exploit this change in oxidation state to support their growth. The reduced nitrogen atoms have a negative oxidation state and bacterial oxidation of these compounds releases electrons which can be moved through the electron transport system to O_2 with the production of ATP. The oxidized nitrogen atoms have a positive oxidation state and bacteria can divert electrons from cellular metabolism to reduce the nitrogen atom. Bacteria have become specialized to obtain energy for growth coupled to the change in oxidation state of nitrogen. The nitrogen cycle consists of a series of steps and the processes associated with these changes of oxidation state of nitrogen are an important aspect in understanding environmental microbiology.

8.5.1 Nitrification

The oxidation of ammonia to nitrate is termed nitrification and is accomplished by bacteria several different physiological groups of the *Proteobacteria*. Members of the *Thaumarchaeota*, undesignated species of archaea, oxidize ammonia to nitrite (Offre et al. 2013). This conversion of nitrogen is as follows:

$$NH_3 \rightarrow NH_2OH \rightarrow NO_2^- \rightarrow NO_3^-$$

Nitrification occurs in two major steps with the first being the oxidation of ammonia to nitrite and the second is the oxidation of nitrite to nitrate. The reactions are as follows:

Ammonia oxidation: $$NH_3 + O_2 + 2H^+ + 2e^- \rightarrow NH_2OH + H_2O$$

$$NH_2OH + H_2O \rightarrow NO_2^- + 5H^+ + 4e^-$$

Nitrite oxidation: $$NO_2^- + H_2O \rightarrow NO_3^- + 2H^+ + 2e^-$$

Hydroxylamine (NH_2OH) is highly instable and occurs in the environment only transiently. For the oxidation of ammonia or nitrite O_2 is the final electron acceptor and with electrons shuttled through the cytochrome system, ATP produced by the oxidative phosphorylation system provides energy for the bacteria. *Nitrosomonas, Nitrosococcus,* and *Nitrosospira* are key bacterium in the oxidation of ammonia to nitrite and these bacteria are present in freshwater, soil, and marine environments. The cells of *Nitrosomonas* and *Nitrosococcus* have become highly specialized in that they have extensive of internal membranes for the metabolism of ammonia. With *Nitrosomonas*, the oxidation of ammonia to hydroxylamine is attributed to ammonia monooxygenase (AMO) and hydroxylamine is oxidized to nitrite by hydroxylamine oxidoreductase (HAO). The oxidation of nitrite to nitrate is attributed to nitrite oxidoreductase which is present on the cytoplasmic side of the plasma membrane and may account for 25% of total protein in the nitrifying bacteria. The oxidation of nitrite to nitrate is attributed to *Nitrobacter, Nitrospina,* and *Nitrococcus* which are present in marine, soil or fresh water environments. No one bacterial species is capable of oxidizing ammonia to nitrate so nitrification is dependent on cooperative processes of distinct physiological types of bacteria.

An agricultural practice where ammonia is injected into the soil in the fall is an excellent example of nitrification. The two-step oxidation of ammonia to nitrate occurs before the planting season and nitrate is released to the growing plants. While plants assimilate nitrate rather than use ammonium, nitrate is slowly released in the soil in the nitrification process. If nitrate fertilizers are used to enhance plant growth, nitrate may not be retained in the root zone but is readily washed away by rain or irrigation.

8.5.2 Denitrification

Bacteria growing anaerobically may employ nitrate or nitrite as electron acceptors. The conversion of nitrate to N_2 by respiring bacteria is known as denitrification with nitrogen reduction proceeding by a step-wise process as indicated below:

$$NO_3^- \rightarrow NO_2^- \rightarrow NO \rightarrow N_2O \rightarrow N_2$$

The function of nitrate is to serve as an electron sink and reoxidize the electron carriers in the bacterial cell. Many of the soil or marine bacteria have all the enzymes for complete conversion of nitrate to dinitrogen. In hot water springs, saline waters and other extreme

Table 8.8 Key enzymes and genes in the nitrogen cycle.[a]

Process	Enzyme[b]	Genes
Dinitrogen fixation	MoFe-nitrogenase reductase	*nifH*
Mineralization (release NH_4^+)	Chitinolysis	*chiA*
	Alkaline proteases	*aprA*
Nitrification	Ammonia monooxygenase	*amoA*[c]
	Hydroxylamine Oxidoreductase	*hao*
	Nitrite oxidoreductase	*norA*
Heterotrophic nitrification ($R\text{-}NH_2 \rightarrow NO_3^-$)	Nitrate reductase (?)	*narH/J* (?)
Dissimilatory		
Nitrate respiration	Nitrate reductase (m)	*narG*
Nitrate reduction	Nitrate reductase (p)	*napA*
Nitrite denitrification	Nitrite reductase (p)	*nirS/nirK*
Nitrite detoxification	Nitrite reductase (c)	*nirBD*
	Nitric oxide reductase	*norB*
	Nitrous oxide reductase	*nosZ*
Ammonification ($NO_3^- \rightarrow NO_2^- \rightarrow NH_3$)	Nitrate reductase (m)	*nap*
	Nitrite reductase (p)	*nrfA*
Assimilation		
Assimilatory reduction	Bacterial nitrate reductase[d] (c)	*nasCA*
	Cyanobacterial nitrate reductase (c)	*narB*
Assimilatory reductase	Bacterial nitrite reductase[d] (c)	*nasB*
	Cyanobacterial nitrite reductase (c)	*nirA*

[a] *Source:* Brankatschk et al. (2011); Zehr and Kudela, (2011); Moreno-Vivián et al. (1999).
[b] Located in the periplasm (p), cytoplasm (c) or plasma membrane (m).
[c] Distinct primers used for ammonia monooxygenase of bacteria (AOB) and archaea (AOA).
[d] Refers to heterotrophic bacteria.

environments, members of the archaea participate in denitrification as well as in other aspects of the nitrogen cycle (Cabello et al. 2004). The enzymes and genes associated with these reductions are listed in Table 8.8. There are two types of denitrifying prokaryotes. Those with a complete denitrification system couple nitrate reduction to anaerobic respiration and use generation of N_2 to contribute to energetic gains of the cell. In incomplete denitrification, not all genes are present for reduction of nitrate to N_2 and this absence of certain genes reflects the adaptation of microbes to specific environments (Moreno-Vivián et al. 1999).

8.5.3 Nitrate Reduction

Nitrate reductase is broadly distributed in bacteria where it may function for assimilation of nitrogen into compounds for growth (see Section 5.7), anaerobic respiration with energy production coupled to electrons moving to nitrate and the dissimilatory use of nitrate as an electron sink to maintain an appropriate redox level in cells (Moreno-Vivián et al. 1999). The nitrate respiratory process in Gram-negative bacteria is induced under anaerobiosis and uses a membrane-bound enzyme to reduce nitrate with nitrite released into the cytoplasm. The reduction of nitrate is achieved with electrons from quinol lipids in the plasma and this is coupled to proton export which is used for energizing the membrane. Nitrite is exported into the periplasm where enzymes reduce nitrite to N_2.

Many Gram-negative bacteria have a dissimilatory nitrate reductase which is located in the periplasm and is not associated with nitrate assimilation or nitrate respiration processes (Moreno-Vivián et al. 1999). While dissimilatory nitrate reduction is characteristically an anaerobic event, several denitrifying bacteria display nitrite reduction under aerobic conditions. The aerobic denitrification process is useful to bacteria growing in microaerobic conditions or at the aerobic-anaerobic interface. The dissimilatory reduction of nitrate functions to use excess reducing equivalents and balance redox activity in the cell.

8.5.4 Nitrite Reductase

Dissimilatory nitrite reductases have been isolated from numerous bacteria and based on the prosthetic group required for the reaction, contain either heme or copper (Barton 2005). Numerous studies have confirmed that nitrate from foods are converted first to nitrite and then to NO by facultative anaerobic bacteria on the dorsal surface of the tongue (Rőszer 2012). When the NO reaches the stomach, it enhances the antibacterial activity of gastric acid and is more effective than when only stomach acid is present. In the oral mucosa, NO accumulation is beneficial in preventing periodontitis attributed to *Porphyromonas gingivalis*.

There a few studies addressing the production of NO by bacteria reducing nitrite. Bacteria in the intestine of rats and humans are capable of generating NO from nitrate or nitrite (Sobko et al. 2005) while lactobacilli and bifidobacteria have greatest NO production with nitrite. The impact of NO generated in the gut is a topic of considerable interest because the formation of *N*-nitroso compounds by nitrate-reducing bacteria may induce human gastric cancer (Moreno-Vivián et al. 1999). In swampy soil, methanotrophic growing at the aerobic-anaerobic interface display dissimilatory reduction of nitrate with release of NO (Ren et al. 2000). In agricultural environments that are transiently flooded, dissimilatory nitrate reducing bacteria produce N_2 and also release NO. Bacteria in terrestrial and aquatic environments release about 40% of the NO global budget (Williams et al. 1992).

8.5.5 Metabolism of NO and N_2O

NO and N_2O are intermediates in denitrification and these compounds may be released because the binding to reductases cannot keep up with the rate at which the nitrogen oxides are being produced. With environmental isolates including cultures of *Pseudomonas denitrificans* and *Rhodobacter sphaeroides*, the amount of NO produced has been reported to range from nanometer to micrometer NO per hour (Choi et al. 2006). NO is known to strongly inactivate metalloproteins and to be toxic to bacteria in the environment where denitrification is occurring (Mancardi et al. 2004). *Bacillus subtilis*, a bacterium that does not produce NO, is able to grow in the presence of denitrifying bacteria because NO induces *B. subtilis* to produce a flavohemoglobin which oxidizes NO to NO_3^- and protects against NO toxicity (Moore et al. 2004). In microenvironments with denitrifying bacteria, NO may influence the microbial community by inhibiting the growth of bacteria lacking genes for inactivation of NO.

According to the Inventory of US Greenhouse Gas Emissions and Sinks: 1990–2012, nitrous oxide is the fourth largest contributor to greenhouse gases while water vapor, carbon dioxide and methane are of greater concern. Although only 300 ppb of the Earth's atmosphere is nitrous oxide, it is important because at a molecular equivalence, the impact of nitrous oxide on global warming is several hundred greater than for carbon dioxide. About 75% of nitrous oxide emissions are attributed to microbial activities in the soil and about 5% from management of solid animal wastes. Thus, bacterial production and utilization or nitrous oxide is highly important. Nitrous oxide is an obligatory intermediate in the denitrification process of respiratory-coupled reduction of nitrite to dinitrogen (N_2). It is assumed that incomplete denitrification as occurs in methanotrophic bacteria (Ren et al. 2000) may account for the release of nitrous oxide into the atmosphere (Ward 2012).

While many bacteria transform nitrous oxide to dinitrogen, the multicopper-containing nitrous oxide reductase from *Pseudomonas* and *Paracoccus* has received the greatest amount of attention (Zumft and Kroneck 2007). *Wolinella succinogenes* contains a nitrous oxide reductase distinct from the one in *Pseudomonas* and *Paracoccus* and an additional type of nitrous oxide reductase may be present in certain archaea. It is proposed by Zumft and Kroneck (2007) that nitrous oxide reductase may have evolved prior to the separation of bacteria and archaea.

8.5.6 Production of NO by NOS

The principle nitrogen oxides in the nitrogen cycle involving microorganisms are NO and nitrous oxide (N_2O). While NO is an intermediate in the denitrification process, another system contributing to the production of NO is associated with cellular metabolism. The production of NO from L-arginine by nitric oxide synthase (NOS) is summarized in the following reaction:

$$\text{L-Arginine} + 2O_2 + 1.5\,\text{NADPH} + 1.5H^+ \rightarrow \text{L-Citrulline} + \text{NO} + 1.5\,\text{NADP}^+$$

As reviewed by Rőszer (2012), the first demonstration of NOS was in a *Nocardia* strain but NOS activity has been reported for strains of *Lactobacillus, Rhodococcus, Salmonella, Bacillus, Deinococcus, Staphylococcus,* and *Streptomyces*. In most cases, the bacterial NOS enzymes are structurally similar to NOS in mammalian cells. This production of NO by the NOS occurs in the cytoplasm of the bacterium and effects vary with the bacterium. In several bacteria, NO production is important for defense against oxidative stress by inhibiting the generation of hydroxyl free radicals and increased production of catalase. In *B. subtilis,* NO production is responsible for resistance to quinolones and β-lactam-containing antibiotics. Cytoplasmic NO is required to produce thaxtomin-A, which is a plant toxin produced by *Streptomyces* sp. Following the exposure of *Deinococcus* to high levels of ionizing radiation, NO enhances recovery of *Deinococcus* to radiation damage. NO released from rhizospheric bacteria may contribute to rhizobia-plant symbiosis and inhibition of bacteria in biofilms. The enzymology for NOS is not present in all prokaryotes and this generation of NO from L-arginine may have evolved after the dissimilatory reduction of nitrite.

8.5.7 Respiratory Ammonification

Some anaerobic bacteria (i.e. *Escherichia coli* K-12, *Shewanella oneidensis* MR-1, *Desulfovibrio desulfuricans* DSM 27774, and *W. succinogenes*) display the reduction of nitrate with the production of ammonium. Respiratory nitrate ammonification is the dissimilatory reduction of nitrate with ammonium as the end product and the reactions are as given below:

$$NO_3^- \rightarrow NO_2^- \rightarrow NH_4^+$$

Nitrate reduction by nitrate reductase produces nitrite which is subsequently reduced by nitrite reductase to ammonium. The six-electron reduction step is catalyzed by a membrane-bound nitrite reductase which charges the plasma membrane and therefore, is an energy yielding process (Simon 2002). This process is distinct form respiratory nitrate reduction where

N_2 is the end product with nitrogen oxides as intermediates. Desulfovibrio gigas and a few other bacteria cannot reduce nitrate but can reduce nitrite to ammonia.

8.5.8 Anammox Reaction

A few anaerobic bacteria are capable of oxidizing ammonia with the reduction of nitrite and oxidation of ammonia according to the following reactions:

$$NO_2^- \rightarrow NO \text{ (nitrite oxidoreductase)}$$

$$NO + NH_4^+ \rightarrow HNNH \text{ (hydrazine hydrolase)}$$

$$N_2H_4 \rightarrow N_2 \text{ (hydrazine dehydrogenase)}$$

This process has been called the anammox reaction and it occurs inside a special cytoplasmic body designated as the anammoxosome (ANS) and it contains a typical membrane of lipid and protein. Unique to the ANS membrane are lipids which contain C20 ladderane alcohols (see Figure 1.2) attached to a glycerol backbone along with fatty acids (Karta et al. 2013). The enzymes for the conversion are either inside the ANS or in the membrane of this specialized unit. At this time, there are 10 anammox bacterial species in the following genera: *Kuenenia, Jettenia, Scalindua, Anammoglobus,* and *Brocadia*. These bacteria have been found in marine sediments, freshwater ecosystems and waste water treatment systems. It is believed that the anammox reaction may account for >40% of N_2 released from oceans.

8.5.9 Assimilation of Nitrogen

In the assimilatory nitrate reduction system, nitrate uptake and reduction is closely regulated by microbial metabolism. For assimilation of nitrate into amino acids, nitrate is initially transported across the plasma membrane. Nitrate is initially transported across the plasma membrane and this is followed by the sequential action of nitrate reductase and nitrite reductase. There are two different classes of assimilatory nitrate reductases found in bacteria and they have different electron donors. The ferredoxin or flavodoxin class is found primarily in cyanobacteria (Table 8.8) and is designated as the "nar system" while the NAD(P)H class is found in bacteria other than cyanobacteria (Moreno-Vivián et al. 1999). If the assimilatory reduction of nitrate to nitrite occurs in an aerobic environment, one of the subunits of the enzyme generates superoxide which is quickly detoxified by SOD.

For the synthesis of amino acids, ammonium ion from assimilatory nitrate reduction or the environment is incorporated into carbon units to produce amino acids. If the concentration of ammonium is greater than 1 mM, glutamate dehydrogenase catalyzes the addition of ammonium onto α-ketoglutarate to produce L-glutamate. If the concentration of ammonium is less than 1 mM, the sequential action of glutamine synthetase and glutamate synthetase is required to produces L-glutamate. Alternate pathways in bacteria for assimilation of ammonium involve the production of L-asparagine or L-alanine.

8.5.10 Dinitrogen Fixation

A select group of prokaryotes have the capability of converting N_2 to NH_3. The triple bond between the two nitrogen atoms is strong and a considerable amount of energy and reducing power is required for fixation of nitrogen (reaction follows).

$$N_2 + 8e^- + 8H^+ 16ATP \rightarrow 2NH_3 + H_2 + 16ADP + 16P_i$$

The enzyme responsible for this fixation of dinitrogen is commonly referred to as nitrogenase and it consists of two subunits. Component I has nitrogenase activity because it separates the dinitrogen molecule and Component II is the dinitrogen reductase which donates electrons to the nitrogenase or Component I. The *nifH* gene encodes for dinitrogen reductase and presence of this gene in an environmental isolate is used to indicate a nitrogen-fixing microorganism. There several physiological groups of organisms with nitrogen-fixing capability and some of these include cyanobacteria (i.e. *Nostoc, Anabaena, Calothrix*), heterotrophic soil bacteria (i.e. *Azospirillum, Azotobacter, Beijerinckia, Derxia*), anaerobic bacteria (i.e. *Clostridium, Desulfovibrio*), plant symbionts (i.e. *Rhizobium, Bradyrhizobium, Frankia*), plant endophytes (i.e. *Azoarcus, Acetobacter, Herbaspirillum*),photosynthetic bacteria (i.e. *Rhodospirillum, Chromatium, Rhodopseudomonas*), and archaea (i.e. *Methanococcus, Methanosarcina*).

Azotobacter is most unique in that it has several different types based on metal ions present in the protein complex. One type contains iron and molybdenum, a second has iron plus vanadium and a third has only iron. This would enable *Azotobacter* to fix dinitrogen under various metal-limiting conditions. When analyzing the nitrogenase protein complex of the various microorganisms, it is apparent that it is highly conserved and that it had an ancient origin.

8.6 Sulfur Cycling

The sulfur cycle is dependent of aerobic and anaerobic microorganisms for sulfur transformations and a close interaction occurs between the microorganisms of the sulfur cycle and the various components in the nitrogen and carbon cycle. Unlike nitrogen, sulfur is readily available in the aquatic and terrestrial environments and the major reservoir of sulfur is the lithosphere. Sulfur has many inorganic compounds with different oxidation states (Table 8.9) and microorganisms have developed metabolic systems to derive energy by coupling electron flow to or from these inorganic compounds. Important enzymes and genes associated with the sulfur cycle are listed in Table 8.10. Some of the unique metabolic activities of microorganisms are associated with the sulfur cycle (Figure 8.8).

Generally, the sulfur cycle is discussed with respect to global accounting for sulfur compounds but Lorens Bass-Becking in 1925 proposed study of a "sulfuretum," which is the ecology of a semi-tropical environment rich in sulfur. The ideal sulfuretum would have an abundance of organic matter, an anaerobic zone that would support sulfate-reducing bacteria, and bacteria capable of oxidizing hydrogen sulfide. Postgate (1979) provides a review of several sulfuretums and provides insight into cooperative interactions between various sulfur metabolizing bacteria.

Table 8.9 Oxidation state of S in the sulfur cycle.

Oxidation state	Compounds
+6	Sulfate (SO_4^{2-})
+4	Sulfite (SO_3^{2-})
0	Elemental sulfur (S^0)
−2	H_2S (S^{2-})
−2	R-SH
−2	Dimethyl sulfide $(CH_4)_2S$

Table 8.10 Key enzymes and genes in the sulfur cycle.

Process	Enzyme	Genes
Dissimilatory sulfate reduction	APS reductase[a]	*aprAB*
Dissimilatory sulfite reduction	Sulfite reductase	*dsrAB*
Dissimilatory thiosulfate reduction	Thiosulfate reductase	*phsAB*
Anaerobic sulfite reduction	Sulfite reductase (*Salmonella*)	*asrAB*
	Sulfite reductase (*Wolinella*)	*mccA*
Reduction of elemental sulfur	Sulfur reductase (*Acidianus*)	*sreAB*
Oxidation of thiosulfate to sulfate	Thiosulfate oxidation	*sox*
Sulfide oxidation to S^0	Sulfide-quinone oxidoreductase	*sqr*

[a] Also called adenylylsulfate pyrophorylase or sulfate adenyltransferase.
Source: Reproduced with permission from Barton et al. (2014).

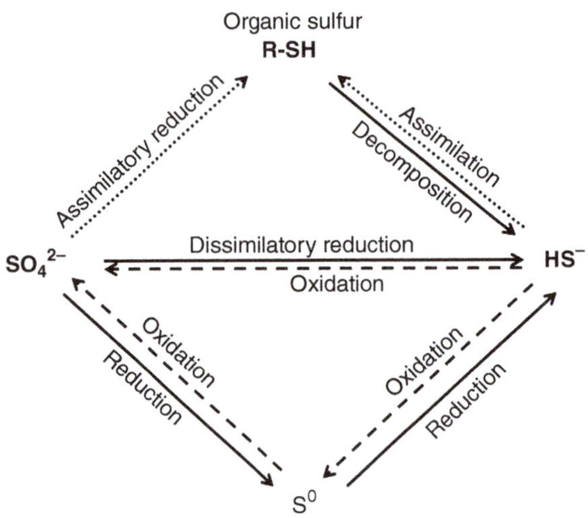

Figure 8.8 Generalized activity in the sulfur cycle involving microorganisms.

8.6.1 Oxidation of Hydrogen Sulfide

The transformations of sulfur occur because bacteria couple oxidation-reduction reactions to obtain favorable energetics. In the case of sulfide oxidation, aerobic bacteria can couple sulfide oxidation to reduction of molecular oxygen as shown below:

$$2H_2S + O_2 \rightarrow 2S^0 + 2H_2O$$

Beggiatoa and *Thioploca* are filamentous sulfide oxidizing bacteria that are broadly dispersed in marine, fresh water or brackish environments (Teske and Nelson 2006). An interesting sulfur transformation process occurs between sulfate-reducing bacteria (i.e. *Desulfovibrio*) and *Beggiatoa*. *Desulfovibrio* in the soil produce hydrogen sulfide by reducing sulfate and *Beggiatoa* will oxidize sulfide to produce intracellular sulfur inclusions. This process detoxifies the soil around rice and other plants. The very large marine cells of *Beggiatoa Thioploca* and

Thiomargarita have central "vacuoles" of nitrate and oxidize intracellular sulfur inclusions to sulfate coupled with nitrate reduction to ammonium. Marine *Beggiatoa* display dissimilatory nitrate reduction with ammonium production according to the following reaction:

$$HS^- + NO_3^- + H^+ + H_2O \rightarrow SO_4^{2-} + NH_4^+$$

The green sulfur photosynthetic bacteria (*Chlorobium*) uses sulfide as the electron donor and elemental sulfur is produced outside the cells. Oscillatoria limnetica, a cyanobacterium, produces extracellular S^0 as it photosynthetically uses hydrogen sulfide as the electron donor for photo system I. These sulfide consuming reactions involve the activities of the carbon and sulfur cycles and are shown below:

$$2H_2S + CO_2 \rightarrow (CH_2O) + 2S^0 + H_2O$$

8.6.2 Oxidation of Elemental Sulfur

Oxidation of elemental sulfur (S^0) is achieved by organisms in both the *Archaea* and *Bacteria* domains (Barton et al. 2014). Sulfur oxidation by organisms of the archaea occurs in *Sulfolobus* but the organism receiving the greatest attention has been the thermoacidophilic aerobic *Acidianus ambivalens*. The enzymes for S^0 oxidation by *A. ambivalens* are in the cytoplasm and would be characterized by the following reaction:

$$2S^0 + 3O_2 + 2H_2O \rightarrow 2SO_4^{2-} + 4H^+$$

Anaerobic bacteria oxidizing elemental sulfur include the green photosynthetic bacteria (*Chlorobaculum tepidum*) and the purple photosynthetic bacteria (*Allochromatium, Rhodobacter, Rhodopseudomonas, Rhodovulum*). There are over 20 genera of purple non-sulfur phototrophic bacteria and numerous species of colorless (lacking photosynthetic pigments) sulfur bacteria that use S^0 as the electron source. Sulfur (S^0) oxidization is found with many bacteria in the *Proteobacterium* phylum and a few examples are listed here: *Alphaproteobacteria* (*Thioclava* and *Starkeya*), *Betaproteobacteria* (*Thiobacillus* and *Sulfuricella*), *Epsilonproteobacteria* (*Sulfurimonas* and *Thiovulvum*) and *Gammaproteobacteria* (*Thiothrix* and *Thiomicrospira*).

8.6.3 Dissimilative S^0 Reduction

When colloidal S^0 is placed in water in the presence of sulfide, polysulfide is formed and this is the substrate for enzymatic reduction resulting in the formation of hydrogen sulfide. Sulfur reductases or polysulfide reductases have been reported for several bacteria including *Desulfovibrio* spp., *Desulfotomaculum* spp., *W. succinogenes, Desulfuromonas acetoxidans*, and *Sulfurospirillum deleyianum* (Barton et al. 2014). Electron transport systems containing sulfur reductases have been reported for several archaea and the best studied are *A. ambivalens, Pyrodictium abyssi*, and *Pyrococcus furiosus*. Electron donors for the reduction of elemental sulfur are usually H_2 or formate and the oxidation of formate with the reduction of elemental sulfur is indicated below:

$$\text{Formate}^- + S^0 \rightarrow CO_2 + HS^-$$

8.6.4 Dissimilative Sulfate Reduction

Dissimilatory sulfate reduction occurs in the anaerobic environments where sulfate is the terminal electron acceptor for a diverse number of prokaryotes and high levels of hydrogen sulfide are produced. There are three genera of archaea (*Archaeoglobus, Caldvirga,* and *Vulcanisaeta*) and over 250 species in 60 genera of bacteria that grow with sulfate as the terminal electron acceptor (Barton et al. 2014). The genomes of over 100 SRP have been sequenced including 36 species of *Desulfovibrio*, the most common genus. These sulfate-reducing prokaryotes (SRPs) are found in numerous environments and many have the capability of bioremediation by oxidation of toxic compounds (Barton and Fauque 2009; Bruschi and Goulhen 2007). Due to the precipitation and immobilization of numerous metals as sulfide precipitates, SRP have the potential for remediation of environments with toxic metals. There is considerable industrial concern that SRP may account for the souring of oil fields by using contaminated drilling machinery or water injected into the oil well. Biocorrosion of ferrous materials is attributed to SRP and this is a problem of global economic importance.

The dissimilatory reduction pathway for conversion of sulfate to hydrogen sulfide is unique. The first reaction is the activation of sulfate by ATP sulfurylase to produce adenosine-5′ – phosphate (APS) and this is equivalent to consumption of two moles of ATP consumed for each mole of sulfate. The conversion of APS to bisulfite (HSO_3^-) is by adenylylsulfate reductase (APS reductase) and finally, there is the reduction of bisulfite to sulfide by dissimilatory sulfite reductase. These reactions are as follows:

$$SO_4^{2-} + 2H^+ + ATP \rightarrow APS + PP_i$$

$$APS + 2H^+ + 2e^- \rightarrow AMP + HSO_3^- + H^+$$

$$HSO_3^- + 6H^+ + 6e^- \rightarrow HS^- + 3H_2O$$

Several bacteria that are not dissimilatory sulfate reducers will reduce oxy-sulfur compounds to hydrogen sulfide. These oxy-sulfur compounds include tetrathionate which is present in soils, thiosulfate found in marine and freshwater sediments and sulfite is an intermediate in dissimilatory sulfite reduction. Respiratory-linked tetrathionate reduction is characteristic of *Salmonella, Citrobacter, Proteus, Pseudomonas,* and several soil bacteria. Environmental bacteria including *Shewanella* and *Salmonella* with other enteric pathogens will reduce thiosulfate by a specific reductase. Dissimilatory sulfite reductase is found in *Methanocaldococcus, Methanopyrus, Methanococcoides, Bilophila, Clostridium, Salmonella,* and *Wolinella.*

8.6.5 Assimilatory Sulfate Reduction

Many organisms use sulfate for the synthesis of cysteine, methionine, and other sulfur-containing organic compounds. Sulfate is activated by ATP by ATP sulfurylase and APS is produced as is the initial reaction in dissimilatory sulfate reduction.

$$SO_4^{2-} \rightarrow APS \rightarrow PAPS \rightarrow SO_3^{2-} \rightarrow HS^- \rightarrow \text{cysteine}$$

8.6.6 Production of H_2S and Dimethyl Sulfide

8.6.6.1 Hydrogen Sulfide

The primary release of hydrogen sulfide (H_2S) into the atmosphere is attributed to the petroleum industry with only about 10^6 tons per year coming from the biogenic activity of saline marshes. Hydrogen sulfide produces an unpleasant odor characteristic of rotten eggs.

Chronic respiratory irritation may occur in humans in long term exposure of 1000–2000 ppm H_2S and death results at concentrations of >2000 ppm There are two distinct physiological groups of microorganisms producing hydrogen sulfide: the dissimilatory sulfate-reducing group (discussed in Section 6.4) and the anaerobic fermenters of sulfur-containing amino acids. For the most part, the production of hydrogen sulfide is restricted to a specific environment where the appropriate nutrients are abundant.

The decomposition of proteins results in production of gases because many bacteria produce the enzyme cysteine desulfhydrase which degrades cysteine to pyruvate, NH_3 and H_2S. Degradation of meat and protein materials with the production of hydrogen sulfide is attributed to many different types of bacteria including *Pseudomonas, Aeromonas, Salmonella, Citrobacter, Proteus,* and *E. coli*. Manure pits or compost piles produce hydrogen sulfide along with methane, carbon dioxide, and ammonia.

There are three additional enzymatic processes that may produce hydrogen sulfide and although these quantities of hydrogen sulfide produced are low, these processes represent the diversity of microbial interactions with the sulfur cycle. Microorganisms will produce sulfite and thiosulfate from partial reduction of sulfate. Many heterotrophic bacteria have sulfite reductase and thiosulfate reductase which reduces sulfite and thiosulfate, respectively, to hydrogen sulfide (Barton et al. 2014). An archaeal isolate has been found to produce HSS^- (hydrogen persulfide) from dissimilatory sulfate reduction and the HSS^- molecule is dismutated to yield hydrogen sulfide and sulfate (Offre et al. 2013). *Desulfovibrio sulfodismutans,* a dissimilatory sulfate reducer, produce hydrogen sulfide from sulfite or thiosulfate by dismutation (Barton et al. 2014). This dismutation process is also called inorganic sulfur fermentation and the reactions are given below:

$$S_2O_3^{2-} + H_2O \rightarrow SO_4^{2-} + HS^- + H^+$$

$$4SO_3^{2-} + H^+ \rightarrow 3SO_4^{2-} + HS^-$$

Several species of the bacterium *Desulfocapsa* is capable of dismutation of elemental sulfur to produce hydrogen sulfide, see below:

$$4S^0 + 4H_2O \rightarrow SO_4^{2-} + 3HS^- + 5H^+$$

Energy associated with dismutation reactions of sulfur, sulfite and thiosulfate enable the bacteria to grow.

8.6.6.2 Dimethyl Sulfide

Various species of bacteria release volatile sulfur compounds (VSCs) from metabolism of cysteine and methionine. Oral malodor (bad breath) is commonly associated with systemic diseases as well as with bacteria growing on the tongue and releasing hydrogen sulfide, methyl mercaptan, and dimethyl sulfide (DMS). The tongue biofilm is supported by nutrients from saliva or food particles and contains several different species of bacteria. Hydrogen sulfide production from *Veillonella, Actinomyces,* and *Prevotella* species are responsible for the oral malodor (Washio et al. 2005). A food chain in the tongue biofilm occurs with *Actinomyces* producing lactic acid from carbohydrates and *Veillonella* utilizing the lactic acid. The tongue biofilm reflects a stable microbial ecosystem.

DMS is the most abundant sulfur compound in the atmosphere released by biological activity. The oxidation of DMS in the atmosphere promotes global climate change by stimulating cloud formation which reduces solar radiation and accounts for the decline in temperature of the Earth. DMS is released by bacterial metabolism of dimethylsulfoniopropionate (DMSP) which is found in marine algae as a stress protectant. DMS is also released by bacteria growing in association with roots of trees in mangroves. The general view of DMS release from DMSP is given below where DMSP is cleaved by a DMSP lyase to give DMS, acrylate and a proton:

$$(CH_3)_2SCH_2CH_2COO^+ \rightarrow (CH_3)_2S + CH_2 = CH_2COO^- + H^+$$

Bacteria that release DMS or metabolize acrylate include *Pseudomonas doudoroff, Desulfovibrio acrylicus, Marinomonas* sp. *Pelagibacter ubique, B. cepacia cepadia* and *Sagittula stellata* (Johnston et al. 2008). Also, DMS is produced by numerous archeal strains of the *Euryarchaeota* including *Halobacteria* (Offre et al. 2013). An interesting ramification of microbial release of DMS in the ocean is that DMS attracts zooplanktonic crustaceans and sea birds. With this influx of animals, various wastes are released which serve to feed bacteria at that site including the DMS producers. This scheme used to describe the attraction of others to contribute to the nutrition of those producing the attractant has been called "the messy eater model" (Johnston et al. 2008).

8.7 Cycling of Trace Elements

8.7.1 Iron

Iron, the fourth most abundant in the Earth's crust, is an essential element for living organisms including bacteria. A cycle indicating major components of bioavailable iron is given in Figure 8.9. In the aerobic environment at neutral pH, Fe(III) is commonly present as insoluble Fe_2O_3, $Fe(OH)_3$ and, $FePO_4$. Biogenic iron-rich minerals may result from the use of Fe(II) as the electron acceptor by nitrate-reducing bacteria or by anaerobic photoautotrophic bacteria (Fortin and Langley 2005). Nanocrystals of biogenic iron oxide may result from iron-oxidizing bacteria coupling the electron flow from Fe(II) to Fe(III) to O_2 or from passive sorption of iron onto extracellular polymeric matrix or the cell surface with the formation of iron oxides by favorable environmental conditions.

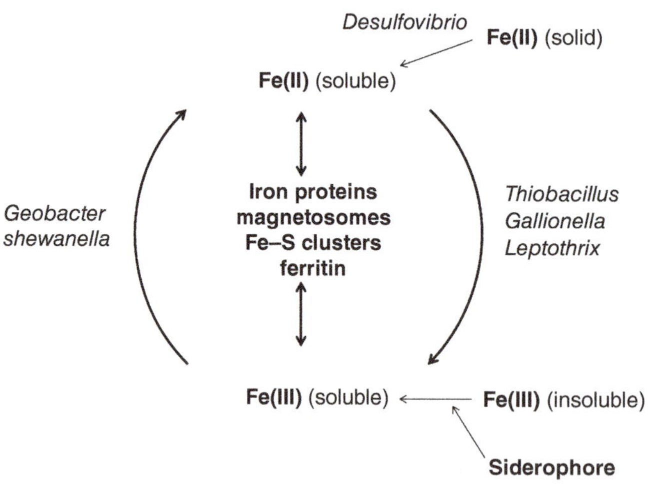

Figure 8.9 Reactions illustrating the oxidation and reduction of iron in the environment.

The banded-iron formations (BIF) are considered to be of biogenic origin and contain about 20 to 40% iron. About 2.5 billion years ago, hydrothermal vents released Fe(II) into the environment and O_2 from cyanobacterial photosynthesis oxidized Fe(II) to Fe(III). See discussion in Section 8.3.1 of this chapter. Primary minerals produced were Fe(III)-containing hydroxides, silicates and carbonates and with iron oxide minerals considered to be of secondary activity (Koehler et al. 2010).

The oxidation-reduction potential of many iron reactions (Table 8.11) are within the ranges of biological range of redox potentials. Microorganisms associated with the iron cycle by having either oxidation or reduction reactions with iron are numerous and have been found in 14 different phyla of Bacteria and two phyla of Archaea (Weber et al. 2006). As recently reviewed (Iibert and Bonnefoy 2013) microorganisms which oxidize iron include the following physiological groups: aerobes including acidophiles and neutrophils; anaerobes including phototrophs and nitrate reducers. Respiratory transfer of electrons from Fe(II) to O_2 at low pH has been examined in bacteria (i.e. *Acidithiobacillus ferrooxidans, Acidithiobacillus ferrivorans, "Thiobacillus prosperus," Leptospirillum* spp. and *Sulfobacillus sibiricus*) and in archaea (*Metallosphaera* spp., *Sulfolobus* sp. and "*Ferroplasma acidarmanus*"). Anaerobic iron oxidation promoting growth of phototrophic bacteria under neutral pH levels has been demonstrated in *Rhodobacter ferrooxidans* and *Rhodopseudomonas palustris* (Iibert and Bonnefoy 2013). Dissimilatory Fe(II) oxidation coupled to nitrate reduction has been shown in a few bacteria and most require an additional electron source or organic carbon for growth. Due to the difference of metabolic components associated with Fe(II) oxidation, Iibert and Bonnefoy (2013) propose multiple origins for iron oxidation with phototrophic anaerobic system developed first, neutrophilic iron oxidation occurred second and acidophilic iron oxidation is the most recent.

The reduction of Fe(III) to Fe(II) is characteristic of several microorganisms and a few species display dissimilatory iron reduction with growth coupled to iron reduction. As reviewed by Lentini et al. (2012), reduction of Fe(III) is accomplished by numerous bacteria; however, relatively few pure cultures of bacteria couple Fe(III) reduction to cell growth. There is a difference in iron mineralogy with Fe(III)-(oxyhydr)oxides existing as over 15 different minerals

Table 8.11 Redox reactions involving iron.

Reaction	Volts[a]
$Fe_{(s)}$/Fe(II)	−0.470[b]
Fe_3O_4 (magnetite)/Fe(II)	−0.314[c]
α-Fe_2O_3(hematite)/Fe(II)	−0.286[d]
α-FeOOH (goethite)/Fe(II)	−0.274[c]
γ-FeOOH (lepidocrocite)/Fe(II)	−0.088[c]
FeOOH/FeCO(siderite)	−0.050[d]
$Fe(OH)_3$/Fe(II)	+0.014[c]
Fe(III)-citrate/Fe(II)-citrate	+0.385[d]
Fe(III)/Fe(II) (pH 2)	+0.770[d]

[a] Data is at pH 7.0 unless indicated.
[b] Data from Venzlaff et al. (2013).
[c] Weber et al. (2006).
[d] Iibert and Bonnefoy (2013).

and each iron oxide mineral has a different oxidation-reduction potential which is dependent on particle size, pH and degree of crystallinity (Levar et al. 2017). Ferrihydrite ($Fe_5HO_8 \cdot 4H_2O$) is one of the most bioactive iron oxide minerals but it has a relatively short residence time in sediments and soil. When bacterial sediment communities were enriched with carbon sources and ferrihydrite, cultures receiving acetate the dominant bacterial genus was *Geobacter* while cultures receiving lactate were enriched for *Shewanella* sp., *Geobacter* sp., and *Geothrix* sp. (Lentini et al. 2012). With goethite as the iron oxide and glucose as the carbon source, the enriched population contained primarily *Spirochaetes, Firmicutes,* and *Enterobacteriaceae.* With goethite and lactate as the carbon source, *Desulfovibrio, Actinobacteria, Aeromonas,* and *Spirochaetes* were dominant in the enriched population (Lentini et al. 2012). With hematite as the iron oxide, glucose enrichments contained high percentages of bacteria within the *Enterobacteriaceae* while lactate enrichments contained high numbers of bacteria associated with *Desulfovibrio* sp., *Actinobacteria* sp., and *Enterobacteriaceae* (Lentini et al. 2012).

Dissimilatory iron reduction is associated with *Clostridium beijerinckii, Geothrix fermentans, Aeromonas hydrophila, S. deleyianum, Geobacter metallireducens, Geobacter sulfurreducens, and Shewanella alga* (Lentini et al. 2012 and references there in). Several bacteria have respiratory coupled oxidation/reduction reactions with bacteria but these organisms' growth is not coupled to the electron flow. Biogenic reduction of iron is also observed by sulfate-reducing bacteria producing hydrogen sulfide which chemically reacts with iron to produce FeS.

Iron limitation in microbial metabolism is most often a lack of available soluble iron and not the presence of iron in the environment. To facilitate Fe(III) uptake into bacterial and fungal cells, organic chelators with a high affinity for Fe(III) are produced under iron limiting conditions. Siderophores are between 500 and 1500 Da and each species of bacteria or fungi produces a specific iron chelator (Hider and Kong 2010).

8.7.2 Manganese

Manganese is required for several biological activities such as the cleavage of water to release electrons in Photosystem II and as cofactors in numerous enzymes. Oxidized manganese occurs as pyrolusite (MnO_2) and birnessite [$(Na,Ca)Mn_7O_{14} \cdot 2.8H_2O$] which are distributed in a non-uniform manner in the environment (Richter et al. 2011). While manganese is commonly considered to be of two oxidation states in nature, i.e. Mn(IV) and Mn(II), Mn(III) may occurs due to chelation by bacterial siderophores and other organic ligands (Tebo et al. 2010; Geszvain et al. 2012). The biogenic manganese cycle is in (Figure 8.10). As a result of bacterial oxidation of Mn(II) to Mn(IV), ferromanganese nodules are produced (Figure 8.11) and bacterial reduction of these nodules releases Mn(II) and Fe(II).

The oxidation and reduction of manganese is achieved by several bacteria and although the reactions are energetically favorable, it remains to be fully established that oxidation or reduction reaction of manganese can be coupled to bacterial growth.

Pure cultures of *Aurantimonas manganoxydans, Pseudomonas putida, Leptothrix discophora, Pedomicrobium* sp. ACM 3067, *Erythrobacter* sp. SD-21, and spores of *Bacillus* sp. SG-1 (Tebo et al. 2010). There are two types of bacterial enzymes responsible for Mn(II) oxidation: calcium-binding heme peroxidases and multicopper oxidases. These enzymes are involved in one electron transfer and deposit granules of MnO_2 on the surface of the bacterial cell. *Thiobacillus thiooxidans* and *Thiobacillus ferrooxidans* couple reduction of Mn(IV) to the oxidation of S^0 but in both instances growth is not supported by this reaction (Lovley 1991). Also, *S. oneidensis* reduces Mn(IV) with electrons from H_2 or formate without growth.

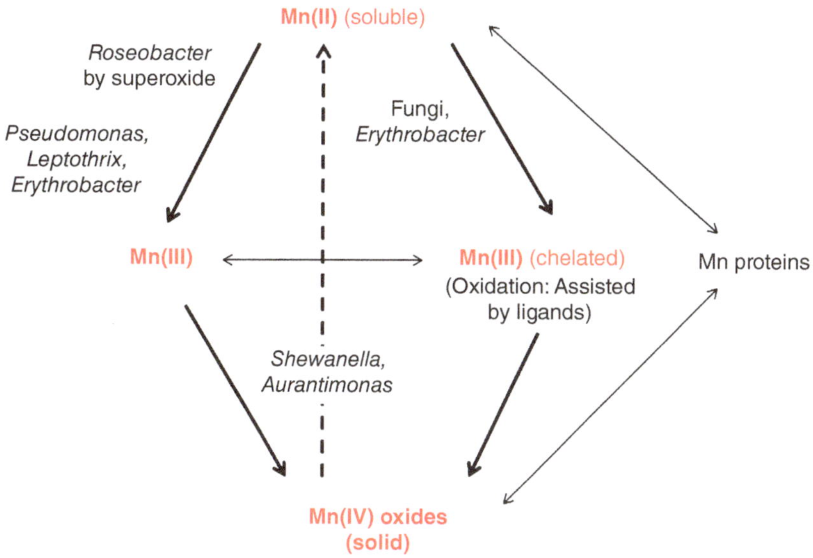

Figure 8.10 Reaction driven by bacteria using manganese in the environment.

Figure 8.11 An example of a ferromanganese nodule collected from the Pacific Ocean floor. Specimen from the Geology Museum, University of Nanjing, China. *Source:* Photo courtesy of Larry L. Barton.

8.8 Phosphorus Cycling

Phosphorus is an essential nutrient for all life forms and for aquatic bacteria, the phosphorus content ranges from 0.01% to 3% dry weight (Godwin and Cotner 2015). This phosphorus content in bacteria is in contrast to plants that are relatively constant at 0.2% phosphorus. An explanation for the range of phosphorus levels in bacteria is the accumulation of phosphate as an internal polyphosphate granule in bacteria under conditions where phosphorus is not limiting.

In soil environments, phosphorus is available as the free inorganic form, organic residues, inorganic form adsorbed onto charged items in the environment and insoluble mineral form. pH is extremely important in soil where maximum bioavailability of phosphate is between pH 6 and 7.5. At pH 4–6, phosphorus is bound to iron or aluminum and in alkaline soils including many deserts, phosphorus is bound to calcium. Soil microorganisms are important in phosphorus cycling (Figure 8.12) by enhancing bioavailability of phosphate (Richardson and Simpson 2011). Mycorrhizal fungi and rhizospheric bacteria stimulate branching or root hair formation in plant roots which amplifies the root surface and enhance phosphate uptake from the environment. Bacteria and fungi in the root zone release organic acids which mobilize phosphate and siderophores to chelate iron and release phosphate from iron phosphate minerals.

Microorganisms release phosphatases and phytases which release orthophosphate from organic structural molecules containing phosphorus, information processing biomolecules (i.e. DNA, RNA), ferritin and other organic phosphorus containing materials. In mineralization of organic matter, phosphate is also obtained from metabolism of phosphite (PO_3^{3-}), hypophosphite ($H_2PO_2^{1-}$) and phosphonate [$C-PO(OH)_2$]. While these unusual phosphorus-containing compounds are of minor abundance in microbial biomass, they reveal the diversity of phosphorus compounds in the biomass. The production of phosphite is proposed to be linked to the reduction of phosphate from phosphoenolpyruvate (Pasek et al. 2014). Anaerobic bacterial metabolism of organic matter and often associated with sewage treatment plants results in the release of small amounts of phosphine (PH_3) gas.

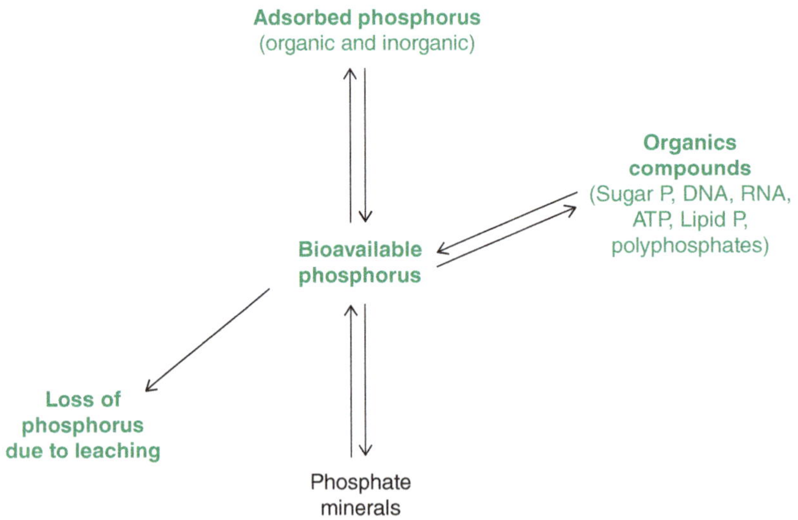

Figure 8.12 The flow of phosphorus in the environment.

Phosphorus cycling involves microorganisms in oceans and freshwater lakes. The source of phosphorus for oceans is river runoff, which annually adds over 1.5 Mt. of dissolved phosphorus and 20 Mt. of suspended phosphorus to the world's oceans (Baturin 2003). While the amount of soluble inorganic phosphorus in lakes is <5%, particulate phosphorus is frequently >70%. The impact of phosphorus runoff on near-shore ecosystems is seen in the coral core samples taken from the Great Barrier Reef (Mallela et al. 2013). The decline of the coral growth and health as observed in the 60 year coral core coincided with the elevated fertilizer discharge between the years of 1949 and 2008.

The dominance of cyanobacteria in freshwater lakes is closely related bioavailable phosphorus with limited levels of nitrogen. The nitrogen-fixing cyanobacteria grow in lakes when phosphorus is present in the lake resulting in eutrophication, Figure 8.13. Productivity of lakes can be established by determining the amount of total phosphorus in the water, see Table 8.12. Phosphorus could be released from decay of organic material or from runoff waters carrying phosphate from agricultural fertilizers.

Figure 8.13 Harmful algae bloom. Bolles Harbor, Monroe, MI, Lake Erie, 22 July 2011. *Source:* NOAA Great Lakes Environmental Research Laboratory, https://www.flickr.com/photos/noaa_glerl/8741970440. Licensed under CC BY-SA 2.0.

Table 8.12 Classification of lake productivity based on phosphorus in lake water.

Status or classification	Total phosphorus ($\mu g\ l^{-1}$)
Oligotrophic	<10
Mesotrophic	10–30
Eutrophic	30–>100

8.9 Selenium Cycling

Selenium is a trace element required for biological systems for a few specific activities but at elevated concentrations selenium is toxic to plants and animals. The oxidation states of selenium are similar to sulfur and the oxidation–reduction reactions in the selenium cycle (Figure 8.14) are driven by different physiological types of bacteria (Staicu and Barton 2017; Stolz et al. 2006). Selenate is the most oxidized state of selenium and once it enters the cell, often using a sulfate uptake system, it may be incorporated into seleno-cysteine or selenomethionine by an assimilatory reduction system. Selenium is used by some bacteria in the formation of xanthene dehydrogenase, hydrogenase, COD, formate dehydrogenase and glycine reductase. Another system involving selenium reduction is associated with detoxification where selenate or selenite is reduced to elemental selenium (Se^0) by mechanisms that are not completely understood. Reduction of selenate or selenite is not linked to bacterial respiration, but most bacteria reduce selenite by reactions involving glutathione reductase or thioredoxin reductase (Staicu and Barton 2017). In a few bacteria, nitrate reductase, nitrite reductase and fumarate reductase can recognize selenate and enzymatically reduce selenate to Se^0. Many bacteria are capable of detoxify selenium environments production of volatile end products including hydrogen selenide, dimethyl selenide, and dimethyldisulfide.

Some bacteria have the capability of dissimilatory selenate reduction with growth coupled to selenate reduction (Staicu and Barton 2017). *Thauera selenatis* couples the oxidation of formate, lactate or acetate to the reduction of arsenate to arsenite with energy conserved to support cell growth. Growth of *Sulfurospirillum barnesii* and *Bacillus arsenicoselenatis* is attributed to energy released in the flow of electrons from lactate to selenate or selenite with Se^0 as the end product. As reviewed by Staicu and Barton (2017), the list of bacteria reported to grow by selenite respiration is expanding and, in addition to the ones listed previously in this section, includes strains of *Enterobacter taylorae, Pseudomonas stutzeri, A. hydrophila, Neisseria mucosa, Rahnella aquatilis, Hafnia alvei, Ferrimonas futtsuensis, Salana multivorans*, and *Citrobacter freundii. Pyrobaculum arsenaticum* and *Pyrobaculum aerophilum* are archaea that have the capability of growing by respiratory reduction of selenate.

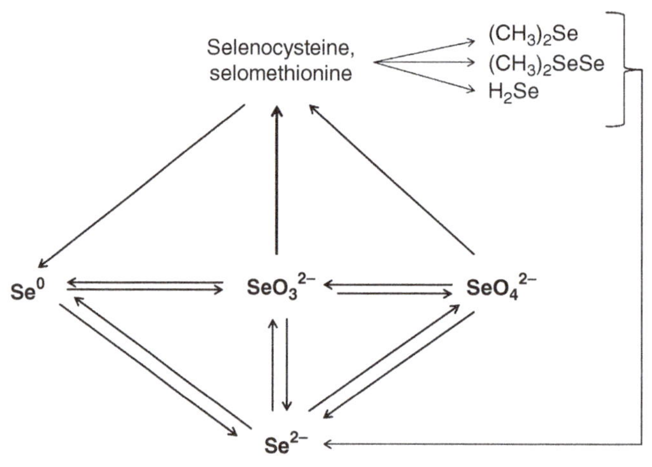

Figure 8.14 The selenium cycle is like the sulfur cycle because both sulfur and selenium have identical oxidation states.

8.10 Cycling Toxic Elements

8.10.1 Mercury

Some of the most toxic chemicals in the environment are associated with mercury. As shown in Figure 8.15, mercury can have several oxidation states and microorganisms are associated with the cycling of mercury through the biosphere. While all terrestrial environments contain small amounts of mercury, elevated levels of mercury are found in certain natural sources or because of commercial and industrial applications. The greatest source of mercury is from the atmosphere, where the primary sources for volatile species of mercury (Hg^0 and methylmercury) are from volcanoes, marine environments, and coal incineration. Other sources of mercury include geological formations, mining of gold, metal processing, and pharmaceutical wastes.

Several different forms of mercury are highly toxic. Elemental mercury (Hg^0) is volatile and readily diffuses through membranes due to its lipid solubility. An early treatment of syphilis was to rub Hg^0 on the back of the afflicted individual; however, the patient most frequently died from mercury toxicity and not the bacterial infection. Mercury amalgam in dental fillings has been a concern due to the extended period that an individual is exposed to mercury, and while mercury amalgam is being replaced in the US, many regions of the world continue to use it. Mercury salts (Hg^{2+}) were used as disinfectants in the early days of medicine but the cumulative exposure to Hg^{2+} was detrimental due to binding of Hg^{2+} to sulfhydryl groups (R-SH) resulting in protein denaturation. To reduce toxicity to tissues, Hg^{2+} is commonly converted to an organomercury compound (i.e. thimerosal or phenylmercuric acetate) and these organomercury compounds may enter the environment as they are discarded. Methylmercury may be produced by various physiological types of bacteria and under certain conditions dimethylmercury, a potent neurotoxin, is also produced. Mono-methylmercury is soluble and may be retained in the tissues of small organisms at the bottom of the food chain. As methyl-mercury compounds are retained in lipids, the tissue concentration increases through trophic levels (i.e. small organisms get eaten by larger ones, etc.), a process called biomagnification. As a result, large fish may contain toxic levels of mercury due to biomagnification of mercury exposure. Thus, while all stages of the mercury cycle may contain chemicals highly toxic to humans, the greatest concern is from ingestion of food containing methylmercury.

Detoxification of mercury by bacteria using the *mer* operon has been studied for many years. The *mer* operon (Figure 8.16) may be on the bacterial chromosome or on a conjugative plasmid and if it is on a plasmid, mercury resistance may be rapidly dispersed throughout the bacterial population by horizontal gene transfer. Various Gram-positive and Gram-negative bacteria have the *mer* operon which consists of genes for regulation (*merR*), binding Hg^{2+} (*merP*), transport of Hg^{2+} across the plasma membrane (*merT*), reduction of Hg^{2+} to Hg^0 (*merA*), and a lyase for hydrolysis of organomercury to produce Hg^{2+} plus an organic compound (*merB*).

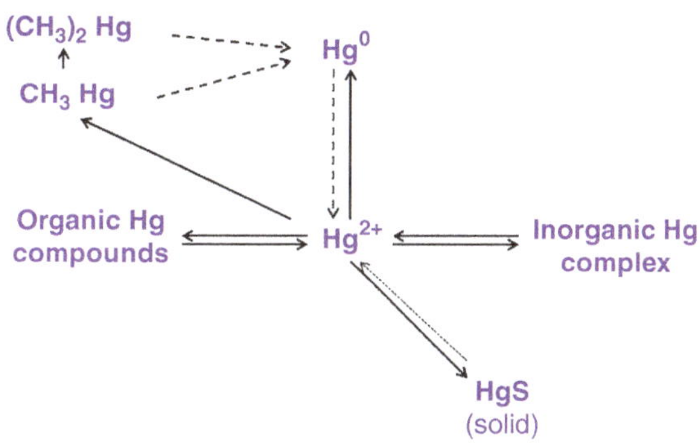

Figure 8.15 Movement of mercury in the biosphere. Elemental mercury, methyl-mercury and dimethyl'mercury are volatile. Dashed line, reactions driven by ultraviolet light. Formation of mercury sulfide is attributed to biotic and abiotic production of hydrogen sulfide.

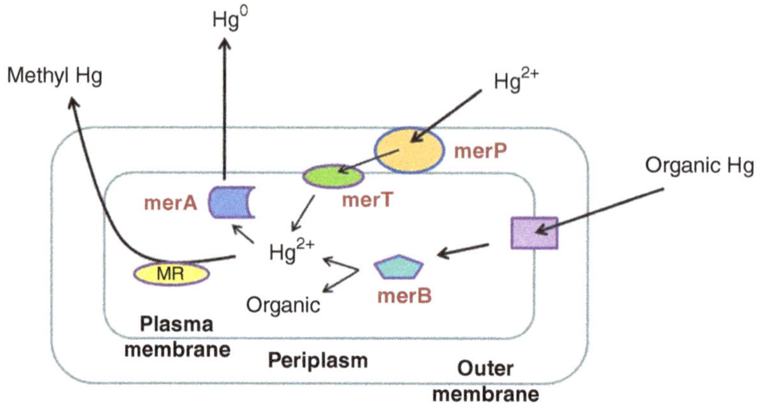

Figure 8.16 Model for bacterial reduction of mercury and production of methyl mercury. Segments of the *mer* operon expressed include: merP, mercury binding protein in the periplasm; merT, uptake transporter for mercury across the plasma membrane; merA, mercury reductase; merB, enzyme releases mercury from organomercury; MR, methylase reaction for mercury.

Anaerobic bacteria and archaea may have a unique system for detoxifying mercury. The Bacterial Methylation Team at the Oak Ridge National Laboratory, U.S. Department of Energy has been exploring the genetic basis of the two gene system for mercury cluster (hgcA and hgcB). The protein expressed from *hgcA* gene is required for mercury methylation and consists of two domains. One domain has a four transmembrane region and the other is a cytoplasmic segment. The HgcA is a corrinoid protein with a putative role as a methyl carrier and the HgcB contains a 2[4Fe-4S] cluster that may be the electron donor for corrinoid reduction (Parks et al. 2013). Studies are ongoing to further identify mechanisms in mercury methylation by the HgcAB system.

While Desulfovibrio desulfuricans ND 132 had long been one of the few sulfate-reducing bacteria displaying mercury methylation, several reports indicate that mercury methylation is found in many but not all *Desulfovibrio*. When testing cultures of sulfate-respiring bacteria, physiological age may be important because cells in stationary phase have the highest methylating activity and fresh environmental isolates may display mercury methylation while laboratory cultures do not. Using a specific gene cluster (*hgcAB*) as a predictor of mercury methylation, microorganisms containing *hgcAB* orthologues were found to methylate mercury and this molecular probe seems useful to survey environmental isolates. From an extensive survey of bacteria, microorganisms methylating mercury are not only sulfate-reducing bacteria (*Desulfovibrio*) and iron-reducing bacteria (*Geobacter*) but include methanogens and bacteria of the *Firmicutes* that display syntrophic, acetogenic or fermentative characteristics (Gilmour et al. 2013).

Other reactions in the mercury cycle merit additional examination. Demethylation of methylmercury is poorly understood but it may be attributed to a methylmercury lyase. The oxidation of Hg^0 to Hg^{2+} has been demonstrated in aerobic bacteria but it is not clear if this is attributed to catalase/peroxidase activity or to another enzyme. Dimethylmercury is highly volatile and is readily released from anaerobic sediments but the mechanisms of the biotic (or abiotic) reactions are not clearly defined.

8.10.2 Arsenic

Iron arsenide sulfide (FeAsS) is the most common arsenic compound in the environment but arsenic is present in over 200 minerals. Adamite (Zn_2AsO_4OH), alarsite ($AlAsO_4$), erythrite [$Co_3(AsO_4)_2 \cdot 8H_2O$] and annabergite [$Ni_3(AsO_4)$] are additional examples of minerals containing arsenic. Volcanic action is responsible for about one-third of atmospheric arsenic and in some regions inorganic arsenic compounds are found at toxic levels in groundwater (Mandal and Suzuki 2002). Some terrestrial and marine species produce dimethylarsene, trimethylarsene, arsenocholine, arsenobetaine, arsenosugars, and arsenolipids and other organic arsenic compounds (Dhuldhaj et al. 2013). A generalized cycle of arsenic is given in Figure 8.17.

In the early nineteenth century, arsenic as copper arsenite was used to color wallpaper green, but bacteria and fungi metabolizing the green pigmented wallpaper in humid environments released toxic trimethylarsene [$(CH_3)_3As$] and so its use of arsenic in wall paper or paints was discontinued. At one time, the greatest use of arsenic was in the timber industry where copper chrome arsenate (CCA) was used to treat lumber against fungal and bacterial decomposition. While CCA is an excellent preservative for lumber, its use increases toxic metals in the environment, so alternative wood preservatives are being considered. Arsenic compounds have been added to feed for chickens, turkeys, and pigs to prevent infections by intestinal parasites. It was originally believed that the arsenic compounds did not cross the intestinal barrier to contaminate the meat; however, subsequent studies showed trace levels in liver and other tissues of these animals. Another problem was the elevated levels of arsenic in plants caused by using manure containing arsenic as fertilizer. The arsenic-containing feed has been discontinued for chickens and pigs but is often still used in the first six weeks of the turkey to prevent the lethal protozoan disease of blackhead or histomoniasis. The toxicity of arsenate is due to its interference with ATP synthesis because arsenate mimics the structure of phosphate.

Arsenic is a metalloid that is commonly found as arsenate (AsO_4^{3-}; As(V), arsenite [AsO_3^{3-}]), and dimethylarsene [$(CH_3)_2AsH$] with oxidation states of 5+, 3+, and 3−, respectively. A biogeochemical cycle of arsenic is provided in Figure 8.17. Because arsenic has multiple oxidation states and the redox reaction of arsenic is within bacterial metabolic capability, bacteria are highly involved in the arsenic cycle.

Initially, only one or two species of bacteria were considered capable of coupling energy from arsenate redox reactions to growth; however, it is now apparent that arsenate is an important biogenetic substrate for bacteria. Using the diversity of arsenic metabolizing enzymes (respiratory arsenate reductase, arsenite oxidase, and alternative arsenite oxidases), the evolution of these arsenic pathways are now being analysed (van Lis et al. 2013). Anaerobic bacteria

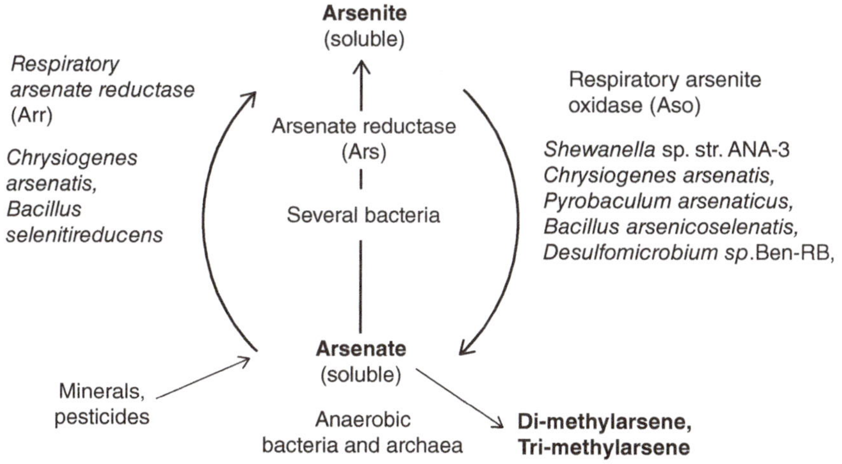

Figure 8.17 In the environment, the major arsenic compounds include the highly toxic arsenate and or arsenite with reduced toxicity. Arsenate is reduced by either heterotrophic bacteria or dissimilatory arsenate reducing bacteria.

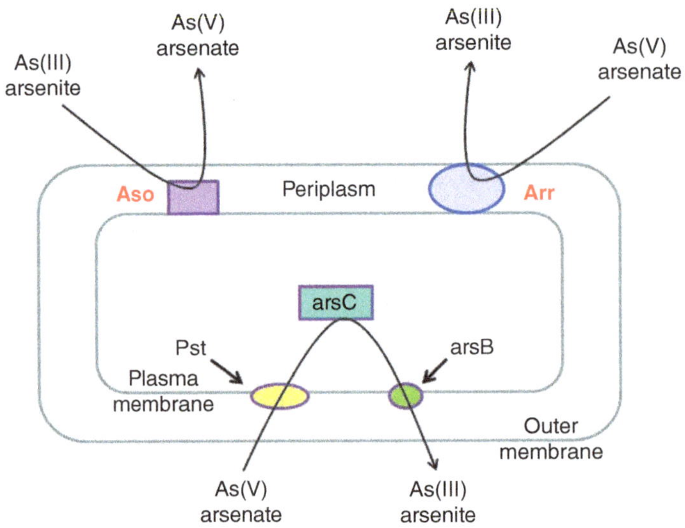

Figure 8.18 Cellular localization of arsenic reactions in bacteria. Pst, the phosphate uptake transporter that transports arsenate; arsC, arsenate reductase; arsB, arsenite export transporter; Aso, arsenite oxidase; Arr, arsenate reductase.

like *Ectothiorhodospira* and *Oscillatoria* that use arsenite (As III) as an electron donor for photosynthesis with the production of arsenate (AsV) have been isolated (Kulp et al. 2008). Several anaerobic organisms couple dissimilatory arsenate reduction to growth and these have been designated as dissimilatory arsenate reducing prokaryotes (DARPs). Examples of DARPs include bacteria (i.e. *Sulfurospirillum arsenophilum, Shewanella* sp. str. ANA-3, *B. arsenicoselenatis* str. E1H, *Bacillus selenitireducens* str.MLS10, *Desulfotomaculum auripigmentum, Desulfomicrobium* sp. str. Ben-RB, *Chrysiogenes arsenatis*) and archaea (*P. arsenaticum* and *P. aerophilum*). Bacteria such as *Anaeromyxobacter* sp. strain PSR-1 (Kudo et al. 2013) can reduce both arsenate and iron (FeIII), which is important for understanding how arsenic absorbed onto iron minerals is released.

Arsenate is commonly found in the aerobic environment and arsenite is found in aerobic and anaerobic environments. The microbial reduction of arsenate may be by a respiratory enzyme located in the plasma membrane or periplasm and a detoxifying enzyme located in the cytoplasm, see Figure 8.18 (Stolz et al. 2006). Respiratory reductase (Arr) is attributed to *arrA* and *arrB,* a two gene cluster in the *arr* operon and primers constructed from *arrA* have been used to search for bacteria with respiratory arsenate reduction activity. Arr enzyme has two subunits: ArrA and ArrB. In *C. arsenatis,* the Arr enzyme is in the periplasm but in *B. selenitireducens* the enzyme is in the membrane. The product of arsenate reduction by ArrAB is arsenite and it is released outside of the plasma membrane.

Many aerobic strains of bacteria detoxify arsenate in the cytoplasm by enzymatic reduction to arsenite, which is exported from the cell or methylated to produce trimethylarsene (Silver and Phung 2005). Cytoplasmic arsenate reductase is encoded on the *arsC* gene and the *ars* operon is found on either a plasmid or incorporated into the bacterial chromosome. Adjacent to the *arsC* gene is *arsB,* which encodes for the arsenite efflux pump from the cell. Electron donors for reduction of arsenate are either glutaredoxin or thioredoxin.

Arsenite is highly toxic to the cell and bacteria will eliminate it from the environment by methylation activities to produce volatile methylarsines. Arsenite, whether from respiratory arsenate reduction or from other sources in the environment, is transported into cells by a glycerol uptake system because of the structural similarity between glycerol and nonionized arsenite [$As(OH)_3$]. Several biological systems will reduce and methylate arsenite by the enzyme, ArsM. Homologs of ArsM are found in many bacteria and archaea (Stolz et al. 2006). In aerobic fungi and animal tissues, the methyl source is S-adenosylmethionine (SAM) while in anaerobic bacteria and archaea methyl cobalamin donates the methyl group for production of dimethyl- and trimethylarsine (Stolz et al. 2006). Reduced glutathione is the electron donor for the SAM-mediated ArsM process and various reduced thiol-containing proteins (i.e. thioredoxin, glutaredoxin) may provide electrons for anaerobes.

Several bacteria use arsenite oxidase to convert the highly toxic arsenite to the less toxic arsenate. The enzyme arsenite oxidase is currently designated as Aio (van Lis et al. 2013) although previously it was referred to as Aox, Aso, or Aro (Stolz et al. 2006). *Rhizobium* sp. NT-26 is a facultative autotroph that couples growth to the transfer of electrons from arsenite to O_2 with the formation of arsenate and water (Warelow et al. 2013). Using Basic Local Alignment Search Tool (BLAST) searches, putative arsenite oxidases have been found in archaea (*Aeropyrum pernix, Sulfolobus tokodaii*), anaerobic phototrophic bacteria (*Chloroflexus aurantiacus, Chlorobium limicola, Chlorobium phaeobacteroides*), and heterotrophic bacteria (*Nitrobacter hamburgensis, Rhodoferax ferrireducens, Agrobacterium tumefaciens, Cenibacterium arsenoxidans* (renamed *Herminiimonas arsenicoxydans*) strain ULPAs1) (Stolz et al. 2006).

8.11 Summary

Oxidation/reduction reactions are important in mineral cycles. In this regard, bacteria could be considered environmental opportunists that exploit thermodynamically favorable inorganic reactions to produce ATP for growth. Bacteria will often use an electron donor from one geochemical cycle and an electron acceptor from another cycle, which contributes to a dynamic chemical environment. The energy flow through geochemical cycles accounts for transformation of toxic compounds in the environment (such as, mercury, arsenic, selenium) to compounds with reduced toxicity. Commonly, a given bacterial species has a narrow substrate range for either electron donor or electron acceptor activity, with global cycling a summation of the various individual chemical transformations.

Discussion Questions

1. Discuss how a microbial population could interact with the carbon, nitrogen, and sulfur cycles.
2. Discuss how you could determine the rate limiting steps in the nitrogen and sulfur cycles.
3. Discuss the advantages and disadvantages bioremediation versus chemical remediation to detoxify a site containing mercury compounds.
4. Discuss the similarities between the sulfur cycle and the selenium cycle.
5. Discuss how the greenhouse gases could be reduced by manipulation of the carbon cycle.
6. Discuss why there are more phases of transition in the nitrogen cycle than in the iron or manganese cycle.

References

Amo, T., Atomi, H., and Imanaka, T. (2002). Unique presence of a manganese catalase in a hyperthermophile archaeron, *Pyrobacterium calidifontis VA1. J. Bacteriol.* 184: 3305–3312.

Anthony, C. (1988). Quinoproteins and energy transduction. In: *Bacterial Energy Transduction* (ed. C. Anthony), 294–316. New York, NY: Academic Press.

Baldori, B., Trinetta, V., Dalgalarrondo, M. et al. (2007). Isolation, taxonomic identification and hydrogen peroxide production by *Lactobacillus delbrueckii* subsp. *lactis* T31, isolated from yogurt: inhibitory activity on food-borne pathogens. *J. Appl. Microbiol.* 103: 584–593.

Barton, L.L. (2005). *Structural and Functional Relationships in Prokaryotes*. New York, NY: Springer Science+Business Media, Inc.

Barton, L.L. and Fauque, G.D. (2009). Biochemistry, physiology and biotechnology of sulfate-reducing bacteria. *Adv. Appl. Microbiol.* 68: 41–98.

Barton, L.L., Fardeau, M.-L., and Fauque, G.D. (2014). Hydrogen sulfide: a toxic gas produced by dissimilatory sulfate reduction and consumed by microbial oxidation. *Met. Ions Life Sci.* 14: 237–278.

Barton, L.L. and Northup, D.E. (2011). *Microbial Ecology*. New York, NY: Wiley-Balckwell.

Baturin, G.N. (2003). Phosphorus cycle in the ocean. *Lithol. Mineral. Resour.* 38: 101–119.

Bayer, E.A. and Lamed, R. (2006). The Cellulosome Saga: early history. In: *Cellulosome* (ed. V. Uversky and I.A. Kataeva), 11–46. New York, NY: Nova Science Publishers, Inc.

Beal, E.J., House, C.H., and Orphan, V.J. (2009). Manganese- and iron-dependent marine methane oxidation. *Science* 325: 184–187.

Blumer, C. and Haas, D. (2000). Mechanism, regulation and ecological role of bacterial cyanide biosynthesis. *Arch. Microbiol.* 173: 170–177.

Boetius, A., Ravenschlag, K., Schubert, C.J. et al. (2000). A marine microbial consortium apparently mediating anaerobic oxidation of methane. *Nature* 407: 623–626.

Brankatschk, R., Töwe, S., Kleineidam, K. et al. (2011). Abundance and potential activities of nitrogen cycling microbial communities along a chronosequence of a glacier forefield. *ISME J.* 5: 1025–1037.

Brennan, P.C. and Feinstein, R.N. (1969). Relationship of hydrogen peroxide production by *Mycoplasma pulmonis* to virulence for catalase-deficient mice. *J. Bacteriol.* 98: 1036–1040.

Bruschi, M. and Goulhen, F. (2007). New bioremediation technologies to remove heavy metals and radionuclides using Fe(III)-, sulfate- and sulfur-reducing bacteria. In: *Environmental Bioremediation Technologies* (ed. S.N. Singh and R.D. Tripathi), 35–56. Heidelberg: Springer.

Burtt, E.H. Jr. and Ichida, J.M. (2004). Gloger's rule, feather-degrading bacteria, and color variation among song sparrows. *The Condor* 106: 681–686.

Burtt, E.H. Jr., Schroeder, M.R., Smith, L.A. et al. (2011). Colorful parrot feathers resist bacterial degradation. *Biol. Lett.* 7: 214–216.

Cabello, P., Roldán, M.D., and Moreno-Vivián, C. (2004). Nitrate reduction and the nitrogen cycle in archaea. *Microbiology* 150: 3527–3546.

Cabiscol, E., Piulats, E., Echave, P. et al. (2000). Oxidative stress promotes specific protein damage in *Saccharomyces cerevisiae. J. Biol. Chem.* 275: 27393–27398.

Chisholm, S.W., Olson, R.J., Zettler, E.R. et al. (1988). A free-living prochlorophyte occurs at high cell concentrations in the oceanic eutrophic zone. *Nature* 334: 340–343.

Choi, P.S., Naal, Z., Moore, C. et al. (2006). Assessing the impact of denitrifier-produced nitric oxide on other bacteria. *Appl. Environ. Microbiol.* 72: 2200–2205.

Coats, J.D., Michaelidou, U., Bruce, R.A. et al. (1999). Ubiquity and diversity of dissimilatory (per) chlorate-reducing bacteria. *Appl. Environ. Microbiol.* 65: 5234–5241.

Das, T. and Manefield, M. (2013). Phenazine production enhances extracellular DNA release via hydrogen peroxide generation in *Pseudomonas aeruginosa*. *Commun. Integr. Biol.* 6: e23570. https://doi.org/10.4161/cib.23570.

Delmotte, N., Knief, C., Chaffron, S. et al. (2009). Community proteogenomics reveals insight into the physiology of phyllosphere bacteria. *Proc. Natl. Acad. Sci. U.S.A.* 106: 16428–16433.

Dhuldhaj, U.P., Yadav, I.C., Singh, S. et al. (2013). Microbial interactions in the arsenic cycle: adoptive strategies and applications in environmental management. *Rev. Environ. Contam. Toxicol.* 224: 1–38.

Ding, S.-Y., Xu, Q., Crowley, M. et al. (2008). A biophysical perspective on the cellulosome: new opportunities for biomass conversion. *Curr. Opin. Biotechnol.* 19: 218–227.

Dridi, B., Fardeau, M.L., Ollivier, B. et al. (2012). *Methanomassiliicoccus luminyensis* gen.Nov., nov., a methanogenic archaeon isolated from human faeces. *Int. J. Syst. Evol. Microbiol.* 62: 1902–1907.

Dubinina, G.A., Grabovich, M.Y., Churikova, V.V. et al. (1990). Hydrogen peroxide production by *Beggiatoa leptomitiformis*. *Mikrobiologiya* 59: 425–431.

Ettwig, K.F., Butler, M.K., Le Paslier, D. et al. (2010). Nitrite-driven anaerobic methane oxidation by oxygenic bacteria. *Nature* 464: 543–548.

Fenchel, T. (2008). The microbial loop – 25 years later. *J. Exp. Marine Biol. Ecol.* 366: 99–103.

Flombaum, P., Gallegos, J.L., Gordillo, R.A. et al. (2013). Present and future global distributions of the marine cyanobacteria *Prochlorococcus* and *Synechococcus*. *Proc. Natl. Acad. Sci. U.S.A.* 110: 9824–9829.

Fortin, D. and Langley, S. (2005). Formation and occurrence of biogenic iron-rich minerals. *Earth-Science Rev.* 72: 1–19.

Gardner M (2004). Ancient lumber from New Zealand and South Carolina, *Fine Woodworking* #179.

Geszvain, K., Butterfield, C., Davis, R.E. et al. (2012). The molecular biogeochemistry of manganese (II) oxidation. *Biochem. Soc. Trans.* 40: 1244–1248.

Gilmour, C.C., Podar, M., Bullock, A.L. et al. (2013). Mercury methylation by novel microorganisms from new environments. *Environ. Sci. Technol.* 47: 11810–11820.

Godwin, C.M. and Cotner, J.B. (2015). Aquatic heterotrophic bacteria have flexible phosphorus content and biomass stoichiometry. *ISEM J.* 9: 2324–2327.

González-Flecha, B. and Demple, B. (1995). Metabolic sources of hydrogen peroxide in aerobically growing *Escherichia coli*. *J. Biol. Chem.* 270: 13681–13687.

Hedderich, R. and Whitman, W.B. (2006). Physiology and biochemistry of the methane-producing archaea. *Prokaryotes* 2: 1050–1079.

Heinhorst, S., Cannon, G.C., and Shively, J.M. (2014). Carboxysomes and their structural organization in prokaryotes. In: *Nanomicrobiology* (ed. L.L. Barton, D.A. Bazylinski and X. Xu), 75–102. New York, NY: Springer.

Hider, R.C. and Kong, X. (2010). Chemistry and biology of siderophores. *Nat. Prod. Rep.* 27: 637–657.

Iibert, M. and Bonnefoy, V. (2013). Insight into the evolution of the iron oxidation pathways. *Biochim. Biophys. Acta* 1827: 161–175.

Janusz, G., Kucharzyk, K.H., Pawlin, A. et al. (2013). Fungal laccase, manganese peroxidase and lignin peroxidase: gene expression and regulation. *Enzyme Microb. Technol.* 52: 1–12.

Jeoung, J.-H., Fesseler, J., Goetzl, S. et al. (2014). Carbon monoxide. Toxic gas and fuel for anaerobes and aerobes: carbon monoxide dehydrogenases. *Met. Ions Life Sci.* 15: 38–69.

Johnston, A.W.B., Todd, J.D., Sun, L. et al. (2008). Molecular diversity of bacterial production of the climate-changing gas, dimethylsulpfide, a molecule that impinges on local and global symbiosis. *J. Expt. Bot.* 59: 1059–1067.

Karta, B., de Almeida, N.M., Maalcke, W.J. et al. (2013). How to make a living from anaerobic ammonium oxidation. *FEMS Rev.* 37: 428–461.

Knittel, K. and Boetius, A. (2009). Anaerobic oxidation of methane: progress with an unknown process. *Ann. Rev. Microbiol.* 63: 311–334.

Knowles, C.J. (1988). Cyanide utilization and degradation by microorganisms. *Ciba Found Symp.* 140: 3–15.
Koehler, I., Konhauser, K., and Kappler, A. (2010). Role of microorganisms in banded iron formations. In: *Geomicrobiology:Molecular and Environmental Perspective* (ed. L.L. Barton, M. Mandel and A. Loy), 309–324. Dordrecht: Springer Science+Business Media.
Koenigs, J.W. (1974). Production of hydrogen peroxide by wood-rotting fungi in wood and its correlation with weight loss, depolymerization and pH changes. *Arch. Microbiol.* 99: 129–145.
Knapp, J.R., Laur, G.L., Vadas, P.A. et al. (2014). Enteric methane in dairy cattle production: quantifying the opportunities and impact of reducing emissions. *J. Dairy Sci.* 97: 3231–3261.
Kremer, R. and Souissi, T. (2001). Cyanide production by rhizobacteria and potential for suppression of weed seedling growth. *Curr. Microbiol.* 43: 182–186.
Kudo, K., Yamaguchi, N., Makino, T. et al. (2013). Release of arsenic from soil by a novel dissimilatory arsenate-reducing bacterium, Anaeromyxobacter sp. strain PSR-1. *Appl. Environ. Microbiol.* 79: 4635–4642:https://doi.org/10.1128/AEM.00693-13.
Kulp, T.R., Hoeft, S.E., Asao, M. et al. (2008). Arsenic (III) fuels anoxygenic photosynthesis in hit spring biofilms from mono Lake, California. *Science* 321: 967–970.
Kutschera, U. (2007). Plant-associated methylobacteria as co-evolved phytosymbionts. *Plant Signal Behav.* 2: 74–78.
Lamed, R., Setter, E., and Bayer, E.A. (1983). Characterization of a cellulose-binding, cellulase-containing complex in Clostridium thermocellum. *J. Bacteriol.* 156: 828–836.
Lentini, C.J., Wankel, S.D., and Hansel, C.M. (2012). Enriched iron (III)-reducing bacterial communities are shaped by carbon substrate and iron oxide mineralogy. *Front. Microbiol.* 3: 404: https://doi.org/10.3389/fmicb.2012.00404.
Levar, C.E., Hoffman, C.L., Dunshee, A.J. et al. (2017). Redox potential as a master variable controlling pathways of metal reduction by *Geobacter sulfurreducens*. *ISME J.* 11: 741–752.
Lovley, D.R. (1991). Dissimilatory Fe(III) and Mn(IV) reduction. *Microbiol. Rev.* 55: 259–287.
Mallela, J., Lewis, S.E., and Croke, B. (2013). Coral skeletons provide historical evidence of phosphorus runoff on the great barrier reef. *PLoS One* https://doi.org/10.1371/journal.pone.0075663.
Mancardi, D., Ridnour, L.A., Thomas, D.D. et al. (2004). The chemical dynamics of NO and reactive nitrogen oxides: a practical guide. *Curr. Mol. Med.* 4: 723–740.
Mandal, B.K. and Suzuki, K.T. (2002). Arsenic round the world: a review. *Talanta* 58: 201–235.
Mazzoli, R., Lamberti, C., and Pessione, E. (2012). Engineering new metabolic capabilities in bacteria: lesions from recombinant cellulolytic strategies. *Trends Biotechnol.* 30: 111–119.
Moore, C.M., Nakano, M.M., Wang, T. et al. (2004). Response of *Bacillus subtilis* to nitric oxide and the nitrosating agent sodium nitroprusside. *J. Bacteriol.* 186: 4655–4664.
Moreno-Vivián, C., Cabello, P., Martinez-Luque, M. et al. (1999). Prokaryotic nitrate reduction: molecular properties and functional distinction among bacterial nitrate reductases. *J. Bacteriol.* 181: 6573–6584.
Mörsdorf, G., Frunzke, K., Gadkari, D. et al. (1992). Microbial growth on carbon monoxide. *Biodegradation* 3: 61–82.
Moy, T.I., Mylonakis, E., Calderwood, S.B. et al. (2004). Cytotoxicity of hydrogen peroxide production by *Enterococcus faecium*. *Infect. Immun.* 72: 4512–4520.
Newnham, R.M., Lowe, D.J., and Green, J.D. (1988). Palynology, vegetation and climate of the Waikato lowlands, North Island, New Zealand, since c. 18,000 years ago. *J. Royal Soc. N. Z.* 19: 127–150.
Noffke, N. and Awramik, S.M. (2013). Stromatolites and MISS – differences between relatives. *Geol. Soc. Am.* 23: 4–6.
Okahashi, N., Nakata, M., Sumitomo, T. et al. (2013). Hydrogen peroxide produced by oral streptococci induces macrophage cell death. *PLOS One*: https://doi.org/10.1371/journal.pone.0062563.

Offre, P., Spang, A., and Schleper, C. (2013). Archaea in biogeochemical cycles. *Annu. Rev. Microbiol.* 67: 437–457.

Parks, J.M., Johs, A., Podar, M. et al. (2013). The genetic basis for bacterial mercury methylation. *Science* 339: 1332–1335.

Pasek, M.A., Sampson, J.M., and Atlas, Z. (2014). Redox chemistry in the phosphorus biogeochemical cycle. *Proc. Natl. Acad. Sci. U.S.A.* 111: 15468–15473.

Postgate, J.R. (1979). *The Sulphate-Reducing Bacteria*. Cambridge, UK: Cambridge University Press.

Pridmore, R.D., Pittet, A.C., Pripian, F. et al. (2008). Hydrogen peroxide by *Lactobacillus johnsonii* NCCNCC 533 and its role in anti-*Salmonella* activity. *FEMS Microbiol. Lett.* 283: 210–215.

Ren, T., Roy, R., and Knowles, R. (2000). Production and consumption of nitric oxide by three methanotrophic bacteria. *Appl. Environ. Microbiol.* 66: 3891–3897.

Richardson, A.E. and Simpson, R.J. (2011). Soil microorganisms mediating phosphorus availability update on microbial phosphorus. *Plant Physiol.* 156: 989–996.

Richter, K., Schicklberger, M., and Gescher, J. (2011). Dissimilatory reduction of extracellular electron acceptors in anaerobic respiration. *Appl. Environ. Microbiol.* 78: 913–921.

Roncel, M., Navarro, J.A., and De La Rosa, M.A. (1989). Coupling of solar energy to hydrogen peroxide production in the cyanobacterium *Anacystis nidulans*. *Appl. Environ. Microbiol.* 55: 483–487.

Rőszer, T. (2012). Nitric oxide is a bioproduct in prokaryotes. In: *The Biology of Subcellular Nitric Oxide*, 19–46. New York, NY: Springer Science+Business Media.

Ryall, B., Lee, X., Ziosnik, J.E.A. et al. (2008). Bacteria of the *Burkholderia cepacia* complex are cyanogenic under biofilm and colonial growth conditions. *BMC Microbiol.* 8: 108: https://doi.org/10.1186/1471-2180-8-108.

Schlesinger, W. (1991). *Biogeochemistry: An Analysis of Global Change*. New York, NY: Academic Press.

Schopf, J.W., Kudyavtsev, A.B., Czaja, A.D. et al. (2007). Evidence of Archean life: Stromatolites and microfissils. *Precambian Res.* 158: 141–155.

Seki, M., Iida, K.-I., Saito, M. et al. (2004). Hydrogen peroxide production in *Streptococcus pyogenes*: involvement of lactate oxidase and coupling with aerobic utilization of lactate. *J. Bacteriol.* 186: 2046–2051.

Sigman, D.M. and Haug, G.H. (2006). The biological pump in the past. In: *Treatise on Geochemistry*, vol. 6 (ed. H.D. Holland and K.K. Turekian), 491–528. Oxford, UK: Permagon Press.

Silver, S. and Phung, L.T. (2005). Genes and enzymes involved in bacterial oxidation and reduction of inorganic arsenic. *Appl. Environ. Microbiol.* 71: 599–608.

Simon, J. (2002). Enzymology and bioenergetics of respiratory nitrite ammonification. *FEMS Microbiol. Rev.* 26: 285–239.

Sobko, T., Reindeers, C.I., Jasson, E.A. et al. (2005). Gastrointestional bacteria generate nitric oxide from nitrate and nitrite. *Nitric Oxide* 13: 272–278.

Staicu, L.C. and Barton, L.L. (2017). Bacterial metabolism of selenium – for survival or profit. In: *Bioremediation of Selenium Contaminated Wastewater* (ed. E.D. van Hullebusch), 1–31. Dordrecht, The Netherlands: Springer International Publishing.

Stolz, J.F., Basu, P., Santini, J.M. et al. (2006). Arsenic and selenium in microbial metabolism. *Annu. Rev. Microbiol.* 60: 107–130.

Strus, M., Gosiewski, T., Fyderek, K. et al. (2009). A role of hydrogen peroxide producing commensal bacteria present in colon of adolescents with inflammatory bowel disease in perpetuation of the inflammatory process. *J. Physiol. Pharmacol.* 60: 49–54.

Tebo, B.M., Geszvain, K., and Lee, S.-W. (2010). The molecular geomicrobiology of bacterial manganese (II) oxidation. In: *Geomicrobiology* (ed. L.L. Barton, M. Mandel and A. Loy), 285–308. Dordrecht: Springer Science+Business Media B.V.

Tenenbaum, D.J. (2004). Underwater logging: submarine rediscovers lost wood. *Environ. Health Perspect.* 112: A892–A895.

Teske, A. and Nelson, D.C. (2006). The genera *Beggiatoa* and *Thioploca*. *Prokaryotes* 6: 784–810.

Tolli, J.D., Sievert, S.M., and Taylor, C.D. (2006). Unexpected diversity of bacteria capable of carbon monoxide oxidation in a costal marine environment and contribution of the *Roseobacter*-associated clade to total CO oxidation. *Appl. Environ. Microbiol.* 72: 1966–1973.

van Lis, R., Nitschke, W., Duval, S. et al. (2013). Arsenics as bioenergetics substrates. *Biochim. Biophys. Acta* 182: 176–188.

Venzlaff, H., Enning, D., Srinivasan, J. et al. (2013). Accelerated cathodic reaction in microbial corrosion of iron due to direct electron uptake by sulfate-reducing bacteria. *Corr. Sci.* 66: 88–96.

Walker, J.C.G. (1980). *The Oxygen Cycle in the Natural Environment and the Biogeochemical Cycles*. Berlin: Springer-Verlag.

Ward, B. (2012). The global nitrogen cycle. In: *Fundamentals of Geobiology* (ed. A.H. Knoll, D.E. Canfield and K.O. Konhauser), 36–48. Chichester, West Sussex, UK: Blackwell Publishing Ltd.

Washio, J., Sato, T., Koseki, T. et al. (2005). Hydrogen sulfide-producing bacteria in tongue biofilm and their relationship with oral malodor. *J. Med. Microbiol.* 54: 889–895.

Warelow, T.P., Oke, M., Schoepp-Cothenet, B. et al. (2013). The respiratory arsenite oxidase: structure and the role of residues surrounding the Rieske cluster. *PLoS One*: https://doi.org/10.1371/journal.pone.0072535.

Weber, K.A., Achenbach, L.A., and Coates, J.D. (2006). Microorganisms pumping iron: anaerobic microbial iron oxidation and reduction. *Nature Rev. Microbiol.* 4: 752–764.

Williams, E.J., Guenther, A., and Fehsenfeld, F.C. (1992). An inventory of nitric oxide emissions from soils of the United States. *J. Geophys. Res.* 97: 7511–7519.

Wilkes D (2014). Surreal seascape revealed by the storms: Ancient oaks and pines from 5000 year old forest rise as Welsh beach is washed away, *The Daily Mail* (On line news) Australia.

Zehr, J.P. and Kudela, R.M. (2011). Nitrogen cycle of the open ocean: from genes to ecosystems. *Annu. Rev. Mar. Sci.* 3: 197–225.

Zhao, J., de Koker, T.H., and Janese, B.J.H. (1998). Comparative studies of lignin peroxidases and manganese-independent peroxidases produced by selected white rot fungi in solid media. *FEMS Microbiol. Lett.* 145: 393–399.

Zumft, W.G. and Kroneck, P.M. (2007). Respiratory transformation of nitrous oxide (N_2O) to dinitrogen by bacteria and archaea. *Adv. Microbiol. Physiol.* 52: 107–227.

Further Reading

Cristina, N., Butterfield, C.N., and Bradley, T.M. (2017). Substrate specificity and copper loading of the manganese-oxidizing multicopper oxidase Mnx from *Bacillus* sp. PL-12. *Metallomics* 2017 (9): 183: https://doi.org/10.1039/C6MT00239K.

Fitzgerald, W.F., Lamborg, C.H., and Hammerschmidt, C.R. (2007). Marine biogeochemical cycling of mercury. *Chem. Rev.* 107: 641–662.

Hug, K., Maher, W.A., Stott, M.B. et al. Microbial contributions to coupled arsenic and sulfur cycling in the acid-sulfide hot spring champagne pool, New Zealand. *Front. Microbiol.* 2014: https://doi.org/10.3389/fmicb.2014.00569.

Lovley, D.R., Holmes, D.E., and Nevin, K.P. (2004). Dissimilatory Fe(III) and Mn(IV) reduction. *Adv. Microbial. Physiol.* 49: 219–286.

Muyzer, G. and Stams, A.J.M. (2008). The ecology and biotechnology of sulphate-reducing bacteria. *Nat. Rev. Microbiol.* 6: 441–454.

9

Bioremediation Using Microorganisms

9.1 Overview and Focus

The addition of undesired chemicals to an environment is called pollution. This chapter focuses on microbial metabolism of chemical pollutants. If the undesired chemical is from a specific site, the source of pollution is referred to as a point source but if the source cannot be determined, the pollution is designated as diffuse or from a non-point source. Because the site of contamination is extremely important, remediation is highly dependent on the physical, chemical, and biological activities in the immediate environment. For the best bioremediation in soil, the following soil parameters are suggested: 10% air filled pore space, pH between 5.5 and 8.5, temperature between 15 and 45 °C, low clay or silt content, with a water holding capacity of 25–28% (Sharma 2012). As indicated in Table 9.1, environmental pollution is extremely broad and continues to be a problem despite considerable focus on curtailing the release of pollutants into the environment. There are many different chemicals associated with pollution and most are in the following categories: pesticides, halogenated industrial compounds, polyaromatic hydrocarbons (PAHs), petroleum products, complex organic molecules, and nitrogen compounds.

Environmental parameters greatly influence microbial activities following chemical pollution. The pollution site will dictate whether bacterial metabolism will be aerobic or anaerobic. Deposition of chemicals in soils generally leads to aerobic activities whereas polluted lakes, rivers, marshes or sites subject to flooding lead to anaerobic activities. Usually, aerobic metabolism produces a greater rate of detoxification, with organic compounds mineralized to carbon dioxide. In general, higher rates of chemical detoxification occur at sites where the existing microbial activity is robust and the type of bacteria present is highly diverse. If the terrestrial site contaminated is barren, the bacterial metabolic activity will be low and dominated by relatively few species. This chapter focuses on the microbial activities contributing to the detoxification of organic or inorganic compounds. The general theme of bioremediation is summarized in Figure 9.1.

9.2 Microbial Bioremediation: Strategies and Applications

Microorganisms in an environment persist because they have adapted to the chemical and physical conditions present. An influx of chemicals changes the environment and usually stresses the bacterial population. Depending on the chemical pollutant and the site, substantial numbers of bacteria may be killed. Chemicals with the highest impact on bacteria are those

Environmental Microbiology and Microbial Ecology, First Edition. Larry L. Barton and Robert J.C. McLean.

Table 9.1 Sources of pollution and types of chemicals associated with specified examples of pollutant.

Examples	Chemicals commonly present
Industrial effluents	Various types of water soluble chemicals containing carbon and nitrogen that promote rapid growth of bacteria. May contain organic compounds that persist for long periods
Sewage effluent	Oxygen depleting substances are present including high levels of nitrogen, phosphorus compounds in additions to carbon compounds. May contain pharmaceuticals and hormones
Oil storage facilities including leaking pipelines and gas storage tanks	Various types of hydrocarbons
Urban storm water runoff from paved areas	Compounds containing nitrogen and phosphorus, heavy metals and salts, soaps and persistent organic pollutants. Hydrocarbons from car and truck traffic
Landfill sites	Various organic compounds including oxygen depleting substances. Lechate may contain high levels of inorganic nitrogen compounds
Mining	Heavy metals and acidic drainage
Livestock feed lots, farm wastes, and fish farming	Various forms of nitrogen and phosphorus
Agricultural fertilizers	Inorganic forms of nitrogen and phosphorus
Pesticide applications	Complex organic compounds that decompose slowly
Power generating systems	Compounds of nitrogen and sulfur. Release of mercury

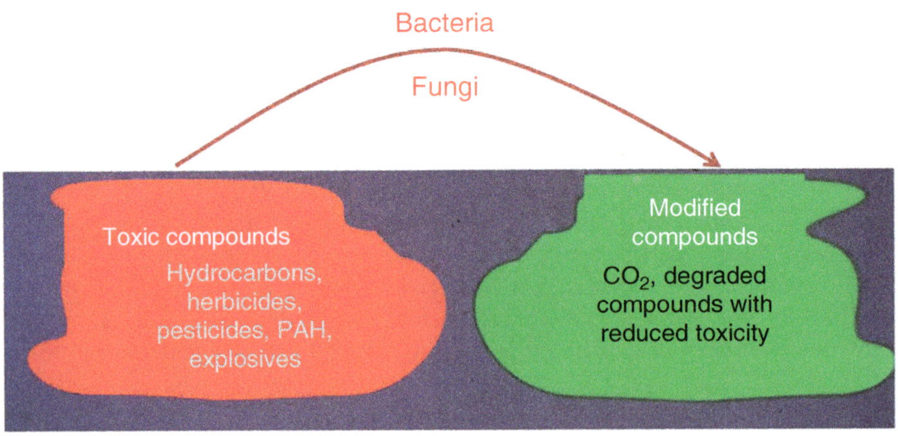

Figure 9.1 Goal of bioremediation of toxic organic compounds.

that are water soluble, destroy membrane function, or inhibit enzymatic processes. Selection of bacterial populations may result from sensory activities that induce genes, and the proteins that are specifically expressed enable the bacteria to resist the toxic effects of the polluting chemical. In some instances, resistance to toxic chemicals is carried on special plasmids in a few of the bacterial genera present and over time these resistance plasmids may be transferred to other bacteria in the environment.

The goal of bioremediation is to use microorganisms, principally bacteria, archaea or fungi, to convert contaminants to non-toxic chemicals or materials of reduced toxicity. While the intent is to convert contaminating organic chemicals to carbon dioxide by aerobic microorganisms, the initial enzymatic reactions of degradation are sufficient to detoxify some hazardous chemicals. With organic contaminants, the type of chemical involved may be extremely important in determining if bioremediation will be successful.

According to the United States Environmental Protection Agency (EPA 2006), there has been considerable bioremediation success at sites containing halogenated and non-halogenated volatile organic compounds (VOCs) and at sites contaminated with non-halogenated semi-volatile organic compounds (SVOCs). Relatively limited bioremediation success has been observed at sites contaminated with PAHs, organic pesticides and herbicides, or polychlorinated biphenyls (PCBs). When using bioremediation to treat soils and sediments contaminated with metals, the best outcome is the reduction of chemical toxicity by altering the oxidation state of the metal or reducing the mobility of the toxic metal (Tabak et al. 2005). Various designs have been used in bioremediation of toxic contaminants in soil and aquatic environments.

When comparing bioremediation with the physical/chemical alternatives, there are several factors to consider. Bioremediation is almost always relatively economical to implement and the pollutant is often completely degraded, so processing of the contaminated site is not required. However, there are some disadvantages to bioremediation. Bioremediation at polluted sites may require months to years, so when chemicals are an extreme health hazard to humans, society can't wait for microorganisms to metabolize the chemicals. Bacterial metabolism of chemicals may also result in the production of toxic byproducts. Thus, a careful evaluation of the chemicals involved and the chemical character of the soil or water environment are needed before deciding whether to use bioremediation or physical/chemical treatment.

9.2.1 Biostimulation

When the quantity of chemical contamination is low at a specific site and the bacterial activity is robust, remediation may be conducted using indigenous (native) bacteria. Millions of bacterial cells representing thousands of species may be present in a gram of soil. Although few bacteria with the metabolic system needed for remediation may be present initially, selective pressures will enable these bacteria to flourish and to remediate the contaminant. Indigenous bacteria are remarkably effective when soil containing low levels of petroleum wastes is applied to rich agricultural soil. This remediation strategy is referred to as land farming and it is not appropriate when the indigenous bacterial numbers are low due to an arid climate or when the site has a low nutritional level.

Addition of phosphate, nitrogen, or carbon compounds that support the growth of the desired bacteria is often used to enhance the rate of bioremediation and this is called biostimulation (Tyagi et al. 2011). Agricultural products including animal manure is often used as an inexpensive nutrient source for surface contamination sites and the additions can be easily incorporated into the soil. However, with subsurface contamination sites, chemicals are added using injection pipes. The high concentration of nutrients at the outlet of the pipe contributes to dense mats of bacterial growth, which reduces the translocation rates for the nutrients.

The rate at which a site can be decontaminated is related to the solubility of the pollutant because bacteria prefer water soluble substrates. Oil, tar, PAHs, and other water insoluble materials require surfactants to enhance the bioavailability of the lipophilic material and, therefore, are needed to increase bioremediation. Commercial surfactants such as Biosolve, S-200, Surfactant AB01039, and Inipol EAP 22 are biodegradable surfactants with Inipol EAP 22 being classified as a fertilizer because it also contains nitrogen and phosphate (Tyagi et al. 2011).

Chemicals tightly bound onto clay or mineral matrices are metabolized by bacteria at a low rate; however, the application of surfactants can solubilize the chemicals and increase their bioavailability. Many bacteria produce surfactants with structures containing fatty acids, neutral lipids, phospholipids, glycolipids or lipopeptides (Georgiou et al. 1999). Surfactants produced by bacteria include glycolipids by *Bacillus* sp. and *Aeromonas* sp., surfactin by *Bacillus subtilis*, and rhamnolipids by *Pseudomonas fluorescens* and *Pseudomonas aeruginosa* (Das and Chandran 2011). Therefore, when augmenting the bacteria at bioremediation sites where the bioavailability of the pollutant is low, surfactant-producing bacteria should be considered. In soils contaminated with heavy metals, bacteria producing surfactants are mostly gram-positive whereas in petroleum contaminated soils they are mostly common gram-negative bacteria (Bodour et al. 2003). The effectiveness of bacterial surfactants for solubilizing hydrocarbons is well known because biosurfactants produced by *B. subtilis* are important in the tertiary stage of the oil recovery process (Ron and Rosenberg 2001). The flavolipids, a new class of biosurfactants produced by *Flavobacterium,* have a structure similar to bacterial siderophores. Flavolipid surfactants bind metal ions and may be important in bioremediation of metal contaminated sites (Bodour et al. 2004). Peptidoglycan and other cell surface components of bacteria are able to bind and concentrate metal ions (reviewed in Beveridge 1989).

9.2.2 Bioaugmentation

9.2.2.1 Indigenous Bacteria

Frequently the contaminated site contains a low population of indigenous bacteria that can detoxify the spilled chemical, so more bacteria need to be introduced into the site; this is called bioaugmentation. Soil containing bacteria from another site contaminated with the same chemical is collected and mixed with the contaminated soil at the new site (Tyagi et al. 2011). In a few instances, bacteria or fungi capable of detoxifying the chemical may be isolated from a natural environment and cultivated in the laboratory under controlled conditions before they are added to the contaminated site. A laboratory study enables scientists to best assess the nutrients that would stimulate the growth of the desired bacterium and perhaps identify chemicals to stimulate the metabolic pathways used for detoxification.

9.2.2.2 Genetically Modified Organisms

Various commercial agencies market mixtures of bacterial species that may be used to promote decomposition and remediation, or engineered species of bacteria may be introduced into the polluted site. Ananda Chakrabarty introduced four distinct plasmids associated with hydrocarbon metabolism into *Burkholderia* sp. Although this organism was engineered to mineralize crude oil, it has never been released into the environment to remediate oil spills. Genetically modified organisms (GMOs) are not used in field bioremediation sites because of their potential to disrupt the natural ecosystem. Generally, bioaugmentation using bacteria from the contaminated site is more likely to have a longer effect in the environment than introduction of GMO.

Genetically modified bacteria have been developed to provide information about specific bioremediation processes. Bacteria have been constructed with an enhanced capability to degrade chlorinated organic compounds, PAHs, insecticides, and insoluble narcotics (Das and Chandran 2011). Laboratory studies with GMO enable scientists to study stress responses, appropriate inducers for metabolic pathways, the rate of decomposition of pollutants, products generated in bioremediation, and assessment of toxicities. Additionally, monitoring systems can be developed in laboratory studies for bioremediation practices in the field.

The concern that plasmids from the GMO can be readily transferred to indigenous bacteria in the environment is well founded. A report showed that when *Pseudomonas putida* containing plasmids for herbicide 2,4-D degradation were released into the environment, *P. putida* containing the plasmids readily disappear from the B horizon. However, 2,4-D degradation plasmids were readily transferred to indigenous soil bacteria in the B horizon (Dejonghe et al. 2000).

Another example of horizontal plasmid transfer in bioremediation is the *pheBA* operon, which encodes for two enzymes in the catabolism of phenol (see review by Perpetuo et al. 2011). *Pseudomonas putida* PaW85 carrying the conjugative plasmid with the *pheBA* operon was released into a river that had been contaminated by phenolic compounds. Analysis of the river water six years later revealed that *P. putida* PaW85 was not present but that the phenol degrading plasmid had been transferred to nine different strains of *Pseudomonas*. In one case, the plasmid carrying the *pheBA* operon was integrated into the chromosome of the new host. Continuous release of phenolics into the river accounted for the persistence of the *pheBA* operon in native bacteria even though the original host, *P. putida* PaW85, could not compete in the open environment.

Future use of genetically engineered bacteria for in situ bioremediation is dependent on the use of biological containment strategies (Mandell et al. 2015; Paul et al. 2005; Sayler and Ripp 2000). There are two major strategies for containment of GMO: (i) modification of the chromosome of the bacterium so that this added organism is unable to compete in the natural ecosystem; (ii) construction of a suicide factor in the plasmid or genome of the engineered bacteria so that in the absence of an inducer (the chemical contaminant), the GMO causes other proteins to be produced which kill it. For example, suicide factors may target membrane structure, electron transport systems, protein synthesis, or DNA replication.

9.2.3 Intrinsic Bioremediation

In situations where there is a small amount of contamination and public health is not an issue, bioremediation treatments are not established. These contamination sites are low risk and treatment attempts to leave the environment undisturbed by using indigenous microorganisms. This treatment strategy is called intrinsic bioremediation or natural attenuation (Box 9.1). A study of soil contaminated with diesel and oil indicated that intrinsic bioremediation was about half as effective as biostimulation (Margesin and Schinner 2001).

9.2.4 Microbial Consortium

Organisms will form associations in the environment that enable greatest rates of growth, metabolism, and species stability. In a microbial consortium, two or more organisms grow in close proximity and perform metabolic activities generally not possible for a single species (Brenner et al. 2008). Microbial consortia display several unique characteristics: (i) populations maintain homeostasis, (ii) horizontal gene transfer readily occurs, and (iii) communication between cells of the same species is possible. In bioremediation, the consortium designates

Box 9.1 Natural Attenuation

The term "natural attenuation" is used in discussing treatment measures to reduce pollution in the environment. Simply stated, natural attenuation is the use of natural processes to reduce the impact of the pollutant. There is a tendency to assume that natural attenuation refers only to the use of microorganisms in the environment to metabolize the contaminant and reduce the concentration of the toxic pollutant. However, there are also nonbiological attenuations including the following: (i) sorption of contaminant to soil particles which prevents pollution of groundwater; (ii) reduction of water soluble toxic pollutants through dilution by groundwater; (iii) some volatile pollutants will slowly evaporate into the air where the UV destroys the organic molecules. Thus, the success of natural attenuation varies with the environment and the type of pollutant.

bacterial species to work collectively to detoxify hazardous chemicals. There are numerous reports of bacterial consortia used in the detoxification of pollutants, including the bioremediation of PAH (Thavamani et al. 2012), fluorobenzene (Carvalho et al. 2002), 4-chloroaniline (Radianingtyas et al. 2003), hydrocarbons (Sathishkumar et al. 2008), tannery wastewater (Sivaprakasam et al. 2008), and metals (Borrok et al. 2004).

It is rare that a single type of chemical is involved in pollution as usually various molecular structures are present, as for industrial wastes, PAHs, oil, and diesel, benzene, toluene, ethylbenzene, and xylene (BTEX) spills. Also, heavy metals are commonly present with toxic organic compounds in many pollution sites. Different bacterial strains are needed for the microbial metabolism of the *m*-, *o*-, and *p*-isomers of aromatic compounds (Tabak et al. 1964) and the metabolic activity several bacterial species are required for the complete mineralization of phenolic compounds.

While consortia are common for bacterial systems, the metabolic activities of fungal consortia are also important. Textile dyes contain high levels of metals and an effective remediation of dye wastes has been proposed using *Rhizopus oryzae, Aspergillus terreus,* and *Aspergillus lentulus* as a fungal consortium (Mishra and Malik 2014). In the environment, many different microbial associations contribute to the detoxification of pollutants.

An interesting consortium involving bacterial and archaeal cells may be involved in the oxidation of methane. Methane oxidation in anaerobic marine environments has been studied by many marine ecologists, but it has been difficult to provide evidence of the microorganisms involved. The isolation of a marine consortium consisting of about 100 cells of a methanogen with numerous sulfate-reducing bacterial cells on the periphery may account for the anaerobic oxidation of methane (Boetius et al. 2000). Certainly, there are numerous instances of a structured consortium between microorganisms.

9.2.5 Co-metabolism

Co-metabolism in bioremediation is the microbial decomposition of an organic molecule (secondary substrate) while the microorganism is growing on another compound (primary substrate) (see Table 9.2). Often, the co-metabolism of a compound does not provide energy for the bacteria to grow but may reduce the level of a chemical pollutant to a safe level. Sometimes, the products of secondary substrate decomposition are not used by the bacteria responsible

Table 9.2 Enzymes associated with co-metabolism of numerous polyaromatic and halogen-containing chemical pollutants (Hazen 2010).

Enzymes	Natural substrate	Bacterial source of enzymes
Methane monooxygenase, Methanol dehydrogenase, Alkene dehydrogenase, Catecol dioxygenase	Methane, methanol, propane	*Methylobacteria*
Ammonia monooxygenase	Ammonia	*Nitrosomonas*
Toluene monooxygenase, Toluene dioxygenase	Toluene, butane	*Pseudomonas, Rhodococcus*
Alcohol dehydrogenase	Methanol	*Pseudomonas, Streptomyces*
Dehalogenase	Glucose, acetate	Methanogens, *Clostridium, Desulfovibrio, Geobacter*

for the breakdown of the secondary substrate. Co-metabolism can be best understood by reviewing some instances where it has been demonstrated.

- *Mycobacterium vaccae* can grow as the result of energy obtained from the oxidization of propane. The enzyme oxidizing propane can also oxidize cyclohexane to produce cyclohexanol. *Pseudomonas* sp. can metabolize cyclohexanol but not cyclohexane (Ryoo et al. 2000). The oxidation of cyclohexane by a propane oxidizing enzyme is co-metabolism.
- *Pseudoxanthomonas* sp. wax can co-metabolize 1,1,1-trichloro-2,2-bis(*p*-chlorophenyl) ethane (DDT) while growing on glucose (Wang et al. 2010).
- Bacterial isolates that metabolize 17β-estradiol will co-metabolize 17α-ethinylestradiol, the essential chemical in the contraceptive pill, to nanogram levels (Pauwels et al. 2008).

Microbial bioremediation is most effective when co-metabolism is used.

9.3 Organic Compounds and Xenobiotics Degraded

Due to the diversity of biological organisms present on Earth, there is a continuous supply of many different types of organic molecules. Although not all biological systems are degraded immediately, all are ultimately degraded because in the ecosphere biomolecules do not accumulate. This degradation of diverse organic molecules requires many enzymes. Organic pollutant molecules are most commonly xenobiotics that are not natural but generated by chemical synthesis and added in large quantities. Xenobiotic molecules may have bonds that few bacteria can break, so bacteria that can mineralize toxic pollutants have to be added to the polluted site. For example, a recent discovery has great potential to reduce the accumulation of plastic in the environment. Some plastics are constructed from polyethylene terephthalate (PET). The bacterium *Ideonella sakaiensis* has been isolated that uses PET as a carbon and energy source for growth (Yoshida et al. 2016). Discovery of this bacterium could have a huge impact on the plastic industry.

9.3.1 Pesticides

In this text, pesticide is used to include any compound that destroys or inhibits pests. The use of pesticides here includes herbicides, fungicides, insecticides, nematicides, and molluscicides. The sale and use of pesticides have been regulated in many countries in recent decades but in developing countries numerous areas contaminated with pesticides remain. The decomposition of pesticides is markedly slow and many persist in the soil for as long as 20 years (Alexander 1994; Madigan et al. 2009). According to the US EPA (2010) review, there are four major groups of pesticides and these include the following, with examples in parenthesis: (i) organophosphate molecules (parathion and malathion), (ii) carbamate molecules [aldicarb, carbofuran, carbaryl (Sevin), fenoxycarb], (iii) organochlorine insecticides (DDT, dicofol, heptachlor, endosulfan, chlordane, aldrin, dieldrin, mirex, pentachlorophenol), and (iv) pyrethroid pesticides (permethrin). The diversity of molecular structures used as pesticides is apparent from the structures of insecticides given in Figure 9.2 and herbicides given in Figure 9.3.

As reviewed by Arif et al. (2012), the remediation of sites by pesticide has been attributed to the use of various bacteria including *Pseudomonas* sp., *Micrococcus* sp., *Sphingobacterium* sp., *Klebsiella jilinsis, Burkholderia cepacia, Paenibacillus polymyxa,* and *Chryseobacterium* spp. Several fungal strains contributing to the metabolism of pesticides include *Panus conchatus, Trametes versicolor, Pleurotus ostreatus, Trichoderma* sp., *Aspergillus oryzae,* and *Ganoderma austral* (Arif et al. 2012). *Spirulina platensis* and *Microcystis novacekii,* two cyanobacteria,

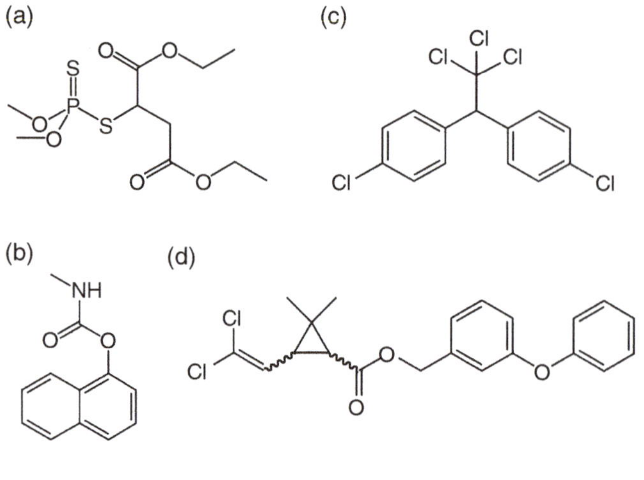

Figure 9.2 Structures of common insecticides: (a) malathion (an organophosphate); (b) carbaryl (a carbamate); (c) dichlorodiphenyltrichloroethane, DDT (an organic molecule containing chlorine); (d) permthrin (a pyrethroid).

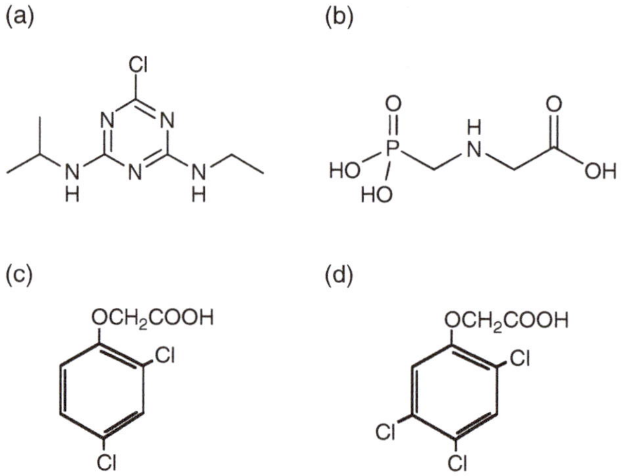

Figure 9.3 Structures of herbicides: (a) atrazine; (b) glyphosphate (Roundup™); (c) 2,4-dichlorophenoxyacetic acid (2,4-D); (d) 2,4,5-trichlorophenoxyacetic acid (2,4,5-T).

degrade methyl parathion and *Haloarcula* sp. and *Haloferax* sp., two archaea, metabolize heptadecane (Arif et al. 2012). Thus, enzymes that are capable of degradation of pesticides are found in many indigenous microorganisms. Recently, *Stenotrophomonas* sp. and *Pseudomonas syringae* have been genetically engineered to degrade organophosphate pesticides (Yang et al. 2010). Microbial detoxification of pesticides results from the following enzymatic activities: (i) by deamination (removal of $-NH_2$) groups, replace chloride, bromide, or iodide with a hydrogen atom, (ii) use of esterases or etherases to reduce the size of the molecule, (iii) addition of hydroxyl or methyl groups, (iv) reduce triple or double bonds to a single bond between carbon and nitrogen atoms, and (v) removal of methyl or methoxy groups from the pesticide (Alexander 1994).

9.3.2 Chlorinated Organic Compounds

There are several chlorinated organic carbon compounds that are of concern because of their toxicity and persistence in the environment. While utilization of chlorine (or other halogens) is not of metabolic importance by environmental groups, the enzymology of forming chlorine–carbon bonds does exist. Small concentrations of chlorinated organic compounds are found in marine algae (Gribble 1996) and 4-chloroindole-3-acetic acid, which is a plant hormone (Magnus et al. 1997). Therefore, it should not be impossible to degrade chlorinated organic compounds but the rates or metabolism may be slow. Some insecticides (malathion, lindane, and dichlorodiphenyltrichloroethane or DDT) an herbicides (atrazine, dalapin, 2,4-dichlorophenoxyacetic acid or 2,4-D, and 2,4,5-trichlorophenoxyacetic acid or 2,4,5-T) also contain chlorine and this may contribute to their resistance to biodegradation.

Toxicities associated with some of the chlorinated pesticides had resulted in their restricted use. A highly effective defoliant used in Vietnam was Agent Orange (a mixture of 2,4-D and 2,4,5-T); human toxicities attributed to this compound resulted in banning the worldwide use of this compound.. Insects have become resistant to 2,4-D and the World Health Organization (WHO) has restrictions on its global use. Lindane, hexachlorcyclohexane, has been associated with damage to the neurological system, liver, and kidney and WHO has restricted the use of lindane to treatment for lice and scabies.

9.3.2.1 Chloroethylenes

Chlorinated volatile organic compounds (CVOCs) associated with chemical pollution and remediation by bacteria include tetrachloroethene or perchloroethene (PCE), trichloroethene (TCE), *cis*-1,2-dichloroethene (*cis*-DCE), vinyl chloride (VC), 1,1,2,2-tetrachloroethane (1,1,2,2-PCA), 1,1,1-trichloroethane (1,1,1-TCA), carbon tetrachloride (CT), chloroform (CF), and dichloromethane or methylene chloride (MC). These CVOCs are metabolized by either oxidation (electron donor) or reduction (electron accepting) reactions and sometimes the chlorinated compound is used as the carbon source.

Reductive dehalogenation of CVOCs of chlorinated compounds occurs when these molecules are used as the electron acceptor and this is frequently called dehalo-respiration. The steps involving dehalogenation of PCE to produce ethene (ethylene) are shown in Figure 9.4.

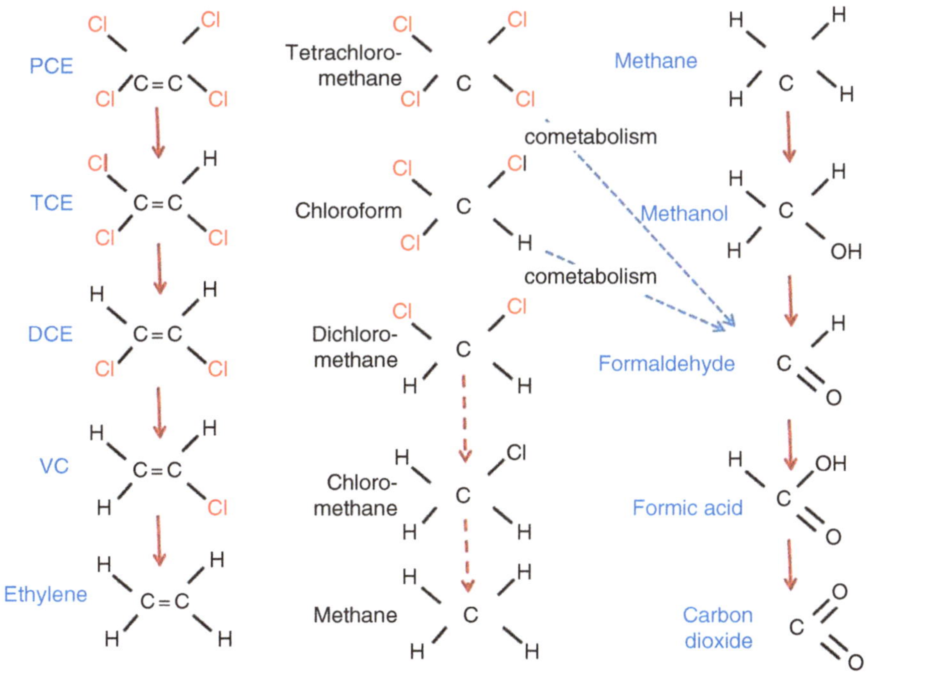

Figure 9.4 Structures of chlorinated volatile organic compounds with pathways for degradation.

Box 9.2 Oxygenase Enzyme Reactions

Oxygenase enzyme reactions require the presence of O_2. Electrons from O_2 are transferred to the substrate and the oxygen atom is incorporated into the product. In contrast, "oxidase" is used to describe any enzyme reaction that accounts for the oxidation of a substrate by catalyzing electron flow from a substrate.

Reduction of PCE and TCE has been reported for *Dehalobacter restrictus* (Maillard et al. 2003), *Desulfitobacterium hafniense* (Gerritse et al. 1999), *Dehalospirillum multivorans* (Neuman et al. 2002), *Desulfuromonas chloroethenica* (Krumholz 1997), and *Dehalococcoides ethenogenes* (Magnuson et al. 2000). In environments where PCE and TCE have been dehalogenated, *cis*-DCE and VC will accumulate unless an organism such as *D. ethenogenes* is present to complete the dechlorination to produce ethene (Maymó-Gatell et al. 2001).

Oxidation of TCE is attributed to low substrate specificity of monooxygenase and dioxygenase enzymes produced by bacteria. These enzymes degrade CVOCs and aromatic hydrocarbons without benefit to the bacteria that produce the oxygenase (Box 9.2), a characteristic of co-metabolism. TCE degradation has been reported for: *Pseudomonas* sp. JS150 (currently classified as *Burkholderia*), which produces a toluene monooxygenase (Johnson and Olsen 1995); *P. putida* expressing a toluene dioxygenase (Wackett and Gibson 1988); *M. vaccae,* which has a propane monooxygenase (Wackett et al. 1989); *Nitrosomonas europaea* using ammonia monooxidase (Arciero et al. 1989); and *Methylosinus trichosporium* using methane monooxygenase (MMO) (Jahng and Wood 1994). While the involvement of monooxygenases and dioxygenases in oxidation of TCE has mostly been ascertained from genetic analysis, the partially purified monooxygenase from *Stenotrophomonas maltophilia* degrades TCE as well as CF, toluene, hexane, and benzene (Mukherjee and Roy 2013). Methanotrophic genera (*Methylobacter, Methylococcus, Methylocystis, Methylosinus,* and *Methylomicrobium*) produce MMO which is either soluble in the cell (sMMO) or associated with the membrane (pMMO). Bacteria expressing sMMO may use a broader spectra of substrates than the bacteria producing pMMO and although sMMO degrade substrates at faster than pMMO, the pMMO producing bacteria may better withstand the chlorinated ethenes accumulating from TCE oxidations (Semrau 2011). Due to the broad distribution of bacteria producing MMOs, the methanotrophs are important indigenous bacteria for in situ bioremediation.

9.3.2.2 Chloromethanes

The oxidation of carbon tetrachloride, CF, dichloromethane, and chloromethane (methyl chloride) are mineralized by reductive dehalogenases produced by bacteria under anaerobic and aerobic conditions. The steps in sequential degradation of carbon tetrachloride to methane and methane oxidation to carbon dioxide are provided in Figure 9.4.

Degradation of tetrachloromethane (carbon tetrachloride) occurs under anaerobic conditions by co-metabolism processes, but microbial growth with tetrachloromethane as the only carbon source has not been demonstrated (Penny et al. 2010). Methanogens (*Methanobacterium thermoautotrophicum* and *Methanosarcina barkeri*), acetogenic bacteria (*Moorella thermoacetica* and *Acetobacterium woodii*), and sulfate reducing bacteria (*Desulfobacterium autotrophicum*) degrade tetrachloromethane by reactions involving metal-bound coenzymes in these anaerobes. The rate of degradation of tetrachloromethane is enhanced by humic acids and other chemicals serving as electron shuttles between the anaerobic cells and tetrachloromethane. Although there are a few reports of using CF as the final electron acceptor, co-metabolism is the major process used for biodegradation of CF (Cappelletti et al. 2012). As for tetrachloromethane, CF degradation is associated with sulfate-reducing, methanogenic, acetogenic, and fermenting microbial cultures.

Unlike tetrachloromethane and CF, dimethyl chloride and methyl chloride can be used by microorganisms as a carbon and energy source for anaerobic growth (Leisinger and Braus-Stromeyer 1995). Additionally, there are numerous examples of the degradation of dichloromethane and methyl chloride by bacteria using co-metabolism (McAnulla et al. 2001a; Semrau 2011). The biodegradation of dichloromethane (and dibromomethane) occurs in both aerobic and anaerobic methylotrophic bacteria with the conversion of dichloromethane to $2Cl^-$ and formaldehyde (Krausova et al. 2003). Dichloromethane dehydrogenase is encoded on the *dcmA* gene which is found in *Methylobacterium dichloromethanicum* (currently *M. extorquens*) and is presumed to be highly mobile due to horizontal transfer (Kayser et al. 2002). Other bacteria known to use dichloromethane as a carbon source include *Methylophilus* sp., *Pseudomonas* sp., and *Hyphomicrobium* sp. Methyl chloride and methyl bromide degradation with assimilation of the methyl carbon to support growth is attributed to the *cmuA* and *cmuB* genes (Miller et al. 2003; McAnulla et al. 2001a) and gene probes have been developed to screen for chloromethane bacteria in the environment (McAnulla et al. 2001b). Bacteria capable of degrading chloromethane are found in many environments and the organisms best studied include *Hyphomicrobium chloromethanicum* and *Methylobacterium chloromethanicum* (McAnulla et al. 2001a).

9.3.2.3 Polychlorinated Biphenyl Compounds

Polychlorinated biphenyl compounds have been produced for various industrial uses including fire retardants, plasticizers, heat transfer, and hydraulic fluids; however, when released into the environment they persist for a long time and their toxicity to the ecosystem and human health is of concern. As indicated in Figure 9.5, chlorine may be attached to any of the carbons of the biphenyl molecule, which provides considerable structural variability for bioremediation. The PCBs bind to clay in the soil, which results in low bioavailability and limits bioremediation. Natural or industrial products used as surfactants release PCBs from the soil and increase their bioavailability (Seeger et al. 2010). The bacterial degradation of PCBs may be by anaerobic dehalogenation or by aerobic degradation.

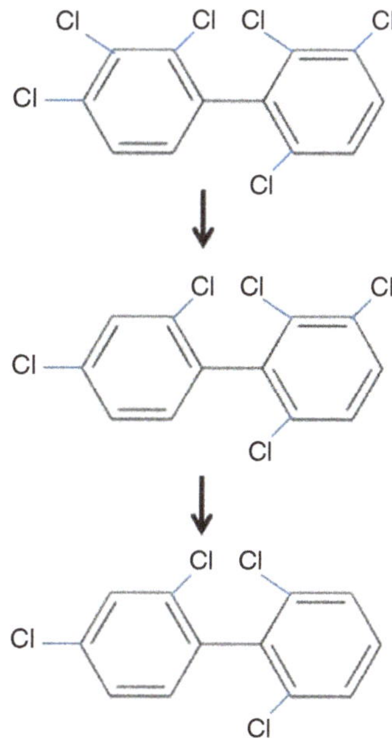

Figure 9.5 Dechlorination patterns of PCB by *Dehalococcoides ethenogenes*.

Anaerobic dehalogenation. PCB molecules can serve as alternate electron acceptors for certain bacteria. Reductive dechlorination of PCB has been associated with metabolism by strains of *D. ethenogenes* and the proposed pathway for degradation is listed in Figure 9.5. The specificity of enzymatic removal of chlorine at *para, meta,* and *ortho* positions has been reviewed by Seeger et al. (2010) and Pieper and Seeger (2008). The bacteria responsible for anaerobic dechlorination of PCB are unable to degrade the chlorinated PCBs that result from reduction processes. This would suggest that another process would be required to complete the degradation process and further degradation may occur by aerobic metabolism.

Aerobic metabolism. Aerobic bacteria (*Pseudomonas, Bacteria, Comamonas, Cupriavidus, Sphingomonas Acidovorax, Rhodococcus,* and *Burkholderia*) are capable of degrading PCBs

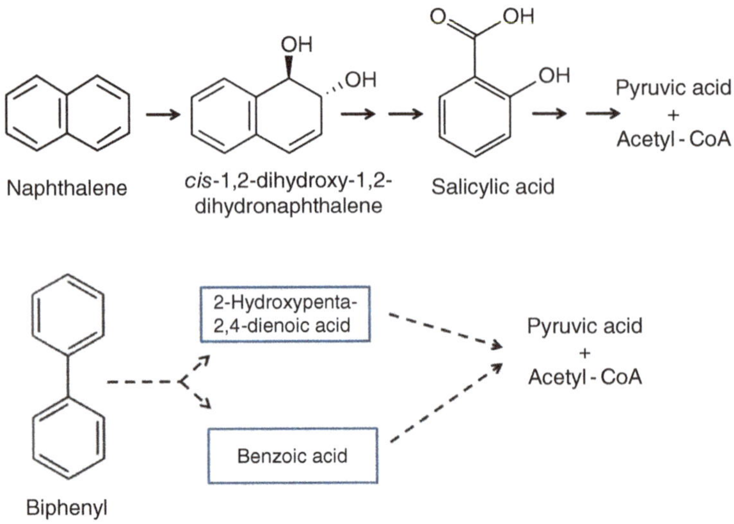

Figure 9.6 Aerobic dioxygenase reactions with naphthalene and biphenyl.

by using the pathway for degradation of biphenyl. The rate of degradation of PCBs is dependent on the enzyme from specific isolated bacteria because each enzyme is specific for chlorines positioned at various sites on the biphenyl (Pieper and Seeger 2008). The dioxygenase and dehydrogenase enzymes for PCB degradation are located on the *bph* gene cluster and the enzymes produced by *Burkholderia xenovorans* LB400 have broad specificity for sites containing chlorides. Like the biphenyl degradation shown in Figure 9.6, one of the phenyl molecules from PCB is degraded to produce benzoate and the other to give 2-hydroxypenta-2,4-dienoic acid. Both products are further metabolized by catabolism using the TCA cycle (Furukawa et al. 2004). Bacteria with PCB-degrading genes are present in the soil but these genes need to be induced for enzyme production. Nontoxic inducers for PCB degradation include terpenes, flavonoids, and other plant exudates (Pieper and Seeger 2008). Due to the structural complexity of PCBs, co-metabolism of mixed cultures is required for complete degradation.

9.3.3 Population Dynamics in Degradation of Hydrocarbons

9.3.3.1 Oil Spills

Crude oil contains hundreds of hydrocarbon compounds and when it contaminates soil or water it causes huge problems for society. The five major oil spills in the world are listed in Table 9.3. The Exxon Valdez oil tanker disaster mobilized society, including the scientific community, to increase their efforts for in situ bioremediation (Atlas and Hazen 2011). There are two general approaches for bioremediation of oil spills: (i) biostimulation by adding chemicals to enhance indigenous bacteria and (ii) bioaugmentation, which is the addition of oil-degrading bacteria from an oil remediation site to supplement the population of indigenous bacteria (Das and Chandran 2011; Wang et al. 2011). As discussed previously, genetically modified bacteria are not added to open environments but may be used in controlled laboratory studies.

The diversity and insolubility of chemical species in crude oil complicates the rate at which bioremediation occurs. The degradation of *n*-alkanes, various aliphatic compounds, and aromatic hydrocarbons in oil requires enzymes from different bacterial species. *Methylobacteria* use the sMMO and pMMO to degrade C_1–C_8 alkanes and cycloalkanes. Various species of *Pseudomonas, Burkholderia,* and *Rhodococcus* have alkane hydroxylases that metabolize

Table 9.3 Largest oil spills in the world.

Size of spill (million barrels)	Name of spill	Country or region	Year
9	Lakeview gusher	California	1910–1911
~4.2	Deepwater horizon	Gulf of Mexico	2010
2–6	Gulf War Oil Spill	Iraq, Kuwait	1991
~3.4	Ixtoc 1	Gulf of Mexico	1979–1980
~0.25–7.5	Exxon Valdez	Alaska	1989

Source: Based on data from (Wang et al. 2011).

C_5–C_{16} alkanes, alkyl benzenes, and cycloalkanes. Dioxygenases produced by *Acinetobacter* sp. degrades C_{10}–C_{30} alkanes. PAHs are markedly insoluble and their metabolism is discussed in Section 9.3.3.3.

Hydrocarbon-degrading bacteria appear to be ubiquitous throughout marine and terrestrial ecosystems; however, they may be a small fraction of the microbial population in unpolluted environments. Therefore, when oil spills occur there is a change in the bacterial population. Using experimental plots, MacNaughton et al. (1999) found that gram-negative bacteria within the *Alphaproteobacteria* increased in quantity following oil addition. In marine environments, a group of obligate hydrocarbonoclastic bacteria (OHCB) are considered important in degradation of hydrocarbons (Yakimov et al. 2007). The OHCB group includes *Alcanivorax, Marinobacter, Thallassolituus, Cycloclasticus,* and *Oleispira,* which are undetected in the environment before an oil spill but their numbers substantially increase after. The members of the OHCB responsible for oil degradation may vary in different geographical regions; the OHCB group have extremely low numbers among the indigenous bacteria in unpolluted seawater. The OHCB with this unique hydrocarbon metabolism may possibly be present in seawater because there is a natural release of oil from the ocean floor.

9.3.3.2 Fuel Hydrocarbons

Saturated hydrocarbons are a major component of petroleum-derived diesel and a smaller fraction consists of naphthalenes, alkylbenzenes, and other aromatic hydrocarbons. In diesel-contaminated groundwater, the dominant bacteria belong to *Proteobacteria, Bacillus,* and *Actinobacteria* phyla (Rodríguez-Martinez et al. 2006). Evaluation of the bacterial population using functional gene array revealed that 270 genes for organic contamination degradation were present, which included enzymes for the metabolism of naphthalene, toluene, octane, biphenyl, xylene, benzene, pyrene, and phenanthrene. Certainly, the decomposition of diesel fuel requires a sizeable bacterial population to provide the genes required. The hydrocarbons of interest in bioremediation of sites contaminated with gasoline are the monoaromatic compounds of BTEX. The degradation of BTEX compounds by aerobic indigenous bacteria in the environment is an attractive treatment technology; however, the high oxygen demands for this biodegradation often necessitates the injection of oxygen releasing compounds into the environment or the use of engineered treatments involving bioventing or biosparging.

Aromatic oxygenases are involved in the initial steps of aerobic biodegradation of BTEX (Figure 9.7). The toluene monooxygenase of *P. putida* F1 is capable of oxidizing >100 compounds including BTEX and the versatility of some of these oxygenases is revealed in their degradation of numerous aromatic substrates. Toluene degradation may involve either a toluene monooxygenase pathway or a toluene dioxygenases path depending on the bacterial

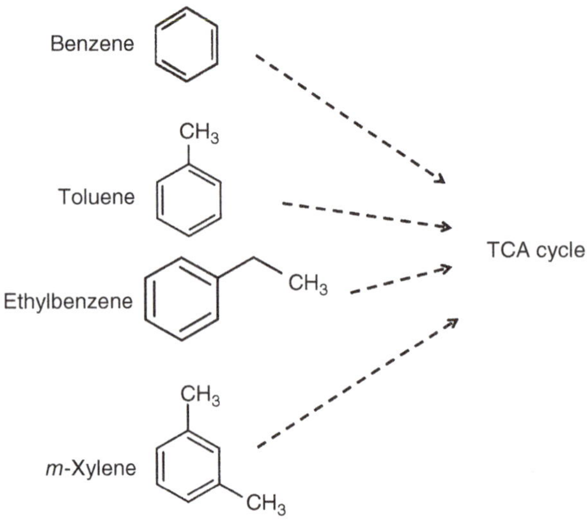

Figure 9.7 Aerobic degradation of BTEX compounds.

system involved. BTEX and other volatile fuel hydrocarbons are ultimately degraded to pyruvate and acetyl, which are metabolized via the TCA cycle.

As summarized by Xie et al. (2010), numerous aerobic bacteria and *Actinobacteria* phyla have been associated with BTEX degradation. Benzene degradation has been reported for members of the *Alphaproteobacteria* and *Betaproteobacteria* phyla (*Pseudomonas, Burkholderia, Ralstonia, Achromobacter, Pseudoxanthom, Hydrogenophaga, Alicycliphilus, Acinetobacter, Polaromonas,* and *Rhodococcus*), members of the *Actinobacteria* phylum (*Rhodococcus* and *Arthrobacter*) and members from the *Acidobacteria* phylum (*Acidobacterium*). The use of terminal restriction fragment length polymorphism (TRFLP) has enabled identification of many aerobic benzene degraders (Xie et al. 2010) and PCR primer sets have detected aromatic monooxygenase and dioxygenase genes contributing to aerobic degradation of BTEX (Baldwin et al. 2009; Nebe et al. 2009; Hendrickx et al. 2006).

In addition to BTEX, other compounds with aromatic rings are cleaved by different bacterial systems using dioxygenases. Biphenyl dioxygenase, the initial enzyme in decomposition of biphenyl and chlorinated biphenyl compounds, is encoded on the *bphA* gene and the *pbh* gene cluster is characteristically associated with the metabolism of biphenyl. The *pbh* gene cluster has been reported in *Pseudomonas pseudoalcaligenes, P. putida* KF715, *Burkholderia* sp. strain LB400, *Ralstonia eutropha* A5, *Rhodococcus* sp. strain M5, and *Comamonas testosterone* B-356 (Furukawa et al. 2004).

Anaerobic degradation of BTEX occurs with numerous physiological groups of bacteria where the BTEX molecules are the electron donor with NO_3^-, Fe^{3+}, Mn^{4+}, ClO_4^-, or SO_4^{2-} as electron acceptors. As reviewed by Weelink et al. (2010), most research uses enrichment cultures and although a few pure cultures are identified, more information is needed about the metabolic pathways of BTEX bioremediation. Anaerobic degradation of BTEX, based on current understanding, is given in Figure 9.8. Bacteria growing on toluene with nitrate as the electron acceptor include those in the genera of *Thauera, Azoarcus,* and *Magnetospirillum* and *Geobacter metallireducens* use NO_3^-, Fe^{3+} or Mn^{4+} as electron acceptors. *Desulfobacula toluolica* and *Desulfotignum toluenicum* use sulfate while *Dechloromonas aromatica* uses perchlorate (ClO_4^-). In anaerobic environments, xylene degradation is influenced by molecular structure with *m*-xylene more rapidly degraded than *o*-xylene or *p*-xylene. *Desulfosarcina* sp. selectively oxidizes *o*-xylene with sulfate as the electron acceptor while *Azoarcus* sp. oxidizes *m*-xylene.

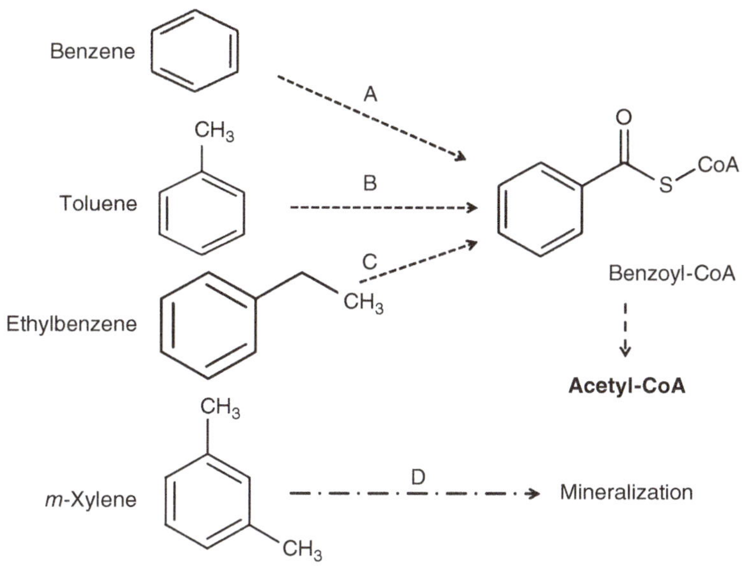

Figure 9.8 Anaerobic degradation of BTEX compounds.

Pure cultures that oxidize *p*-xylene have not yet been obtained. Aside from *Dechloromonas* and *Azoarcus*, benzene oxidation has been best achieved with mixed bacterial cultures. Ethyl benzene degradation occurs with enrichments for nitrate-reducing and sulfate-reducing bacteria.

Since breaking the benzene ring is difficult for anaerobic bacteria, the initial reaction leading to degradation of benzene is thought to be either (i) methylation, (ii) hydroxylation, or (iii) carboxylation (Vogt et al. 2011). A central intermediate in anaerobic benzene degradation is benzoyl-CoA, which is degraded to complete the process of mineralization. The gene associated with the initial anaerobic metabolism is *abcA* which encodes for benzene carboxylase. The *bssA* gene, which encodes for benzyl-succinate synthase, is the initial step in anaerobic degradation of toluene and is assumed to be important for all BTEX compounds Figure 9.6. Recently, the degradation of benzene by *G. metallireducens* has been shown to be by the pathway of phenol metabolism and the genes for this decomposition of benzene are designated as *GMET* (Zhang et al. 2014).

9.3.3.3 Polyaromatic Hydrocarbons

Polyaromatic hydrocarbon (PAH) compounds contain polyaromatic rings which consist entirely of C, O, and H and do not carry substituents or side chains. Although over 100 PAHs have been associated with crude oil, coal, and tars, 16 PAH compounds (see Figure 9.9 for 15 PAH compounds; the 16th for naphthalene is given in Figure 9.7) have been identified by United States Environmental Protection Agency (US EPA) as pollutants of major concern. Numerous reviews are available on bacterial bioremediation of PAHs and the reader is referred to these for additional information (Arun and Eyini 2011; Haritash and Kaushik 2009).

Naphthalene. The degradation of naphthalene has attracted the greatest amount of research because it is the smallest and most water soluble of the PAH compounds. Several strains of *Pseudomonas* produce naphthalene dioxygenase, which is the first enzyme in the degradation of naphthalene and is capable of degrading numerous substrates (Parales et al. 2000). Other bacterial strains reported to produce naphthalene dioxygenase include *Burkholderia* sp. strain LB400, *Comamonas testosteroni* B-356, *Sphingomonas yanoikuyae* B8/36, *Pseudomonas stutzeri* AN10

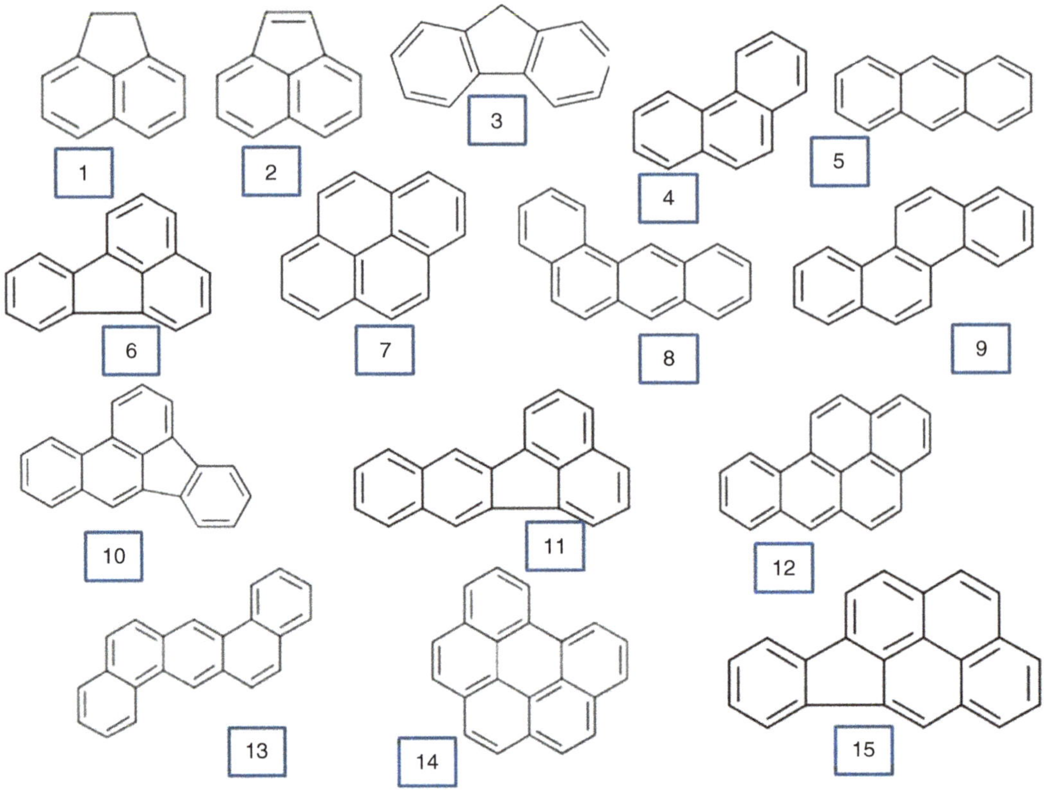

Figure 9.9 PAHs designated by US EPA as priority pollutants: (1) acenaphthylene; (2) acenaphthene; (3) fluorene; (4) phenenthrene; (5) anthracene; (6) fluoranthene; (7) pyrene; (8) benz[a]anthracene; (9) chrysene; (10) benzo[b]fluoranthene; (11)benzo[k]fluoranthene; (12) benzo[a]pyrene; (13) dibenz[a, h]anthracene; (14) benzo[g, h, i]perylene; (15) indeno[1,2,3 cd]pyrene). Naphthalene (see Figure 9.7) is also designated by US EPA as a significant pollutant.

and *P. putida* G7. An abbreviated pathway proposed for some *Pseudomonas* sp. to use naphthalene as a carbon and energy source is given in Figure 9.6.

Three and four ring PAHs. Heavy aromatic compounds of three or more ring constructions (see Figure 9.9) are markedly insoluble in water and surfactants are required (see Section 9.2.1) to enhance their bioavailability. Studies on the degradation of heavier PAHs has focused on the aerobic bacteria involved in the process, the metabolic pathways used by aerobic bacteria for PAH degradation, and the molecular techniques used to evaluate PAH catabolic genes (Aguawal and Shahi 2015; Sihag and Pathak 2014; Haritash and Kaushik 2009; Mrozik et al. 2003; Arun et al. 2010). Some of the bacteria capable of degrading PAH compounds are listed in Table 9.4.

Current research indicates that several aerobic bacterial strains catabolize PAHs but the focus of enzyme attack may be different for each bacterial species (Mrozik et al. 2003). For example, the initial cleavage of the five membered ring of fluorene by *Pseudomonas* sp. F274 produces 2,3-dihydroxy-biphenyl which is metabolized via the biphenyl pathway. However, *Pseudomonas cepacia* F297, *Brevibacterium* sp. DPO1361, *Staphylococcus auriculans* DBF63, and *Arthrobacter* sp. F101 each use different catabolic pathways to degrade fluorene (Mrozik et al. 2003). Bacteria that are associated with PAH catabolism are presented in Table 9.4. With the tricyclic aromatic hydrocarbons, anthracene and phenanthrene, there are several different pathways for bacterial degradation. Genes encoding catabolic PAH enzymes can be on

Table 9.4 A partial listing of bacteria decomposing PAHs.

Chemical	Bacteria
Phenanthrene	*Pseudomonas* sp., *Massilia* sp., *Sphingomonas* sp., *Brevibacillus* sp.
Anthracene	*Microbacterium* sp., *Martelella* sp., *Ochrobactrum* sp., *Rhodococcus opacus*
Fluoranthene	*Herbaspirillum chlorophenolicum*
Pyrene	*Klebsiella oxytoca, Bacillus vallismortis, Diaphorobacter* sp.

plasmids or located on the bacterial chromosome. With catabolism of PAHs by aerobic bacteria, monooxygenase or dioxygenase enzymes seem to be involved, with the products of catabolism entering the TCA cycle for complete mineralization.

Bacterial decomposition of PAHs has been studied in polluted soil and marine sediments. Using a denaturing gradient gel electrophoresis technique revealed a shift in bacterial communities in a creosote-contaminated soil as decomposition progressed (Viñas et al. 2005). Anaerobic degradation of PAHs occurs under denitrifying conditions (Minati and Marsan 2005) and under sulfate-reducing conditions (Coates et al. 1997); the specific anaerobic bacteria and the metabolic pathways for anaerobic degradation remain unknown.

9.3.3.4 Azo Dyes

There are several classes of molecules that account for the colors of dyes, with azo and anthraquinone chromophores present in over half of the synthetic dyes produced. The industrial application of dyes is in the production of textiles, leather, and cosmetics and in food processing and paper production. There are approximately 100 000 dyes produced, with azo dyes making up about 60–70% of the commercial market. Azo dyes have one or more $R—N{=}N—R^1$ functional groups and produce a rainbow of colors (see Figure 9.10 for examples of azo dye structures). Azo dye molecules have a low toxicity; an average person would need to consume over 100 kg in a day to show any ill effects. In the case of azo dye pollution, millions of gallons containing many tons of dye had been released into the environment. The breakdown products of azo dyes could be harmful to humans and so in 2002 the European Union banned the addition of azo dyes to food (Lehto et al. 2017) and the presence of certain azo dyes in consumer goods (Piccinini et al. 2008).

In countries with textile mills, azo dyes are a major problem when they are released in industrial wastewater because they alter the pH, clarity, and biological demand of the water at the release site (Sudha et al. 2014). Bacteria isolated from sites containing azo dyes include *Morganella* sp., *Sphingomonas* sp., *Proteus hauseri, Aeromonas caviae,* and *Rhodococcus globerulus* (Aguawal and Shahi 2015). Anaerobic bacteria, including *Desulfovibrio* spp., will reduce and decolorize dyes using the end products of metabolism or reducing enzymes produced by the bacteria (Barton and Fauque 2009). Various anaerobic intestinal bacteria produce toxic aromatic amines and aromatic nitroso compounds by reducing azo dyes, so the consumption of the dyes by humans is a health concern (Chung et al. 1992).

9.3.4 Explosives

9.3.4.1 Trinitrotoluene

There are only a few naturally produced nitroaromatic compounds, which is why 2,4,6-trinitrotoluene (TNT) is not readily decomposed in the environment. The microbial decomposition of TNT occurs slowly under either aerobic or anaerobic conditions (see Figure 9.11). Under ligninolytic

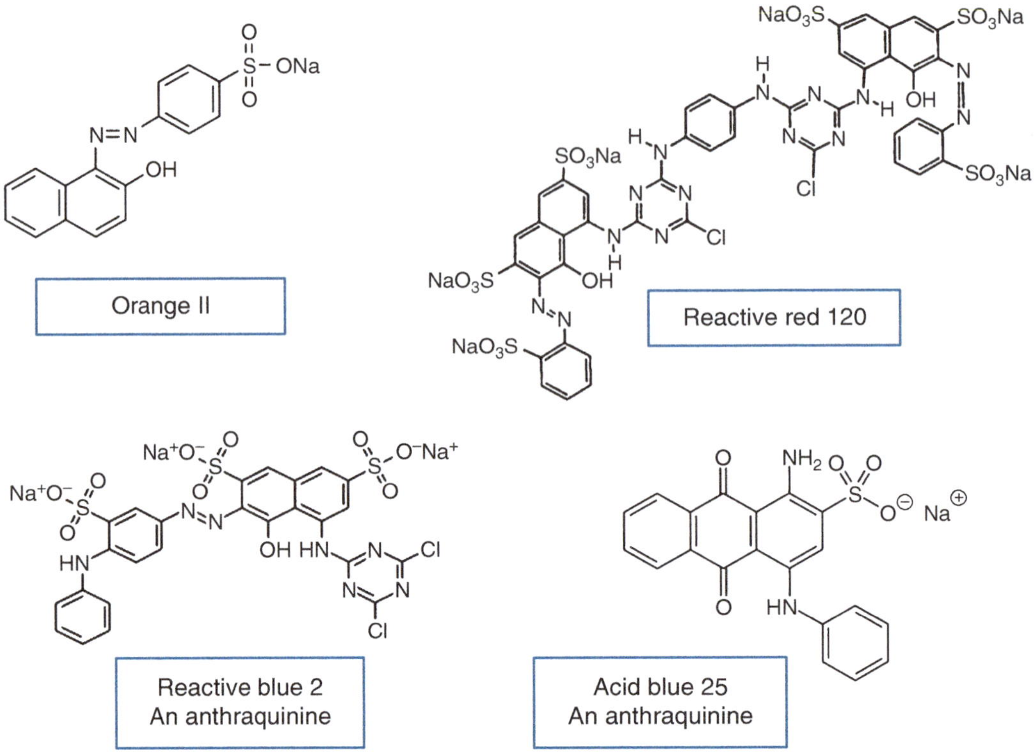

Figure 9.10 Structures of dyes decolorized by bacteria.

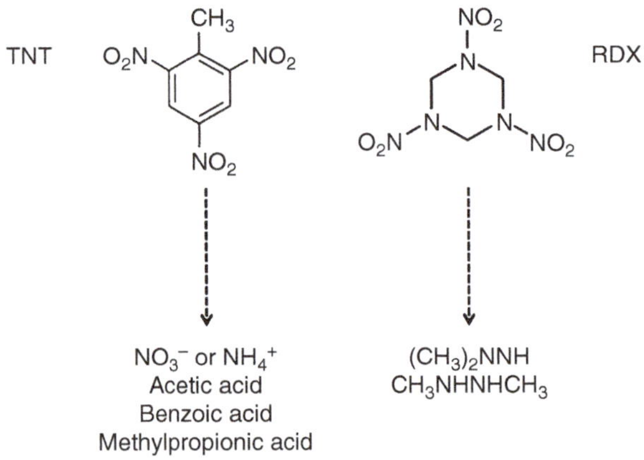

Figure 9.11 Degradation of TNA and RDX by bacteria.

conditions, *Phanerochaete chrysosporium* (a fungus) mineralizes TNT using lignin and manganese peroxidases. Aerobic bacteria (*Pseudomonas* sp., *Bacillus* sp., *Staphylococcus* sp., *Actinomycetes* sp.) remove nitro groups from TNT using nonspecific nitroreductases resulting in the formation of toluene (Esteve-Núñez et al. 2001). Anaerobic bacteria (*Clostridium pasteurianum, Desulfovibrio* sp., and *Moorella thermoacetica*) reduce TNT to 2,4-diamino-6-nitrotoluene by a reductase (Esteve-Núñez et al. 2001; Blehert et al. 1999). Under conditions of limiting sulfate, TNT is reduced by *Desulfovibrio* by reactions that include hydrogenase and dissimilatory sulfite reductase (French et al. 2001). Often it is most desirable to use a bacterial consortium for the decomposition of

TNT because intermediates are toxic and should not be allowed to accumulate.

9.3.4.2 RDX and HMX

Other explosives include hexahydro-1,3,5-trinitro-1,3,5-triazine (RDX), and octahydro-1,3,5,7-tetranitro-1,3,5,7-tetrazocine (HMX). The biodegradation of HMX and RDX by bacteria has been reported and the reader is advised to consult the review by Singh and Singh (2012) for proposed pathways of degradation.

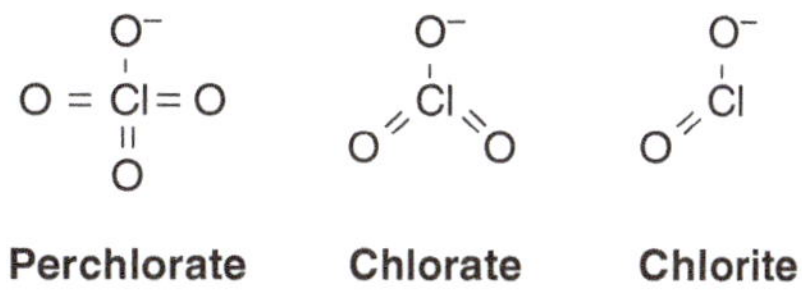

Figure 9.12 Important propellant compounds in the environment that are degraded by bioremediation.

9.3.4.3 Perchlorate

Perchlorate is used as a propellant for rockets, missiles, and explosives (Figure 9.12). There are numerous sites contaminated with this explosive and bioremediation is a viable treatment. The dissimilatory reduction pathway that couples energy production with electron transport by bacteria is as follows:

$$\text{Perchlorate} \rightarrow \text{Chlorate} \rightarrow \text{Chlorite} \rightarrow O_2 + E^-$$

Bacteria capable of perchlorate reduction have been isolated in many different environments and are assigned to the five taxonomic subclasses of *Proteobacteria* and of *Firmicutes* (Bardiya and Bae 2011). A few of the physiological differences of the perchlorate reducing bacteria are as follows: *Dechloromonas* and *Azospira* (formerly *Dechlorosoma*) are capable of reducing both perchlorate and chlorate; *Dechloromonas* PC1 and *Dechlorosoma* sp. reduce both perchlorate and chlorate but chlorate accumulates in the process; *Ideonella dechloratans, Pseudomonas chloritidismutans,* and *Alicycliphilus denitrificans* reduce chlorate but not perchlorate; *Rhodobacter capsulatus, Rhodobacter sphaeroides, Haloferax denitrificans,* and *Paracoccus halodenitrificans* are denitrifying microorganisms that can reduce perchlorate without receiving energy from the process. The final reaction in the anaerobic dissimilatory reduction of perchlorate (chlorite dismutase) is extremely dangerous because it generates O_2.

9.3.5 Bioremediation and Detoxification of Metal(loid)s

There are numerous instances where soil and water environments have been contaminated with heavy metals or metalloids at levels that far exceed safe health limits. The heavy metal elements of concern include lead, chromium, zinc, cadmium, copper, mercury, and nickel. Selenium, a metalloid, and uranium, an actinide element, are of interest in specific environments where these elements serve as pollutants. The element arsenic is neither a metal nor a metalloid but it is included in discussions concerning metal pollution. Unlike with organic pollutants where the goal of bioremediation is to mineralize carbon compounds to CO_2, in situ bacterial metabolism of heavy metals does not remove them from the site but transforms them to a less toxic form (Figure 9.13) (Wuana and Okieimen 2011; Gadd 2010; Barton and Fauque 2009). For pollution by chromate (CrO_4^{3-}), arsenate (AsO_4^{2-}), and uranyl (UO_4^{2-}), bacterial reduction to Cr^{3+}, AsO_3^{2-} and UO_2, respectively, results in a less toxic material but these elements remain at the site. There is always the possibility that these reduced toxic materials could be reoxidized by bacterial or chemical processes in the environment. Removal of metals from the site would require physically collecting the contaminated soil, moving it to another area, washing it, and chemically treating it to remove the toxic elements. The chemical process for treating soil contaminated with heavy metals could be replaced by a microbial treatment,

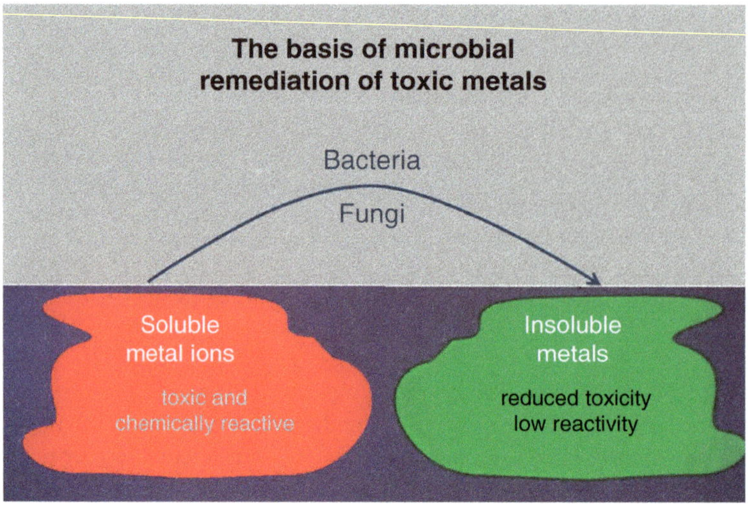

Figure 9.13 Goal for remediation of toxic metals.

possibly by binding the metal ions to bacterial or fungal cells; however, the expense of generating the bacterial or fungal biomass has favored the use of resins to bind metals.

9.3.5.1 Dissimilatory Metal(loid) Reduction

Bacteria have developed unique mechanisms to grow or persist in the presence of toxic metals or metalloids. Several of these resistance patterns are indicated in Figure 9.14 and the activities

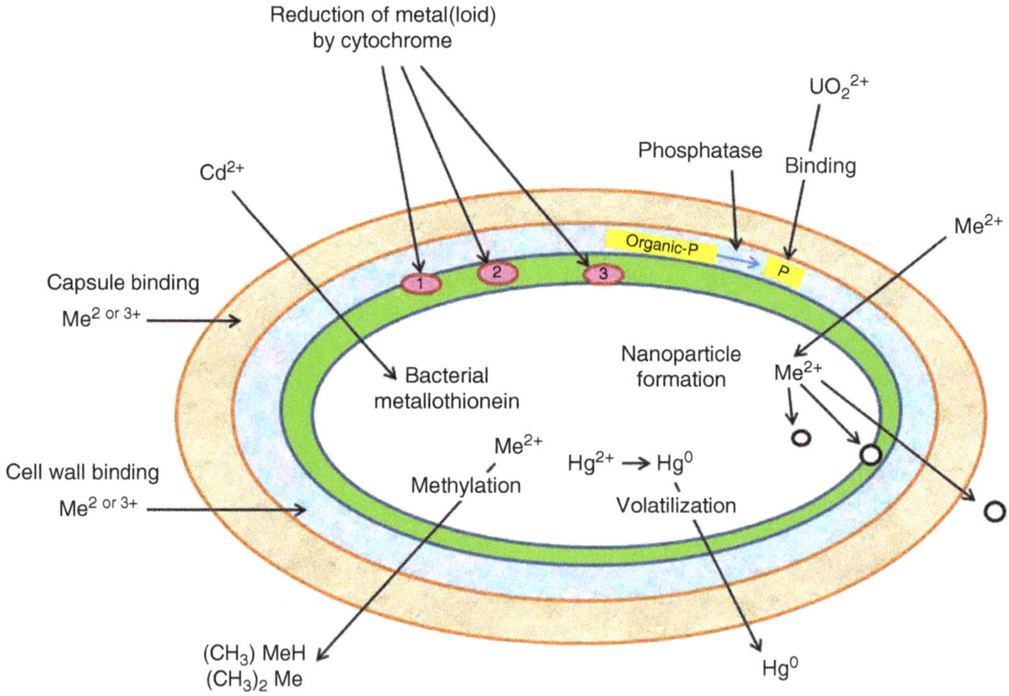

Figure 9.14 Model of bacterial interaction with toxic metals.

Table 9.5 Characteristics of metal resistance in bacteria.

Metal resistance	Characteristics
Precipitation	The formation of insoluble metal salts could result from metal sulfide production from hydrogen sulfide released from sulfate-reducing bacteria. Examples would include CdS and ZnS. Insoluble metal phosphates may be produced on the cell surface following enzymatic release of phosphate from organophosphates. Examples of precipitation of toxic elements incudes uranium phosphate and cadmium phosphate.
Enzymatic reduction	Some mechanisms of reduction of metal(loid)s is by highly specific enzymes (i.e. mercury reductase, arsenare reductase) but reduction of several (i.e. selenium, chromium, tellurium, uranium, vanadium) oxidized elements is by cytochromes or respiratory enzymes (Barton et al. 2015).
Binding to cell walls	Gram-positive bacteria and fungi have many acidic groups in their cell walls, which bind cationic metals.
Binding to capsules	Cyanobacteria, *Pseudomonas*, *Bacillus*, and numerous other bacteria produce acidic capsules, which are attached to the bacterial cell and bind cationic metals.
Immobilization by metallothionein	Cysteine-rich peptides are produced by cyanobacteria and a few other bacteria, which bind intracellular cadmium.
Volatilization	Many bacteria and fungi methylate metalloids and metals, which results in the release of the toxic elements into the atmosphere. Additionally, many bacteria will reduce divalent cationic mercury to elemental mercury, which is volatile and escapes into the atmosphere.
Nanoparticle formation	Many bacteria, fungi, and archaea reduce metal(loid)s to the elemental form with the production of metallonanoparticles. This form of detoxification is relatively new and has not been explored for bioremediation (Barton et al. 2014).

of these mechanisms are given in Table 9.5. Many bacterial species reduce metals by using electrons derived from cellular metabolism to produce nanoparticles of metals or metalloids (Barton et al. 2014, 2015). A few of the bacteria display dissimilatory reduction, which is the coupling of growth to the reduction of metals or metalloids. Over the years, an interest has been developed for bioremediation of soil sites contaminated with various metals and radionuclides. The reduction kinetics for Co(III), U(VI), Cr(VI), and Tc(VII) have been measured using pure cultures of *Shewanella putrefaciens* CN32, *Shewanella alga* BrY, *Shewanella oneidensis* MR-1, and *G. metallireducens* GS-15 and H_2 functioned as a better electron donor than lactate (Liu et al. 2002). Using an immobilized culture of *Desulfovibrio desulfuricans,* reduction kinetics have been calculated for U(VI), Mo(VI), Se and Cr(VI) (Tucker et al. 1998, 2000); however, *D. desulfuricans* is unable to grow with metals as the electron acceptor. The reduction of U(VI) to U(IV) in situ by indigenous bacteria has been examined at several US Department of Energy sites (Rifle, Colorado; Oak Ridge, Tennessee; Hanford, Washington) and insoluble uraninite [U(IV)] is produced with the addition of acetate as the carbon source (Newsome et al. 2014). Long-term bioremediation of uranium is difficult because it is not possible to prevent the later release of soluble uranium due to microbial activity or chemicals in groundwater.

9.3.5.2 Methylation Reactions

The removal of toxic heavy metals or metalloids from an environment is due to the formation of methylated metal(loid) species, which are released into the atmosphere where ultraviolet light oxidizes the methyl group, with the released metal slowly settling over the biosphere. While methylation of transition metals does not occur under normal environmental conditions,

Table 9.6 Methylated elements in the environment produced by microorganisms.

Metals	Metaloids
Arsenic	Selenium
Antimony	Tellurium
Bismuth	
Cadmium	
Germanium	
Lead	
Mercury	
Thallium	
Tin	

Source: Reproduced with permission from Mason (2013) and references within this review.

methyl groups are added onto many metals and metalloids by microorganisms. The metal most commonly associated with methylation is mercury (see Section 8.10.1); however, there are numerous metal(loid)s volatilized by addition of methyl groups. Some of these methylated species are listed in Table 9.6. The presence of atmospheric heavy metals near the North and South polar regions is attributed to the microbial methylation of mercury, lead, and cadmium (Pongratz and Heumann 1999). Mason (2013) reviews methylation reactions of various elements that give a carbon–metal bond and the primary reactions for methylation by microorganisms that involve *S*-adenosylmethionine, methylcobalamin, or *N*-methyltetrahydrofolate as the methyl donor. The methylation of Hg^{2+} by sulfate-reducing bacteria and iron-reducing bacteria requires the methyl carbanion (CH_3^-), which can only be supplied by methylcobalamin. The methyl group released by *S*-adenosylmethionine and *N*-methyltetrahydrofolate is the carbocation (CH_3^+) moiety.

For arsenic, methylation of the reduced element is coupled to the oxidation of As(III) to produce As(V)-CH_3; however to add another methyl group, arsenic must be reduced to As(III)-CH_3 to produce As(V)-$(CH_3)_2$. It has been suggested that methylation of selenium is similar to the reaction of arsenic (Mason 2013). Many different species of bacteria and fungi are able to methylate metal(loid)s and different environments may be characterized by evaluating the methylation products being released.

9.4 Design of Systems for Bioremediation

9.4.1 In Situ vs Ex Situ

Each approach to bioremediation has some advantages and some disadvantages (EPA 2006; Carberry and Wik 2001). Ecologists and engineers will need to assess the contaminating chemicals as well as the site of contamination. With in situ bioremediation, a large area of contaminated soil can be treated and it is generally less expensive than ex situ because there is no need to haul the soil away. The prospect of remediation is excellent for sandy soil but it is difficult for high clay soil because air or nutrients cannot readily penetrate to the site of contamination. Generally ex situ is a fast bioremediation process and the uniform treatment can be readily monitored. Ex situ bioremediation includes solid-phase treatment such as land farming or slurry-phase treatment including bioreactors. Natural attenuation, permeable reactive barriers, biosparging, and bioventing are commonly used techniques with in situ bioremediation.

9.4.2 Bioreactors

Slurry-phase bioreactors use 5–40% contaminated soil mixed with water. Nutrients and bacteria are added to the reactor, which is maintained at optimal temperature, aeration, and pH. It is ideally used for petrochemicals, insecticides, PAH, and soil contaminated by explosives (Robles-González et al. 2008). Usually, remediation can be achieved in a few weeks. The disadvantages of slurry bioreactors are that they expensive to operate, the water must be removed from the soil at the end of the process, and the reactor water needs to be treated before disposal.

9.4.3 Biofarming

Contaminated sludge soil or sediments may be tilled into the upper soil zone where indigenous bacteria and fungi metabolize the organic chemical contaminant (Song et al. 1990). Biofarming is an in situ process that has been used to detoxify hydrocarbons, oily sludge refinery wastes, and pesticides. Small organic compounds are readily degraded but polycyclic hydrocarbons with numerous rings, and chlorinated or nitrated compounds require a considerable time to detoxify. Creosote, pentachlorophenol, and contaminants containing metal ions are generally remediated by a process other than biofarming. An interesting extension of this land use is the distribution of PAH contaminated soil in forest areas where fungal activity is high. The process used by wood rot fungi to decompose lignin involves the enzymatic release of hydrogen peroxide and this will also degrade PAH.

9.4.4 Permeable Reactive Barriers

Treatment of contaminated groundwater by placing a permeable reactive barrier across the flowing plume was initially used about 25 years ago (Obiri-Nyarko et al. 2014). These permeable reactive barriers can be used to remove heavy metals, organic solvents, and pesticides. This technology can be applied in any region where the groundwater moves slowly and is close to the surface. To make a barrier, a trench is dug perpendicular to the flow of groundwater; straw, manure, and the desired environmental bacteria fill the trench. Bacteria grow in the trench and metabolize the organic contaminants, detoxify heavy metals by reduction activities, and bind charged by-products into the barrier matrix. Monitoring of the site indicates when the barrier needs to be removed and reconstructed.

9.4.5 Groundwater and Lagoon Treatment

Contamination of groundwater is a result of a contaminant plume reaching subsurface water. Bioremediation of groundwater is widely used because it is relatively inexpensive and it can be modified to fit site-specific conditions (EPA 2003). Organic pollutants can be removed by enhancing bacterial growth through aerobic respiration, anaerobic respiration, or fermentation. Stimulation of aerobic metabolism is achieved by the addition of oxygen directly to the groundwater by adding injection pipes that deliver O_2 or oxygen-releasing compounds to the groundwater. Nutrients are introduced into the groundwater to stimulate indigenous bacteria or specific strains of bacteria are added to promote co-metabolism of contaminants. For organic contaminants, aerobic metabolism is desirable because the end products of oxidation would be CO_2 and water but when the contaminating organic compound contains chlorine, the initial metabolism is anaerobic, followed by aerobic metabolism. Lagoons are often used to collect

BTEX, PAHs, heavy metals, and other industrial wastes. These shallow waste sites can be aerated to enhance aerobic metabolism or mixed to redistribute chemicals in the sediment.

9.4.6 Bioventing

Bioventing is injecting or pumping air into contaminated water or soil. The added oxygen enhances the metabolic activity of indigenous bacteria and brings the volatile contaminant to the surface, where it can be collected or processed through a special bioreactor. Excellent success has been achieved when using bioventing for bioremediation of petroleum hydrocarbons, solvents, and various organic compounds (Lee and Swindoll 1993). Monitoring of volatile gases is required and dry soil may limit the metabolic activity of the bacteria.

9.5 Summary

Xenobiotics, hydrocarbons and explosives that have been designated as toxic or hazardous to humans can be degraded by bacterial metabolic systems. Often the enzymology of degradation is sufficiently complex that several species of bacteria cooperate to metabolize a given pollutant. To ensure that there are enough bacteria for a bioremediation process, the site can be augmented with bacteria that have the desired metabolic capability and are isolated from the environment. Bioremediation of organic matter involves bacterial metabolism where the end product is a non-toxic compound or, for aerobes, the organic pollutant is oxidized completely to carbon dioxide. Occasionally, reductive dehalogenation by anaerobic bacteria may be used to initiate the bioremediation event. With bioremediation of toxic heavy metals and metalloids, considerable interest has been given to metabolic reduction by anaerobic bacteria. The reduced metallic elements are less toxic and more insoluble than their oxidized counterparts. The release of hydrogen sulfide by sulfate-reducing bacteria is especially effective in immobilizing toxic elements as insoluble metal sulfides. With mercury and selenium, detoxification may be attributed to methylation reactions with the removal of the toxic element from the site of pollution. Systems for bioremediation vary with the types of chemical pollutants present and with the site involved.

Discussion Questions

1 Discuss the main issues that must be addressed when using in situ bioremediation of an area contaminated with diesel fuel.

2 Discuss some of the main issues confronting scientists when using bacterial bioremediation to clean up a lake with a toxic level of selenite.

3 Discuss some of the future applications for using genetically engineered bacteria for bioremediation. Are there some situations where the use of genetically modified bacteria may be preferred?

4 Are there any organic pollutants that bacteria cannot degrade? Defend your answer.

5 When bioremediation is used for a marine oil spill, would you expect to see a succession of microbial populations? If so, what physiological types of bacteria would be involved?

References

Aguawal, N. and Shahi, S.K. (2015). An environmental cleanup strategy – microbial transformation of xenobiotic compounds. *Int. J. Curr. Microbiol. Appl. Sci.* 4: 429–461.

Alexander, M. (1994). *Biodegredation and Bioremediation*. New York, NY: Academic Press.

Arciero, D., Vannelli, T., Logan, M. et al. (1989). Degradation of trichloroethylene by the ammonia-oxidizing bacterium *Nitrosomonas europaea*. *Biochem. Biophys. Res. Commun.* 159: 640–643.

Arif, I.A., Bakir, M.A., and Khan, H.A. (2012). Microbial remediation of pesticides. In: *Pesticides: Evaluation of Environmental Pollution* (ed. H.S. Rathore and N. LML), 131–146. Boca Raton, FL: CRC Press, Taylor and Francis Group, LLC.

Arun, K., Ashok, M., and Rajesh, S. (2010). Crude oil PAH constitution, degradation pathway and associated bioremediation microflora: an overview. *Int. J. Environ. Sci.* 1: 1427–1446.

Arun, A. and Eyini, M. (2011). Comparative studies on lignin and polycyclic aromatic hydrocarbons degradation by basidiomycetes fungi. *Bioresour. Technol.* 102: 8063–8070.

Atlas, R.M. and Hazen, T.C. (2011). Oil biodegradation and bioremediation: a tale of the two worst spills in US history. *Environ. Sci. Technol.* 45: 6709–6715.

Bardiya, N. and Bae, J.-H. (2011). Dissimilatory perchlorate reduction: a review. *Microbiol. Res.* 166: 237–254.

Baldwin, B.R., Nakatsu, C.H., Nebe, J. et al. (2009). Enumeration of aromatic oxygenase genes to evaluate biodegradation during multi-phase extraction at a gasoline-contaminated site. *J. Hazard. Mater.* 163: 524–530.

Barton, L.L. and Fauque, G.D. (2009). Biochemistry, physiology and biotechnology of sulfate-reducing bacteria. *Adv. Appl. Microbiol.* 68: 43–78.

Barton, L.L., Tomei-Torres, F.A., Xu, F. et al. (2014). Nanoparticles formed by microbial metabolism of metals and minerals. In: *Nanomicrobiology* (ed. L.L. Barton, D.A. Bazylinski and H. Xu), 145–176. New York, NY: Springer.

Barton, L.L., Tomei-Torres, F.A., Xu, F. et al. (2015). Metabolism of metals and metalloids by the sulfate-reducing bacteria. In: *Bacterial-Metal Interactions* (ed. D. Saffarini), 57–84. New York, NY: Springer International Publishing.

Beveridge, T.J. (1989). Role of cellular design in bacterial metal accumulation and mineralization. *Annu. Rev. Microbiol.* 43: 147–171.

Blehert, D.S., Fox, B.G., and Chambliss, G.H. (1999). Cloning and sequence analysis of two pseudomonas flavoprotein xenobiotic reductases. *J. Bacteriol.* 181: 6254–6263.

Boetius, A., Ravenschlag, K., Schubert, C.J. et al. (2000). A marine microbial consortium apparently mediating anaerobic oxidation of methane. *Nature* 407: 623–626.

Bodour, A.A., Guerrero-Barajas, C., Jiorle, B.V. et al. (2004). Structure and characterization of flavolipids, a novel class of biosurfactants produced by *Flavobacterium* sp. strain MTN11. *Appl. Environ. Microbiol.* 70: 114–120.

Bodour, A.A., Drees, K.P., and Maier, R.M. (2003). Distribution of biosurfactant-producing bacteria in undisturbed and contaminated arid southwestern soils. *Appl. Environ. Microbiol.* 69: 3280–3287.

Borrok, D.M., Fein, J.B., and Kilpa, C.F. Jr. (2004). Cd and proton adsorption onto bacterial consortia grown from industrial wastes and contaminated geological settings. *Environ. Sci. Technol.* 38: 5656–5664.

Brenner, K., You, L., and Arnold, F.H. (2008). Engineering microbial consortia: a new frontier in synthetic biology. *Trends Biotechnol.* 26: 483–489.

Cappelletti, M., Frascari, D., Zannoni, D. et al. (2012). Microbial degradation of chloroform. *Appl. Microbiol. Biotechnol.* 96: 1395–1409.

Carberry, J.B. and Wik, J. (2001). Comparison of ex situ and in situ bioremediation of unsaturated soils contaminated by petroleum. *J. Environ. Sci. Health. A Tox. Hazard Subst. Environ.* 38: 1491–1503.

Carvalho, M.F., Alves, C.C.T., Ferreira, M.I.M. et al. (2002). Isolation and initial characterization of a bacterial consortium able to mineralize fluorobenzene. *Appl. Environ. Microbiol.* 68: 102–105.

Chung, K.T., Stevens, S.E. Jr., and Cerniglia, C.E. (1992). The reduction of azo dyes by the intestional microflora. *Crit. Rev. Microbiol.* 18: 175–190.

Coates, J.D., Woodward, J., Allen, J. et al. (1997). Anaerobic degradation of polycyclic aromatic hydrocarbons and alkanes in petroleum-contaminated marine harbor sediments. *Appl. Environ. Microbiol.* 63: 3589–3593.

Das, N. and Chandran, P. (2011). Microbial degradation of petroleum hydrocarbon contaminants: an overview. *Biotechnol. Res. Inter.* 2011 (2011): 941810. https://doi.org/10.4061/2011/941810.

Dejonghe, W., Goris, J., El Fantroussi, S. et al. (2000). Effect of dissemination of 2,4-D degradation plasmids on 2,4-D degradation and on bacterial community structure in two different soil horizons. *Appl. Environ. Microbiol.* 66: 3297–3304.

EPA, United States Environmental Protection Agency (2006). *In Situ and Ex Situ Biodegradation Technologies for Remediation of Contaminated Sites*, EPA/625-R-06-015 October 2006 www.epa.gov

EPA, United States Environmental Protection Agency (2003). *Introduction to In Situ Bioremediation of Groundwater*, EPA/542-R-13-018 December 2013 www.epa.gov

Esteve-Núñez, A., Caballero, A., and Ramos, J.L. (2001). Biological degradation of 2,4,6-trinitrotoluene. *Microbiol. Mol. Biol. Rev.* 65: 335–352.

French, C.E., Rosser, S.J., and Bruce, N.C. (2001). Biotransformations of explosives. *Biotechnol. Genet. Eng. Rev.* 18: 171–217. https://doi.org/10.1080/02648725.2001.10648013.

Furukawa, K., Suenaga, H., and Goto, M. (2004). Biphenyl dioxygenases: functional versatilities and directed evolution. *J. Bacteriol.* 186: 5189–5196.

Gadd, G.M. (2010). Metals, minerals and microbes: geomicrobiology and bioremediation. *Microbiology* 156: 609–643.

Georgiou, G., Lin, S.C., and Sharma, M.M. (1999). Surface-active compounds from microorganisms. *Biotechnology* 10: 60–65.

Gerritse, J., Drzyzga, O., Kloetstra, G. et al. (1999). Influence of different electron donors and acceptors on dehalorespiration of tetrachloroethene by *Desulfitobacterium frappieri* TCE1. *Appl. Environ. Microbiol.* 65: 5212–5221.

Gribble, G.W. (1996). Naturally occurring organohalogen compounds – a comprehensive review. *Prog. Chem. Org. Nat. Prod.* 68: 1–423.

Haritash, A.K. and Kaushik, C.P. (2009). Biodegradation aspects of polycyclic aromatic hydrocarbons (PAHs): a review. *J. Hazar. Mater.* 169: 1–15.

Hazen, T.C. (2010). Cometabolic bioremediation. In: *Handbook of Hydrocarbon and Lipid Microbiology* (ed. K.N. Timmis), 2502–2514. Berlin: Springer-Verlag.

Hendrickx, B., Junca, H., Vosahlova, J. et al. (2006). Alternative primer sets for PCR detection of genotypes involved in bacterial aerobic BTEX degradation: distribution of the genes in BTEX degrading isolates and in subsurface soils of a BTEX contaminated industrial site. *J. Microbiol. Methods* 64: 250–265.

Jahng, D. and Wood, T.K. (1994). Trichloroethylene and chloroform degradation by a recombinant pseudomonad expressing soluble methane monooxygenase from *Methylosinus trichosporium* OB3b. *Appl. Environ. Microbiol.* 60: 2473–2482.

Johnson, G.R. and Olsen, R.H. (1995). Nucleotide sequence analysis of genes encoding a toluene/benzene monooxygenase from *Pseudomonas* sp. strain JS150. *Appl. Environ. Microbiol.* 61: 3336–3346.

Kayser, M.F., Ucurum, Z., and Vuilleumier, S. (2002). Dichloromethane metabolism and C1 utilization genes in *Methylobacterium* strains. *Nat. Rev. Microbiol.* 148: 1915–1922.

Krausova, V.I., Robb, F.T., and González, J.M. (2003). Bacterial degradation of dichloromethane in cultures and natural environments. *J. Microbiol. Methods* 54: 419–422.

Krumholz, L.R. (1997). *Desulfuromonas chloroethenica* sp. nov. uses tetrachloroethene and trichloroethylene as electron acceptors. *Int. J. Syst. Bacteriol.* 47: 1262–1263.

Lee, M.D. and Swindoll, C.M. (1993). Bioventing for in situ remediation. *Hydrolog. Sci. J.* 38: 273–282.

Lehto, S., Buchweitz, M., Klimm, A. et al. (2017). Comparison of food colour regulations in the EU and the US: a review of current provisions. *Food Addit. Contam. Part A Chem. Anal. Control Expo. Risk Assess.* 34: 335–355.

Leisinger, T. and Braus-Stromeyer, S.A. (1995). Bacterial growth with chlorinated methanes. *Environ. Health Perspect.* 103: 33–36.

Liu, C., Gorby, Y.A., Zachara, J.M. et al. (2002). Reduction kinetics of Fe(III), Co(III), U(VI), Cr(VI) and Tc(VII) in cultures of dissimilatory metal-reducing bacteria. *Biotechnol. Bioeng.* 80: 637–649.

Madigan, M.T., Martinko, J.M., Dunlap, P.V. et al. (2009). *Brock Biology of Microorganisms*, 12e. Glenville, IL: Pearson Education.

MacNaughton, S.J., Stephen, J.R., Venosa, A.D. et al. (1999). Microbial population changes during bioremediation of an experimental oil spill. *Appl. Environ. Microbiol.* 65: 3566–3574.

Magnus, V., Ozga, J.A., Reinecke, D.M. et al. (1997). 4-chloroindole-3-acetic and indole-3-acetic acids in *Pisum sativum*. *Phytochemistry* 46: 675–681.

Magnuson, J.K., Romine, M.F., Burris, D.R. et al. (2000). Trichloroethene reductive dehalogenase from *Dehalococcoides ethenogenes* sequence of *tceA* and substrate range characterization. *Appl. Environ. Microbiol.* 66: 5141–5147.

Maillard, J., Schumacher, W., Vazuquez, F. et al. (2003). Characterization of the corrinoid iron-sulfur protein tetrachloroehtene reductase dehalogenase of *Dehalobacter restrictus*. *Appl. Environ. Microbiol.* 69: 4628–4638.

Mandell, D.J., Lajole, M.J., Mee, M.T. et al. (2015). Biocontainment of genetically modified organisms by synthetic protein design. *Nature* 518: 55–60.

Margesin, R. and Schinner, F. (2001). Bioremediation (natural attenuation and biostimmulation) of diesel-oil-contaminated soil in an alpine glacier skiing area. *Appl. Environ. Microbiol.* 67: 3127–3133.

Mason, R.P. (2013). *The Methylation of Metals and Metalloids in Aquatic Systems*. INTECH https://doi.org/10.5772/51774.

Maymó-Gatell, X., Nijenhuis, I., and Zinder, S.H. (2001). Reductive dechlorination of cis-1,2-dichloroethene and vinyl chloride by "*Dehalococcoides ethenogenes*". *Environ. Sci. Technol.* 35: 516–521.

McAnulla, C., McDonald, I.R., and Murrell, J.C. (2001a). Methyl chloride utilizing bacteria are ubiquitous in the natural environment. *FEMS Microbiol. Lett.* 201: 151–155.

McAnulla, C., Woodall, C.A., and McDonald, I.R. (2001b). Chloromethane utilization gene cluster from *Hyphomicrobium chloromethanicum* strain CM2 and development of functional gene probes to detect halomethane-degrading bacteria. *Appl. Environ. Microbiol.* 67: 307–316.

Miller, L.G., Warner, K.L., Baesman, S.M. et al. (2003). Degradation of methyl bromide and methyl chloride in soil microcosms: use of stable C isotope fractionation and stable isotope probing to identify reactions and the responsible microorganisms. *Geochim. Cosmochim. Acta* 68: 3271–3282.

Minati, J.L. and Marsan, F.A. (2005). Anaerobic PAH degradation in soil by a mixed bacterial consortium under denitrifying conditions. *Chemosphere* 60: 1231–1236.

Mishra, A. and Malik, A. (2014). Novel fungal consortium for bioremediation of metals and dyes from mixed waste stream. *Bioresour. Technol.* 171: 217–226.

Mrozik, A., Piotrowska-Soget, Z., and Labużek, S. (2003). Bacterial degradation and bioremediation of polycyclic aromatic hydrocarbons. *Pol. J. Environ. Stud.* 12: 15–25.

Mukherjee, P. and Roy, P. (2013). Copper enhanced monooxygenase activity and FT-IR spectroscopic characterization of biotransformation products in trichloroethylene degrading bacterium: *Stenotrophomonas maltophilia* MP102. *BioMed Res. Inter.* 2013 (2013): 723680. https://doi.org/10.1155/2013/723680.

Nebe, J., Baldwin, B.R., Kassab, R.L. et al. (2009). Quantification of aromatic oxygenase genes to evaluate enhanced bioremediation of oxygen releasing materials at a gasoline-contaminated site. *Environ. Sci. Technol.* 43: 2029–2034.

Neuman, A., Siebert, A., Trescher, T. et al. (2002). Tetrachloroethene reductive dehalogenase of *Dehalospirillum multivorans*: substrate specificity of the native enzyme and its corrinoid cofactor. *Arch. Microbiol.* 177: 420–426.

Newsome, L., Morris, K., and Lloyd, J.R. (2014). The biogeochemistry and bioremediation of uranium and other priority radionuclides. *Chem. Geol.* 363: 164–184.

Obiri-Nyarko, F., Grajales-Mesa, S.J., and Malina, G. (2014). An overview of permeable reactive barriers for *in situ* sustainable groundwater remediation. *Chemosphere* 111: 243–259.

Parales, R.E., Lee, K., Resneck, S.M. et al. (2000). Substrate specificity of naphthalene dioxygenase: effect of specific amino acids at the active site of the enzyme. *J. Bacteriol.* 182: 1641–1649.

Paul, D., Pandey, G., Pandey, J. et al. (2005). Accessing microbial diversity for bioremediation and environmental restoration. *Trends Biotechnol.* 23: 135–142.

Pauwels, B., Wille, K., Noppe, H. et al. (2008). 17α- ethinylestradiol cometabolism by bacteria degrading estrone, 17β- estradiol and estriol. *Biodegradation* 19: 683–693.

Penny, C., Vuilleumier, S., and Bringel, F. (2010). Microbial degradation of tetrachloromethane: mechanisms and perspectives for bioremediation. *FEMS Microbiol. Ecol.* 74: 257–275.

Perpetuo, F.A., Souza, C.B., and Nascimento, C.A.O. (2011). Engineering bacteria for bioremediation. In: *Progress in Molecular and Environmental Bioengineering – From Analysis and Modeling to Technology Applications* (ed. Carpi). INTECH Open, August 1, 2011 under CC BY-NC-SA 3.0 license.

Piccinini, P., Senaldi, C., and Buriova, E. (2008). *European Survey on the Presence of Banned Azodyes in Textiles.* Luxembourg: Office for Official Publications of the European Communities.

Pieper, D.H. and Seeger, M. (2008). Bacterial metabolism of polychlorinated biphenyls. *J. Mol. Microbiol. Biotechnol.* 15: 121–138.

Pongratz, R. and Heumann, K.G. (1999). Production of methylated mercury, lead and cadium by marine bacteria as a significant natural source for atmospheric heavy metals in polar regions. *Chemosphere* 39: 89–102.

Ron, E.Z. and Rosenberg, E. (2001). Natural roles of biosurfactants. *Environ. Microbiol.* 3: 229–236.

Radianingtyas, H., Robinson, G.K., and Bull, A.T. (2003). Characterization of a soil-derived bacterial consortium degrading 4-chloroaniline. *Microbiology* 149: 3279–3287.

Robles-González, I.V., Fava, F., and Poggi-Varaldo, H.M. (2008). A review on slurry bioreactors for bioremediation of soils and sediments. *Microb. Cell Fact.* 7: 5. https://doi.org/10.1186/1475-2859-7-5.

Rodríguez-Martinez, E.M., Pérez, E.X., Schadt, C.W. et al. (2006). Microbial diversity and bioremediation of a hydrocarbon-contaminated aquifer (Vega Baja, Puerto Rico). *Int. J. Environ. Res. Public Health* 3: 292–300.

Ryoo, D., Shim, H., Canada, K. et al. (2000). Aerobic degradation of tetrachlorothylene by toluene-o-xylene monooxoygenase of *Pseudomonas stutzeri* OX1. *Nat. Biotechnol.* 18: 775–778.

Sathishkumar, M., Binupriya, A.R., Baik, S.-H. et al. (2008). Biodegradation of individual bacterial strains and a mixed bacterial consortium isolated from hydrocarbon contaminated areas. *Clean Soil Air Water* 36: 92–96. https://doi.org/10.1002/clen.200700042.

Sayler, G.S. and Ripp, S. (2000). Field application of genetically engineered microorganisms for bioremediation processes. *Curr. Opin. Biotechnol.* 11: 268–269.

Seeger, M., Hernández, M., Méndez, V. et al. (2010). Bacterial degradation and bioremediation of chlorinated herbicides and biphenyls. *J. Soil Sci. Plant Nutr.* 10: 320–332.

Semrau, J.D. (2011). Bioremediation via methanotrophy: overview of recent findings and suggestions for future research. *Front. Microbiol.* 2 (209): https://doi.org/10.3389/fmicb.2011.00209.

Sharma, S. (2012). Bioremediation: features, strategies and application. *Asian J. Pharmacy Life Sci.* 2: 202–213.

Sihag, S. and Pathak, H. (2014). A review: biodegradation of polyaromatic hydrocarbons. *Am. J. Earth Sci.* 1: 9–20.

Singh, R. and Singh, A. (2012). Biodegradation of military explosives RDX and HMX. In: *Microbial Degradation of Xenobiotics* (ed. S.N. Singh). Berlin: Springer-Verlag.

Sivaprakasam, S., Mahadevan, S., Sekar, S. et al. (2008). Biological treatment of tannery wastewater by using salt-tolerant bacterial strains. *Microb. Cell Fact.* 7: 15. https://doi.org/10.1186/1475-2859-7-15.

Song, H.G., Wang, X., and Bartha, R. (1990). Bioremediation potential of terrestrial fuel spills. *Appl. Environ. Microbiol.* 56: 652–656.

Sudha, M., Saranya, A., Selvakumar, G. et al. (2014). Microbial degradation of azo dyes: a review. *Int. J. Curr. Microbiol. App. Sci.* 3: 670–690.

Tabak, H.H., Chambers, C.W., and Kabler, P.W. (1964). Microbial metabolism of aromatic compounds. I. Decomposition of phenolic compounds and aromatic hydrocarbons by phenolic-adapted bacteria. *J. Bacteriol.* 87: 910–919.

Tabak, H.H., Lens, P., van Hullebusch, E.D. et al. (2005). Developments in bioremediation of soils and sediments polluted with metals and radionuclides – 1. Microbial processes and mechanisms affecting bioremediation of metal contamination and influencing metal toxicity and transport. *Rev. Environ. Sci. Biotechnol.* 4: 115–156.

Thavamani, P., Meghara, M., and Naidu, R. (2012). Bioremediation of high molecular weight polyaromatic hydrocarbons co-contaminated with metals in liquid and soil slurries by metal tolerant PAHs degrading bacterial consortium. *Biodegradation* 23: 823–835.

Tucker, M.D., Barton, L.L., and Thomson, B.M. (1998). Reduction of Cr, Mo, Se and U by *Desulfovibrio desulfuricans* immobilized in polyacrylamide gels. *J. Ind. Microbiol. Biotechnol.* 20: 13–19.

Tucker, M.D., Barton, L.L., and Thomson, B.M. (2000). Removal of U and Mo from water by immobilized *Desulfovibrio desulfuricans* in column reactors. *Biotechnol. Bioeng.* 60: 88–96.

Tyagi, M., da Fonseca, M.M., and de Carvalho, C.C. (2011). Bioaugementation and biostimulation strategies to improve the effectiveness of bioremediation strategies. *Biodegradation* 22: 331–341.

US EPA (2010). *Types of Pesticides*. Washington DC: US Environmental Protection Agency http://www.epa.gov/pesticides/about/types.htm.

Viñas, M., Sabaté, J., Espuny, M.J. et al. (2005). Bacterial community dynamics and polycyclic aromatic hydrocarbon degradation during bioremediation of heavy creosote-contaminated soil. *Appl. Environ. Microbiol.* 71: 7008–7018.

Vogt, C., Kleinsteuben, S., and Richnow, H.-H. (2011). Anaerobic benzene degradation by bacteria. *Microb. Biotechnol.* 4: 710–724.

Wackett, L.P., Brusseau, G.A., Householder, S.R. et al. (1989). Survey of microbial oxygenases: trichloroethylene degradation by propane-oxidizing bacteria. *Appl. Environ. Microbiol.* 55: 2960–2964.

Wackett, L.P. and Gibson, D.T. (1988). Degradation of trichloroethylene by toluene dioxygenase in whole-cell studies with *Pseudomonas putida* F1. *Appl. Environ. Microbiol.* 54: 1703–1708.

Wang, G., Wang, Z.L., Liang, B. et al. (2010). Co-metabolism of DDT by the newly isolated bacterium, *Pseudoxanthomonas* sp. wax. *Braz. J. Microbiol.* 41: 431–438.

Wang, Q., Zhang, S., Li, Y. et al. (2011). Potential approaches to improving biodegradation of hydrocarbons for bioremediation of crude oil pollution. *J. Environ. Protection* 2: 47–55.

Weelink, S.A.B., van Eckert, M.H.A., and Stams, A.J.M. (2010). Degradation of BTEX by anaerobic bacteria: physiology and application. *Rev. Environ. Sci. Biotechnol.* 9: 359–385.

Wuana, R.A. and Okieimen, F.E. (2011). Heavy metals in contaminated soils: a review of sources, chemistry, risks and best available strategies for remediation. *ISNR Ecol.* 2011 (2011): 402647. https://doi.org/10.5402/2011/402647.

Xie, S., Sun, W., Luo, C. et al. (2010). Novel aerobic benzene degrading microorganisms identified in three soils by stable isotope probing. *Biodegradation* https://doi.org/10.1007/s10532-010-9377-5.

Yakimov, M.M., Timmis, K.N., and Golysin, P.N. (2007). Obligate oil-degrading marine bacteria. *Curr. Opin. Biotechnol.* 18: 257–266.

Yang, C., Song, C., Mulchandani, A. et al. (2010). Genetic engineering of *Stenotrophomonas* strain yc-1 to possess a broader substrate range for organophosphates. *J. Agric. Food. Chem.* 58: 6762–6766.

Yoshida, S., Hiraga, K., Takehana, T. et al. (2016). A bacterium that degrades and assimilates poly(ethylene terephthalate). *Science* 351: 1196–1199.

Zhang, T., Tremblay, P.-L., Charurasia, A.K. et al. (2014). Identification of genes specifically required for the anaerobic metabolism of benzene in *Geobacter metallireducens*. *Front Microbiol.* 5 (245): https://doi.org/10.3389/fmicb.2014.00245.

Further Reading

Adams, G.O., Fufeyin, P.T., Okoro, S.E. et al. (2015). Bioremediation, biostimulation and bioaugmentation: a review. *Int. J. Environ. Biorem. Biodegrad.* 3: 28–39.

Biswas, K., Paul, D., and Sinha, S.N. (2015). Biological agents of bioremediation: a concise review. *Front. Environ. Microbiol.* 1 (3): 39–43. https://doi.org/10.11648/j.fem.20150103.11.

Dixit, R., Wasiullah, Malaviya, D. et al. (2015). Bioremediation of heavy metals from soil and aquatic environments: an overview of principles and criteria of fundamental processes. *Sustainability* 7: 2189–2212.

Hazen, T.C. (2016). Marine Oil Biodegradation. *Environ. Sci. Technol.* 50: 2121–2129.

10

Biocorrosion and Geomicrobiology

10.1 Overview and Focus

Microorganisms interface with various metallic cations and anions and also with metals in the elemental state. For biocorrosion, single cells initiate the bacterial–metal interaction but the formation of biofilms consisting of thousands of bacterial cells enhances the corrosive process. The biocorrosion of metal surfaces is interesting because electrons from elemental metals are moved into the bacteria to support growth, which brings up the subject of extracellular metabolism and how it is achieved by bacteria. Corrosion of nonferrous materials by microorganisms is of concern when dealing with concrete, paintings, and art objects and these corrosion activities include metabolic products of various bacteria and fungi. Another area of industrial importance is bioleaching, which is used by several countries for biomining, with the most common elements extracted being gold, copper, cobalt, nickel, zinc, and uranium (Schippers et al. 2014).

10.2 Microbially Influenced Corrosion (MIC) of Ferrous Metals

The damage from microbially influenced corrosion (MIC) of ferrous and nonferrous metals is considerable and the cost of replacing damaged metal pipes is enormous. While only 10–30% of metal corrosion is attributed to biocorrosion, the annual cost to the petroleum industry from metal corrosion in the USA exceeds one hundred million dollars (Beech and Sunner 2007). Iron alloys containing chromium, nickel, and molybdenum produce corrosion resistant stainless steel but its expense precludes its use in pipes, so carbon steel is commonly used around the world.

10.2.1 Current Theories of Biocorrosion

The biocorrosion of ferrous pipes was recorded over 100 years ago; however, biocorrosion mechanisms are poorly understood. The initial working concept by von Wolzogen Kühr and van der Vlugt was that electrons were liberated from iron at a region designated as the anode and this resulted in the release of Fe^{2+} with the formation of a pit in the metal. At a region referred to as the cathode, H_2 formed from the electrons from the anode and the H_2 collected at the surface of iron (cathode) and so reduced the rate of pit formation. Bacteria were envisioned as removing H_2 and by depolymerization of the cathode, the rate at which biocorrosion of the iron was accelerated. This proposed theory of biocorrosion was called cathodic depolymerization. However, because of newly isolated strains of bacteria applied in lithotrophic

Environmental Microbiology and Microbial Ecology, First Edition. Larry L. Barton and Robert J.C. McLean.

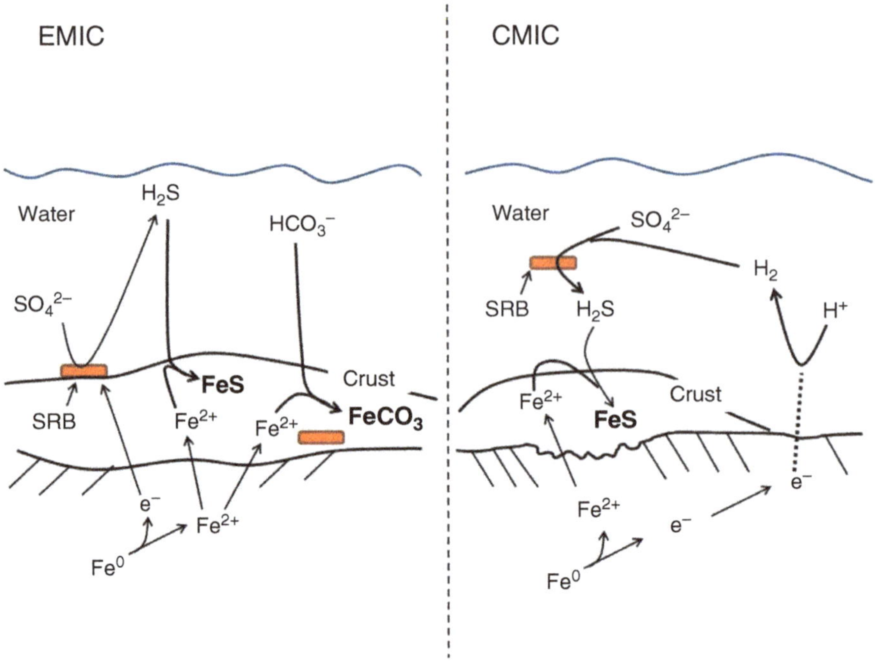

Figure 10.1 Iron corrosion based on the CMIC and EMIC models. SRB = Sulfate-reducing bacteria.

reactions where iron (Fe^0) is the electron donor, newer theories are currently being considered and these are discussed below.

The questions associated with ferrous biocorrosion that interest engineers and scientists are as follows:

- How are electrons extracted from Fe^0 in the oxidation process and what is the fate of these electrons?
- What is the fate of Fe^{2+} that is released from the metal?
- Which microorganisms have a primary role in the ferrous corrosion process and what are the mechanisms involved?
- Which groups of microorganisms enhance the activities of the primary iron oxidizers and what is their role?
- What is the role of biofilms and crust formation in the iron corrosion process?

To some extent, these questions remain unanswered despite considerable research. Starting with the abiotic reaction with Fe^0 oxidation (Figure 10.1), there is a slow process of chemical interaction as shown in the following equation:

$$4Fe^0 + SO_4^{2-} + 4H_2O \rightarrow FeS + 3Fe^{2+} + 8OH^-$$

The hallmark feature of the microbe – related corrosion is an enhancing of the rate at which the oxidation of Fe^0 occurs. Models that have emerged to describe the role of bacteria and archaea in biocorrosion of ferrous materials include the “electrical microbially influenced corrosion” (EMIC) and “chemical microbially influenced corrosion” (CMIC) (Enning and Garrelfs 2014). In EMIC, electrons are moved from the zero-valent metal to the bacterial cell with the release of Fe^{2+}, whereas with CMIC, corrosion of iron is attributed to the bacterial production of H_2S with the formation of FeS as the active corrosive agent. Models of EMIC and CMIC are given in Figure 10.1.

10.2.1.1 EMIC

In EMIC, the oxidation of iron is initiated by an anode reaction:

$$Fe^0 \leftrightarrow Fe^{2+} + 2e^- \quad E_0' = -0.47\ V \tag{10.1}$$

The electron acceptors are protons from the dissociation of water, with the formation of H_2 by the following abiotic, cathodic reaction:

$$2e^- + 2H^+ \leftrightarrow H_2 \qquad E_0' = -0.41\ V \tag{10.2}$$

Summation of the overall reactions produces the following:

$$Fe^0 + 2H^+ \rightarrow H_2 + Fe^{2+} \tag{10.3}$$

It is thought that MIC is due to microorganisms with an active hydrogenase that diverts electrons from hydrogen oxidation into cells, thereby contributing to biocorrosion. However, it has been demonstrated that bacteria and archaea that acquire electrons directly from iron (Reaction (10.1)) have a higher corrosion rate than those organisms using H_2 (Enning et al. 2012). Note that protons are limiting in the environment and this would not provide adequate substrate levels for Reaction (10.3). Therefore, the reduction of sulfate by sulfate-reducing bacteria (SRB) would be attributed to electrons from Reaction (10.1). Additionally, those microorganisms that grow lithotrophically using iron as the electron donor, have high rates of biocorrosion. Microorganisms using lactate or other organic electron donors have rates of ferrous biocorrosion that are no greater than sterile controls.

Not all strains of SRB can use Fe^0 as an electron donor for respiration and growth (Enning et al. 2012). Since electrons cannot move through water but require a conductive material, SRB would require a unique electron conductive component on the surface of the cell and currently there is no evidence of such a specific material. Two recent SRB isolates identified as *Desulfovibrio ferrophilus* and *Desulfopila corrodens* grow on electrons from iron (Enning et al. 2012) and numerous strains of other recently isolated SRB can acquire electrons directly in the corrosion process (EMIC) (Enning and Garrelfs 2014).

Several other electrogenic anaerobes can use electrons released from iron to support growth. The newly isolated acetogenic strain of *Sporomusa* sp. and the established strain *Sporomusa sphaeroides*, grow with Fe^0 as the electron donor and have enhanced biocorrosion activity (Kato et al. 2015). *Methanococcus maripaludis*, an electromethanogenic archaeon, can grow with oxidation of Fe^0 as the source of electrons, using extracellular enzymes (Deutzmann et al. 2015). Cellular enzymes such as hydrogenase or formate dehydrogenase are released on cell lysis, become bound onto the cell surface, and participate in the removal of electrons from the surface of the corroding ferrous material. It not known if the lysis of *M. maripaludis* is due to stress factors in the biofilm or due to a subpopulation that is being sacrificed for the greater benefit. Strains of *Shewanella* and *Geobacter* can interface with iron surfaces and acquire electrons by using cytochromes in the outer membrane or through nanowire connections (Pirbadian et al. 2014; Okamoto et al. 2011). For more about EMIC in the context of other extracellular electron transport by microorganisms, see Section 11.6.

10.2.1.2 CMIC

Respiratory coupled reduction of sulfate by SRB produces H_2S, which chemically reacts with Fe^{2+} to give insoluble FeS:

$$Fe^{2+} + H_2S \rightarrow FeS + 2H^+ \tag{10.4}$$

For corrosion to occur at an optimal rate, Fe^{2+} must be removed either by precipitation as FeS or by interaction with readily available bicarbonate (HCO_3^-) in the environment to form insoluble $FeCO_3$. Additionally, the bioproduction of H_2S leads to the formation of FeS as follows:

$$Fe^0 + H_2S \rightarrow FeS + H_2 \tag{10.5}$$

The hallmark activity of the CMIC theory (Figure 10.1) is the oxidation of iron as in Reaction (10.5) and the overall reaction of SRB is given below:

$$3\ \text{Lactate} + 2Fe^0 + 2SO_2^{2-} + H^+ \rightarrow 3\ \text{Acetate} + 2FeS + 3HCO_3^- + 2H_2O \tag{10.6}$$

The biogenic generation of FeS, Reactions (10.5) and (10.6), contributes to the sulfide stress cracking (SSC) of metallic iron (Enning and Garrelfs 2014). SSC contributes to a change in the properties of metallic iron and further stimulates the erosion of iron. Usually, when organic carbon is the electron source in biocorrosion studies, the oxidation of iron is due to CMIC and not to EMIC.

10.2.1.3 Iron Sulfide Crusts

FeS crusts occur on the surface of corroding iron. Some scientists have suggested that deposits of insoluble FeS on the iron surface may be a contributing factor in the diminishing oxidation rate of metallic iron. Others propose that electrons may move through the semiconductive ferrous sulfide to facilitate electron transfer as iron is oxidized by the SRB (Enning et al. 2012).

10.2.1.4 Biofilms and Extracellular Matrix

Bacteria associated with EMIC may secrete corrosive exopolymers containing uronic acids, proteins, and nucleic acids, all of which contribute to the biocorrosion event. The biofilm is not a pure culture of SRB but contains cells of various bacterial species and an extracellular polymeric matrix. The matrix may serve to entrap metabolic end products of SRB and other bacteria in the environment, bind various metal ions, entrap metallic nanoparticles, and contribute to the conductivity of the environment surrounding the SRB (Beech and Sunner 2007). It is strongly suggested that biofilms containing EMIC are resistant to biocides, and biofilms may promote biocorrosion because EMIC organisms use end products of other bacteria in the biofilm as nutrients. Biofilms in aerobic environments may harbor bacteria that consume oxygen thus rendering the interior of the biofilm anaerobic and favoring SRB. Additionally, bacteria associated with EMIC may use quorum sensing to promote a collective response to produce antibiotics or compounds deleterious to other bacteria. Specific details are needed to explain the role of biofilms and the extracellular polymeric matrix to biocorrosion.

As indicated in Table 10.1, there a several physiological groups of microorganisms that are thought to be associated with microbial-related biocorrosion. In addition to the SRB that oxidize iron, there is a group of methanogenic archaea that oxidize H_2 as it is formed. Pure cultures of *Methanobacterium* strain IM1 and *M. maripaludis* strain KA1 have been shown to produce methane with concomitant oxidation of H_2 (Kato 2016). It is known that pure cultures of microorganisms corrode iron less effectively than environmental samples

Table 10.1 Microorganisms associated with corrosion of ferrous metals.

Physiological type	Phyla	Genus and species
A Iron oxidizers		
Sulfate reducers	*Deltaproteobacteria* *Deltaproteobacteria*	*Desulfopila corrodens* strain Is4 *Desulfovibrio ferrophilus* strain IS5
B Hydrogen oxidizers		
Methanogens	*Euryarchaeota* *Euryarchaeota*	*Methanobacterium* sp. strain 1M1 *Methanococcus maripaludis* strain KA1
C Acetogens	*Clostridiales* *Clostridiales* *Clostridiales*	*Acetobacterium* sp. *Sporomusa spaeroides* *Syntrophomonas* sp.
D Nitrate reducers	*Firmicutes* *Bacteroidetes*	*Bacillus licheniformis* ATCC 14580 *Prolixibacter* sp. strain MIC-1
E S^0 reducers	*Thermotogales* *Euryarchaeota* *Euryarchaeota* *Euryarchaeota*	*Geotoga* sp. *Methanobacterium* *Methanosarcina* *Methanolobus*
F Iron reducers	*Gammaproteobacter* *Gammaproteobacter* *Deltaproteobacter* *Euryarchaeota*	*Pelobacter* *Pseudomonas* *Desulfovibrio* *Methanococcus maripalusia*

Sources: Based on data from Kato (2016); Vigneron et al. (2016); Enning et al. (2012).

containing numerous physiological groups of microorganisms. Microbial isolates from field corrosion sites contain acetogenic, sulfur oxidizing, iron oxidizing, and nitrate reducing microorganisms. These organisms (Table 10.1) may be in the biofilm of the corroding metal and their growth may be coupled to the acquisition of electrons as the corrosion proceeds. Some microorganisms participating in the corrosion event may use extracellular electron carriers and couple these electrons to manganese oxidation and iodide oxidation.

10.2.2 Biocorrosion of Nonferrous Materials

In bacterial system involving nonferrous metals, the process of biocorrosion is due to fermentative bacteria and acetogenic bacteria that produce acetic acid, which chemically reacts with copper, aluminum, or zinc. Biocorrosion is mainly due to bacteria immobilized in a biofilm and not to planktonic or swimming bacteria because immobilized bacteria focus their activities on the metal.

Tanks and other metallic chambers used for storing hydrocarbons have heterotrophic bacteria and fungi growing at the water–oil interface that secrete organic acids, which will pit copper, copper-nickel alloys, aluminum, and aluminum alloys (Little and Lee 2007). Blue water may result from the action of microorganisms present in biofilms when surface water is transported in copper tubing. Usually, the oxides on copper and aluminum protect against corrosion; however, the acidic end products of microbial metabolism and the secretion of an extracellular polymeric matrix diminish this protective effect. If the environment favors the growth of SRB, the production of sulfide will also erode the protective oxide layer on metals. Metals not subject to biocorrosion included titanium and stainless steel, an iron alloy containing chromium, and molybdenum.

Biocorrosion is commonly localized at the site where metals have been welded. The weld surface is irregular with a grain (crystallite of metal) size greater than the metal surface and the weld provides a region for biofilm development. The heat from the weld process changes some physical characteristics of the metal, including surface oxidation and stress on the adjacent metal surface. Damage due to biocorrosion of brass (copper and zinc alloy) materials is most often at the weld site.

10.2.3 Control of Biocorrosion

Various approaches have been used to prevent biocorrosion of metals by microorganisms; however, industrial applications always include a concern for cost. Iron pipes or storage tanks may be coated with a plastic liner to prevent bacterial attack on the iron surface. Compounds (i.e. fillers, plasticizers, and stabilizers) present in the plastics are used as nutrients for bacteria and fungi in the environment and this microbial activity enhances corrosion (Little and Lee 2007). Aeration has been used deter the viability of anaerobic bacteria; however, aeration pulses often dislodge biofilms or encourage increased growth of heterotrophic bacteria present. Chromate (CrO_3^-) salts have been used to inhibit the growth of bacteria and especially SRB; however, many different species of bacteria will reduce chromate to Cr^{3+} and insoluble chromium hydroxide [$Cr(OH)_3$] is produced at neutral pH. Various chemicals have been used as biocides (chlorine, chlorine dioxide, ozone, formaldehyde, cetylpyridinium chloride, ethanol, and methylene cyanate) and the use of these chemicals is often dependent on dose level and frequency, toxicity, application process, safety. and chemical compatibility (Jack and Westlake 1995).

Some effort has been given to inhibit bacteria associated with MIC by modifying the environment so SRB growth is not favored or to use biofilms to inhibit SRB (Little and Lee 2007). As an alternative to the use of biocides, adding nitrate and/or nitrite to natural environments shifts the bacterial population from SRB to nitrate-reducing bacteria (NRB). This use of nitrate to inhibit SRB works only if there is a limiting amount of carbon as the electron donor. Nitrite is inhibitory to the sulfur reductases of the SRB, while NRB outcompete the SRB for the carbon in the environment. Using bacteria to inhibit bacteria is reminiscent of probiotics used to influence microbial populations in animals. Laboratory tests suggest that it may be feasible to induce a biofilm on ferrous material and inhibit SRB from establishing in the biofilm. In one instance, a biofilm produced by *Bacillus brevis* was established and this organism secreted gramicidin S, which inhibited *Desulfovibrio vulgaris* growth (Jayaraman et al. 1999).

10.3 Bioalteration of Rocks, Monuments, and Other Surfaces

10.3.1 Biofilms on Rocks and Buildings

Desert varnish or rock varnish is a stable 1mm thick layer found on rock surfaces located worldwide from deserts to high mountain regions. This orange, brown, or black layer consists of clays with oxides and hydroxides of iron and manganese. Although the origin of rock varnish is unresolved, strong evidence indicates that the origin of black rock varnish is microbial (Dorn and Oberlander 1981). Often prehistoric persons would remove the desert varnish or patina with a sharp tool to expose a lighter colored rock and produce images which are referred to as petroglyphs (see Figure 10.2).

Microorganisms have been reported to create undesired appearances as they grow on house siding or roof shingles. In humid environments, algae and cyanobacteria produce streaks on asphalt shingles and on vinyl or wooden house siding. Although microbial growth is not detrimental to the house structure, efforts are often made to remove it because it is unsightly.

(a) (b)

Figure 10.2 Petroglyphs resulting from removal of desert varnish on rock surfaces: (a) "Star being" from Petroglyph National Monument in Albuquerque, Rinconada section; (b) a petroglyph of a caravan of bighorn sheep near Moab, Utah, USA. *Source:* (a) Jerry Willis, https://www.flickr.com/photos/glyphwalker/2680378470. Licensed under CC BY-SA 2.0; (b) Jim Bouldin, https://commons.wikimedia.org/wiki/File:MtnSheepPetroglyph.jpg. Licensed under CC BY-SA 3.0.

10.3.2 Biodegradation of Art Objects

The preservation of historic materials, including art, is important for the documentation of cultural heritage. The biodegradation of paintings, glass, ceramics, statues, stone buildings, photographic film, paper, and leather is attributed to microorganisms. While each site has specific physical and chemical characteristics that influence microbial activity, the degradation of items in the environment is due to microbial colonization followed by a succession of populations. A broad range of bacteria, algae, yeast, and fungi that are physiologically capable of degrading proteins, cellulose, and lipids, have been isolated from items of organic or inorganic objects of cultural heritage (Ranalli et al. 2009). Biodegradation is commonly initiated by phototrophic bacteria, algae, and lithotrophic bacteria. Heterotrophic bacteria have been isolated from art objects and it was thought that heterotrophic bacteria grew by using organic material released from dead autotrophic microorganisms; however, it is now obvious that heterotrophs absorb organic material from the air and this supports their growth (Ranalli et al. 2009). The following are some case studies involving biodegradation of important art objects or historic structures.

10.3.2.1 Marble Statues in Italy

Algae have been found to colonize the surfaces of marble statues in the Boboli Gardens in Florence, Italy (Lamenti et al. 2000). Within two years after cleaning and restoring the ornamental marble statues, the green alga *Coccomyxa* was found on the surface at a density of 300 cells cm^{-2}. Cyanobacteria along with *Coccomyxa* established a green biofilm and by the sixth year following cleaning, the density of photosynthetic cells was 30 000 cells cm^{-2}. The cyanobacteria produced considerable exopolysaccharide, which stabilized the biofilm and enhanced the adherence of the cyanobacteria onto the marble surface.

10.3.2.2 Paintings in the Lascaux Cave in France

There are about 600 paintings in the Lascaux Cave in southwestern France that date back to 15 000–13 000 BCE. The cave containing paleolithic art was opened to visitors in 1948 but closed in 1963 after it became apparent that the paintings were being damaged. The lighting system introduced to provide safety, the disturbing of cave soil by workers and the elevated

CO_2 levels due to respiration by visitors contributed to an environment that favored the growth of algae, cyanobacteria, bacteria, and fungi. One of the contaminants on the walls of the Lascaux Cave was a green algal organism *Bracteacoccus minor* that produced a green patina over the paintings (Ciferri 1999). In 2001, the fungus *Fusarium solani* was found throughout the cave and by 2006, extensive growth of another fungus *Ochroconis lascauxensis* covered many of the paintings with a black stain (Martin-Sanchez et al. 2014). To control the rapid growth of these fungi, benzalkonium chloride was employed as a biocide and as a result, members of the genera *Ralstonia* and *Pseudomonas* became dominant bacteria in the cave environment (Bastian et al. 2009).

10.3.2.3 Mogao Grottoes in China

The caves of Mogao Grottoes are in the Gansu Province of China located on the Silk Road, which connects Europe and Asia. These caves were built from about 400–1300 AD and about 450 of these caves with mural areas remain. Although the climate where these caves were built is dry, considerable damage to the pigmentation has occurred. While considerable diversity of fungal populations occurs in these caves, members of the *Ascomycota, Zygomycota*, and *Basidiomycota* phyla were the most frequently isolated. Using molecular biology techniques, bacterial species belonging to the *Firmicutes* and *Proteobacteria* phyla were most frequently isolated (Ma et al. 2015). Thus far, no suitable means of controlling the biodegradation of murals in the Mogao Grottoes have been developed.

10.3.2.4 Damage to Frescoes

Medieval frescoes were painted using various metals (i.e. arsenic, mercury, lead, iron, aluminum, and cadmium) to provide rich colors and these toxic elements initially prevented microbial growth on the paint. The paints and support layer for the painting were often mixed with egg proteins, casein, fats, plant oils, and fibers, which are excellent substrates for microbial growth. However, from numerous surveys, the types of microorganisms commonly associated with frescoes include fungi, cyanobacteria, soil bacteria, and eukaryotic algae (Ciferri 1999). Fungi growing on frescoes are responsible for staining and discoloration as well as forming patinas and efflorescence. Physical damage attributed to fungi is the release paint flakes following growth between the paint layer and the supporting matrix. Bacteria degrade frescoes by forming a biofilm over the paint surface (Milanesi et al. 2006). Frescoes in crypts and tombs are subject to deterioration by *Streptomyces* and *Nocardia,* with over one million bacteria per gram of sample removed from frescoes in various Italian hypogean sites (Ciferri 1999).

An interesting study was conducted at the chapel of the Holy Nail, Siena, Italy (Milanesi et al. 2009). Frescoes painted by Lorenzo di Pietro in 1446–1449 depict biblical scenes and in the tenth century the chapel served as a hospital. Using DNA extraction and bacterial 16S rRNA gene cloning, Milanesi et al. (2009) isolated and identified bacteria from many different phylogenetic groups from the frescoes. Of considerable interest were potentially pathogenic bacteria from the *Neisseriaceae, Burkholderiales*, and *Streptococcus* groups. In 1966, the acrylic resin Paraloid B 72 b was applied to the frescoes to conserve the art; however, it has been found that fungi can deteriorate epoxy resins (Milanesi et al. 2009; Pangallo et al. 2015).

10.3.3 Biotechnology for Restoration of Artworks and Historic Stones

In areas where air pollution is persistent and the temperature and relative humidity is high, deterioration of architectural heritage and art objects are of great concern (Dakal and Cameotra 2012). Algae grow on the surfaces of many historic buildings (Figure 10.3) and distort their appearance. Restoration efforts include the use of bacteria for biocleaning damaged surfaces

Figure 10.3 Lichen on a historic building. The image of the building is in the inset and the enlarged image reveals lichens on the wall. While structural damage to the historic building is a relatively minor problem, the major concern is the discoloration which detracts from the initial structure. This picture of a historic building in Northern Spain is provided by Larry Barton.

(Bosch-Roig and Ranalli 2014). The restoration of altered surfaces is required every few years because deterioration is ongoing and cannot be permanently stopped. While each site has unique characteristics contributing to biodeterioration, the focus here will be on treating frescoes and stone surfaces.

- The fourteenth century "Conversion of S. Efisio and Battle" fresco at the Pisa Camposanto Monumentale, Italy, was restored but the fresco became altered by the animal glues used. Later application of a culture of *Pseudomonas stutzeri* restored the fresco to its original appearance (Ranalli et al. 2005).
- The Ajanta caves in India date back to the second century BC and contain Indian wall paintings. Over the years, conservators applied shellac, a secretion from an Indian insect containing monoesters and polyesters of hydroxy aliphatic and sesquiterpenoid acids, as a fixative on mural paintings. With time, the shellac became yellow and darkened and the murals required restoration. Laboratory tests simulating paints covered with shellac indicated that three isolates (a *Pseudomonas* sp., an *Achromobacter* sp., and an *Acinetobacter* sp.) used shellac as a carbon source and were potentially useful in removing shellac covering the murals (Barbabietola et al. 2012).
- An interaction between organic pollutants in the atmosphere and mineral deposits causes black crusts to form on stone surfaces. Sulfate deposits come from acid rain. Organic compounds in smog particles, especially polyaromatic hydrocarbons, from the atmosphere are

adsorbed in the mineral matrix to produce black crusts (Saiz-Jimenez 1997). A cleaning process to remove the black crust, which uses an aerotolerant strain of *D. vulgaris* – a sulfate-reducing bacterium – to metabolize gypsum ($CaSO_4$) without the production of hydrogen sulfide (Cappitelli et al. 2007).

- Normally, carbonate is deposited on the stone surfaces by bacterial activity and the process of calcite formation by microbes is considered desirable in retarding rates of biodecomposition (Dhami et al. 2014). Weathered building stone often contains small pores or cracks, especially if the building is constructed of limestone. The formation of calcite reduces water penetration of the stone surface and also prevents the deposition of nutrients needed for bacterial growth.

10.4 Biodeterioration of Concrete

Concrete is a mixture of inorganic alkaline precipitates (cement) and aggregates (stones and rocks). While MIC refers to the corrosion of metallic structures, the degradation of plain concrete or metallic reinforced concrete is designated as microbial influenced degradation (MID). Concrete structures in wastewater treatment plants, sewage systems, barn floors, and marine ports are deteriorated primarily by the combined activities of SRB and sulfur oxidizing bacteria (Wei et al. 2013). Freshly poured concrete has a highly alkaline surface because of the calcium hydroxide formed in the hydration process. The alkaline surface of fresh concrete inhibits colonization by bacteria but microcracks eventually develop as the concrete solidifies and the final internal structure is achieved. The moist surface of concrete often supports the growth of photosynthetic microorganisms (Figure 10.4) and heterotrophic microorganisms grow on the carbon compounds secreted by the phototrophs. Thus, algae and cyanobacteria may initiate the biodeterioration of concrete and stone surfaces.

Colonization by autotrophic thiobacilli causes the oxidation of reduced sulfur forms to produce sulfuric acid (Li et al. 2017). As the surface alkalinity declines to about pH 9,

Figure 10.4 Algae growing on concrete is an initial step in concrete biodeterioration. *Source:* Courtesy of Nathaniel L. Ritz.

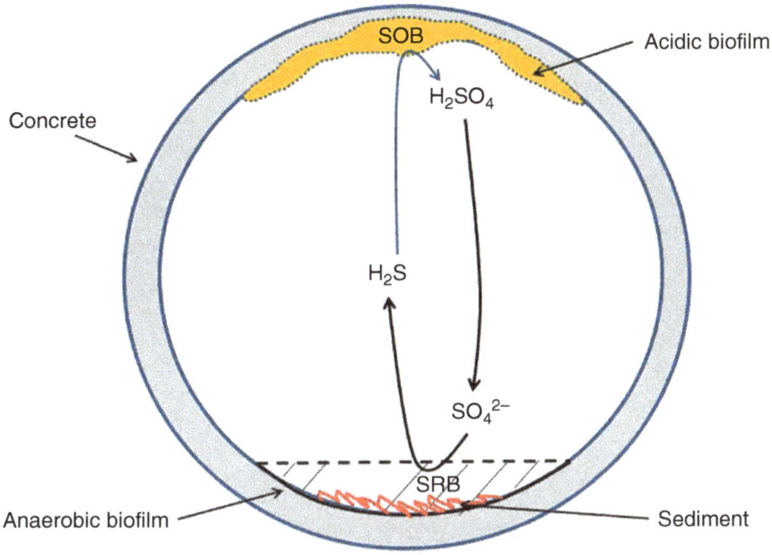

Figure 10.5 Deterioration of concrete as a result of acid formation by aerobic bacteria. With the loss of surface material, internal pores in the concrete are exposed and invasion by bacteria into these pores results in a weakened concrete structure. SRB = Sulfate-reducing bacteria. SOB = Sulfur-oxidizing bacteria.

microcolonies of cyanobacteria, heterotrophic bacteria, and SRB establish a mixed microbial mat. Under appropriate nutritional conditions, SRB grow on the sulfate produced by the thiobacilli and release H_2S, which neutralizes the alkaline concrete surface. The first species of thiobacilli that colonize on concrete are *Thiobacillus thioparus, Thiobacillus novellus, Thiobacillus intermedius,* and *Thiobacillus neopolitanus* because they can grow at pH levels near neutrality. Once the surface of concrete reaches about pH 5, *Acidithiobacillus thiooxidans* appears and the neutrophilic thiobacilli are inhibited by the high acidic level. As the metabolism of the sulfur oxidizing bacteria and SRB display nutritional cooperation in running a mini-sulfur cycle, biodeterioration of concrete continues (Figure 10.5).

Algae are important organisms in MID of metallic reinforced concrete and several events are thought to be important (Javaherdashti et al. 2009). The initial deterioration involves an attack on concrete by algae extracting magnesium, calcium, and other elements from the cement paste of the cement, which produces minute cavities and voids in the concrete. Cracks eventually develop that connect the internal cavities of the cement. Algal activity along with sulfur-oxidizing bacteria will further deteriorate the concrete and contribute to the corrosion of the steel used as reinforcement. Algal activity includes the production of organic acids that damage the integrity of the concrete. Due to the photosynthetic production of oxygen and the respiration of algae to consume oxygen, electrochemical cells are developed that contribute to regions of varying oxygen concentrations and this favors the growth of MIC organisms and the corrosion of steel.

An attempt to control concrete biodeterioration includes the incorporation of commercial biocides in the concrete mixture. Useful inexpensive additives that inhibit concrete biodeterioration include zinc oxide, copper slag, sodium bromide, and fly ash (Wei et al. 2013). Submerged concrete breakers found along the coast of some of the Roman ports remain in good condition but the mystery of how these structures have avoided biodeterioration remains unresolved.

10.5 Mineral Interaction and Biomineralization

The formation of mineral deposits by microorganisms is called biomineralization. Mineral precipitation due to microbial metabolic or respiratory end products is designated as biological induced mineralization because once the chemicals are released from microorganisms, the cell has no control of the mineral product. Another form of microbial activity is biologically controlled biomineralization (Box 10.1), where the development of precipitated minerals is under cellular control and the minerals have an important physiological role (Konhauser 2007). Of the over 70 minerals produced by bacteria, the most common are iron, calcium and magnesium complexed with carbonate, phosphate, sulfate and sulfide. Microbial cell surfaces and biofilms can also bind metal ions and thus promote (i.e. nucleate) the formation of a number of minerals (reviewed in Fortin et al. 1997). A few minerals produced by microbial activity will be discussed in the following paragraphs (Dunn et al. 1997).

10.5.1 Iron Hydroxides

Bacteria provide sites of nucleation for minerals for precipitation of ferric hydroxide. When water containing reduced iron (Fe^{2+}) comes in contact with oxygen (O_2), abiotic formation of insoluble $Fe(OH)_3$ results. The anionic bacterial cell wall or extracellular polymeric material may serve as a site for accumulation of amorphous ferric hydroxide (reviewed in Fortin et al. 1997). This passive iron mineralization encrusts single cells or entire bacterial communities and produces ochre, which is a red to brown formation. Iron hydroxide is converted under abiotic conditions to the crystalline minerals of goethite (FeOOH) or hematite (Fe_2O_3) and the accumulation of these minerals around the bacterial cell, encrusts the cell in iron oxides.

Some bacteria oxidize Fe^{2+} to $Fe^{3+}$and use energy from this reaction for growth. *Chlorobium ferrooxidans,* a green sulfur bacterium, *Rhodobacter ferrooxidans,* a purple nonsulfur bacterium, and *Thiodictyon* sp., a purple sulfur bacterium, are examples of anaerobic photoautotrophic bacteria that use Fe^{2+} as an electron donor with the release of Fe^{3+}. Anaerobic phototrophic bacteria can use siderite ($FeCO_3$) or iron sulfide (FeS) as electron sources but they cannot obtain electrons from iron in crystalline vivianite [$Fe_3(PO_4)_2$], magnetite (Fe_3O_4) or pyrite (FeS_2). *Acidithiobacillus ferrooxidans* and *Leptospirillum ferrooxidans* are chemolithotrophic bacteria growing at acidic pH with the oxidation of Fe^{2+} to Fe^{3+}. Under anaerobic

Box 10.1 Mineralization and Biomineralization

Mineralization is a term which has a specific meaning in different scientific areas. In biology, mineralization is used to describe the precipitation of inorganic materials inside or outside of the individual, including the formation of bones and exoskeleton. In microbiology, this includes the production of magnetic granules and calcium crystals. Additionally, the decomposition of natural materials with the release of inorganic compounds and other small molecules is often called mineralization. In soil science, mineralization is used to describe the enrichment of soil with plant-available compounds from decomposition or nitrogen fixation. In geology, mineralization is the formation of insoluble ores and crystals. Because mineralization in geology may be by chemical or biological processes, the designation of biomineralization was initially used to designate the involvement of microbial activities in this process. However, many scientists now attach "bio" to many of the processes involving bacteria or fungi. Thus, the context in which mineralization is used is required to identify the process being addressed.

conditions or in acidic environments, Fe^{3+} remains soluble while insoluble ferric hydroxide forms only in aerobic water with a near neutral pH.

Members of the *Sphaerotilus, Crenothrix,* and *Metallogenium* genera are heterotrophic bacteria that grow at neutral pH and contribute to iron mineralization by metabolizing the organic structures that chelate Fe^{3+}. These heterotrophs do not obtain energy from oxidation of Fe^{2+} and have sheaths where the cells are transiently present inside the sheath cover. The sheaths are made of complex sugar polymeric material, which binds Fe^{3+} as the iron hydroxide and prevents the bacterial cells from being encrusted inside the iron precipitate. *Gallionella ferruginea,* an iron oxidizing bacterium, grows at neutral pH in waters with moderately reduced oxygen. Attached to the bacterial cell are long filamentous ribbons. $Fe(OH)_3$ precipitates along the ribbons and not on the cells, thereby, preventing the bacterial cells from becoming encrusted in iron and unable to divide.

10.5.2 Magnetic Mineral Crystals

Certain bacteria display specialized activities in response to magnetic fields. Magnetotactic prokaryotes are distinguished by having internal organelles containing crystals of magnetite (Fe_3O_4) or greigite (Fe_3S_4) (Lefèvre and Bazylinski 2013a). The internal granules, termed magnetosomes, are aligned to orientate the cell along the Earth's magnetic field. The magnetosomes are enclosed in a membrane containing either magnetite or greigite and range from 35 to 120 nm (Lefèvre and Bazylinski 2013b). Magnetite formation may be according to the following reaction:

$$Fe^{2+} + 2OH^- + 2Fe(OH)_3 \rightarrow Fe_3O_4 + 4H_2O$$

Bacteria producing magnetosomes include *Magnetospirillum magnetotacticum, Magnetospirillum* marinus, *Magnetospirillum blakemore, Magnetospirillum magneticum, Magnetospirillum gryphiswaldense,* and *Desulfovibrio magneticus* RS-1 (Lefèvre and Bazylinski 2013b; Sakaguchi et al. 2002). Several bacteria with Fe^{3+} reducing activity or sulfate-reducing capability produce magnetite (Fe_3O_4), greigite (Fe_3S_4), and pyrrhotite (Fe_7S_8). *Geobacter metallireducens* GS-15, *Geobacter sulfurreducens,* and *Shewanella putrefaciens* produce magnetite from reduction of iron in ferric hydroxide, with nanoparticles of magnetite produced outside the cell (Vali et al. 2004; Konhauser 2007; Corker et al. 2009).

10.5.3 Manganese Oxides

In aquatic and soil systems, hydroxides of manganese as Mn^{2+} and Mn^{4+} occur at the aerobic–anaerobic interface in a manner similar to formation of iron hydroxides. The coprecipitation of manganese oxides and iron oxides results in ferromanganese nodules, which are mineral deposits found in soil and sediments in lakes and seas. The formation of aquatic ferromanganese nodules is the result of a series of microbiological activities. Initially, cyanobacteria and algal cells concentrate soluble manganese and iron on cell surfaces or capsular materials. Upon death of these phototrophic organisms, the cells with metals attached descend to the sediment where the cellular organic material is digested and the metals are liberated. The reduction of $Mn^{3+}/Mn^{4+}/Fe^{3+}$ oxyhydroxides present in the river occurs and as the particulate minerals move upward in the water column, a new collection of cells with minerals becomes attached to the particulate surface. This cycling of particulate material containing oxidized and reduced manganese and iron produces a layer effect on the ferromanganese nodule. Bacteria have been isolated on the surface inside the ferromanganese nodule and it is assumed that these bacteria contribute to the ferromanganese nodule formation.

The molecular features of Mn^{2+} oxidation by bacterial systems have been described and under appropriate conditions, microbial oxidation of Mn^{2+} to Mn^{4+} produces Mn^{3+} as a stable intermediate (Tebo et al. 2010). Endospores of *Bacillus* sp. strain SG-1 are capable of oxidizing Mn^{2+} by a multicopper oxidase (MCO). The oxidation of Mn^{2+} by *Leptothrix discophora* and *Pedomicrobium* sp. ACM 3067 may also be attributed to an enzyme of the MCO family. The manganese oxidizing activity of *Aurantimonas manganoxydans* and *Erythrobacter* sp. strain SD-21 is associated with a heme containing peroxidase, like that found in *Phanerochaete chrysosporium* for lignin oxidation. The peroxidase produces hydrogen peroxide which oxidizes manganese oxide.

10.5.4 Carbonates

Microorganisms have a vital role in the formation of carbonate minerals because they either increase the alkalinity of the area surrounding the cell or secrete substances that promote nucleation of minerals (Konhauser 2007). The environment determines which physiological type of bacteria is involved in the carbonate formation. Siderite ($FeCO_3$) nanocrystals are produced by *Acidiphilium* sp. strain PM, an iron reducing bacterium (Sánchez-Román et al. 2014). Factors controlled by bacteria that enable siderite formation in an acidic environment (pH 3.5) include the following. Fe^{2+} collects at the cell surface as ammonia and carbon dioxide are released from the cell by metabolic activities. Ammonia promotes alkalinisation (pH 6–7) according to the following reaction and globules of $FeCO_3$ on the surface of the *Acidiphilium* sp.:

$$NH_3 + H_2O \rightarrow NH_4^+ + OH^-$$

Many different microbial species produce calcium and manganese carbonates in environments containing high concentrations of salt. As reviewed by Rivadeneyra et al. (2006), over 200 soil bacteria or 22 halophilic bacteria can produce $CaCO_3$, indicating that carbonate precipitation is a general characteristic of bacteria. The ratio of Mg^{2+} to Ca^{2+} is thought to influence the formation of calcite and aragonite, as both have the same chemical formula of $CaCO_3$ but different crystalline structures. Since a high concentration of Mg^{2+} inhibits the bacterial production of $CaCO_3$, Ca^{2+} is concentrated on the cell surface by ionic interaction with organic acids and this high density of Ca^{2+} enables the formation of calcium carbonate to occur.

In anaerobic environments, the formation of dolomite [$CaMg(CO_3)_2$] is influenced by SRB. Laboratory experiments with pure cultures indicate that several strains of SRB, including *Desulfovibrio brasiliensis* and *D. hydrothermalis*, produce dolomite and the environmental activities of SRB induce dolomite formation in hypersaline lagoons in Brazil (Lith et al. 2003). The contribution of SRB to dolomite precipitation include changes in the environment such as increased pH levels and alkalinity and depletion of sulfate. Additionally, charges on the cell surface and charges in extracellular polymeric substances concentrate Mg^{2+} and Ca^{2+}, which result in the binding of HCO_3^- and the ultimate formation of $CaMg(CO_3)_2$.

10.5.5 Phosphates

The accumulation of phosphate by planktonic phototrophs and heterotrophic bacteria is associated with microbial activity in marine environments. Because phosphate is often a limiting nutrient in the environment, bacteria have developed highly active systems for collecting phosphate and retaining it on the cell surface or internally as polyphosphate granules. Thus, with the distribution of dead phototrophs throughout the water column, heterotrophic bacteria actively collect phosphate for their growth. Additionally, inorganic phosphate released from organic materials may be removed from the water column when iron phosphate precipitates are formed.

Phosphorus-rich nodules or grains, called phosphorites, are found along coastal regions where upwelling of phosphate-rich waters occurs. Phosphorites are an organic matrix containing at least 10% phosphate minerals. Bacteria have been implicated in the formation of phosphorite for many years (Nathan et al. 1993). Fossilized phosphoritic sediments are of interest because they may help explain some ancient activities in the development of the Earth (Bailey et al. 2013). Fossilized phosphorite material is seen in stromatolites, which are mats of cyanobacterial cells, that contain calcium fluorapatite $[Ca_{10}(PO_4)_{6-x}(CO_3)_xF_{2+x}]$ (Konhauser 2007).

Struvite ($MgNH_4PO_4{\cdot}6H_2O$) is a mineral associated with the decomposition of organic matter and bacterial activities. A high percentage of bacteria present in wastewater treatment systems can precipitate phosphorus as struvite and most of the bacteria with this capability are of the *Alphaproteobacteria* group (Rivadeneyra et al. 2014). Struvite has also been found is in anaerobic digesters, where there are high levels of ammonium, phosphate, and magnesium ion.

10.5.6 Sulfates

Gypsum is precipitated by cyanobacteria and is found in ancient stromatolites, indicating that it is not a recently evolved process. Nucleation is along the S-layer of the cyanobacterial cell and in addition to binding Ca^{2+}, the S-layer also binds Sr^{2+} to form celestite ($SrSO_4$) (Schultze-Lam and Beveridge 1994). The role of heterotrophic bacteria in the formation of sulfate minerals has not been fully explored except for a report that *Myxococcus xanthus* can form barite ($BaSO_4$) (Konhauser 2007) and that gypsum is produced by *Arthrobacter* sp. (Cockell et al. 2010).

10.5.7 Sulfides

Copious quantities of hydrogen sulfide are produced in the metabolism of SRB. The sulfide reacts with Fe^{2+} in the environment to produce amorphous iron sulfide. It has long been speculated that bacteria have a role in the process of pyrite (FeS_2) formation from FeS; however, evidence of this has not yet been presented. Several of the following sections address the metabolism of metal sulfides.

10.5.8 Clays

Several different strains of bacteria have been reported to reduce Fe^{3+} in clay crystals (Kostka et al. 1999; Kostka et al. 2002; Li et al. 2004; Kashefi et al. 2007; O'Reilly et al. 2005; Zhang et al. 2007). Bacteria that can reduce solid phase Fe^{3+} in clay crystals under anaerobic conditions include *Shewanella oneidensis* strains MR-1, *Thermoanaerobacter ethanolicus, Geothermobacter ehrlichii, Geobacter* sp. *Desulfitobacterium hafniense, Desulfovibrio* sp. strain G-11, and *Pseudomonas* spp. A few bacteria reduce Fe^{3+} in clay crystals only under microaerophilic conditions. Also, it appears that in the presence ferruginous clays, methanogens will reduce Fe^{3+} but not produce methane. Model compounds of ferruginous smectite that are used for bacterial studies include SWa-1, Swy-1, NAu-1, and NAu-2. As seen from the chemical formula, $[Na_{0.87}(Si_{7.29}Al_{0.62})(Fe^{3+}{}_{2.67}Fe^{2+}{}_{0.01}Al_{1.08}Mg_{0.23})O_{20}(OH)_4]$, SWa-1 contains an appreciable quantity of Fe^{3+} (Kashefi et al. 2007). The clay crystals are highly organized and in some cases the reduction of Fe^{3+} results in decay of the crystal and in some instances the insertion of K^+ adjacent to the reduced iron atom stabilizes the clay crystal.

10.5.9 Uranium Precipitate and Crystals

The atomic age has given interest to uranium in the environment. To provide industry with uranium, uranium mining has been promoted in several regions and the fallout of this activity is environmental contamination where uranium milling occurs. Uranium as (UO_2^{2+}, uranyl ion) produces a light-yellow solution with water and many bacteria have been reported to reduce uranyl ion to water insoluble, black uraninite (UO_2). Oxygen remains in close association with uranium as uranyl or uraninite, where the element is U(VI) and U(IV), respectively. Depending on the bacterial species, the deposition of uraninite by bacteria may be in the periplasm of gram-negative bacteria, on the cell surface, in the cytoplasm, or extracellular. Cytochromes in the outer membrane of *Shewanella* (Marshall et al. 2006) and *Geobacter* (Shelobolina et al. 2007) provide electrons for extracellular uranyl reduction while nanowires, previously called conductive pili, provide electrons for uranyl by *G. sulfurreducens* (Cologgi et al. 2011). A list of 30 strains from the many bacteria that can reduce uranyl ion to uraninite are given in a recent review (Barton et al. 2014). *Cupriavidus metallidurans* and *Rhodopseudomonas palustris* have tremendous capability to metabolize uranium and the potential for these strains for uranium bioremediation are presented by Llorens et al. (2012).

The stability of uraninite is important for bioremediation of uranium. Uraninite can be oxidized by a chemical reaction to uranyl by an end product from microorganisms. Fe^{3+} from oxidation of Fe^{2+} by *A. ferrooxidans, L. ferrooxidans,* species of *Sulfolobus* or other iron reducing microorganisms provide the electron acceptor for oxidation of uraninite. The following chemical reaction could accomplish this oxidation of uraninite:

$$UO_2 + Fe_2(SO_4)_3 \rightarrow UO_2^{2+} + SO_4^{2-} + 2FeSO_4$$

In some situations, biominerals of uranium are produced by bacteria. The bioprecipitation of UO_2^{2+} by *Citrobacter* sp. with the formation of a $NH_4UO_2PO_4$ crystal is due to the release of phosphate from organic matter (Macaskie et al. 2000). *Thermoterrabacterium ferrireducens* can reduce pale green uramphite (U(VI); $[(NH_4)(UO_2)(PO_4)_2 \cdot 3H_2O]$ to brown-green ningyoite (U(IV); $[CaU(PO_4)_2 \cdot H_2O]$ (Khijniak et al. 2005).

10.5.10 Gold Grains

Historically, gold has been considered to be metabolically inert; however, the presence of biofilms containing various bacteria on gold grains has prompted scientists to examine gold mineralization with the formation of "bacterioform" gold (Reith et al. 2013). There are three oxidation states for gold (Au^0, Au^+, Au^{3+}) and the enzymatic reduction of Au^+ or Au^{3+} to Au^0 has been demonstrated in numerous microorganisms, including *Thermotoga maritima* and *Pyrobaculum islandicum* (Reith et al. 2007). Reduction of ionic gold may contribute to cell toxicity by free radical formation because this reduction is associated with a single electron transport process. Additionally, ionic gold has strong binding affinity to carboxyl, thiol, and other charged groups found in bacterial cell walls. In addition to passive interaction with bacteria, cationic gold is actively metabolized by some bacteria. *A. ferrooxidans* and *A. thiooxidans* will oxidize sulfide minerals with release of gold. Thiosulfate may be produced from bacterial sulfide oxidation (or added to mining operations) to form soluble gold complexes and metabolism of thiosulfate by *Desulfovibrio* sp. could result in formation of gold deposits.

Thus far, a few of the bacterial interactions with Au^+ and Au^{3+} have been reported and the diversity of gold-bacteria interactions reflects varied stress response (Reith et al. 2007; Reith et al. 2013). *C. metallidurans* CH34 responds to Au^+ and Au^{3+} by activating gold-induced genes (*gig*) which induces the uptake of Au^{3+}, the cytoplasmic reduction of Au^{3+} to Au^+ including possible methylation of gold and export of Au^+ to the periplasm where it is reduced to Au^0. Also involved in the exposure to ionic gold is the induction of an oxidative stress response and genes for resistances genes including *cop, ars, cup,* and *mer*. *Salmonella typhimurium* has a system specific for ionic forms of gold and since *S. typhimurium* is found in the intestine of birds and animals, this may reflect resistance to ingestion of gold containing plant or animal material. In the presence of Au^{3+}, *Delftia acidovorans* secretes delfibactin (Johnson et al. 2013) which selectively chelates the gold ion protecting the cell and contributing to the formation of Au^0 in the extracellular polysaccharide (EPS) matrix. Cyanobacteria, purple nonsulfur phototrophs and *Stenotrophomonas* sp. also form nanoparticles of Au^0 in the EPS. *Micrococcus luteus* has ionic gold in the metal center of a methane oxidase (Levchenko et al. 2001). Certainly, the area of gold interaction with bacteria is an extremely interesting area of research.

Acidithiobacillus ferrooxidans and acid tolerant species of *Sulfolobus* are capable of leaching gold from arsenopyrite ore (Lindström et al. 1992). In this activity, the microorganisms release gold (Au) with the oxidation of iron and arsenic according to the following reaction:

$$2FeAsS(Au) + 7O_2 + 2H_2O + H_2SO_4 \rightarrow Fe_2(SO_4)_3 + 2H_3AsO_4 + (Au)$$

Gold is solubilized with cyanide according to mining procedures. In the final stages of the gold recovery process, cyanide is detoxified by bacterial metabolism to produce carbon dioxide and urea.

10.6 Interactions with Transition and Rare Earth Elements

10.6.1 Transition Elements

Microorganisms are found in most mineral deposits and are considered by many to be important in the cycling of transition elements including Fe, Mn, Co, Mo, Cu, and Ni. Although Zn is not a transition metal, it will be included in this discussion. Numerous metals are associated with laterite ores and fungal leaching of laterite solubilizes several transition elements. *Aspergillus niger* and *Penicillium simplicissimum* have been used to leach Ni, Co, Cu, Zn, and Mn from laterite by the secretion of oxalic and citric acids (Konhauser 2007). Both oxalic and citric acids are chelators of metal ions and their affinity for transition metal cations (Me^{2+}) is enough to extract the metal ions from the ore matrix.

Bacteria such as *A. ferrooxidans* have been used in bioleaching to extract transition metals from various minerals. *A. ferrooxidans* can oxidize sulfide in several sulfide ores and this produces a soluble transition metal (Table 10.2). The oxidation of molybdenum in ore (presumed to be molybdenite) to give molybdate is also attributed to bacteria (Lyalikova and Lebedeva 1984). Zn^{2+}, Ni^{2+} and Cu^{2+} and are released from sphalerite, millerite, and bornite, respectively, by *A. ferrooxidans* (Balci 2010; Razzell and Trussell 1963). Many acidophilic iron-oxidizing prokaryotes including *A. ferrooxidans* and *L. ferrooxidans* can oxidize pyrite and marcasite. The mineralization of pyrite progresses through several phases and this process serves as a useful model for microorganisms metabolizing other solid minerals (Konhauser 2007). Initially,

Table 10.2 Examples of sulfide minerals metabolized by bacteria.

Mineral	Reactions
Pyrite (cubic crystal structure)	$FeS_2 + 3.5O_2 + H_2O \rightarrow FeSO_4 + H_2SO_4$
Marcasite (orthorhombic crystal structure)	$FeS_2 + 3.5O_2 + H_2O \rightarrow FeSO_4 + H_2SO_4$
Sphalerite	$ZnS + 2O_2 \rightarrow ZnSO_4$
Millerite	$NiS + 2O_2 \rightarrow NiSO_4$
Molybdenite	$2MoS_2 + 9O_2 + 6H_2O \rightarrow 2H_2MoSO_4 + 4H_2SO_4$

bacteria migrate to the pyrite and the bacterial cells interface with the surface of the iron mineral. The cell attaches to the mineral by electrostatic charges using pili, filaments, or an EPS matrix. The bacteria grow to form a microcolony and, if nutrients are available, a mixed biofilm is established.

The role of *A. ferrooxidans* in leaching Cu^{2+} from copper minerals may be by either a reaction requiring O_2 or by an anaerobic reaction. Bacterial oxidation of Cu^{+} in chalcocite to covellite is followed by the oxidation of sulfide in covellite to sulfate by the following reactions:

$$2Cu_2S + 2H_2SO_4 + O_2 \rightarrow 2CuS + 2CuSO_4 + 2H_2O$$

$$CuS + 2O_2 \rightarrow CuSO_4$$

In another oxidative reaction that is oxygen independent, chalcocite is oxidized to covellite and copper sulfate with Fe^{3+} as the electron acceptor:

$$Cu_2S + Fe_2(SO_4)_3 \rightarrow CuSO_4 + CuS + 2FeSO_4$$

The sulfide in covellite is oxidized to elemental sulfur (S^0), with Fe^{3+} reduced to Fe^{2+} as in the following reaction:

$$CuS + Fe_2(SO_4)_3 \rightarrow CuSO_4 + S^0 + 2FeSO_4$$

Elemental sulfur is oxidized by *A. ferrooxidans* to sulfate.

$$2S^0 + 3O_2 + 2H_2O \rightarrow 2H_2SO_4$$

Although these reactions may occur slowly in heap leaching, they are effective and the process is relatively inexpensive.

10.6.2 Rare Earth Elements

Rare earth elements (REEs) include both the lanthanide and actinide series of elements. The REE are receiving increased interest due to their role in numerous technological applications. Elements in the lanthanide series include cerium, praseodymium, neodymium, promethium, samarium, europium, gadolinium, terbium, dysprosium, holmium, erbium, thulium, ytterbium, lutetium, lanthanum, and often yttrium. Uranium, plutonium, and thorium are elements of interest in the actinide series and uranium interaction was discussed in Section 10.5.9. Most studies have focused on REE binding to carboxyl or phosphate groups at the periphery of bacteria and cells in natural microbial mats, which enhances the concentration of REE from the environment (Takahashi et al. 2005). In general, a greater accumulation of REE occurs with gram-positive bacteria than with gram-negative bacteria. Ferric ion chelators of low binding affinity for REE elements are more conducive for biosorption to cells than if chelators of high binding affinity are used. In at least one case, *Methylacidiphilum fumariolicum* requires lanthanides for growth and in this case REE are required as a metal cofactor for methanol dehydrogenase (Poi et al. 2014).

REE activities associated with gram-positive bacteria are as follows. *Arthrobacter luteolus* takes up samarium and scandium into cell material and this uptake is probably driven by a ferric ion catecholate siderophore or low molecular weight organic acids (Emmanuel et al. 2012). Bacillus cereus adsorbs cerium and neodymium (Emmanuel et al. 2011). *Mycobacterium smegmatis* accumulates lanthanide, europium, and ytterbium of the lanthanide series as well as uranium and thorium of the actinide series (Andres et al. 1993).

Activities associated with gram-negative bacteria would include the following. Lanthanum, europium, and terbium are concentrated in the periplasm of growing cultures of Escherichia coli (Bayer and Bayer 1991). Lanthanum, cerium, and promethium are adsorbed by *Variovorax paradox* and *Comamonas acidovorans* (Emmanuel et al. 2011). Lanthanum is adsorbed onto cells of *Pseudomonas* sp. (Kazy and Susanta 2006) and *M. xanthus* (Merroun et al. 2003).

In the actinide series, thorium (Th^{4+}) is bound onto the cells of microorganisms and this research has focused on bioremediation of thorium. *Pseudomonas* sp., *M. smegmatis, Altermonas* sp., *Vibrio* sp., and *Flavobacterium* sp. adsorb thorium from experimental solutions and bind thorium onto cell walls (Hirose and Tanoue 2001; Sar and D'Souza 2002). The biosorption and desorption of thorium onto *Rhizopus arrhizus* and *A. niger* has been examined in simulated process streams (Gadd and White 1992).

10.7 Toxic Elements

10.7.1 Mercury

Mercury is frequently found in the environment and is toxic to biological systems at low concentrations. Mercury sulfide (red cinnabar and black metacinnabar) is found in waste piles of inoperative mines worldwide and is highly stable under aerobic and anaerobic conditions. However, bacteria in acid mine drainage from sites readily release mercury from cinnabar. Analysis of an acid mine drainage system in New Idria, California, USA, revealed that about 55% of the bacteria present in the top 1 cm of the sediment were sulfur-oxidizing bacteria of the species *Thiomonas*. Also prominent in this microbial community were iron-oxidizing bacteria that are members of the orders *Ferritrophicales* and *Gallionellas* (Jew et al. 2014).

Metacinnabar is an important source of Hg^{2+} in the aquatic environment. In water, the solubility of cinnabar and metacinnabar is enhanced in the presence of polysulfide (Jay et al. 2000). Polysulfide is produced when sulfide interacts with elemental sulfur and at high pH the dominant form of mercury is $Hg(S_x)_2^{2-}$. Sulfide for the formation of polysulfide could be produced by SRB. Using *T. thioparus* in pure culture, Hg^{2+} is released from metacinnabar at neutral pH and additions of thiosulfate enhance the rate of Hg^{2+} solubilized by formation of a mercury-thiosulfate complex with elevated release of volatile Hg^0 (Vázquez-Rodriguez et al. 2015).

10.7.2 Chromium

The primary source of chromium in nature is chromite [$FeCr_2O_4$] with Cr^{3+} found in ultramafic rocks and ophiolite complexes. As chromite is weathered, Cr^{3+} is released and bacterial resistance to Cr^{3+} is broadly distributed throughout many different physiological groups of bacteria. Chromate (Cr^{6+}) is found in certain groundwaters perhaps due the oxidation of chromite by birnessite, a manganese mineral containing both Mn^{3+} and Mn^{4+} (Oze et al. 2007).

Chromate has become deposited in the environment because it was used as a cleaning agent, as an inhibitor of biofilm formation in reactor cooling towers, and in numerous industrial processes. Many bacteria are resistant to chromate and reduce chromate to Cr^{3+}, as found in a study of chromium contaminated sludge (Francisco et al. 2002). After analysis of cellular proteins and fatty acid methyl esters, 10 different clusters of bacteria were obtained and these groups varied in chromate resistance and reduction (Francisco et al. 2002). Cationic chromium readily binds to cellular materials and it readily forms chromium hydroxide, which is insoluble and precipitates from water.

The reduction of chromate (Cr^{6+}) to Cr^{3+} presents an interesting problem for bacteria because reduction is not a multiple of two electrons. Usually, respiring cells process two electrons at a time and single electrons in aerobic conditions may produce free radicals. Glutathione reductase, lipoyl dehydrogenase, and ferredoxin-NADP oxidoreductase are flavoproteins that reduce Cr^{6+} to Cr^{5+} and produce cytotoxic reactive oxygen species (ROS) (Ackerley et al. 2004; Thatol et al. 2014). ChrR and YieF are enzymes with chromate reduction activity with minimal ROS production because Cr^{3+} is formed (Ackerley et al. 2004). The proteins with chromate reduction activity can be localized in the cytoplasm or bound to the cell membranes of aerobic or anaerobic bacteria.

10.7.3 Arsenic

Arsenopyrite (FeAsS), realgar (As_4S_4), orpiment (As_2S_3), enargite (Cu_3AsS_4), and tennantite [$(Cu,Fe)_{12}As_4S_{13}$] are the most common arsenic sulfide minerals. Release of arsenic from arsenic-bearing minerals may be attributed to microorganisms that can oxidize iron, sulfide, or selenide or sequester iron from the mineral using siderophores produced by indigenous bacteria (Lukasz et al. 2014). Also, depending on the chemistry of the suspending water, chemical oxidation of arsenide could be coupled to chemical (abiotic) activity. As a result, arsenate (As^{5+}), arsenite (As^{3+}), elemental As (As^0), and selenide (Se^{3-}) are found in the environment. At least 24 prokaryotes, Bacteria and Archaea, have the capability of dissimilatory arsenate reduction. The bacterium *Alcalilimnicola ehrlichii* has the capability of oxidizing arsenite with the reduction of nitrate and it appears that arsenite-oxidizing bacteria/NRB are distributed broadly in nature (Stolz et al. 2006).

In biological systems, arsenate interferes with phosphate metabolism and bacteria have developed a highly specific system to detoxify arsenate (Stolz et al. 2006). Bacteria resistant to arsenic reduce the toxic arsenate to arsenite by a cytoplasmic or membrane bound enzyme encoded on the *ars* operon. Although arsenite has reduced toxicity as compared to arsenate, arsenite is exported from the cell by a specific efflux pump.

10.7.4 Selenium

Selenium is broadly distributed and is present in copper deposits, antimony deposits, uranium-vanadium deposits, coal deposits, volcanic tuffs, and sedimentary iron ores. Naumannite (Ag_2Se), tiemannite (HgSe), eucairite (CuAgSe), bellidoite (Cu_2Se), and covellite (CuSe) are common minerals containing selenium. *Thiobacillus ferrooxidans* oxidizes selenide in covellite according to the following reaction:

$$CuSe + 2H^+ + 0.5O_2 \rightarrow Cu^{2+} + Se^0 + H_2O$$

Additionally, aerobic bacteria including *T. ferrooxidans* can produce selenate (SeO_4^{2-}) and selenite (SeO_3^{2-}) from elemental selenium under aerobic conditions.

Selenium is a chalcophile element with a high affinity for sulfur, which accounts for the association of selenium with sulfur deposits. Seleniferous soils in the Western United States support the growth of selenium-requiring plants including *Astragalus, Stanleya,* and *Xylorrhiza*. These selenium-accumulating plants and selenium-containing aquatic environments may be toxic to grazing animals and water fowl. In animals and bacteria, selenium (as selenate, SeO_4^{2-}, or selenite, SeO_3^{2-}) is required at trace levels but at elevated concentrations it is toxic.

Many bacteria can reduce selenate or selenite to elemental selenium and numerous groups of bacteria can oxidize elemental selenium (Barton et al. 2014). Some interest has been shown in elemental selenium in the environment. Se^0 microparticles (>20 µm long) have been reported in weathered sideritic mudstones in the Czech Republic and their origin remains unresolved but microbial origin is a possibility (Matýsek and Skupien 2015).

10.8 Metallic and Metalloid Nanoparticles of Microbial Origin

Bacteria have been the primary group of microorganisms examined for the production of metallic nanoparticles (Box 10.2). These studies have focused on the bacterial conversion of soluble elements to insoluble forms. Initially the formation of bionanoparticles was of academic interest with potential applications in the remediation of sites contaminated with toxic metal(loid)s. However, the recent thrust of this research has been for industrial applications because bacterial production of nanoparticles is inexpensive and does not affect the environment. Nanoparticles of bacterial origin are 5–60 nm in diameter and may aggregate into larger particles that reach 200 nm. As recently reviewed (Barton et al. 2014), the metallic nanoparticles produced by bacteria may be a single element, a metal oxide, or contain several elements (Table 10.3). There are multiple processes that may account for bacterial formation of nanoparticles of metal(loid) elements and this varies with the element as well as with the bacterial cell (Figure 10.6). In some instances, redox active metallic or metalloid ions are reduced by cellular processes to the elemental form by cytochromes or nonspecific enzymes on the surface or in the periplasm of the cell. In some bacteria, the metal or metalloid is transported into the periplasm or cytoplasm, reduced, and may or may not be exported from the cell. Nanoparticles are collected in the EPS matrix by various types of charge interactions. It appears that many of the elemental nanoparticles have a protein covering that may either influence the size of the developing nanoparticle or some may be exported from the cell. The mechanisms of formation for many of the nanoparticles and especially those containing mixed elements, remain to be established.

It was thought that formation of these metallic or metalloid nanoparticles was a defense mechanism of the bacteria enabling them to persist in toxic environments. In some instances, toxic metals induce a response in bacteria that activates genes to be expressed that are normally

Box 10.2 Nanoparticles

Discrete granules that are 1–100 nm in diameter are referred to as nanoparticles. In some instances, nanoparticles may aggregate to produce structures that are 200–300 nm and these aggregates are also referred to as nanoparticles. While metallic nanoparticles may be produced by chemical or biological processes and the term "bionanoparticle" designates nanoparticles that are produced by bacteria, fungi, or other biological systems. There is considerable interest in using bacteria and fungi for the production of nanoparticles because it is a green process that, unlike chemical processes, does not contribute to environmental pollution.

Table 10.3 Examples of metal(loid) nanoparticles produced by microorganisms.[a]

Single element	Metallic oxide	Metal(loid) sulfide	Metallic carbonate	Mixed elements
Au^0	CuO	AsS	$CdCO_3$	Au/Ag
Ag^0	Fe_3O_4	As_2S_3	$PbCO_3$	$BaTiO_3$
u^{0b}	MnO_2	As_4S_3	$SrCO_3$	Cd/Se
Fe^{0c}	Sb_2O_3	As_4S_4		$CoFe_2O_4{}^d$
Ni^0	TiO_2	CdS		Se/S
Pd^0	UO_2	CuS		Si/GeO
Pt^0	ZrO_2	FeS		Si/SiO_2
Se^0		ZnS		
Te^0				
Hg^0				

[a] References for nanoparticle formations are found in Barton et al. (2014) with the following additions.
[b] Hofacker et al. (2015).
[c] Rahman and Bastola (2014).
[d] Corker et al. (2009).

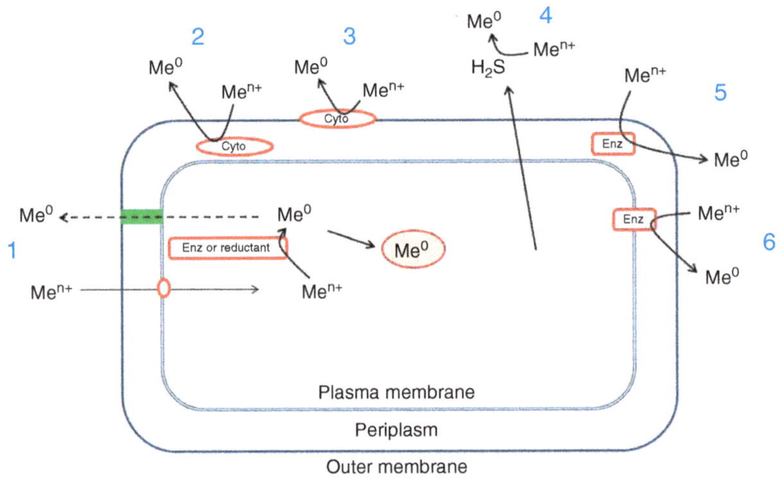

Figure 10.6 Model indicating regions of a gram-negative cell where nanoparticles may be produced. (1) Metal is internalized by an uptake transporter and reduction in the cytoplasm is dependent on an enzyme or reductant (reduced ferredoxin or reduced glutathione). The nanoparticle may remain inside the cytoplasm or is exported from the cell by a multi-facilitator system. (2) The metal is reduced by cytochrome in the periplasm. (3) The metal ion is reduced by a cytochrome in the outer membrane. (4) Hydrogen sulfide is a product of cellular metabolism and it freely crosses the cell membranes with reduction of metal ions outside of the cell. (5) An enzyme in the periplasm with exposed reduced metallocenters interacts with the metal ion and reduces it. (6) A metalloenzyme in the plasma membrane reduces the metal ions.

silent. Almost all the nanoparticle-producing bacteria are common inhabitants of marine or terrestrial environments and once the metallic nanoparticles are formed, they do not remain indefinitely in the environment. The charge on elemental nanoparticles is subject to a variety of chemical reactions in the environment as well as oxidative metabolic events of bacteria. The next phase of investigation concerning elemental nanoparticles is their activity in the environment. That is, to what extend do they participate in geochemical changes since it is evident that they are involved in mercury, iron, and selenium geochemical cycles (see Chapter 9). An interesting question is "What is the impact of metal and metalloid nanoparticles on plant-root activity and on human metabolism?" It appears that bacteria–metal(loid) interactions with living systems is introducing a new field of research.

10.9 Summary

MIC of ferrous and nonferrous metals is a process involving anaerobic bacteria that solubilize the solid metal by removal of electrons and transport these electrons to support bacterial growth. The EMIC and CMIC are two models to describe MIC. Electrons are moved from Fe^0 bacterial cells in EMIC whereas with CMIC, corrosion of iron is attributed to bacterial production of H_2S. Biocorrosion is enhanced by biofilms and extracellular polymeric materials collecting at the site of corrosion. Biocorrosion is controlled by using chromium-steel alloys, the addition of biocides, and the depletion of electron donor substances in the area by selective support of NRB. Microorganisms are often responsible for eroding or changing the surface of buildings and artistic materials. Algae, lichens, and sulfide-oxidizing, acid-producing bacteria erode the surfaces of concrete and stone. Desert varnish on rock surfaces is attributed to microbial production of manganese and other metal oxides. Historic paintings may be degraded by metal resistant bacteria and fungi growing on materials added to the initial paints. While some microorganisms harness electron flow to or from ionic iron and manganese to support growth, a few bacteria produce magnetic nanogranules and other bacteria form crystal and clay complexes containing iron. Soluble anions of carbonate, sulfate, phosphate, and sulfide are immobilized by the metabolic activity of several bacteria. Ionic forms of uranium and gold are precipitated by the nonspecific redox activities of several bacteria. Transition elements are solubilized from minerals in small concentrations to support microbial growth. REEs are not required for microbial growth but may be acquired by nonspecific uptake mechanisms. Microorganisms have developed detoxification mechanisms to enable them to grow in toxic environments containing soluble forms of the elements mercury, chromium, arsenic, and selenium. Occasionally, bacteria use energy from reduction processes to grow. Many microorganisms nonspecifically convert soluble toxic metals and metalloid ions to nanoparticles with reduced toxicity.

Discussion Questions

1. Explain how would you experimentally distinguish between chemical corrosion and microbially influenced corrosion.
2. As the biodegradation of paintings has been attributed to the growth and metabolism of bacteria, why is it reasonable to propose that bacteria can be used to restore artworks and historic stones?
3. Several examples of mineral transformation are given in this chapter. Is it possible for microorganism to mineralize all inorganic compounds found on Earth? If not, what would be some of the limiting factors preventing bacterial or archaeal metabolism.
4. Describe the mechanisms that enable certain bacteria and archaea to grow in the presence of toxic levels of arsenic, selenium, mercury, and chromium.

References

Ackerley, D.F., Gonzalez, C.F., Park, C.H. et al. (2004). Chromate-reducing properties of soluble flavoproteins from *Pseudomonas putida* and *Escherichia coli*. *Appl. Environ. Microbiol.* 70: 873–882.

Andres, Y., MacCordick, H.J., and Hubert, J.C. (1993). Adsorption of several actinide (Th,U) and lanthanide (La, Eu, Yb) ions by *Mycobcterium smegmatis*. *Appl. Microbiol. Biotechnol.* 39: 413–417.

Bailey, J.V., Corsetti, F.A., Greene, S.E. et al. (2013). Filamentous sulfur bacteria preserved in modern and ancient phosphatic dediments: implications for the role of oxygen and bacteria in phosphorogenesis. *Geobiology* 11: 397–405.

Balci, N.C. (2010). Effect of bacterial activity on trace metals release from oxidation of sphalerite at low pH (<3) and implications for AMD environment. *Environ. Earth Sci.* 60: 485–493.

Barbabietola N, Tasso F, Grimaldi M, et al. (2012). Microbe-based technology for a novel approach to conservation and restoration, *NEAE Energia Ambiente e Innovazione Speciale II-2012 Knowledge, Diagnostics and Preservation of Cultural Heritage.*

Barton, L.L., Tomei-Torres, F.A., Xu, H. et al. (2014). Nanoparticles formed by microbial metabolism of metals and minerals. In: *Nanomicrobiology* (ed. L.L. Barton, D.A. Bazylinski and H. Xu), 145–176. New York, NY: Springer.

Bastian, F., Alabouvette, C., Jurado, V. et al. (2009). Impact of biocide treatments on the bacterial communities of the Lascaux cave. *Naturwissenschaften* 96: 863–868.

Bayer, M.E. and Bayer, M.H. (1991). Lanthanide accumulation in the periplasmic space of *Escherichia coli* B. *J. Bacteriol.* 173: 141–149.

Beech, I.B. and Sunner, J.A. (2007). Sulphate-reducing bacteria and their role in corrosion of ferrous materials. In: *Sulphate-Reducing Bacteria – Environmental and Engineered Systems* (ed. L.L. Barton and W.A. Hamilton), 459–482. Cambridge, UK: Cambridge University Press.

Bosch-Roig, P. and Ranalli, G. (2014). The safety of biocleaning technologies for cultural heritage. *Front. Microbiol.* 5: 155. https://doi.org/10.3389/fmicb.2014.00155.

Ciferri, O. (1999). Microbial degradation of paintings. *Appl. Environ. Microbiol.* 65: 879–885.

Cappitelli, F., Toniolo, L., Sansonetti, A. et al. (2007). Advantages of using microbial technology over traditional chemical technology in removal of black crusts from stone surfaces of historical monuments. *Appl. Environ. Microbiol.* 73: 5671–5675.

Cockell, C.S., Osinski, G.R., Banerjee, N.R. et al. (2010). The microbe-mineral environment and gypisum neogenesis in a weathered polar evaporate. *Geobiology* 8: 293–308.

Cologgi, D.L., Lampa-Pastirk, S., Speers, A.M. et al. (2011). Extracellular reduction of uranium via *Geobacter* conductive pili as a protective cellular mechanism. *Proc. Natl. Acad. Sci. U.S.A.* 108: 15248–15252.

Corker, V.S., Telling, N.D., van der Laan, G. et al. (2009, 3). Harnessing the extracellular bacterial production of nanoscale cobalt ferrite with exploitable magnetic properties. *ASC NANO* (7): https://doi.org/10.1021/nm900293d.

Dakal, T.C. and Cameotra, S.S. (2012). Microbially induced deterioration of architectural heritages: routes and mechanisms involved. *Environ. Sci. Europe* 24: 36. https://doi.org/10.1186/2190-4715-24-36.

Deutzmann JS, Sahin M, Spormann AM (2015). Extracellular enzymes facilitate electron uptake in biocorrosiol and bioelectrosynthesis, *mBio* 6(2):e00496–15. https://doi.org/10.1128/mBio.00496-15.

Dhami, N.K., Reddy, M.S., and Mukherijee, A. (2014). Application of calicificying bacteria for remediation of stones and cultural heritages. *Front. Microbiol.* 5: 304. https://doi.org/10.3389/fmicb.2014.00304.

Dorn, R.I. and Oberlander, T.M. (1981). Microbial origin of desert varnish. *Science* 213: 1245–1247.

Dunn, K.A., McLean, R.J.C. McLean., Upchurch, Jr., G.R.. et al. (1997). Enhancement of leaf fossilization potential by bacterial biofilms. *Geology.* 25: 1119–1122.

Emmanuel, E.S.C., Vignesh, V., Anandkumar, B. et al. (2011). Bioaccumulation of cerium and neodymium by *Bacillus cereus* isolated from rare earth environments of Chavara and Manavalakurichi, India. *Indian J. Microbiol.* 51: 488, 1–495.

Emmanuel, E.S.C., Ananthi, T., Anandkumar, B. et al. (2012). Accumulation of rare earth elements by siderophore-forming *Arthrobacter luteolus* isolated from rare earth environment of Chavara, India. *J. Biosci.* 37: 25–31.

Enning, D. and Garrelfs, J. (2014). Corrosion of iron by sulfate-reducing bacteria: new views of an old problem. *Appl. Environ. Microbiol.* 80: 1226–1236.

Enning, D., Venzlaff, H., Garrelfs, J. et al. (2012). Marine sulfate-reducing bacteria cause serious corrosiol of iron under electroconductive biogenic mineral crust. *Environ. Microbiol.* 14: 1772–1787.

Fortin, D., Ferris, F.G., and Beveridge, T.J. (1997). Surface-mediated mineral development by bacteria. *Rev. Mineral. Geochem.* 35: 161–180.

Francisco, R., Alpoim, M.C., and Morais, P.V. (2002). Diversity of chromium-resistant and –reducing bacteria in a chromium-contaminated activated sludge. *J. Appl. Microbiol.* 92: 837–843.

Gadd, G.M. and White, C. (1992). Removal of thorium from simulated acid process streams by fungal biomass: potential for thorium desorption and reuse of biomass and desorbent. *J. Chem. Technol. Biotechnol.* 55: 39–44.

Hirose, K. and Tanoue, E. (2001). Strong ligands for thorium complexation in marine bacteria. *Mar. Environ. Res.* 51: 95–112.

Hofacker, A.E., Behrens, S., Voegelin, A. et al. (2015). *Clostridium* species as metallic copper-forming bacteria in soil under reducing conditions. *Geol. J.* 32: 130–139.

Jack, T.R. and Westlake, D.W.S. (1995). Control in industrial settings. In: *Sulfate-Reducing Bacteria* (ed. L.L. Barton), 265–292. New York, NY: Plenum Press.

Javaherdashti, R., Nikraz, H., Borowitzka, M. et al. (2009). On the impact of algae on accelerating the biodeterioration/biocorrosion of reinforced concrete: a mechanistic review. *Eur. J. Sci. Res.* 36: 394–406.

Jay, J.A., Morel, F.M.M., and Hemond, H.F. (2000). Mercury speciation in the presence of polysulfides. *Environ. Sci. Technol.* 34: 2196–2200.

Jayaraman, A., Hallock, P.J., Carson, R.M. et al. (1999). Inhibiting sulfate-reducing bacteria in biofilms on steel with antimicrobial peptides generated in situ. *Appl. Microbiol. Biotechnol.* 52: 267–275.

Jew, A.D., Behrens, S.F., Rytuba, J.J. et al. (2014). Microbially enhanced dissolution of HgS in an acid mine drainage system in the California coast range. *Geology* 12: 20–33.

Johnson, C.W., Wyatt, M.A., Li, X. et al. (2013). Gold mineralization by a metallophore from a gold-associated microbe. *Nat. Chem. Biol.* 99: 241–243.

Kato, S. (2016). Microbial extracellular electron transfer and its felevance to iron corrosion. *Microb. Biotechnol.* 9: 141–148.

Kato, S., Yumoto, I., and Kamagata, Y. (2015). Isolation of acetogenic bacteria that induce biocorrosion by utilizing metallic iron as the sole electron donor. *Appl. Environ. Microbiol.* 81: 67–73.

Kashefi, K., Shelobolina, E.S., Elliott, W.C. et al. (2007). Growth of thermophilic and hyperthermophilic Fe(III)-reducing microorganisms on a ferruginous smectite as the sole electron acceptor. *Appl. Environ. Microbiol.* 74: 251–258.

Kazy, K. and Susanta, K. (2006). Lanthanum biosorption by a *Pseudomonas* sp. equilibrium studies and chemical characterization. *J. Ind. Microbiol. Biotechnol.* 33: 773–783.

Khijniak, T.V., Slobodkin, A.I., Coker, V. et al. (2005). Reduction of uranium (VI) phosphate during growth of the thermophilic bacterium *Thermoterrabacteriu ferrireducens*. *Appl. Environ. Microbiol.* 71: 6423–6426.

Konhauser, K. (2007). *Introduction to Geomicrobiology*. Malden, MA: Blackwell Publishing.

Kostka, J.E., Wu, J., Nealson, K.H. et al. (1999). The impact of structural Fe(III) reduction by bacteria on the surface chemistry of smecitite clay minerals, *Geochim. Cosmochim. Acta* 63: 3705–3713.

Kostka, J.E., Dalton, D.D., Skelton, H. et al. (2002). Growth of iron (III)-reducing bacteria on clay minerals as the sole electron acceptor and comparison of growth yields on a variety of oxidized iron forms. *Appl. Environ. Microbiol.* 68: 6256–6262.

Lamenti, G., Tiano, P., and Tomaselli, L. (2000). Biodeterioration of ornamental marble statues in the Boboli gardens (Fluorence, Italy). *J. Appl. Phycol.* 12: 427–433.

Lefèvre, C.T. and Bazylinski, D.A. (2013a). Ecology, diversity and evolution of magnetotactic bacteria. *Microbiol. Mol. Biol. Rev.* 77: 497–526.

Lefèvre, C.T. and Bazylinski, D.A. (2013b). The bacterial magnetosome: a unique prokaryotic organelle. *J. Mol. Microbiol. Biotechnol.* 23: 63–80.

Levchenko, L.A., Sadkov, A.P., Lariontseva, N.V. et al. (2001). Methane oxidation catalyzed by the au-protein from *Micrococcus luteus*. *Dokl. Biochem. Biophys.* 377: 123–124.

Li, X., Kappler, U., Jiang, G. et al. (2017). The ecology of acidophilic microorganisms in the corroding concrete sewer environment. *Front. Microbiol.* 8: 683. https://doi.org/10.3389/fmcb.2017.00683.

Li, Y.-L., Vall, H., Sears, S.K. et al. (2004). Iron reduction and alteration of nontronite NAu-2 by a sulfate-reducing bacterium, *Geochim. Cosmochim. Acta* 68: 3251–3260.

Lindström, E.B., Gunneriusson, E., and Tuovinen, O.H. (1992). Bacterial oxidation of refractory sulfide ores for gold recovery. *Crit. Rev. Biotechnol.* 12: 133–155.

Little, B.J. and Lee, J.S. (2007). *Microbiologically Influenced Corrosion*. New York, NY: Wiley.

Llorens, I., Untereiner, G., Jaillard, D. et al. (2012). Uranium interaction with two multi-resistant environmental bacteria: *Cupriavidus metallidurans* CH34 and *Rhodopseudomonas palustris*. *PLoS One* 7 (12): e51783. https://doi.org/10.1371/journal.pone.00511783.

Lukasz, D., Liwia, R., Aleksandra, M. et al. (2014). Dissolution of arsenic minerals mediated by dissimilatory arsenate reducing bacteria: estimation of the physiological potential for arsenic mobilization. *Biomed. Res. Int.* 2014 (2014): 841892. https://doi.org/10.1155/2014/841892.

Lyalikova, N.N. and Lebedeva, E.V. (1984). Bacterial oxidation of molybdenum in ore deposits. *Geomicrobiol. J.* 3: 307–318.

Ma Y, Zhang H, Du Y, et al. (2015). The community distribution of bacteria and fungi on ancient wall paintings of the Mogao grottoes. *Sci. Rep.* 7752, doi:https://doi.org/10.1038/srep07752

Macaskie, L.E., Bonthrone, K.M., Yong, P. et al. (2000). Enzymatically mediated bioprecipitation of uranium by a *Citrobacter* sp.: a concerted role for extracellular lipopolysaccharide and assorted phosphatase in biomineral formation. *Microbiology* 146: 1855–1867.

Marshall, M.J., Beliaev, A.S., Dohnalkova, A.C. et al. (2006). *c*-type cytochrome-dependent formation of U(IV) nanoparticles by *Shewanella oneidensis*. *PLoS Biol.* 4: 1324–1333.

Martin-Sanchez, P.M., Jurado, V., Porca, E. et al. (2014). Airborne microorganisms in Lascaux Cave (France). *Int. J. Speleol.* 43: 295–303.

Matỳsek, D. and Skupien, P. (2015). The origin of native selenium microparticles during the oxidation of sideritic mudstones in the Veřovice formation (outer western Carpathians). *Geol. Carpath.* 66: 303–310.

Merroun, M.L., Ben Chekroun, K., Arias, J.M. et al. (2003). Lanthanum fixation by *Myxococcus xanthus*: cellular location and extra cellular polysaccharide observation. *Chemosphere* 52: 113–120.

Milanesi, C., Baldi, F., Sara, B. et al. (2006). Biodeterioration of a fresco by biofilm forming bacteriaconca. *Int. Biodeter. Biodegrad.* 57: 168–173.

Milanesi, C., Baldi, F., Borin, S. et al. (2009). Deterioration of medieval painting in the chapel of the holy nail, Siena (Italy) partially treated with Paraloid B72. *Int. Biodeter. Biodegrad.* 63: 844–850.

Nathan, Y., Bremner, J.M., Loewenthal, R.E. et al. (1993). Role of bacteria in phosphorite genesis. *Geomicrobiol. J.* 11: 69–76.

Okamoto, A., Nakamura, R., and Hashimoto, K. (2011). In-vivo identification of direct electron transfer from *Shewanella oneidensis* MR-1 to electrodes via outer-membrane OmcA-mtrCAB protein complexes. *Electrochim. Acta* 56: 5526–5531.

O'Reilly, S.E., Watkins, J., and Furukawa, Y. (2005). Secondary mineral formation associated with respiration of nontronite, NAu-1 by iton-reducing bacteria. *Geochem. Trans.* 6 (67): https://doi.org/10.1186/1467-4866-6-67.

Oze, C., Bird, D.K., and Fendorf, S. (2007). Genesis of hexavalent chromium from natural sources in soil and groundwater. *Proc. Natl. Acad. Sci. U.S.A.* 104: 6544–6549.

Pangallo, D., Bučková, M., Kraková, L. et al. (2015). Biodeterioration of epoxyresin: a microbial survey through culture-independent and cultural-dependent approaches. *Environ. Microbiol.* 17: 462–479.

Pirbadian, S., Barchinger, S.E., Leung, K.M. et al. (2014). *Shewanella oneidensis* MR-1 nanowires are outer membrane and periplasmic extensions of the extracellular electron transport complex components. *Proc. Natl. Acad. Sci. U.S.A.* 111: 12883–12888.

Poi, A., Barends, T.R.M., Dietl, A. et al. (2014). Rare earth metals are essential for methanotrophic live in volcanic mudpots. *Appl. Microbiol.* 16: 255–264.

Rahman, P.K.S.M. and Bastola, S. (2014). Biological reduction of iron to the elemental state from ochre deposits of Skelton Beck in Northeast England. *Front. Environ. Sci.* 2014: https://doi.org/10.3389/fenvs.2014.00022.

Ranalli, G., Zanardini, E., and Sorlini, S. (2009). Biodeterioration – including cultural heritage. In: *Encyclopedia of Microbiology*, 3e (ed. M. Schaechter), 191–205. Oxford, UK: Elsevier Inc.

Ranalli, G., Alfano, G., Belli, C. et al. (2005). Biotechnology applied to cultural heritage: biorestoration of frescoes using viable bacterial cells and enzymes. *J. Appl. Microbiol.* 98: 73–83.

Razzell, W.E. and Trussell, P.C. (1963). Microbial leaching of metallic sulfides. *Appl. Microbiol.* 11: 105–110.

Reith, F., Brugger, J., Zammit, C.M. et al. (2013). Geomicrobiological cycling of gold: from fundamental processes understanding to exploration solutions. *Minerals* 3: 367–394.

Reith, F., Lengke, M.F., Falconer, D. et al. (2007). The microbiology of gold. *ISME J.* 1: 567–584.

Rivadeneyra, M.A., Delgado, R., Párraga, J. et al. (2006). Precipitation of minerals by 22 species of moderately halophilic bacteria in artificial marine salt media: influence of salt concentration. *Folia Microbiol.* 51: 445–453.

Rivadeneyra, A., Gonzalez-Martinez, A., Gonzalez-Lopez, J. et al. (2014). Precipitation of phosphate minerals by microorganisms isolated from a fixed-biofilm reactor used for the treatment of domestic wastewater. *Int. J. Environ. Res. Public Health* 11: 3689–3704.

Saiz-Jimenez, C. (1997). Biodeterioration vs biodegradation: the role of microorganisms in the removal of pollutants deposited on historic buildings. *Int. Bioteter. Biodegrad.* 40: 225–232.

Sakaguchi, T., Arakaki, A., and Matsunaga, T. (2002). *Desulfovibrio magneticus* sp. nov., a novel sulfate-reducing bacterium that produces intracellular single-domain-sized magnetite particles. *Int. J. Syst. Evol. Microbiol.* 52: 215–221.

Sánchez-Román, M., Fernández-Remolar, D., Amils, R. et al. (2014). Microbial mediated formation of Fe-carbonate minerals under extreme acidic conditions. *Sci. Rep.* 4: 4767. https://doi.org/10.1038/srep04767.

Sar, P. and D'Souza, S.F. (2002). Biosorption of thorium (IV) by a *Pseudomonas* biomass. *Biotechnol. Lett.* 24: 239–243.

Schippers, A., Hedrich, S., Vasters, J. et al. (2014). Biomining: metal recovery from ores with microorganisms. *Adv. Biochem. Eng. Biotechnol.* 141: 1–47.

Schultze-Lam, S. and Beveridge, T.J. (1994). Nucleation of celestite and strontianite on a cyanobacterial S-layer. *Appl. Environ. Microbiol.* 60: 447–453.

Shelobolina, E.S., Coppi, M.V., Korenevsky, A.A. et al. (2007). Importance of *c*-type cytochromes for U(VI) reduction by *Geobacter sulfurreducens*. *MBC Microbiol.* 2007 (7): 16. https://doi.org/10.1186/1471-2180-7-16.

Stolz, J.F., Basu, P., Santini, J.M. et al. (2006). Arsenic and selenium in microbial metabolism. *Annu. Rev. Microbiol.* 60: 107–130.

Tebo, B.M., Geszvain, K., and Lee, S.-W. (2010). The molecular geomicrobiology of bacterial manganese (II) oxidation. In: *Geomicrobiology: Molecular and Environmental Perspective* (ed. L.L. Barton, M. Mandl and A. Loy), 285–308. Dordrecht: Springer.

Takahashi, Y., Châtellier, X., Hattori, K.H. et al. (2005). Adsorption of rare earth elements onto bacterial cell walls and its implication for REE sorption onto natural microbial mats. *Chem. Geol.* 219: 53–67.

Thatol, H., Das, S., Mishra, J. et al. (2014). Bacterial chromate reductase, a potential enzyme for bioremediation of hexavalent chromium: a review. *J. Environ. Manage.* 146: 383–399.

Van Lith, Y., Warthmann, R., Vasconcelos, C., and McKenzie, J.A. (2003). Sulphate-reducing bacteria induce low-temperature Ca-dolomite and high mg-calcite formation. *Geology* 1: 71–79.

Vali, H., Weiss, B., Li, Y.-L. et al. (2004). Formation of tabular single domain mangnetite induced by *Geobacter metallireducens* GS-15. *Proc. Natl. Acad. Sci. U.S.A.* 101: 16121–16126.

Vigneron, A., Alsop, E.B., Chambers, C. et al. (2016). Complementary microorganisms in highly corrosive biofilms from an offshore oil production facility. *Appl. Environ. Microbiol.* 82: 2545–2554.

Vázquez-Rodriguez, A., Hansel, C.M., Zhang, T. et al. (2015). Microbial- and thiosulfate-mediated dissolution of mercury sulfide minerals and transformation to gaseous mercury. *Front. Microbiol.* 2015. https://doi.org/10.3389/fmicb.2016.00596.

Wei, S., Jiang, Z., Liu, H. et al. (2013). Microbiologically induced deterioration of concrete – a review. *Braz. J. Microbiol.* 44: 1001–1007.

Zhang, G., Dong, H., Kim, J. et al. (2007). Microbial reduction of structural Fe^{3+} in nontronite by a thermophilic bacterium and its role in promoting the smectite to illite ratio. *Am. Minerol.* 92: 1411–1419.

Further Reading

Ehrlich, H.L. and Newman, D.K. (2009). *Geomicrobiology*, 5e. Boca Raton, FL: CRC Press.

Gadd, G.M. (2010). Metals, minerals and microbes: geomicrobiology and bioremediation. *Microbiology* 156: 609–643.

Gutarowska B, Celikkol-Aydin S, Bonifay V, et al. (2015). Metabolomic and high-throughput sequencing analysis—modern approach for the assessment of biodeterioration of materials from historic buildings, *Front. Microbiol.*; 6: 979. doi:https://doi.org/10.3389/fmicb.2015.00979

11

Microbial Communities and Metabolic Networks

11.1 Overview and Focus

Microbial communities are highly complex consisting of an enormous array of species and even though the genome of these species has similar housekeeping genes, each species has unique genetic components. It is considered that there are two distinct ecological strategies for microbial success (Freilich et al. 2009). In one situation the niche is unique, the growth rate is relatively slow, and the environment is restrictive for most other microorganisms – these are specialist microorganisms. Specialist microorganisms are often environmentally dependent and are found in thermal pools, brine ponds, environments containing chemical toxins, or areas where chemolithotrophy is required for bacterial growth. Another strategy, which is characteristic of populations of considerable species diversity, is the metabolic adaptability with considerable nutrient exchange that enables bacteria to grow rapidly with a minimal amount of competition – these could be considered generalists. Many generalists produce a broad array of secondary metabolites. In both strategies, the operative word is growth. Because over 99% of most microbial populations cannot be cultivated, analysis of metabolic networks is carried out using proteomics, metagenomics, the identification of genes expressed by evaluation of mRNA (transcriptomics), and imaging mass spectrometry (Ponomarova and Paitil 2015). In microbial ecology with complex populations, there is a move toward systems microbiology, where the individuals within a population are considered as a single unit and the subunits are integrated (Röling et al. 2010).

11.2 Examples of Succession of Populations

In open environments, microbial populations have become established over the years and the abundance of various species and the total population size is relatively stable. When nutrients or chemical inhibitors are added into the environment, the populations rapidly adjust, with the most capable microorganisms becoming dominant. In mixed cultures, the bacteria that respond the fastest will become dominant; however, dominant populations often decline to become replaced with a marked increase in the cell number of another group of bacteria. This decline and replacement by a new population continues for several cycles and gradually there is the return to a "steady state," where the population size and diversity matches the environment before the additions were made. The decline in a population may be due to the release of acidic end products, production of toxic peptides, and antibiotics while in some instances it is the consumption of an electron donor or an electron acceptor. Recent interest has been in

Environmental Microbiology and Microbial Ecology, First Edition. Larry L. Barton and Robert J.C. McLean.

bacterial viruses that lyse the dominant bacterial populations. This destruction of dominant cells in a population reflects an approach reminiscent of "killing the winner"; however, bacteria have numerous strategies for gene induction, which enable them to adapt to stress environments but lack flexible systems that would enable them to remain dominant.

11.2.1 Development of Coral Black Band Disease

Black band disease (BBD) is extremely important for maintaining coral reefs around the world and it may involve over 500 different bacterial species. Photosynthetic filamentous cyanobacteria, sulfur-oxidizing bacteria and sulfate-reducing bacteria are the principal physiological types of bacteria in BBD (Bourne et al. 2011). Several species of cyanobacteria have been isolated from the microbial mat growing on the coral and the dark red color in the mat is attributed to phycoerythrin, an ancillary photopigment in cyanobacteria. In the microbial biomat, *Beggiatoa* spp. (sulfur-oxidizing bacteria) and *Desulfovibrio* spp. (sulfate-reducing bacteria) are commonly present. Hydrogen sulfide produced by the sulfate reducers is toxic to the living tissue of the coral.

Studies of the Great Barrier Reef near Pelorus Island reveal that a succession of bacterial populations on *Montipora hispida,* the reef coral, is characteristic of BBD (Sato et al. 2010). Initially there are patches of cyanobacteria consisting of *Blennothrix* sp. and this is followed by thick growth of *Oscillatoria* spp. Numerous species of *Alphaproteobacteria* are present in these cyanobacterial patches. With the onset of BBD lesions, members of the *Gammaproteobacteria* become dominant along with *Beggiatoa* spp. and *Desulfovibrio* spp. As the sulfur-cycling bacteria colonize on the coral at the site of the cyanobacterial mat, the pathogenicity of the BBD reaches its greatest level.

11.2.2 Population Succession in Production of Dairy Products

In a healthy cow, milk is free of bacteria but becomes contaminated as the milk is exposed to the environment. Sources for bacterial contamination include the skin and surface of the animal as well as airborne organisms. Milk contains lactose, which is readily metabolized by many microorganisms. One of the first bacteria to dominate in the spoilage of raw milk is *Streptococcus lactus,* which uses homofermentation to produce lactic acid. As the lactic acid accumulates, the pH drops from an initial level of 6.5 to about 5.5, killing the streptococci. Lactobacilli, also homofermenters, are the second bacterial population to dominate and these bacteria continue producing lactic acid until the milk has a pH of about 4. This high acidity kills the lactobacilli and coagulates the milk protein, producing a curdling effect. Yeast and fungi tolerate the acidic environment and rapidly use the lactic acid, with the pH being slowly shifted to about 6. Proteolytic bacteria such as the spore-forming *Bacillus* grow and represent the final population in the spoilage of milk. This entire process can occur in only a few days.

The production of cheese from milk uses the succession of microbial populations in cheese ripening to produce the desired product. While the process of making cheese has been handed down from generation to generation, the product is highly dependent on the bacteria present and cheese making has been considered a family secret. With the development of molecular biology techniques, changes in the cheese microbiome can be evaluated and a process can be selected that will impact the product quality. To understand the role of temperature in the ripening phase of cheese production, consult the article by De Filippis et al. (2016) concerning the regulation of bacterial gene expression in proteolysis, lipolysis, and amino acid metabolism during cheese maturation. To illustrate the population dynamics of bacteria in cheese production, the following paragraphs discuss two examples of cheese production in Italy.

In a study of cheese production in the Italian Dolomites, raw milk was collected from two different areas (Poznanski et al. 2004). Milk from the Passo Rolle area contained many mesophilic bacterial strains including a few enterococci and the thermophilic *S. thermophiles* whereas milk from the Vanoi area contained numerous strains of mesophilic bacteria including *Lactobacillus lactus, Lactobacillus cremoris,* and *Lactobacillus paracasei.* A natural starter culture was used that contained thermophilic cocci and a few enterococci. *Enterococcus faecium* and *Streptococcus macedonicus* were dominant in the Vanor cheese during the first two days and at the 15th day, *L. paracasei* was most abundant. Thermophilic *S. macedonicus* and *Streptococcus thermophilus* were initially dominant in the Rolle cheese and mesophilic cocci were in secondary abundance. By the 15th day, the mesophilic cocci and lactobacilli were dominant. Using 16S rRNA sequence analysis, raw milk at both areas contained many different bacterial species but by the end of the ripening process, only strains of *S. macedonicus, Lactobacillus plantarum, Lactobacillus fermentum, L. paracasei,* and *E. faecium* were found in both the Vanor and Rolle cheeses.

A change in bacterial populations is also seen in the ripening of pasta filata Caciocavallo Pugliese cheese (De Pasquale et al. 2014). Pyrosequencing analysis of bacterial DNA isolated from raw cow's milk indicated the presence of strains of bacteria associated with *Firmicutes* (53%), *Proteobacteria* (39%), *Bacteroides* (7.8%), *Actinobacteria* (0.06%), and *Fusobacteria* (0.03%), with the starter culture predominately *S. thermophiles*. By the end of the first month, the abundance of *S. thermophilus* decreased while *Lactobacillus casei* and *L. paracasei* become dominant. Even though the numbers of *S. thermophiles* were decreasing in the middle-ripening stage, the presence of free amino acids and volatile compounds (alcohols, esters, ketones, and furans) were correlated with the presence of *S. thermophiles.* There is a quick turnover in this cheese production with this pasta filata usually marketed after two weeks of ripening or after four months if it is to be grated.

11.2.3 Population Dynamics in Fermentation of Non-dairy Foods

11.2.3.1 Kimchi

Kimchi is a traditional Korean fermented food prepared from Chinese cabbage or white radish, with additives such as red peppers, garlic, and fermented sea food. The fermentation does not use a starter culture but is dependent on indigenous bacteria on the vegetables. The addition of salt (table or marine) to the initial fermentation inhibits many of the food spoilage bacteria but salt is tolerated by the important fermenting bacteria. In an extensive metagenomic analysis, it was revealed that the most prominent species of bacteria in this 30-day fermentation process were *Leuconostoc, Lactobacillus,* and *Weissella* (Jung et al. 2011). The initial dominant population was *Leuconostoc* with populations of *Lactobacillus* and *Weissella* abundant throughout the fermentation but the *Leuconostoc* population was dominant at the 30th day. Metagenomic sequences revealed that kimchi fermentation was accomplished by many different heterofermentative lactic acid bacteria and displayed high metabolic diversity. Metabolism of mono-, di- and oligosaccharides produced a variety of end products including lactic acid, acetic acid, ethanol, and mannitol. Numerous phage DNA sequences were obtained in the molecular analysis of the kimchi fermentation and while the role of these viruses in the fermentation were not discussed, it is probable that bacterial viruses participate in control of susceptible bacterial populations.

11.2.3.2 Coffee

Coffee plants are colonized by fungi, yeasts and bacteria that occur as endophytes or are attached to the plant surfaces and account for the coffee microbiome (Vaughan et al. 2015). Over 215 species of fungi have been isolated from coffee fruit and seed tissues, with the most common isolates belonging to the genera of *Aspergillus* and *Penicillum*. The presence of *Aspergillus ochraceus* and *Aspergillus carbonarius* are important because they can produce ochratoxin A, a carcinogen for animals and humans. Yeast genera commonly present include *Saccharomyces, Pichia,* and *Candida*. Over one hundred bacterial species belonging to *Proteobacteria, Firmicutes, Actinobacteria,* and *Bacteroides* phyla have been isolated from coffee beans. Coffee beans are coated with a gelatinous matrix, which is the substrate for bacteria in the initial stages of the fermentation. It is commonly found that 80% of the bacteria are gram-positive and most are *Bacillus* sp. (Silva et al. 2008). *Serratia, Enterobacter,* and *Acinetobacter* are examples of gram-negative bacteria present in the initial phases of the fermentation. Since only 6% of the coffee fruit mass is marketed as coffee grounds, there is a great quantity of coffee pulp byproduct. Microbial processes have been proposed to remove caffeine from the coffee byproduct so that it can be used for animal feed or to produce α-amylase, xylanase, β-fructofuranosidase, or other useful enzymes.

11.2.3.3 Cocoa

Chocolate is produced from the microbial fermentation of cocoa beans, which are cocoa seeds. There are two steps in cocoa bean fermentation: (i) the digestion of the pulp containing the beans, and (ii) the hydrolytic reactions occurring inside the bean (da Veiga Moreira et al. 2013). Yeast are the initial colonizers of the cocoa pulp with strains of *Saccharomyces, Pichia,* and *Hanseniaspora* present is significant numbers. After a few hours, the dominant yeast are replaced by bacteria that produce lactic acid. Initially, *L. plantarum* is dominant but in the later stages of fermentation, *L. fermentum* becomes predominant. After 48 hours of fermentation by lactic acid bacteria, the final phase is initiated by obligate aerobic acetic acid bacteria that oxidize the alcohol produced by the yeast to acetic acid. Acetic acid bacteria isolated from the final stage of fermentation often include the following: *Acetobacter aceli, Acetobacter pasteurianus, Acetobacter ghanensis, Acetobacter senegalensis, Acetobacter lovaniensis, Acetobacter fabarum, Gluconobacter oxydans,* and *Gluconobacter xylinus* (da Veiga Moreira et al. 2013). The second part of cocoa fermentation is biochemical. Ethanol and lactic acid produced by fermentation diffuse into the cocoa beans, killing the seed embryo and activating hydrolyases that degrade the internal components of the cocoa bean, releasing flavor and color precursors.

11.2.3.4 Chinese Soy Sauce

The traditional production of Chinese soy sauce does not use starter cultures but relies on environmental bacteria to conduct the fermentation. The soy sauce microbiome has been examined using metagenomic DNA analysis and the metabolic pathways used by the fermenting bacteria have been elucidated (Sulaiman et al. 2014). In soy fermentation, a brine solution is used that inhibits the growth of some environmental bacteria and yeast. *Salmonella, Listeria* and other pathogenic bacteria that may be accidentally introduced into the fermentation process are inhibited by antimicrobial peptides produced by lactic acid bacteria. The first phase of soy sauce fermentation is dominated by lactic acid bacteria, which decrease the pH level as a consequence of lactic acid production. One group of lactic-acid producing bacteria includes the homofermenters (*Lactococcus, Enterococcus,* and *Pediococcus*), which use the Embden–Meyerhof–Parnas pathway to produce lactic acid. Another group of lactic acid producers are heterofermenters that use the pentose or the hexose monophosphate pathway and these include *Leuconostoc, Weissella,* and some lactobacilli. The heterofermenters are the dominant bacteria in Chinese soy sauce fermentation and end products of these bacteria include carbon dioxide, ethanol, and acetic acid in addition to lactic acid. Hydrolytic enzymes produced by bacteria degrade polymeric plant materials with the release of reducing sugars. At about the fourth month, the yeast population becomes abundant, with most of the yeast belonging to the genera of *Candida, Starmerella,* and *Wickerhamiella*. Unlike the Japanese Koikuchi shoyu, alcohol producing yeast (i.e. *Zygosaccharomyces* and *Saccharomyces*) are absent in Chinese soy sauce fermentation. Toward the end of fermentation, gene analysis of the fermentation mash indicated the production of branched chain amino acids (isoleucine, leucine, and valine). Branched-chain amino acids are important for the production of hydrolases by lactic acid bacteria and for the production of flavor compounds in the soy sauce by *Candida*. In the final stage of fermentation, acid tolerant bacteria from the *Propionibacteriaceae* and *Acidobacteriaceae* families appear. While *Propionibacterium* does not contribute to the initial stages of fermentation, it is important in contributing aromatic compounds to Chinese soy sauce (Sulaiman et al. 2014).

11.2.4 Composting Plant Material

Composting of plant materials is perhaps one of the most obvious examples of the succession of microbial populations and it involves specific microbial populations at the three stages or phases of decomposition. The initial phase involves mesophilic bacteria present on the plant material and reflects the bacterial population in top soil. As the bacteria rapidly consume sugars, fats, starch, and proteins, the compost temperature increases due to the rapid metabolic activity of the microorganisms and spore-forming *Bacillus* become dominant. The temperature increases to about 55 °C and this thermophilic phase lasts for several weeks. During this time, thermophilic *Bacillus* and other bacteria maintain an elevated temperature due to the high metabolism, which destroys mesophilic bacteria, including animal pathogens. While fly larvae and weed seeds are destroyed at these elevated temperatures, pathogenic protists may survive and is why animal wastes is not added to compost for garden use. The compost must be appropriately aerated to maintain aerobic activity. The bacteria present produce high levels of extracellular enzymes that degrade cellulose and other complex carbohydrates. If the compost becomes anaerobic, heterotrophic anaerobic bacteria will grow, with the production of volatile end products that contribute to the odor of mixed acid bacterial fermentation. As the high energy substrates for the bacteria are used up, the metabolic rate diminishes, and the compost gradually cools. Important microorganisms in this third phase of composting include actinomycetes and mesophilic fungi, which continue to degrade the cellulosic matter, bark, and woody stems.

11.3 Impact of Climate Change on Microorganisms

From 1906 to 2005, the global temperature increased 0.74 ± 0.18 °C (IPCC, 2007), with the increase of atmospheric CO_2, a principle greenhouse gas, becoming a considerable concern. Since a change in global temperature is tightly associated with increased CO_2 production, there is concern that an increase of temperature may initiate dramatic changes in microbial populations. A discussion of carbon dioxide capture and storage is given in Chapter 3, Section 3.5.3. CO_2 and other drivers of climate change could have an impact on host–parasite relationships, symbiotic interactions, and cell–cell communication of microorganisms in marine and in terrestrial environments. Climate change produces slow ecological changes and many of these changes can be examined with the appropriate use of long-term models.

11.3.1 Marine Environment

Ocean acidification (OA) resulting from increased atmospheric CO_2 concentrations will have a pronounced effect on the pH of marine water as well as change marine carbonate chemistry (O'Brien et al. 2016). Coral reefs account for about one quarter of marine biodiversity and, with elevated CO_2 concentrations in the oceans, appropriate carbonate levels are not available to enable corals and other organisms with calcium carbonate skeletons to grow optimally. Sponges, molluscs, echinoderms, and cnidarians depend on the coral reefs to provide a complex array of microorganisms to grow. The holobiont, an association between scleractinian corals and associated microorganisms, has been the focus of numerous studies (Rosenberg et al. 2007). In conjunction with this coral are cyanobacteria, bacteria, and archaea associated with nutrient cycles, fungi, and viruses with *Symbiodinium*, an endosymbiotic photosynthetic dinoflagellate, serving as the major provider of the energy source. Coral-associated bacteria enhance the nitrogen level by fixation of dinitrogen and dimethylsulfoniopropionate – produced by *Symbiodinium* and the coral host – is metabolized by bacteria to produce antioxidants that protect the coral against free radicals. Sponges harbor a rich microbial population and it is assumed that the sponge population will greatly increase because of their resistance to the effects of climate change and OA. Acidophilic archaea with chemosynthetic capabilities may be important in associating with bivalves with CO_2 conversion to CH_4. With an increased activity in the oceans, bacteria may produce fewer antibiotics that protect corals and some bacteria, such as *Vibrio tubiashii*, may become more pathogenic as shown by the increased pathogenicity of *V. tubiashii* against blue mussels under slightly acidic conditions. Members of the phylum *Bacteroidetes* and certain fungi displayed an increase in virulence genes when marine water was acidified to 1.4 pH units. It is difficult to generalize about the effect of OA on bacteria because pH changes commonly occur in surface waters, most of the cultures involved are currently uncultivable, bacteria are highly adaptable, and they have persisted throughout time. It is reasonable to expect that the composition of populations will change with OA and that a new dynamic state will be produced in the marine ecosystem. A scenario for climate change is given in Figure 11.1.

11.3.2 Soil Environment

A change in global temperature may initiate dramatic changes in microbial populations. Changes may be expected in host–parasite relationships, symbiotic interactions, and cell–cell communication of microorganisms in marine and terrestrial environments. To illustrate the types of direct and indirect changes that are predicted with global warming, models have been developed for the impact of climate change on soil microbes and root–microbe interactions. It is challenging to predict the impact of climate change on microorganisms due to the adaptability of bacteria and the complexity of interactions within the terrestrial food web.

With about 120 Gt of carbon consumed and released annually by terrestrial ecosystems, the forest ecosystem is important because it contains about 40% of the global carbon pool. The release of CO_2 by soil microorganisms from the decomposition of organic matter accounts for about nine times the amount of CO_2 released from burning fossil fuels. Global warming may increase the level of CO_2 in the atmosphere due to microbial activity in the soil, with this rate of CO_2 release influenced by top down or bottom up feedback regulatory systems. Predation and grazing are important in the top down system, whereas resources such as nutrients limit the bottom up feedback system. As detailed in the review by Classen et al. (2015), an increase of 5 °C in a temperate forest enhances bacterial growth and increases the bacteria : fungi ratio in the forest soil. Microbial respiration increased with an elevated CO_2 level and there was increased enzyme activity associated with mineralization of plant material. Significant changes occur with warming that lead to a lengthened growing season, an expanded plant range, and increased plant productivity. However, with an increase in temperature there is frequently a decrease in soil moisture, which increases the relative abundance of *Acidobacteria* but

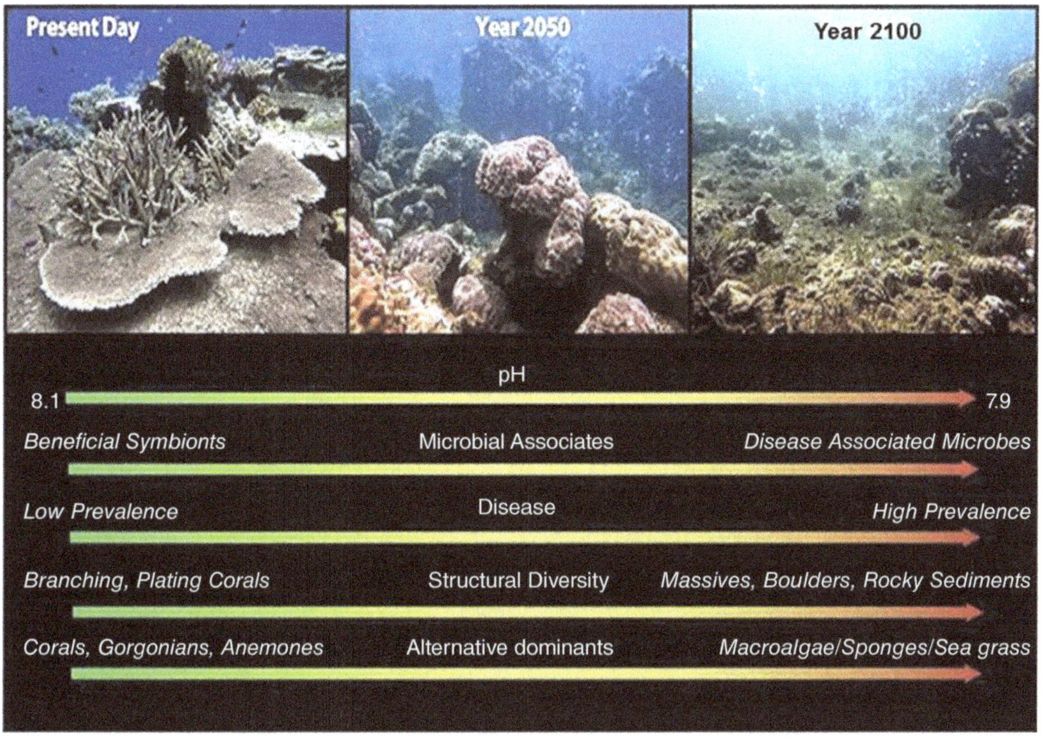

Figure 11.1 The impact of climate changes on the environment. Photos, taken along a *p*CO2/pH gradient in Papua New Guinea (year 2014), and text depict three potential scenarios for coral reefs in present and in future (years 2050 and 2100) ocean acidification conditions if continue the current predicted CO_2 emissions trajectories continue. Present Day illustrates a healthy coral reef with high structural complexity and diversity that is likely colonized by beneficial microbial associates and experiences a low incidence of disease. As pCO_2 increases, scenarios for years 2050 and 2100 illustrate the successional degradation of coral reef heterogeneity (structural diversity), destabilization of microbial associations, and increase in disease, transitioning to an alternative stable state composed of competitive dominants (e.g. sponges, macroalgae) and/or sediment/rubble. Arrows from green to red indicate the positive to negative changes a coral reef might experience over time. *Source:* O'Brien et al. (2016). https://www.frontiersin.org/articles/10.3389/fmars.2016.00047/full. Licensed under CC BY 4.0.

markedly diminishes the abundance of *Proteobacteria*. A change in precipitation has a dramatic effect on the composition of the fungal soil clone 1 (Castro et al. 2009), recently established as *Archaeorhizomycetes* (Rosling et al. 2011). Members of soil clone 1 are unusual ascomycetes that grow on root surfaces by using organic compounds secreted by tree roots, but these fungi are not mycorrhizal. If nitrogen in the environment becomes a limiting factor, plants with symbiotic nitrogen-fixing bacteria are favored over plants without nitrogen-fixing systems. Under the conditions discussed, bottom up regulation would be important in the quantity of CO_2 released from the soil.

There are two types of grazing activities (top down control) that are relevant to climate change (Classen et al. 2015; Crowther et al. 2015). One is the amount of grazing above ground by native livestock, because with more plant mass removed by grazing there is less plant detritus above ground and in the soil. However, the quantity of plant material removed by grazing may be diminished in new plant ranges due to anti-herbivore chemicals produced by epiphytic and endophytic bacteria and fungi. Often the root–mycorrhizal association is not an obligate symbiotic interaction but with the new stress on the plant–mycorrhizal system, the stressed mycorrhizal fungi could become pathogenic. Grazing soil invertebrates are highly significant in consumption of saprophytic soil fungi (Crowther et al. 2015). Soil fungi grow in linear "strands" or "cords" in the soil and isopods switch their feeding habit from consuming plant litter to eating fungal cords. With a reduction of fungal biomass (fungal cords), the quantity of CO_2 released from the soil is also decreased. If nitrogen is not limiting, fungal grazing by isopods is a major factor in influencing the release of CO_2 from the soil.

11.4 Syntrophy and Co-metabolism

The dependency of one species for nutrients supplied by another species is often referred to as microbial syntrophy (Schink and Stams 2006) or cross-feeding (see Box 11.1). It is distinguished from symbiosis because symbiotic associations can be based on protection from mechanical or chemical stress in addition to metabolism (Morris et al. 2013). This nutritional dependence has been viewed by microbiologists as an obligate mutualistic form of metabolism in which reducing equivalents that may be low molecular weight organic compounds or H_2 are transferred from one microbial species to another. The production of H_2 and acetate by one species and consumption of these chemicals by another has been designated interspecies electron transfer (IED) or more specifically, direct interspecies electron transfer (DIET) (Dolfing 2014). The first use of the term "syntrophy" described cross-feeding with *Salmonella typhi* secreting tryptophan, which supported the co-culture growth of a tryptophan-requiring mutant strain of *Escherichia coli* (Fildes 1956). The classic example of bacterial syntrophism is the co-culture of "S" organism, which fermented ethanol to acetate and hydrogen, with *Methanobacterium bryantii* strain M.o.h, which used acetate and hydrogen to produce methane (Bryant et al. 1967). The reactions are as follows:

$$\text{"S" organism} \quad 2C_2H_5OH + 2H_2O \rightarrow 3CH_3COOH + 4H_2$$

$$\text{M.o.h.strain} \quad 4H_2 + CO_2 \rightarrow CH_4 + 2H_2O$$

The anaerobic methanogenic environment has numerous examples of interspecies syntrophy between bacteria and archaea. There are two types of syntrophic organisms: facultative and obligatory. Initially, facultative syntrophic bacteria were isolated where the bacteria involved in oxidizing acetate or formate could grow without a second partner and many of these bacteria belong to the *Syntrophomonas* genus. Several bacteria have been isolated that require a second living organism for growth and this syntrophy is important in the anaerobic carbon cycle (Schink

Box 11.1 Syntrophy and Co-metabolism

The word syntrophy is from Greek where "syn" means "together" and "trophy" is "nourishment". Commonly this is taken to mean that with two metabolically unrelated microbial species, one lives off the products of the other. Often the consumption of the product of the first species by a second species increases the thermodynamics of the first reaction. With co-metabolism, an enzyme degrades the primary substrate and the product of that reaction is also the substrate for the enzyme that produced it. The secondary product is not generally of nutritional value to the microorganism producing the enzyme and the second end product may be toxic to that microorganism. Monooxygenase, which has many different substrates, is an enzyme commonly associated with co-metabolism.

1997; McInerney et al. 2009). *Syntrophomonas erecta* degrades fatty acids of 4–8 carbon atoms when co-cultured with a methanogen (Wu et al. 2006a). Long chain fatty acids are used by *Syntrophomonas zehnderi* grown with *Methanobacterium formicicum* (Solusa et al. 2007), *Syntrophomonas palmitatica* in co-culture with *Methanospirillum hungatei* (Hatamoto et al. 2007), and *Syntrophomonas cellicola* grown with *Methanobacterium formicicum* (Wu et al. 2006a). Other fatty acid oxidizing bacteria that show obligate syntrophic oxidation of fatty acids in co-culture with *Methanospirillum hungatei* or other hydrogenotrophic methanogens, include *Pelotomaculum schinkii* (de Bok et al. 2005), *Pelotomaculum propionicium* (Imachi et al. 2007), and *Syntrophomonas bryantii* (Wu et al. 2006b). Genomic analysis reveals that bacteria can grow syntrophically with methanogens due to several genetic characteristics, including the presence of an extra-cytoplasmic formate dehydrogenase as one important trait (Worm et al. 2014).

Microbial syntrophy is broadly distributed in the environment and several examples of syntrophy are highlighted here. Sulfur syntrophy is displayed by "*Chloropseudomonas ethylica,*" which is a co-culture of *P. aestuarii* and *Desulfuromonas acetoxidans* (Olson 1978). With acetate as the electron donor, *D. acetoxidans* uses elemental sulfur as the electron acceptor, with the production of H_2S, which is toxic to *D. acetoxidans.* The phototroph *Prosecochloris aestuarii* uses sulfide as the electron donor and produces elemental sulfur. Another example is the co-culture of *Chlorobium*, a green sulfur photosynthetic bacterium, and *Desulfuromonas* (Warthmann et al. 1992). *Chlorobium* uses sulfide as the electron donor with the production of elemental sulfur (S^0) and *Desulfuromonas* oxidizes acetate with electrons going to S^0 with the production of sulfide. Since sulfide is inhibitory to *Desulfuromonas,* the removal of sulfide by *Chlorobium* enhances the growth of *Desulfuromonas.*

Anaerobic bacteria show syntrophic growth on various organic compounds and the mechanisms accounting for these metabolic interactions are fully resolved. The conversion of 3-chlorobenzoate to methane involves a consortium consisting of a dechlorinating bacterium, a benzoate degrading bacterium, and *Methanospirillum* sp. (Dolfing and Tiedje 1986). The degradation of amino acids by bacteria may use a pathway in which one amino acid serves as an electron donor and another amino acid as the electron acceptor. In environments where the amino acids are limiting, syntrophic degradation of amino acids may occur (Schink and Stams 2006). Sugar fermentation in deep lake sediments or in the rumen is due to a syntrophic process where H_2 production by one bacterial type may support methane production by archaea (Dore and Bryant 1990). The degradation of oil occurs more rapidly in mixed cultures compared with single species cultures. *Desulfatibacillum alkenivorans* can complete oxidation of *n*-alkanes when cultured with *Methanospirillum hungatei* (Westerholm et al. 2011). It is possible that corrosion of ferrous metals is by syntrophic activity because mixed cultures have greater rates of corrosion than single species cultures.

Microbial aggregates as granules or biofilms have been associated with microbial syntrophic growth. As reviewed by Kouzuma et al. (2015), co-culture of *Pelotomaculum thermopropionicum* and *Methanobacter thermautotrophicus* are connected by microfilaments that extend from the cells of *P. thermopropionicum* that appear to be flagellum. The function of flagella in *P. thermopropionicum* is for attachment to *Methanobacter thermautotrophicus* to maintain a close association and not for motility. Another physically close relationship is observed with "*Chlorochromatium aggregatum,*" which consists of *Chlorobium chloromatii* and an unidentified bacterium of the *Betaprotobacteria* (Mueller and Overmann, 2011). The phototrophic *Chlorobium chloromatii* tightly surround the bacterial cell and provide organic nutrients for bacterial growth and the bacterial cell uses a single flagellum to move the aggregated cell system in the water column for optimal growth. In the hyperthermophilic archaea, *Pyrococcus furiosus* produces an extensive network of fibers that entrap cells of *Methanopyrus kandleri* (Schopf et al. 2008) and the H_2 released by *P. furiosus* probably supports the production of methane by *Methanopyrus kandleri.*

Geobacter and other anaerobes may be specially suited for interspecies extracellular electron transport (EET) because they have cytochromes in the outer membrane. Syntrophy involving *Geobacter metallireducens* and *Geobacter sulfurreducens* accounts for the oxidation of ethanol and reduction of fumarate, which neither *Geobacter* species can perform without a partner (Summers et al. 2010). Acetate oxidation by *Geobacter* can be coupled to nitrate reduction or methane production with *Thiobacillus* or *Methanosarcina* as organisms in co-culture, respectively (Kato et al. 2012; Cruz Viggi et al. 2014). In these examples, *Geobacter* interfaces with nanoparticles of magnetite to facilitate the acetate–nitrate or acetate–methane reactions.

11.5 Ecosystem Created by Hydraulic Fracturing in Shale

The technology of producing natural gas by injection of water under high pressure into shale formations to release hydrocarbons trapped in small pore spaces is called "fracking" and accounts for a major segment of the natural gas produced in the USA. After hydraulic fracturing of shale in the in the Appalachian Basin, microorganisms in the water of a 2.5 km deep well was examined for 328 days and was found to contain a many unique genomes, including a novel bacterium identified as *Frakibacter* (Daly et al. 2016). These unique bacteria were not introduced with the input fluid but were present in the shale at extremely low numbers that were undetected by traditional sampling. When salts dissolved in the shale matrix, the shale fluid became highly saline and approached a brine condition, with halotolerant bacteria present. Bacterial growth was supported by organic compounds in the input fluids, including trace levels of choline, 0.3 mM sucrose, 3.6 mM ethylene glycol, 1.2 mM monomethylamine, and 331 μM methanol. By the end of the seventh day, sucrose was aerobically oxidized by *Vibrio, Pseudomonas,* and *Marinobacter,* ethylene glycol was used by *Marinobacter* and *Vibrio,* and choline consumed by various bacteria including *Marinobacter*. From the genome analysis of many bacteria isolated from shale fluid, it was apparent that many species could metabolize hydrocarbons. The principle osmoprotectant chemical produced by the bacteria was glycine betaine, which accumulated in the extracellular fluid due to the lysis of bacterial cells by bacterial viruses. The degradation of glycine beatine by *Halanaerobium* and *Frackibacter* produced trimethylamine, which was rapidly consumed by *Methanolobus* and *Methanohalophilus* to produce methane. The monomethylamine and methanol in the input fluid also supported methanogenesis. This study reveals the active role of viral predation and the immunity of bacterial hosts functioning in a contained environment.

11.6 Extracellular Electron Transport

An interesting feature of bacterial syntrophy is the interspecies EET between cells by electric current using conductive material involving nanowires or abiotic material (Kouzuma et al. 2015). There are several ways of approaching this topic including: (i) direct contact involving electron-carrying proteins, (ii) shuttle molecules or compounds that interface with cells, (iii) electron-conductive structures such as pili, and (iv) extracellular polymeric substances of biofilms. This transfer of electrons from cell-to-cell may be between the same species of bacteria or between different species, as for bacterial syntrophy. EET research is rapidly evolving with newer approaches because it is highly applicable to bioenergy and biofuel production. Although the mechanism for electron movement through ferrous materials is not understood, models of biocorrosion often rely on cathodic release of electrons (see discussion in Chapter 10). Additionally, biogeobatteries and long distance electron transfer are related to EET (see Section 11.7.2).

11.6.1 Membrane-bound Proteins

Geobacter and *Shewanella* are especially suited for EET because these gram-negative bacteria have cytochromes in the outer membrane. The initial observations that bacteria could interface with solid minerals containing Fe^{3+} or Mn^{4+} as electron acceptors (Lovley et al. 2004), stimulated the search for the mechanism that would explain this phenomenon. On the outer membranes of *Geobacter sulfurreducens* and *Shewanella oneidensis* MR-1 are multiheme cytochrome *c* molecules facilitating the transfer of electrons across the outer membrane (Choi and Sang 2016). *G. sulfurreducens* has about 30 outer membrane cytochromes and of these, OmcB, OmcE, OmcS, OmcT, and OmcZ are used for the reduction of Fe^{3+} or electrode reduction (Richter et al. 2012). *S. oneidensis* has MtrC, an outer membrane cytochrome which is involved in most EET reactions, and OmcB, an outer membrane cytochrome essential for ferric citrate or ferric oxide reduction (Richter et al. 2012). Additionally, cytochromes are proposed to participate in transmembrane electron transport by the gram-positive bacterium *Thermincola ferriacetica* and the gram-negative bacteria *Sporomusa ovate* and *Moorella thermoacetica* (Kracke et al. 2015 and references therein). Archaea of the genus *Methanosaeta* and *Methanosarcina* have membrane-bound cytochromes (Thauer et al. 2008) and these electron carriers could participate in EET. Although it has not been established, there is the hypothesis that ferredoxin:NAD^+-oxidoreductases or other reductases would be important in transmembrane electron transport by the gram-positive *Acetobacterium woodii* and *Clostridium ljungdahlii* (Nevin et al. 2011). The co-culture of *G. metallireducens* and *G. sulfurreducens* requires the presence of cytochrome *c* for EET associated with the oxidation of ethanol and reduction of fumarate (Summers et al. 2010). Neither *Geobacter* species can perform this metabolism in pure culture.

11.6.2 Electron Shuttling

A process that employs EET with magnetite as an intermediary providing direct contact between two bacteria has been reported. Acetate oxidation by *G. sulfurreducens* can be coupled to nitrate reduction by *Thiobacillus denitrificans* in co-culture, provided nanoparticles of magnetite are present to facilitate interspecies electron transport (Kato et al. 2012). Another example of EET is the oxidation of ethanol by *G. metallireducens,* with methane produced by *Methanosaeta harundinacea* when these two species are grown as a co-culture in a medium containing magnetite nanoparticles (Rotaru et al. 2014). When magnetite was added to anaerobic digesters, the oxidation of acetate by *Geobacter* coupled to methane production by *Methanosarcina* was enhanced with the addition of nanoparticles of magnetite (Kato et al. 2012; Cruz Viggi et al. 2014). A practical application of EET is the stimulation of trichloroethene bioremediation by syntrophic microbial cultures with the addition magnetite nanoparticles (Aulenta et al. 2013). Other electron conductive materials that were found effective in replacing magnetite in this direct transfer of electrons between bacteria included nanoparticles of hematite, graphite, or activated carbon (Kouzuma et al. 2015).

Electrons may be carried from the cell surface to the environment by redox active molecules that function as extracellular electron carriers or electron shuttle molecules. Several strains of *Shewanella* secrete riboflavin (vitamin B2) and flavin mononucleotide (FMN) and these redox active molecules enhance EET in *Shewanella* with reduction of Fe^{3+} oxide (ferrihydrite) and Mn^{4+}oxide (birnessite) by interacting with the outer membrane cytochrome MtrC (Marsili et al. 2008; Kotloski and Gralnick 2013). Other molecules secreted by bacteria that enhance EET include pyocyanin, phenazine-1-carboxamide (Rabaey et al. 2005), and quinones (Newman and Kolter 2000). Humic acid and anthraquinone-2,6-disulfonate, a model quinone for humic acid, serve as electron shuttles for *G. sulfurreducens* and other dissimilatory metal reducing bacteria that use humic substances as the electron acceptor (Scott et al. 1998; Voordeckers et al. 2010).

11.6.3 Nanowires

The term nanowire was used to characterize the electron conductive extensions from cells of *G. sulfurreducens* (Reguera et al. 2005) and *S. oneidensis* (Gorby et al. 2006) to iron oxides. These nanowires participate in EET, are electrically conductive, and can transfer electrons from a donor bacterial cell to a particulate electron acceptor over a distance of several micrometers (Gorby et al. 2006). Nanowires are important for other taxa of microorganisms in addition to *Geobacter* and *Shewanella*. Nanowires may facilitate electron transfer between anaerobic methanotrophic archaea (ANME) and sulfate-reducing bacteria (Wegener et al. 2015). Electron transfer via nanowires may contribute to electric currents in sediments and this activity is similar to a biogeobattery (Revil et al. 2010). Not only are nanowires involved in dissimilatory metal reduction but they also participate in carbon dioxide metabolism by *Synechocystis* PCC6803, an oxygenic phototroph, and by *P. thermopropionicum*, a fermentative bacterium (Gorby et al. 2006).

For some time, it was considered that the nanowires of *G. sulfurreducens* were specialized types of bacterial type IV pili with electrical conductance attributed to aromatic amino acids appropriately stacked in the pilus protein (Vargas et al. 2013); however, this hypothesis has received some criticism (Strycharz-Glaven and Tender 2012; Snider et al. 2012). Studies indicate that the nanowires in the gram-negative *S. oneidensis* are constructed from outer membrane vesicles (OMVs) (Pirbadian et al. 2014). The formation of nanowires produced by *S. oneidensis* MR-1 revealed that MtrC and Omc, the cytochromes in the outer membrane, are present in OMVs and these vesicles fuse to produce the nanowire. Electrical conductivity of the nanowires of *S. oneidensis* MR-1 could be attributed to the multiheme cytochromes making up the nanowire.

11.6.4 Extracellular Electron Movement in Biofilms

The formation of a three-dimensional cellular structure as is the case with biofilms may be highly structured, with cells immersed in a matrix. This formation of a dense cell population resembles a microcolony and may consist of single or multiple species. In addition to the bacterial cells present in the biofilm, there is an extracellular polymeric substance (EPS) consisting of proteins, lipids, polysaccharides, and environmental DNA (eDNA) from bacterial cells lysed by prophages (Gödeke et al. 2011). Cao et al. (2011) determined that EPS of a biofilm consisting of *Shewanella* sp. contained over 50 proteins, many of which included redox active molecules (i.e. iron-sulfur proteins and heme proteins) that could participate in EET (Choi and Sang 2016). While interspecies EET may occur by way of nanowires, it is highly probable that interspecies transfer of electrons in a biofilm is due to H_2 and formate (Rotaru et al. 2012; Boone et al. 1989). Many of the syntrophic activities discussed in Section 11.5 may be involved in transfer of electrons by end product–substrate exchanges and not by direct electrical connections.

11.7 Cross-talk: Interkingdom Signaling

There is also growing evidence that interkingdom signaling occurs between microorganisms and higher life forms. Signal molecules produced by the host and the bacteria are the mediators of interkingdom communication and this system may have evolved because of coevolution. Depending on the system, the impact of the interkingdom signaling could be either beneficial or detrimental.

11.7.1 Microbial Endocrinology

The recognition of communication between bacteria and humans can be traced to experiments that documented the stimulated growth of enteric bacteria by adrenaline, noradrenaline, and catecholamine. In 1993, Mark Lyte termed this activity as "microbial endocrinology" and it has become an important research field to evaluate the control of infectious diseases (Lyte, 2016).

Communication between cells to coordinate population behavior is seen throughout biology, with bacteria using quorum sensing and mammals using hormones. Numerous chemicals in mammals can function as hormones and these chemicals are proteins or peptides, steroids, or amines that are derived from amino acids. Apart from the report that estrone, estriol, and estradiol (estrogen steroids) inhibit quorum sensing in *Pseudomonas aeruginosa*, relatively little is known about the role of steroid hormone interaction with bacteria. In humans, several peptide hormones serve as signals to enhance virulence in pathogenic microorganisms and examples of interactions are given in Table 11.1.

Many papers are available on the topic of microbial interaction with amine hormones that include the catecholamines of adrenaline, noradrenaline, and dopamine. These catecholamines are related to stress and are often referred to as the "fight and flight" hormones. Synthesis of these catecholamines follows the general scheme presented as follows:

$$\text{L-dopa} \rightarrow \text{dopamine} \rightarrow \text{noradrenaline} \rightarrow \text{adrenaline}$$

L-dopa is found in the diet, with substantial levels in bananas and seeds of the broad leaf bean. Adrenaline and noradrenaline may occur in the human intestine at micromolar levels and stimulate the growth of many gram-negative bacteria, including *E. coli*, *Salmonella*, *Helicobacter*, *Listeria*, *Campylobacter*, and *Yersinia* (Sandrini et al. 2015; Kendall and Sperandio 2016; Freestone 2013). Bacteria have a considerable influence on the gut–brain axis and stress response, including behavioral changes (Foster et al. 2017) and perhaps this is related to a response of catecholamine hormones. Stress has an impact on individuals, with elevated catecholamine hormones enhancing bacterial activities in the mouth, respiratory tract, and on the skin (Freestone 2013). As this field of research develops, there are sure to be some interesting associations.

With respect to quorum sensing, there is a distinction with gram-negative bacteria employing *N*-acetyl homoserine lactones (AHLs) and gram-positive bacteria using autoinducing peptides

Table 11.1 Human hormones function as signals for microorganisms.

- *Helicobacter pylori* promotes an imbalance of stomach acidity by interacting with gastrin and somatostatin.
- The virulence of *Burkholderia pseudomallei* is enhanced by binding insulin and is a factor in producing melioidosis.
- The cytotoxicity, modification of lipopolysaccharide and biofilm formation of *Pseudomonas aeruginosa* and *P. fluorescens* is increased with exposure to natriuretic peptides (hormones involved in osmoregulation of blood).
- Virulence of *E. coli*, *Pseudomonas aeruginosa*, *Salmonella enterica*, and *Shigella* is increased by interacting (binding?) cytokines such as interleukins (ILs) and tumor necrosis factors.
- Quorum sensing in *Pseudomonas aeruginosa* is activated by the opioid dynorphin which is released from the intestine follow intestinal surgery or damage to intestine.
- Interaction of the yeast *Candida albicans* with chorionic gonadotropin and luteinizing hormones promotes a phase change to the hyphal form with increased virulence. This may explain the increased susceptibility of *Candida* infections of pregnant women.

Source: Based on data from Kendall and Sperandio (2016); Freestone (2013).

(AIPs). The AHLs and AIPs are species specific and enable bacteria in a mixed culture to function as a unit to produce toxins or other virulence factors. Additionally, many gram-positive and gram-negative bacteria produce a chemical (AI-2) which is not species specific but functions in interspecies communication (Freestone 2013). A considerable amount of research has been focused on the role of AHLs in coordinating bacterial gene expression and establishing an infection. Bacterial AHL are taken up by human cells and inhibit host immune responses. The effect of AHL on the host includes promotion of inflammatory effects by downregulation of specific cytokines and initiation of apoptosis of macrophages, neutrophils, fibroblasts, and vascular endothelial cells. Some mammalian cells disrupt the bacterial signals using enzymes belonging to a family of paraoxonases to degrade the AHLs (Hughes and Sperandio 2008; Freestone 2013).

11.7.2 Cross-signaling in Nonhuman Systems

Several signal molecules have been identified that can facilitate interkingdom communication between bacteria and higher organisms. Examples include signals between *Rhizobium* sp. and legumes that function during nodule formation in plant roots (Venturi and Fuqua, 2013); see discussion in Chapter 6. Interkingdom sensing is also important in host–pathogen relationships (Kendall and Sperandio 2016; Hughes and Sperandio 2008). As a response to AHL released by bacteria as part of their quorum sensing activity, *Medicago truncatula* (alfalfa) secretes compounds with a structure that mimics the bacterial AHL, floods the environment, and prevents the bacterial association occurring. In a similar manner, *Delisea pulchra* (marine alga) produces halogenated furanones that interfere with bacterial AHL signaling (Manefield et al. 1999). To prevent crown gall tumors in plants due to *Agrobacterium tumefaciens*, plants secrete γ-amino butyric acid, which induces *A. tumefaciens* to degrade its own AHL.

11.8 Evolving Systems of Interest

11.8.1 Polyploidy in Bacteria

Despite the long-held generalization that bacteria have a single copy of genomic DNA, it is now apparent that many bacterial and archaeal cells have multiple genome copies. While multiple genome copies are transiently present in cells growing logarithmically, cells in the stationary phase may be polyploid. Most bacteria studied are haploid with one genome copy per cell but if the prokaryote cell has over 10 genome copies, it is known as polyploidy. A few large bacteria have several hundred genome copies per cell and it has been suggested that this represents a prokaryote with chromosomes. There are reports from numerous laboratories using different techniques that many prokaryotic taxa have multiple genome copies, with some researchers suggesting that the presence of one genome copy in bacteria may be relatively rare except in very slow growing bacteria (Pecoraro et al. 2011). The number of gene copies for several bacteria is listed in Table 11.2 and it is apparent that polyploidy is found in several taxonomic groups. The greatest number of genome copies is found in the large bacterium *Epulopiscium* sp., which is estimated to contain several thousand genome copies (Mendell et al. 2008). However, ploidy is not associated with cell size because the endocellular symbiont *Buchnera* sp. is polyploid. The free-living soil bacterium *Azotobacter vinelandii* is polyploid when growing rapidly but during slow growth it has a single genome (Maldonado et al. 1994). The polyploid character extends to several of the *Archaea*, with members of the *Euryarchaeota* displaying polyploidy whereas members of the *Crenarchaeota* are monoploid (Table 11.3). With polyploidy being recently established in microorganisms, there is an interest in determining the role of

Table 11.2 Genome copy number for selected bacteria.

Phylum	Bacteria	Copies of genome	References
Proteobacteria			
Alpha-	*Azotobacteria vinelandii*	80	Punita et al. 1989
	Caulobacter crescentus	1.7–2.3	Pecoraro et al. 2011
Beta-	*Neisseria lactamica*	1	Tobiason and Seifert 2010
	Neisseria gonorrhoeae	3	Tobiason and Seifert 2010
	Neisseria meningitidis	4	Tobiason and Seifert 2010
Gamma-	*Buchnera* sp.	120	Shigenobu et al. 2000; Komacki and Ishikawa 2000
	Escherichia coli	2.5	Pecoraro et al. 2011
	Pseudomonas pudita	19.7	Pecoraro et al. 2011
	Achromatium oxaliferum	polyploidy[a]	Ionescu et al. 2017
Delta-	*Desulfovibrio gigas*	9	Postgate et al. 1984
	Desulfovibrio vulgaris	4	Postgate et al. 1984
Epsilon-	*Wolinella succinogenes*	0.9	Pecoraro et al. 2011
Deinococcus-	*Deinococcus radiodurans*	5–8	Zerulla and Soppe 2014
Thermus	*Thermus thermophiles*	4–5	Ohtani et al. 2010
Firmicutes	*Bacillus megaterium.*	12	Böttinger et al. 2014
Cyanobacteria	*Synechocystis* sp.	40	Griese et al. 2011

[a] Genomes vary in this bacterium and multiple genomes are not identical, see the text.

Table 11.3 Status of genome copy number in selected archaea.

Phylum	Genus	Copies of genome	References
Crenarchaeota	*Acidianus hospitalis*	1–2	Lehman 2003
	Aeropyrum pernix	1–2	Lehman 2003
	Pyrobaculum aerophilum	1–2	Lehman 2003
	Pyrobaculum calidifontis	1–2	Lehman 2003
	Sulfolobus tokodai	1–2	Lehman 2003
	Sulfolobus acidocaldaritus	1–2	Archibald and Rogers 2002
	Silfolobus solfataricus	1–2	Archibald and Rogers 2002
Euryarchaeota	*Halobacterium cutirubrum*	6.3–10.6	Chant et al. 1986
	Halobacterium salinarum	15	Breuert et al. 2006
	Haloferax volcanii	10	Breuert et al. 2006
	Methanocaldococcus jannaschii	1–5	Mackwan et al. 2007
	Methanococcus maripaludis	30	Hildenbrand et al. 2011
	Methanosarcina acetivorans	5–16	Hildenbrand et al. 2011
	Methanothermobacter thermoautotrophicus	2	Lundgren et al. 2008

polyploidy over monoploidy. In general, polyploidy provides an advantage for growth and survival. With multiple genome copies, the microbial cell would have a low mutation rate, high resistance to ionizing radiation, and the extra DNA could serve as a source of phosphate in phosphorus-limited environments (Zerulla and Soppa 2014). Future studies are needed to further examine the environmental advantage of being polyploid to a bacterial cell.

Examination of the genomes from the polyploid *Epulopiscium* sp. and *Candidatus* Marithrix sp. indicated that these cells contain multiple copies of the same genome. A recent report suggests the interesting prospect that individual bacterial cells may be polyploid but harbor several different genomes and not multiple copies of the same genome. Multiple, mixed genomes are present in *Achromatium oxaliferum*, the largest freshwater bacterium, and this raises the potential for intracellular gene transfer with an increase in diversity (Ionescu et al. 2017). Additionally, it is unknown if each of the genomes in *A. oxaliferum*, have the same or different sites for initiation of replication. This complex genomic arrangement is indeed interesting and future research is needed to understand this contribution to biology.

11.8.2 Impact of Viruses and CRISPR-*cas* Systems

Microbial viruses are found wherever microorganisms are present and the abundance of microbial viruses is equal to or several times greater than the number of bacterial and archaeal host cells (Weinbauer 2004). Examples of bacterial provirus abundance and distribution are given in Table 11.4. The impact of bacterial viruses on the marine environment has been discussed (Anderson et al. 2011) and includes the following: (i) as a result of cellular lysis attributed to viruses, carbon and nitrogen compounds are released and through the "virus loop" alter the biogeochemical cycles; (ii) the cells destroyed by viruses are most abundant in the environment and this is the "kill the winner" (Thingstad and Lignell 1997) event; (iii) the DNA virus

Table 11.4 Abundance of provirus and CRISPR loci in bacteria and archaea.[a]

	Organisms	Number of CRISPR loci	Number of provirus
Domain: ***BACTERIA***			
Thermophiles	**Phylum:** *Deferribacteres*		
	Deferribacter desulfuricans	4	1
	Phylum: *Deinococcus-Thermus*		
	Marinithermus hydrothermalis	5	2
	Phylum: *Proteobacteria*		
	Photobacterium profundum	2	3
	Phylum: *Firmicutes*		
	Thermosediminibacter oceani	6	7
Mesophiles	**Phylum:** *Proteobacteria*		
	Nitratifractor salsuginis	2	3
Psychrophiles	**Phylum:** *Proteobacteria*		
	Shewanella piezotolerans	1	5
	Shewanella violacea	1	1
	Candidatus Venteria ishoeyi	18[b]	?
Domain: ***ARCHAEA***			
Extreme thermophiles	**Phylum:** *Euryarchaeota*		
	Aciduliprofundum boonei	2	2
	Methanocaldococcus jannaschii	20	0
	Methanocaldococcus sp.	23	0
	Methanopyrus kandkeri	5	0
	Methanothermococcus okinawensis	7	1
	Thermococcus gammatolerans	3	3

[a] *Source:* Based on data from Anderson et al. (2011).
[b] *Source:* Based on data from Fonseca et al. (2017).

(i.e. lysogenic or temperate virus) carries genetic material from the previous host and new DNA provides the host with new metabolic capabilities.

The interaction between prokaryotic viruses and prokaryotes is highly dynamic, with host cells moderating virus attack by changing cell surface receptors or cytoplasmic nucleases. The bacterial and archaeal viruses also adjust their cell defenses to ensure continuous infection. While there are great differences of virus–host specificity, with some viruses having broad host range and some limited to a given species, there is also a difference in the lethality of viruses. Following bacterial virus infection, the replication of viruses and release of hundreds of new infectious particles occurs at host lysis and this is designated as "lytic virus" activity. A medical application for using bacterial viruses to control infectious bacterial diseases has been considered since the discovery of the bacterial phage (Sulakvelidze et al. 2001) and, because of an increase in antibiotic resistance in bacteria, the use of bacterial viruses to combat bacterial infections in humans continues (Lin et al. 2016). An application of bacterial viruses to control biofilms in the food industry is also being considered (Gutiérrez et al. 2016).

Another bacterial viral activity is the incorporation of DNA from the bacterial virus and previous host into the recipient host and this is achieved by "temperate" or "lysogenic" viruses. The state of lysogeny occurs with the new host bacterial cell incorporating the DNA via viral transfer, and the bacterial virus is identified as a "provirus" and is not expressed as cells replicate and grow. Bacterial cells in the state of lysogeny release infectious viruses upon environmental (i.e. UV or chemicals) or spontaneous mutations. Occasionally, bacteria develop "cryptic" (i.e. hidden) proviruses after prolonged integration into the host DNA and these become a permanent segment of the host genome. Scientists are focusing on lysogenic virus activity to assess the impact this activity has on microbial ecology. Lysogeny improves host fitness because the genes introduced by the lysogenic virus are expressed in the host and are upregulated in low nutrient environments with suboptimal conditions for growth of the host. The number of proviruses in a microbial genome is given in Table 11.4. Thus, lysogeny is important for bacteria and archaea growing in deep sediments, hydrothermal vents, and other regions where the number of potential host cells for viruses is low (Nigro et al. 2017; Anderson et al. 2011). Using cultivation techniques, the incidence of lysogeny in soil bacteria was reported to be about 30% and for Antarctic soils the presence of provirus in cultivated bacteria ranged from 4% to 20%. The isolation of viral DNA segments from the marine environment may indicate that microbial viruses are important for producing microbial diversity in the marine biosphere (Williamson et al. 2008). In ice-covered Antarctic lakes, 89.5% of the bacteria have proviruses (Lisle and Priscu 2004) and in bacteria isolated from soil using the buried-bead technique, 80–89% of the bacteria contained proviruses (Ghosh et al. 2008). Biofilms are found to develop at a higher rate if the bacteria involved carry proviruses. While it is not known how the prophage-containing bacteria contribute to biofilm formation, it has been proposed that the lysis of a small population of bacteria would provide eDNA, which may be critical for cell aggregation in formation of the biofilm (Gödeke et al. 2011). While the presence of proviruses can be identified using programs for searching metagenomic data (Bose and Barber 2006), many of the activities of lysogenic viruses remain to be explored in both archaeal and bacterial cultures.

Clustered regularly interspaced short palindromic repeats (CRISPR) is family of DNA sequences that along with *cas* (CRISPR-associated) genes provide a type of immunity to bacterial virus or plasmid by degrading DNA that enters the cell. About 10% of 1724 sampled microorganisms have genes for the CRISPR–*cas* system (Burstein et al. 2016). The CRISPR contains an AT-rich leader sequence followed DNA repeats of 23–55 bp and the spacers are about 25–75 bp. When foreign DNA enters a cell as a plasmid or bacteriophage, a segment of DNA is synthesized to match a region of the foreign DNA and this segment is inserted into the CRISPR locus. If the foreign DNA enters the cell later, the invading DNA activates the *cas* genes and the foreign DNA is degraded. The CRISPR–*cas* system is regarded as a type of

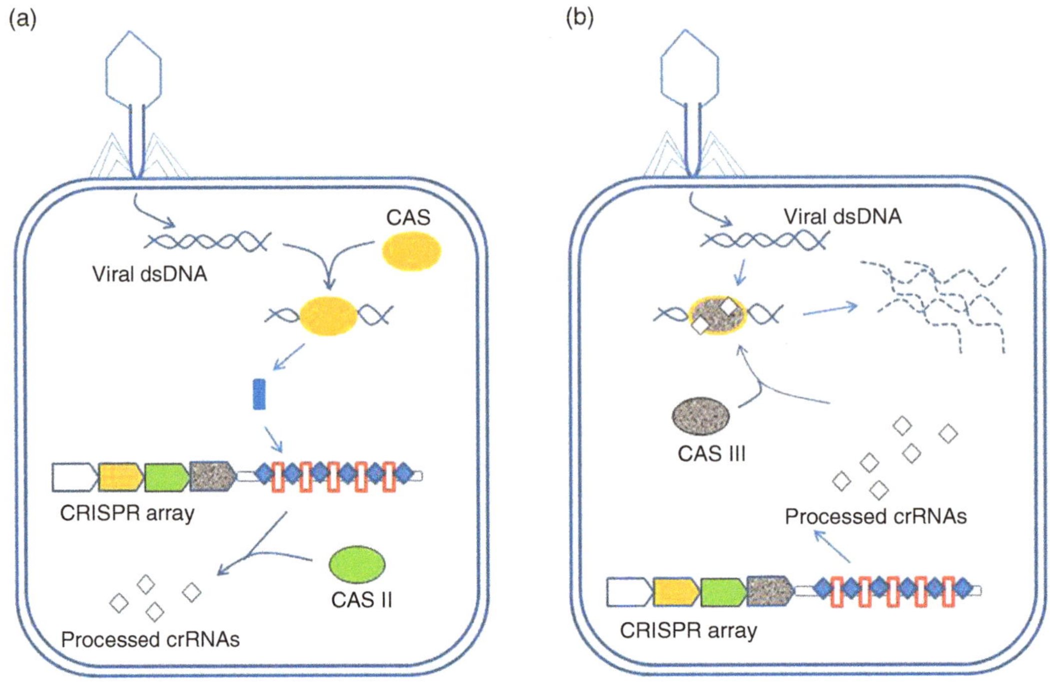

Figure 11.2 The CRISPER-associated (*cas*) proteins maintains a "memory" of previously encountered bacterial viruses and serves to "immunize" the cell against future provirus infections: (a) initial exposure to a provirus results in formation of spacers; (b) the processed crRNAs target homologous provirus DNA and prevent a second infection by the same virus.

viral immunity (Figure 11.2). The composition of the spacer varies with the bacteriophage or plasmid entering the microbial cell (Horvath and Barrangou 2010; Labrie et al. 2010). The presence of the CRISPR–*cas* system is a record of previous virus–cell interactions and the potential for viral-mediated gene transfer, and computer-based programs are available to evaluate metagenomic data files for the presence of CRISPR systems in bacteria and archaea (Grissa et al. 2007a, b). Some bacteria and archaea with high CRISPR loci per genome are found in thermophilic organisms (Table 11.4) and this suggests a high level of gene flow between organisms in hydrothermal vent systems (Anderson et al. 2011). Several groups of microorganisms do not contain the CRISPR–*cas* system and this includes obligate symbionts, extremely slow growing microorganisms that could not rapidly defend against activity resulting from entry of foreign DNA into a cell, and very small cells that could not support a genome containing nonessential DNA (Burstein et al. 2016). It remains to be determined if there are systems in microorganisms other than the CRISPR–*cas* system to prevent virus and plasmid acquisition by microorganisms.

11.8.3 Impact of Outer Membrane Vesicles

For some time, it has been known that gram-negative bacteria produce small membrane units called OMVs (Figure 11.3) and these structures are associated with a range of activities. The OMVs are right-side out structures and some of these OMVs are tethered to the cell (Figure 11.4). Recent reviews (Barton 2014; Kulkarni and Jagannadham 2014; Schwechheimer and Kuehn 2015; Jan 2017) show that the most frequently studied bacteria for OMV production are enteric pathogens, soil bacteria, and dissimilatory metal-reducing bacteria, but OMV production

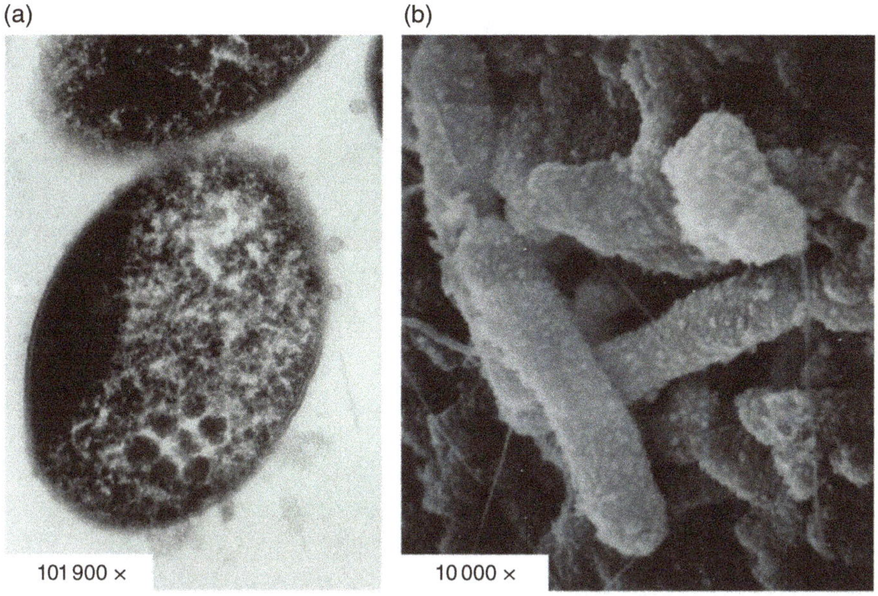

Figure 11.3 Outer membrane vesicles (OMV) on bacteria: (a) thin sections of *Desulfovibrio gigas*; (b) soil bacteria showing presence of OMV on surface of the cell. *Source:* Reproduced with permission from Larry L. Barton.

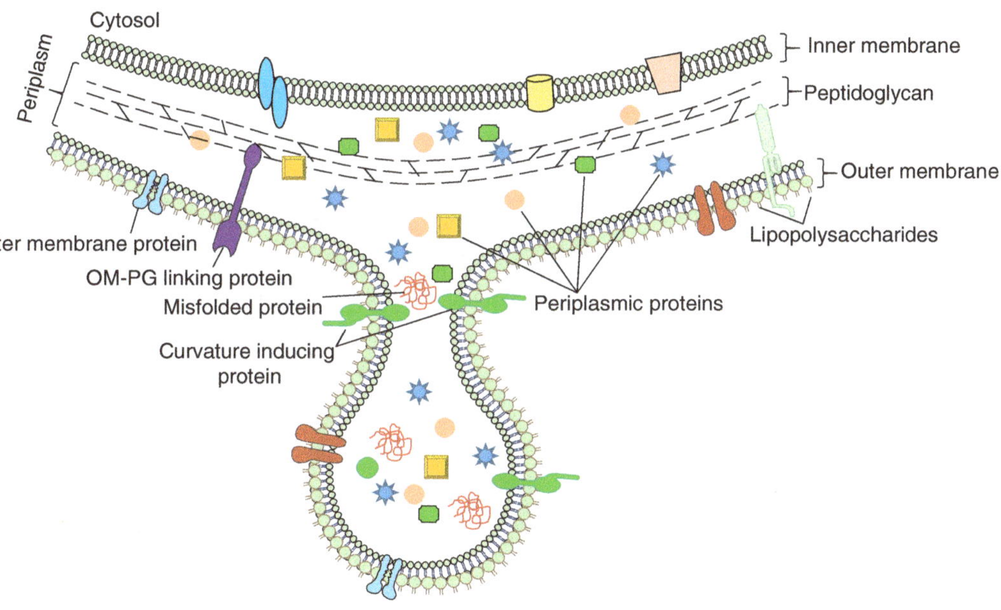

Figure 11.4 Model for biogenesis of outer membrane vesicles in gram-negative bacteria. *Source:* Jan (2017). https://www.frontiersin.org/articles/10.3389/fmicb.2017.01053/full. Licensed under CC BY 4.0

by plant pathogens and by archaeal organisms has also been reported. Scientists have also recognized that membrane vesicles are secreted by gram-positive bacteria and fungi. The structure of the OMVs usually consists of a lipid bilayer (Figure 11.5) but sometimes the vesicle has a single lipid layer. Table 11.5 enumerates some of the microorganisms reported to have OMV and the activities associated with these vesicles.

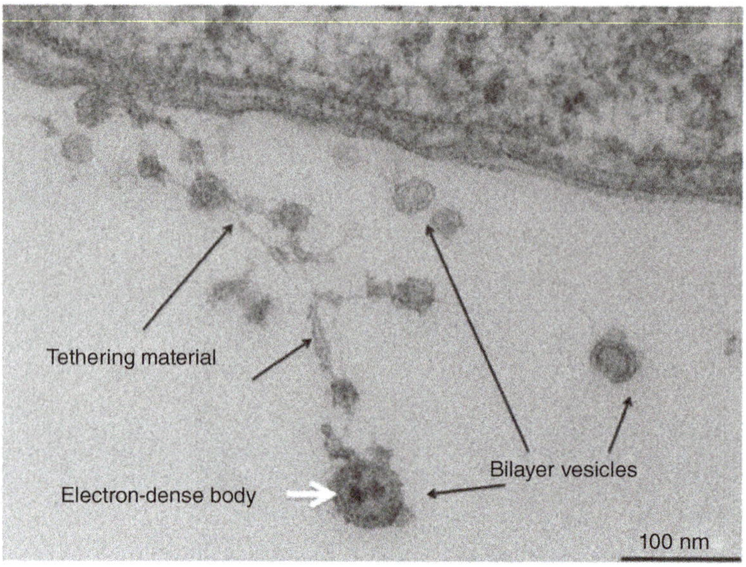

Figure 11.5 Thin section of the NpdA mutant of *Delftia* sp. Cs1–4 showing extracellular vesicles attached to the cell surface. *Source:* Shetty et al. (2011). http://journals.plos.org/plosone/article/figure?id=10.1371/journal.pone.0020725.g005. Licensed under CC BY 4.0.

Table 11.5 Some of the activities associated with outer membrane vesicles.

Bacterial pathogens	Virulence factors and related activities
Escherichia coli	Shiga toxin, enterotoxin, and cytolycin A
Helicobacter pylori	Vacuolating toxin, adherence to cells and cytotoxic activity
Pseudomonas aeruginosa	Alkaline phosphatase, hemolysin, protease, phospholipase C
Borrelia burgdorferi	Adhesion to host cells
Bordetella pertussis	Hemolysin, whooping cough (pertussis) toxin
Xanthomonas campestris	cellulose, type-3 secretion protins
Legionella pneumoniae	Acid phosphatase, protease, chininase, adhesion to host cell

Source: Based on data from Jan (2017).

Bacteria other than gram-negatives also produce membrane vesicles and in general these vesicles contain toxins or compounds that serve to enhance pathogenicity. Membrane vesicles are produced by *Staphylococcus aureus, Mycobacterium tuberculosis, Mycobacterium ulcerans,* and *Bacillus anthrasis, which* contribute to the pathogenesis of human diseases (Brown et al. 2014; Deatherage and Cookson 2012). *B. subtilis* serves as a model system for membrane vesicle formation in gram-positive bacteria. Hydrolytic enzymes enable the membrane vesicle to make its way from the plasma membrane through the peptidoglycan layers of gram-positive bacteria (Toyofuku et al. 2017a). Biosynthesis of the membrane vesicle in *B. subtilis* is dependent on a mutation in the *spf* gene (Brown et al. 2014). In *B. subtilis,* the *spf* gene is involved in the biosynthesis of the lipoprotein surfactin and the siderophore bacillibactin. In strains of bacteria where the *spf* gene is intact and functions appropriately, the surfactin degrades the membrane vesicles as they are formed. Thus, in bacteria carrying a mutation in the *spf* gene, membrane vesicles remain intact after they are produced. This is an interesting case of an exporter system where the cargo carried inside the membrane vesicle can be released by a protein also produced by the same bacterium.

The activities of OMVs in biogeochemical cycling are addressed here from an environmental and ecological perspective. In the marine environment, *Prochlorococcus*, a cyanobacterium, releases lipid vesicles containing proteins, DNA, and RNA that may be used by heterotrophic bacteria as nutrients and, thereby, are involved in oceanic carbon flux (Biller et al. 2014). Isolated OMVs from *Shewanella putrefaciens* CN32 contain *c*-type cytochromes and hydrogenase (Gorby et al. 2008) and this is significant because nanowires of *Shewanella* are constructed from OMVs (Pirbadian et al. 2014); see discussion in Section 11.6.3. Hydrogenase and formate dehydrogenase have been reported for the sulfate-reducing bacterium *Desulfovibrio gigas* (Haynes et al. 1995). A polyaromatic hydrocarbon-degrading bacterium *Delftia* sp. Cs1–4 not only produces OMVs but also produces a straw-like structure called a nanopod (Figure 11.6).

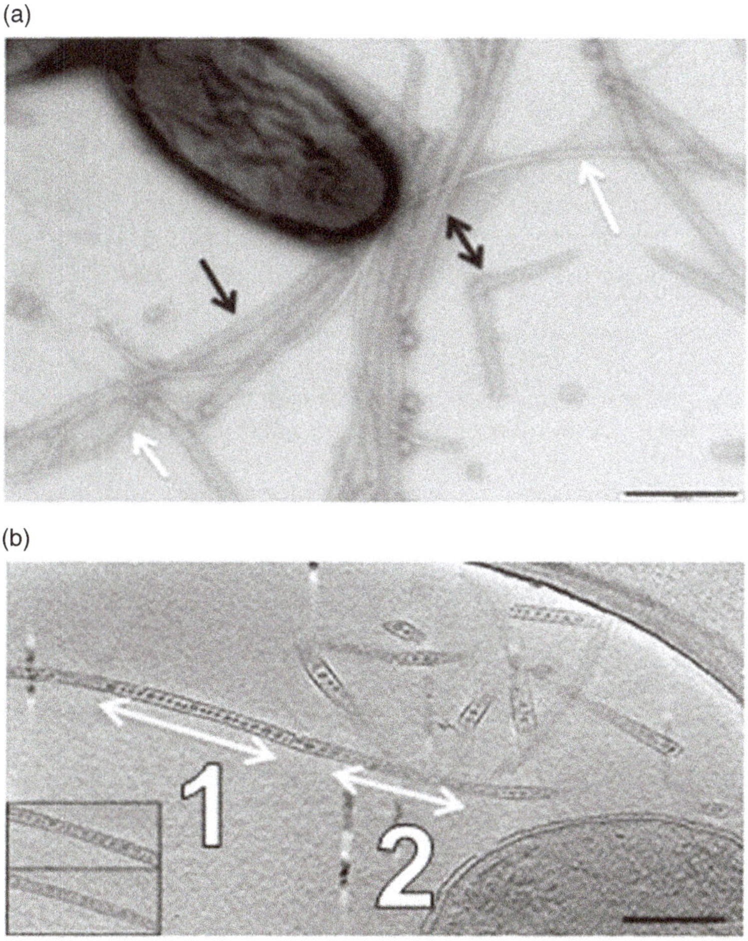

Figure 11.6 Images of nanopods in phenanthrene-grown culture of strain Cs1–4: (a) negatively stained wet mount of a phenanthrene-grown culture of strain Cs1–4 showing nanopods (black arrows) and flagella (white arrows), scale bar = 500 nm; (b) central slice through a 3D electron cryotomographic reconstruction of a phenanthrene-grown *Delftia* sp. Cs1–4 culture showing nanopods next to a cell. Arrows follow an individual nanopod, and illustrate its undulating structure. Region "1" is the exposed interior with vesicles; region "2" is the outer surface. Insets: Tomographic slices just above and below region "1" showing the surface and the crystalline array. Scale bar = 200 nm. *Source:* Shetty et al., (2011). http://journals.plos.org/plosone/article/figure?id=10.1371/journal.pone.0020725.g001. Licensed under CC BY 4.0.

OMVs are inside the nanopod, which protects the OMV as it moves several microns away from the cell (Shetty et al. 2011). The archaeon *Sulfolobus solfataricus,* a sulfur-oxidizing thermophile, secretes a membrane vesicle that is 90–270 nm wide that contains membrane and S-layer from the cell surface. This membrane vesicle contains sulfolobicin, which is toxic to other species of *Sulfolobus* and enables the producing cell to have an advantage in competition for nutrients (Prangishvili et al. 2000).

With respect to communication activity associated with OMVs, it is suggested that *N*-acyl homoserine lactones (AHLs) with long fatty acid chains are water insoluble and use OMVs as a vehicle to carry these quorum sensing molecules (Toyofuku et al. 2017b). The long chain AHLs produced by *Burkholderia pseudomallei, Rhizobium leguminosarum, Paracoccus denitrificans, Sinorhizobium meliloti,* and *Rhodobacter capsulatus* are lipophilic and could be appropriately carried by OMVs. *Pseudomonas aeruginosa* uses OMVs to carry the hydrophobic signaling molecule 2-heptyl-3-hydroxy-4-quinolone for communication within a population (Mashburn and Whiteley, 2005). The role of the OMVs in marine environments and other natural settings continues to be an important area of research.

The interest in human diseases has stimulated investigations into the role of OMVs in pathogen – host interactions. As reviewed by Jan (2017), OMVs carry toxins in diseases attributed to *E. coli, Helicobacter pylori, Borrelia burgdorferi, Shigella dysenteriae, Bordatella pertussis, Vibrio cholerae,* and *Camplyobacter jejuni.* Additionally, the OMVs of pathogens also carry hydrolytic enzymes, adhesion proteins, and immune compromising proteins. The membrane vesicles of the gram-positive *B. anthrasis* carry the anthrax toxin (Brown et al. 2014). The membrane vesicles of *S. aureus* and *Mycobacterium* spp. are known to contain numerous proteins and enhance virulence, but the mechanism of vesicle formation remains unresolved (Brown et al. 2015). Membrane vesicles are reported for numerous eukaryotic organisms including *Cryptococcus neoformans, Histoplasm, Candida, Sporothrix,* and *Leishmania donovani* and are assumed to contribute to their virulence (Deatherage and Cookson 2012). Although there are few observations concerning OMVs in plant diseases, the virulence and signal sensing by the plant pathogen *Xylella fastidiosa* is enhanced by OMVs (Katsir and Bahar 2017).

11.8.4 Atmospheric Microbiology

Aeromicrobiology is the study of microorganisms suspended in the atmosphere. As indicated in Table 11.6, seasonal winds out of Africa may mobilize surface microorganisms and disperse them over great distances. It has been well documented that dust from deserts contains iron minerals that enrich microbial growth downwind. The best example is the movement of dust containing iron from China's Gobi Desert to the Atlantic Ocean, where algal blooms result.

Examination of airborne bacteria in South Korea during an Asian dust event indicated an increase in bacteria of the genera *Bacillus* and *Modestobacter,* with a decrease in the abundance of the *Escherichia-Shigella* group (Cha et al. 2016). Localized extreme climatic conditions can produce intense dust clouds in arid lands and these are referred to as a "haboob", an Arabic word that means "blowing furiously" (see Figure 11.7). Intense dust storms are attributed to tornadoes (Figure 11.7) where soil and other surface particles are carried into the air. Hurricanes and tropical cyclones or typhoons also carry materials from the terrestrial and marine surface into the air currents. These terrestrial disturbances disperse microorganisms into the air where they can be carried for hundreds of kilometers inland by atmospheric currents. Transcontinental winds are of increasing concern because of their capability of carrying viable pathogenic bacteria, fungi, and viruses. While the winds are known to aerosol pathogens for humans, animals, and plants, the extent to which winds are responsible for epidemics and their impact on global

Table 11.6 Winds of the world.

Location/name	Activities
Winds of Provence/ Mistrial and others	A north wind blowing from the Rhone Valley across southern France to the Mediterranean.
African winds	Blowing winds from Sahara Desert, Namib Desert and African savanna to Atlantic Ocean, South America, Caribbean Islands, and southern states of the United States.
Asian dust events	Dust from the Gobi Desert and Taklamakan Desert is blown over the Pacific Ocean, China, Korea, and Japan.
North African winds	Summer: Dust is transported across the Mediterranean to Europe toward the Black Sea and the Middle East
	Spring: Dust is carried across the Atlantic to the Caribbean.
Deserts/Haboob	Found in arid regions around the world. Strong wind results from atmospheric changes associated with the end of a thunderstorm

(a)

(b)

Figure 11.7 Examples of haboob over arid environments: (a) haboob blowing into Ahwatukee, Phoenix, Arizona, on 22 August 2003; (b) category F5 tornado (upgraded from initial estimate of F4) viewed from the southeast as it approached Elie, Manitoba on Friday, June 22nd, 2007. *Source:* (a) English Wikipedia; (b) Justin Hobson. https://commons.wikimedia.org/wiki/File:F5_tornado_Elie_Manitoba_2007.jpg. Licensed under CC BY-SA 3.0.

public health has not been well studied. The abundance of fungi in the haze of Beijing was examined and it was observed that airborne particulate materials carried fungi. On days of haze, the air carried increased levels of *Alternaria, Epicoccum, Aspergillus,* and *Malassezia* (Yan et al. 2016).

Aside from the concern for dispersion of bacteria by winds, considerable interest is being given to microorganism in the air. While clouds around the Earth contain only about 0.03% of the Earth's fresh water, they harbor about 10^{19} microorganisms. In contrast, lakes and rivers contain 10^{26} microorganisms and oceans have 10^{29} microorganisms (Amato 2012). Another way to express bacterial abundance is that 10^3–10^5 are present in a liter of cloud water. Only a fraction of bacteria present in cloud samples can be cultivated and in one account, only three genera of bacteria were cultivated from fog water and they were *Pseudomonas, Bacillus,* and *Acinetobacter,* common soil bacteria. Clouds are a source of nutrients for microorganisms (Table 11.7) and examination of the rRNA of atmospheric bacteria show that bacterial communities in the clouds may be important for cycling of certain organic carbon compounds (Klein et al. 2016). For example, atmospheric acetic acid, lactic acid, formic acid, methanol, and formaldehyde decomposition is attributed to atmospheric bacteria (Amato et al. 2006). It is likely that members of the *Methylobacteriaceae* and *Oxalobacteriaceae* families found in the troposphere degrade oxalic acid (DeLeon-Rodriguez et al. 2013).

Cloud sampling has produced some interesting results. At the Puy de Dôme observatory in France, 45 fungal and 15 yeast isolates were cultivated from microorganisms found in tropospheric clouds (Box 11.2) (Amato et al. 2006). The fungi traveled as spores in cloud water. A highly diverse population of bacteria was present in tropospheric clouds and their abundance was influenced by the seasons (Table 11.8). Gram-negative bacteria were most abundant in the summer and many of the bacteria can grow at cold temperatures. Cloud samples taken at the Mt. Bachelor Observatory, Oregon, USA, contained high levels of *Rhodospirillales* and *Actinomycetales* (Klein et al. 2016). Bacteria of the *Actinomycetale* are commonly found in high altitude atmosphere samples and this may reflect the longevity of their small spores in the atmosphere and their abundance in soil and water. High numbers of the *Rhodospirillales* are found in marine samples and have been reported air samples 8–10 km above the sea. Also, hurricanes have been known to elevate the numbers of atmospheric *Rhodospirillales.* One of the more prominent isolates of the *Rhodospirillales* is *Acidisphaera rubrifaciens,* which has carotenoids to protect it from ultraviolet light.

Table 11.7 Chemicals present in clouds that may be used as nutrients for bacteria.[a]

Ions	Organic compounds	Minerals
Nitrate	Succinic acid[b]	Phosphate
Sulfate	Fumaric acid[b]	Iron
Chloride	Lactic acid[b]	Copper
Ammonium	Malic acid[b]	Magnesium
Sodium	Pyruvic acid[b] Gluconic acid[b] Aldehydes Alcohols	

[a] *Source:* Based on data from Amato (2012).
[b] *Source:* Based on data from Klein et al. (2016).
Succinic acid is in considerable abundance following forest or range fires.

Box 11.2 Tropospheric Clouds

The closest layer of the Earth's atmosphere is the troposphere, which contains most of the water vapors and aerosols. The troposphere is the area where the weather conditions develop and in the polar region, the troposphere is 7 km deep while in the tropics it is 20 km. There are 10 different cloud types in the troposphere, which are grouped into several different altitude levels.

Table 11.8 Abundance of bacteria in tropospheric clouds at the Puy de Dôme observatory in France.

Phyla or Class	Winter (%)	Summer (%)
Proteobacter		
Gammaproteobacter	8	37
Betaproteobacter	4	0
Alphaproteobacter	12	11
Bacteroidetes	2	11
Firmicutes	21	5
Actinobacter	35	37
Unknown	19	0

Source: Based on data from Amato et al. (2006).

When the microbiome of the troposphere was examined before, during, and after the tropical hurricanes of Karl and Earl in the Caribbean Sea, it was revealed that each hurricane aerosolized considerable quantities of bacteria as they moved across the Earth's surface (DeLeon-Rodriguez et al. 2013). Most of the bacteria aerosolized were from the marine surface but when the hurricanes moved over populated land, there was a dramatic increase in the number of *Escherichia* and *Streptococcus* in the air. Bacteria swept into the troposphere by hurricane winds remain viable, can be carried great distances, and remain aloft for days to weeks. Bacteria in the upper troposphere where the temperature is cool may greatly influence cloud formation and serve as nucleation sites in the formation of water droplets. Future studies involving models involving global cloud movement should include the impact by microorganism.

11.8.5 Long-distance Electron Transfer

As reviewed by Müller and et al. (2016), the unique electron transport system for filamentous bacteria evolved from the report of bacteria present in marine sediments (Pfeffer et al. 2012) that could produce electric currents (Nielson et al. 2010). Each filament consists of thousands of bacterial cells with a sheath-like covering, with about 50 ridges in the filament, which gives the appearance of a cable. The distance over which the electrons move can range from 1 to 2 cm. Bacteria that function in this electron transfer are a group within the *Desulfobulbaceae* family including six recently identified species belonging to the new genera of *Candidatus* Electrothrix and *Candidatus* Electronema (Trojan et al. 2016). A model of the long-distance electron transfer characterizing sulfur oxidation is presented in Figure 11.8. Half-cell reactions are separated along the filament and regions of anodic and cathodic activity are present. With acidification occurring in the sediment, this activity could have an important effect on the phosphorous, iron, and manganese cycles. This new area is of considerable interest to geologists

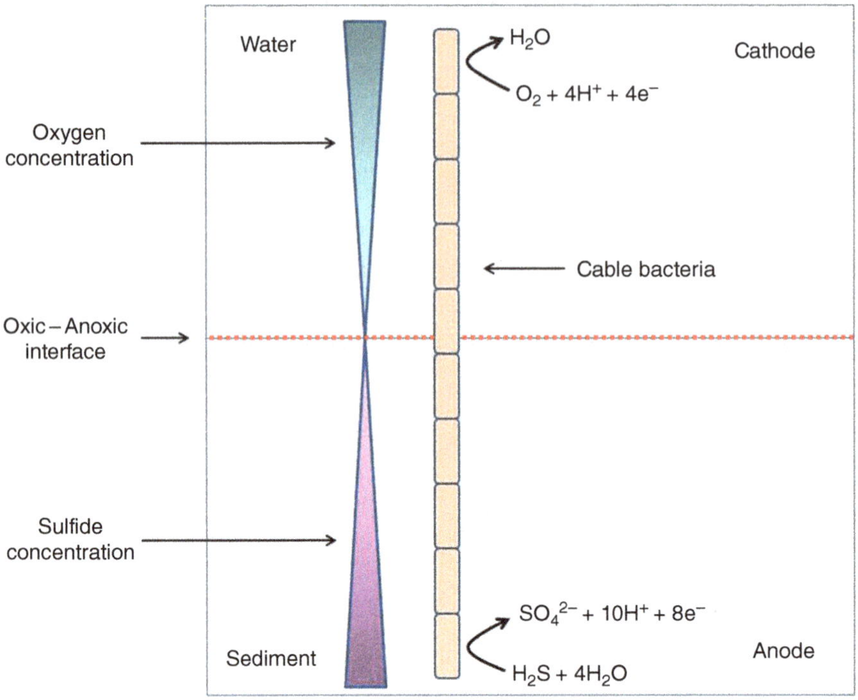

Figure 11.8 Long distance electron transport by cable bacteria.

and microbiologists and there are numerous questions to be addressed. Long-distance electron transfers via bacteria have been demonstrated in fresh water and in marine sediments (Burdorf et al. 2017). The number of filaments in freshwater sediments has been estimated at 4×10^7 per m^2 (Risgaard-Petersen et al. 2015), which suggests that more is to be discovered about the activities of cable bacteria in the environment.

11.9 Summary

Metabolic end products from microbial fermentation and decomposition of plant materials in an environment are the collective expression of the individual species of microorganisms present. Due to the diverse microbial population and their regulation of metabolism, the population of microorganisms and their metabolic end products are continuously changing. This dynamic activity results in changes to the environment. A succession of populations is observed in the progression of detrimental coral BBD as well as in the production of beneficial changes in foods. Microbial fermentation involving bacteria and yeast produce the flavor enhancing qualities found in Kimchi and Chinese soy sauces. Microbial fermentation is part of the natural processing of coffee and cocoa beans. Composting of plant material – another example of the dynamics of a changing microbial population – involves mesophilic cellular soil bacteria, actinomycetes, and thermophilic fungi. As a greenhouse gas, CO_2 acidifies the environment and even though the pH change in soil and marine waters is slow, this will eventually change microbial populations. Syntrophic activities between bacteria is a unique cooperation that enables them to extend the environment where they grow; it will be interesting to see what syntrophic processes may result from the continuous fracture of shale. Electron transfer is a hallmark characteristic of life, and bacteria have developed several electron movement processes outside the cell using outer membrane cytochromes, electroconductive pili, and electron shuttle molecules. Some electron movement processes, such as in biocorrosion and long-distance movement of electrons in bacterial filaments, remain to be established. Microbial endocrinology and cross-talk between bacteria and hosts provides a basis for understanding the interactions between bacteria and human hosts. Some processes that hold potential to provide new insight into environmental microbiology and microbial ecology includes an understanding of genetics with respect to the role of polyploids and the CRISPER–*cas* systems. The activities of OMVs reveal how an OMV can function as a specialized "organelle" to extend the impact of these bacteria on their environment. Numerous bacteria and fungi have been found in the air and a goal of aeromicrobiology is to determine if the troposphere supports bacterial growth and if this represents another biosphere.

Discussion Questions

1. Describe a molecular model that would enable electrons to move from cell-to-cell in the system of long-distance electron transfer.

2. Discuss some specific impacts of climate change on microorganisms in a symbiotic relationship. Which symbiotic microorganisms (bacteria or fungi) would you expect to be most affected by climate change?

3. Would you expect to find the unique bacteria growing near high voltage power lines resistant to electromagnetic forces or capable of using electron flow from the wires to support growth?

4. Microbial growth is a highly dynamic event. Besides the processes discussed in this chapter, in what other areas would a succession of microorganisms be found?

References

Amato, P. (2012). Clouds provide atmospheric oases for microbes. *Microbe* 7: 119–123.

Amato, P., Parazols, M., Sancelme, M. et al. (2006). Microorganisms isolated from the water phase of trophospheric clouds at the Puy de Dôme: major groups and growth abilities at low temperatures. *FEMS Microbiol. Ecol.* 59: 242–254.

Anderson, R.E., Brazelton, W.J., and Baross, J.A. (2011). Is the genetic landscape of the deep subsurface biosphere affected by viruses? *Front. Microbiol.* 2: 219.

Archibald, J.M. and Rogers, A.J. (2002). Gene conversion and the evolution of euryarchaeal chaperonins: a maximum likelihood-based method for detecting conflicting phylogenetic signals. *J. Mol. Evol.* 55: 232–245.

Aulenta, F., Rossetti, S., Amalfitano, S. et al. (2013). Conductive magnetite nanoparticles accelerate the microbial reductive dechlorination of trichloroethene by promoting interspecies electron transfer processes. *ChemSusChem.* 6: 433–436.

Barton, L.L. (2014). Bacterial organization at the smallest level: molecular motors, nanowires and outer membrane vesicles. In: *Nanomicrobiology* (ed. L.L. Barton, D.A. Bazylinski and H. Xu), 103–125. Heidelberg: Springer.

Biller, S.J., Schubotz, F., Roggensack, S.E. et al. (2014). Bacterial vesicles in marine ecosystems. *Science* 343: 183–186.

de Bok, F.A.M., Harmsen, H.J.M., Plugge, C.M. et al. (2005). The first true obligately syntrophic propionate-oxidizing bacterium, *Pelotomaculum schinkii* sp. nov., co-cultured with *Methanospirillum hungatei*, and amended description of the genus *Pelotomaculum*. *Int. J. Syst. Evol. Microbiol.* 55: 1697–1703.

Boone, D.R., Johnson, R.L., and Liu, Y. (1989). Diffusion of the interspecies electron carriers H_2 and formate in methanogenic ecosystems, and implications in the measurement of K_m for H_2 or formate uptake. *Appl. Environ. Microbiol.* 55: 1735–1741.

Bose, M. and Barber, R. (2006). Prophage finder: a prophage loci prediction tool for prokaryotic genome sequences. *In Silico Biol.* 6: 0020.

Böttinger, B., Zerulla, K., and Soppa, J. (2014). Ploidy of *Bacillus subtilis, Bacillus megaterium,* and three new isolates of *Bacillus* and *Paenibacillus*. *PeerJ Prepr.* 2: e306v1. https://doi.org/10.7287/peerj.preprints.306v1.

Bourne, D.G., Muirhead, A., and Sato, Y. (2011). Changes in sulfate-reducing bacterial populations during the onset of black band disease. *ISEM J.* 5: 559–564.

Breuert, S., Allers, T., Spohn, G. et al. (2006). Regulated polyploidy in halophilic archaea. *PLoS One* 1 (1): e92. https://doi.org/10.1371/ journal.pone.0000092.

Brown, L., Kessler, A., Cabezas-Sanchez, P. et al. (2014). Extracellular vesicles produced by the gram-positive *bacterium Bacillus subtilis* are disrupted by the lipopeptide surfactin. *Mol. Microbiol.* 93: 183–198.

Brown, L., Wolf, J.M., Prados-Rosales, R. et al. (2015). Through the wall: extracellular vesicles in gram-positive bacteria, mycobacteria and fungi. *Nat. Rev. Microbiol.* 13: 620–630.

Bryant, M.P., Wolin, E.A., Wolin, M.J. et al. (1967). *Methanobacillus omelianskii*, a symbiotic association of two species of bacteria. *Arch. Mikrobiol.* 59: 20–31.

Burdorf, L.D.W., Tramper, A., Seitaj, D. et al. (2017). Long-distance electric transport occurs globally in marine sediments. *Biogeosci.* 14: 683–701.

Burstein D, Sun CL, Brown CT, et al. (2016). Major bacterial lineages are essentially devoid of CRISPR-Cas viral defense systems, *Nature Commun.* 7:10613 | https://doi.org/10.1038/ncomms10613

Cao, B., Shi, L., Brown, R.N. et al. (2011). Extracellular polymeric substances from *Shewanella* sp. HRCR-1 biofilms: characterization by infrared spectroscopy and proteomics. *Environ. Microbiol.* 13: 1018–1031.

Castro, H.F., Classen, A.T., Austin, E.E. et al. (2009). Soil microbial community responses to multiple experimental climate change drivers. *Appl. Environ. Microbiol.* 76: 999–1007.

Cha, S., Lee, D., Jang, J.H. et al. (2016). Alterations in the airborne bacterial community during Asian dust events occurring between February and March 2015 in South Korea. *Sci. Rep.* 6: 37271. https://doi.org/10.1038/srep37271.

Chant, J., Hui, I., De Jong-Won, D. et al. (1986). The protein synthesizing machinery of archaebacterium *Halobacterium cutirubum*: molecular characterization. *Syst. Appl. Microbiol.* 7: 106–114.

Choi, O. and Sang, B.-I. (2016). Extracellular electron transfer from cathode to microbes: application for biofuel production. *Biotechnol. Biofuels* 20169: 11. https://doi.org/10.1186/s13068-016-0426-0.

Classen, A.T., Sundqvist, M.K., Henning, J.A. et al. (2015). Direct and indirect effects of climate change on soil microbial and soil microbial-plant interactions: What lies ahead? *Ecosphere* 6: 1–21.

Crowther, T.W., Thomas, S.M., Maynard, D.S. et al. (2015). Biotic interactions mediate soil microbial feedbacks to climate change. *Proc. Nat. Acad. Sci. USA* 112: 7033–7038.

Cruz Viggi, C., Rossetti, S., Fazi, S. et al. (2014). Magnetite particles triggering a faster and more robust syntrophic pathway of methanogenic propionate degradation. *Environ. Sci. Technol.* 48: 7536–7543. https://doi.org/10.1021/es5016789.

Daly, R.A., Borton, M.A., Wilkins, M.J. et al. (2016). Microbial metabolism in a 2.5 km deep ecosystem created by hydraulic fracturing in shales. *Nat. Microbiol.* 1: 16146. https://doi.org/10.1038/nmicrobiol.2016.146.

De Filippis, F., Genovese, A., Ferranti, P. et al. (2016). Metatranscriptomics reveals temperature-driven functional changes in microbiome impacting cheese maturation rate. *Sci. Rep.* 6: 21871. https://doi.org/10.1038/srep21871.

De Pasquale, L., Di Cagno, R., Buchin, S. et al. (2014). Microbial ecology dynamics reveals a succession in the core microbiota involves in the ripening of pasta filata Caciocavallo Pugliese cheese. *Appl. Environ. Microbiol.* 80: 6243–6255.

Deatherage, B.L. and Cookson, B.T. (2012). Membrane vesicle release in bacteria, eukaryotes and archaea: a conserved yet underappreciated aspect of microbial life. *Infect. Immun.* 80: 1948–1957.

DeLeon-Rodriguez, N., Lathem, T.L., Rodriguez-R, L.M. et al. (2013). Microbiome of the upper troposphere: species composition and prevalence, effects of tropical storms, and atmospheric implications. *Proc. Natl. Acad. Sci. USA* 110: 2575–2580.

Dolfing, J. (2014). Syntrophy in microbial fuel cells. *ISEM J.* 8: 4–5.

Dolfing, J. and Tiedje, J.M. (1986). Hydrogen cycling in a three-tiered food web growing on the methanogenic conversion of 3-chlorobenzoate. *FEMS Microbiol. Ecol.* 2: 293–298.

Dore, J. and Bryant, M.P. (1990). Metabolism of one-carbon compounds by the ruminal acetogen *Syntrophococcus sucromutans*. *Appl. Environ. Microbiol.* 56: 984–989.

Fildes, P. (1956). Production of tryptophan by *Salmonella typhi* and *Escherichia coli*. *J. Gen. Microbiol.* 15: 636–643.

Fonseca, A., Ishoey, T., Espinoza, C. et al. (2017). Genomic features of "*Candidatus* Venteria ishoeyi", a new sulfur-oxidizing macrobacterium from the Humboldt Sulfuretum off Chile. *PLoS One* 12 (12): e0188371. https://doi.org/10.1371/journal.pone.0188371.

Foster, J.A., Rinaman, L., and Cryan, J.F. (2017). Stress & the gut-brain axis: regulation by the microbiome. *Neurobiol. Stress* 7: 124–136.

Freestone, P. (2013). Communication between bacteria and their hosts. *Scientifica* 2013: 361073. 15 pages. http://dx.doi.org/https://doi.org/10.1155/2013/361073.

Freilich, S., Kreimer, A., Borenstein, E. et al. (2009). Metabolic-network-driven analysis of bacterial ecological strategies. *Genome Biol.* 10 (6): R61. https://doi.org/10.1186/gb-2009-10-6-r61.

Ghosh, D., Roy, K., Williamson, K.E. et al. (2008). Prevalence of lysogeny among soil bacteria and presence of 16S rRNA and trzN genes in viral-community DNA. *Appl. Environ. Microbiol.* 74: 495–502.

Gödeke, J., Paul, K., Lassak, J. et al. (2011). Phage-induced lysis enhances biofilm formation in *Shewanella oneidensis* MR-1. *ISME J.* 5: 613–626.

Gorby, Y.A., Yanina, S., McLean, J.S. et al. (2006). Electrically conductive bacterial nanowires produced by *Shewanella oneidensis* strain MR-1 and other microorganisms. *Proc. Natl. Acad. Sci. USA* 103: 11358–11363.

Gorby, Y., McLean, J., Korenevsky, A. et al. (2008). Redox-reactive membrane vesicles produced by *Shewanella*. *Geomicrobiol.* 6: 232–241.

Griese, M., Lange, C., and Soppa, J. (2011). Ploidy in cyanobacteria. *FEMS Microbiol. Lett.* 323: 124–131.

Grissa I, Vergnaud G, Pourcel C (2007a). CRISPR finder: a web tool to identify clustered regularly interspaced short palindromic repeats, *Nucleic Acids Res.* 35, W52–W57. https://doi.org/10.1093/nar/gkl1118

Grissa I, Vergnaud G, Pourcel C (2007b). The CRISPR db database and tools to display CRISPRs and to generate dictionaries of spacers and repeats, *BMC Bioinformatics* 8: 172.https://doi.org/10.1186/1471-2105-8-172

Gutiérrez, D., Rodríguez-Rubio, L., Martínez, B. et al. (2016). Bacteriophages as weapons against bacterial biofilms in the food industry. *Front. Microbiol.* 7: 2016, 825. https://doi.org/10.3389/fmicb.2016.00825.

Hatamoto, M., Imachi, H., Fukayo, S. et al. (2007). *Syntrophomonas palmitatica* sp. nov., an anaerobic, long-chain fatty acid oxidizing bacterium isolated from methanogenic sludge. *Int. J. Syst. Evol. Microbiol.* 57: 2137–2142.

Haynes, T.S., Klemm, D.J., Ruocco, J.J. et al. (1995). Formate dehydrogenase activity in cells and outer membrane blebs of *Desulfovibrio gigas*. *Anaerobe* 1: 175–182.

Hildenbrand, C., Stock, T., Lange, C. et al. (2011). Genome copy numbers and gene conversion in methanogenic archaea. *J. Bacteriol.* 193: 734–743.

Horvath, P. and Barrangou, R. (2010). CRISPR/Cas, the immune system of bacteria and archaea. *Science* 327: 167–170.

Hughes, D.T. and Sperandio, V. (2008). Inter-kingdom signaling: communication between bacteria and their hosts. *Nat. Rev. Microbiol.* 6: 111–120.

Imachi, H., Sakai, S., Ohashi, A. et al. (2007). *Pelotomaculum propionicicum* sp. nov., an anaerobic, mesophilic, obligately syntrophic, propionate-oxidizing bacterium. *Int. J. Syst. Evol. Microbiol.* 57: 1487–1492.

Ionescu, D., Bizic-Ionescu, M., De Maio, N. et al. (2017). Community-like genome in single cells of the sulfur bacterium *Achromatium oxliferum*. *Nat. Commun.* 8: 455. https://doi.org/10. 1038/s41467-017-00342-9.

IPCC (2007). Climate change 2007: the physical science basis. In: *Contribution of Working Group to the Fourth Assessment Report of the Intergovernmental Panel on Climate Change* (ed. S. Solomon, D. Qin, M. Manning, et al.). Cambridge, UK: Cambridge University Press.

Jan, A.T. (2017). Outer membrane vesicles (OMVs) of gram-negative bacteria: a perspective update. *Front. Microbiol.* 8: 1053. https://doi.org/10.3389/fmicb.2017.01053.

Jung, J.Y., Lee, S.H., Kim, J.M. et al. (2011). Metagenomic analysis of kimchi, a traditional Korean fermented food. *Appl. Environ. Microbiol.* 77: 2264–2274.

Kato, S., Hashimoto, K., and Watanabe, K. (2012). Microbial interspecies electron transfer via electric currents through conductive minerals. *Proc. Natl. Acad. Sci. USA* 109: 10042–10046.

Katsir, L. and Bahar, O. (2017). Bacterial outer membrane vesicles at the plant-pathogen interface. *PLoS Pathog.* 13 (6): e1006306. https://doi.org/10.1371/journal.ppat.1006206.

Kendall, M.M. and Sperandio, V.M. (2016, 2016). What a dinner party! Mechanisms and functions of interkingdom signaling in host-pathogen associations. *mBio* 7: e01748–e01715.

Klein, A.M., Bohannan, B.J.M., Jaffe, D.A. et al. (2016). Molecular evidence for metabolically active bacteria in the atmosphere. *Front. Microbiol.* 7: 772. https://doi.org/10.3389/fmicb.2016.00772.

Komacki, K. and Ishikawa, H. (2000). Genomic copy number of intracellular bacterial symbionts of aphids varies in response to developmental stage and morph of their host. *Insect Biochem. Mol. Biol.* 30: 253–258.

Kotloski, N.J. and Gralnick, J.A. (2013). Flavin electron shuttles dominate extracellular electron transfer by *Shewanella oneidensis*. *mBio* 4 (1): e00553–e00512. https://doi.org/10.1128/mBio.00553-12.

Kouzuma, A., Kato, S., and Watanabe, K. (2015). Microbial interspecies interactions: recent findings in syntrophic consortia. *Front. Microbiol.* 6: 477. https://doi.org/10.3389/fmicb.2015.00477. eCollection 2015.

Kracke, F., Vassilev, I., and Krömer, J.O. (2015). Microbial electron transport and energy conservation – the foundation for optimizing bioelectrochemical systems. *Front. Microbiol.* 11: https://doi.org/10.3389/fmicb.2015.00575.

Kulkarni, H.M. and Jagannadham, M.V. (2014). Biogenesis and multifaceted roles of outer membrane vesicles from gram-negative bacteria. *Microbiol.* 160: 2109–2121.

Labrie, S.J., Samson, J.E., and Moineau, S. (2010). Bacteriophage resistance mechanisms. *Nat. Rev. Microbiol.* 8: 317–327.

Lehman, N. (2003). A case for the extreme antiquity of recombination. *J. Mol. Evol.* 56: 770–777.

Lin, D.M., Koskella, B., and Lin, H.C. (2016). Phage therapy: an alternative to antibiotics in the age of multi-drug resistance. *World J. Gastrointest. Pharmacol. Ther.* 8: 162–173.

Lisle, J.T. and Priscu, J.C. (2004). The occurrence of lysogenic bacteria and microbial aggregates in the lakes of the McMurdo Dry Valleys, Antarctica. *Microb. Ecol.* 47: 427–439.

Lovley, D.R., Holmes, D.E., and Nevin, K.P. (2004). Dissimilatory Fe(III) and Mn(IV) reduction. *Adv. Microb. Physiol.* 49: 219–286.

Lundgren, M., Malandrin, L., Eriksson, S. et al. (2008). Cell cycle characteristics of crenarchaeota: unity among diversity. *J. Bacteriol.* 190: 5362–5367.

Lyte, M. (2016). *Microbial Endocrinology Interkingdom Signaling in Infectious Disease and Health.* New York, NY: Springer.

Mackwan, R.R., Carver, G.T., Drake, J.W. et al. (2007). An unusual pattern of spontaneous mutations recovered in the halophylic archaeon *Haloferax volcanii*. *Genetics* 176: 697–702.

Maldonado, R., Jimenez, J., and Casadesus, J. (1994). Changes of ploidy during the *Azotobacter vinelandii* growth cycle. *J. Bacteriol.* 176: 3911–3919.

Manefield, M., de Nys, R., Naresh, K. et al. (1999). Evidence that halogenated furanones from Delisea pulchra inhibit acylated homoserine lactone (AHL)-mediated gene expression by displacing the AHL signal from its receptor protein. *Microbiology.* 145: 283–291. https://doi: 10.1099/13500872-145-2-283.

Marsili, E., Baron, D.B., Shikhare, I.D. et al. (2008). *Shewanella* secretes flavins that mediate extracellular electron transfer. *Proc. Natl. Acad. Sci. USA* 105: 3968–3973.

Mashburn, L.M. and Whiteley, M. (2005). Membrane vesicles traffic signals and facilitate group activities in a prokaryote. *Nature.* 437: 422–425. https://doi.org/10.1038/nature03925.

McInerney, M.J., Sieber, J.R., and Gunsalus, R.P. (2009). Syntrophy in anaerobic global carbon cycles. *Curr. Opin. Biotechnol.* 20: 623–632.

Mendell, J.E., Clements, K.D., Choat, J.H. et al. (2008). Extreme polyploidy in a large bacterium. *Proc. Natl. Acad. Sci. USA* 105: 6730–6734.

Morris, B.E., Henneberger, R., Huber, H. et al. (2013). Microbial syntrophy: interaction for the common good. *FEMS Microbiol Rev.* 37: 384–406.

Mueller, J. and Overmann, J. (2011). Close interspecies interactions between prokaryotes from sulfureous environments. *Front. Microbiol.* 2: 146.

Müller, H., Bosch, J., Griebler, S. et al. (2016). Long-distance electron transfer by cable bacteria in aquifer sediments. *ISME J.* 10: 2010–2019.

Nevin, K.P., Hensley, S.A., Franks, A.E. et al. (2011). Electrosynthesis of organic compounds from carbon dioxide is catalyzed by a diversity of acetogenic microorganisms. *Appl. Environ. Microbiol.* 77: 2882–2886.

Newman, D.K. and Kolter, R. (2000). A role for excreted quinones in extracellular electron transfer. *Nature* 405: 94–97.

Nielsen, L.P., Risgaard-Petersen, N., Fossing, H. et al. (2010). Electric currents couple spatially separated biogeochemical processes in marine sediment. *Nature* 463: 1071–1074.

Nigro OD, Jungbluth SP, Lin H-T, et al. (2017). Viruses in the oceanic basement, *mBio* vol. 8 no. 2e02129–16: https://doi.org/10.1128/mBio.02129-167

O'Brien, P.A., Morrow, K.M., Willis, B.L. et al. (2016). Implications of ocean acidification for marine microorganisms from the free-living to the host associated. *Front. Marine Sci.* 14: https://doi.org/10.3389/fmars.2016.00047.

Ohtani, N., Tomita, M., and Itaya, M. (2010). An extreme thermophile, *Thermus thermophilus,* is a polyploid bacterium. *J. Bacteriol.* 192: 5499–5505.

Olson, J.M. (1978). Confused history of *Chloropseudomonas ethylica* 2K. *Int. J. Syst. Bacteriol.* 28: 128–129.

Pecoraro, V., Zerulla, K., Lange, C. et al. (2011). Quantification of ploidy in proteobacteria revealed the existence of monoploid, (mero-)oligoploid and polyploid species. *PLoS One* 6 (1): e16392. https://doi.org/10.1371/journal.pone.0016392.

Pfeffer, C., Larsen, S., Song, J. et al. (2012). Filamentous bacteria transport electrons over centimetre distances. *Nature* 491: 218–221.

Pirbadian, S., Barchinger, S.E., Leung, K.M. et al. (2014). *Shewanella oneidensis* MR-1 nanowires are outer membrane and periplasmic extensions of the extracellular electron transport components. *Proc. Natl. Acad. Sci. USA* 111: 12883–12888.

Ponomarova, O. and Paitil, K.R. (2015). Metabollic interactions in microbial communities: untangling the Gordian knot. *Curr. Opin. Microbiol.* 27: 37–44.

Postgate, J.R., Kent, H.M., Robson, R.I. et al. (1984). The genomes of *Desulfovibrio gigas and D. vulgaris*. *J. Gen. Microbiol.* 130: 1597–1601.

Poznanski, E., Cavazza, A., Cappa, F. et al. (2004). Indigenous raw milk microbiota influences the bacterial development in traditional cheese from an alpine park. *Int. J. Food Microbiol.* 92: 141–151.

Prangishvili, D., Holz, I., Stieger, E. et al. (2000). Sulfolobicins, specific proteinaceous toxins produced by strains of the extremely thermophilic archaeal genus *Sulfolobus*. *J. Bacteriol.* 182: 2985–2988.

Punita, J.S.J., Reddy, M.A., and Das, H.K. (1989). Multiple chromosomes of *Azotobacter vinelandii*. *J Bacteriol.* 171: 3133–3138.

Rabaey, K., Boon, N., Höfte, M. et al. (2005). Microbial phenazine production enhances electron transfer in biofuel cells. *Environ. Sci. Technol.* 39: 3401–3408.

Reguera, G., McCarthy, K.D., Mehta, T. et al. (2005). Extracellular electron transfer via microbial nanowires. *Nature* 435: 1098–1101.

Revil, A., Mendonça, C.A., Atekwana, E.A. et al. (2010). Understanding biogeobatteries: where geophysics meet microbiology. *J. Geophys. Res. Biogeosci.* 115 (G1): G00G02: https://doi.org/10.1029/2009.JG001065.

Richter, K., Schicklberger, M., and Gescher, J. (2012). Dissimilatory reduction of extracellular electron acceptors in anaerobic respiration. *Appl. Environ. Microbiol.* 78: 913–921.

Risgaard-Petersen, N., Kristiansen, M., Frederiksen, R.B. et al. (2015). Cable bacteria in freshwater sediments. *Appl. Environ. Microbiol.* 81: 6003–6011.

Röling, W.F.M., Ferrer, M., and Golyshin, P.N. (2010). Systems approach to microbial communities and their functioning. *Curr. Opin. Biotechnol.* 21: 532–538.

Rosenberg, E., Koren, O., Reshef, L. et al. (2007). The role of microorganisms in coral health, disease and evolution. *Nat. Rev. Microbiol.* 5: 355–362.

Rosling, A., Cox, F., Cruz-Martinez, K. et al. (2011). *Archaeorhizomycetes*: unearthing an ancient class of ubiquitous soil fungi. *Science* 333: 876–879.

Rotaru, A.-E., Shrestha, P.M., Liu, F. et al. (2012). Interspecies electron transfer via hydrogen and formate rather than direct electrical connections in cocultures of *Pelobacter carbinolicus* and *Geobacter sulfurreducens*. *Appl. Environ. Microbiol.* 78: 7645–7651.

Rotaru, A.-E., Shrestha, P.M., Liu, F. et al. (2014). Direct interspecies electron transfer between *Geobacter metallireducens* and *Methanosarcina barkeri*. *Appl. Environ. Microbiol.* 80: 4599–4605.

Sandrini, S., Aldriwesh, M., Alruways, M. et al. (2015). Microbial endocrinology: host-bacteria communication with the gut microbiome. *J. Endocrinol.* 225 (2): R21–R34: https://doi.org/10.1530/JOE-14-0615.

Sato, Y., Willis, B.L., and Bourne, D.G. (2010). Successional changes in bacterial communities during the development of black band disease on the reef coral, *Montipora hispida*. *ISEM J.* 4: 203–214.

Schink, B. (1997). Energetics of syntrophic cooperation in methanogenic degradation. *Microbiol, Mol. Biol. Rev.* 61: 262–280.

Schink, B. and Stams, A.J. (2006). Syntrophism among prokaryotes. In: *The Prokaryotes* (ed. M. Dworkin, S. Falkow, E. Rosenberg, et al.), 309–335. New York, NY: Springer.

Schopf, S., Wanner, G., Rachel, R. et al. (2008). An archaeal bi-species biofilm formed by *Pyrococcus furiosus* and *Methanopyrus kandleri*. *Arch. Microbiol.* 190: 371–377.

Scott, D.T., McKnight, D.M., Blunt-Harris, E.L. et al. (1998). Quinone moieties act as electron acceptors in the reduction of humic substances by humics-reducing microorganisms. *Environ. Sci. Technol.* 32: 2984–2989.

Shetty, A., Chen, S., Tocheva, E.I. et al. (2011). Nanopods: a new bacterial structure and mechanism for deployment of outer membrane vesicles. *PLoS One* 6 (6): e20725.

Shigenobu, S., Watanabe, H., Hattori, M. et al. (2000). Genome sequence of the endocellular bacterial symbiont of aphids *Buchnera* sp. *APS Nature* 407: 81–86.

Silva, C.F., Batista, L.R., Abreu, L.M. et al. (2008). Succession of bacterial and fungal communities during natural coffee (*Coffea arabica*) fermentation. *Food Microbiol.* 25: 951–957.

Snider, R.M., Strycharz-Glaven, S.M., Tsoi, S.D. et al. (2012). Long-range electron transport in *Geobacter sulfurreducens* biofilms is redox gradient-driven. *Proc. Natl. Acad. Sci. USA* 109: 15467–15472.

Solusa, D.Z., Smidt, H., Alves, M.M. et al. (2007). *Syntrophomonas zehnderi* sp. nov., an aerobe that degrades long-chain fatty acids in co-culture with *Methanobacterium formicicum*. *Int. J. Syst. Evol. Microbiol.* 57: 609–615.

Strycharz-Glaven, S.M. and Tender, L.M. (2012). Reply to the 'Comment on "on electrical conductivity of microbial nanowires and biofilms"' by N. S. Malvankar, M. T. Tuominen and D. R. Lovley. *Energy Environ. Sci.* 5: 6250–6255.

Sulaiman, J., Gan, H.M., Yin, W.-F. et al. (2014). Microbial succession and the functional potential during the fermentation of Chinese soy sauce brine. *Front. Microbiol.* 5: 556: https://doi.org/10.3389/fmicb.2014.00556.

Sulakvelidze, A., Alavidze, Z., and Morris, J.G. Jr. (2001). Bacteriophage therapy. *Antimicrob. Agents Chemother.* 45: 649–659.

Summers, Z.M., Fogarty, H.E., Leang, C. et al. (2010). Direct exchange of electrons within aggregates of an evolved syntrophic coculture of anaerobic bacteria. *Science* 330: 1413–1415.

Thauer, R.K., Kaster, A.-K., Seedorf, H. et al. (2008). Methanogenic archaea: ecologically relevant differences in energy conversation. *Nat. Rev. Microbiol.* 6: 579–591.

Thingstad, T. and Lignell, R. (1997). Theoretical models for the control of bacterial growth rate, abundance, diversity and carbon demand. *Aquat. Microbiol. Ecol.* 13: 119–127.

Tobiason, D.M. and Seifert, H.S. (2010). Genomic content of *Neisseria* species. *J. Bacteriol.* 192: 2160–2168.

Toyofuku, M., Cárcamo-Oyarce, G., Yamamoto, T. et al. (2017a). Prophage-triggered membrane vesicle formation through peptidoglycan damage in *Bacillus subtilis*. *Nature Commun.* 8: 481:https://doi.org/10.1038/s41467-017-00492-w.

Toyofuku, M., Morinaga, K., Hashimoto, Y. et al. (2017b). Membrane vesicle-mediated bacterial communication. *ISME J.* 11: 1504–1509.

Trojan, D., Schreiber, L., Bjerg, J.T. et al. (2016). A taxonomic framework for cable bacteria and proposal of the candidate genera *Electrothrix* and *Electronema*. *Syst. Appl. Microbiol.* 39: 297–306.

Vargas, M., Malvankar, N.S., and Tremblay, P.-L. (2013). Aromatic amino acids required for pili conductivity and long range extracellular electron transport in *Geobacter sulfurreducens*. *mBio* 4 (2): e00105–e00113.

Vaughan, M.J., Mitchell, T., and McSpadden Gardener, B.B. (2015). What's inside that seed we brew? A new approach to mining the coffee microbiome. *Appl. Environ. Microbiol.* 81: 6518–6527.

da Veiga Moreira, I.M., Pedrozo, d.C., Miguel, M.G. et al. (2013). Microbial succession and the dynamics of metabolites and sugars during the fermentation of three different cocoa (*Theobroma cacao* L.) hybrids. *Food Res. Internat.* 54: 9–17.

Venturi, V. and Fuqua, C. (2013). Chemical signaling between plants and plant-pathogenic bacteria. *Ann. Rev. Phytopath.* 51: 17–37.

Voordeckers, J.W., Kim, B.-C., Izallalen, M. et al. (2010). Role of *Geobacter sulfurreducens* outer surface c-type cytochromes in reduction of soil humic acid and anthraquinone-2,6-disulfonate. *Appl. Environ. Microbiol.* 76: 2371–2375.

Warthmann, R., Cypionka, H., and Pfennig, N. (1992). Photoproduction of H_2 from acetate by syntrophic cocultures of green sulfur bacteria and sulfur-reducing bacteria. *Arch. Microbiol.* 157: 343–348.

Wegener, G., Krukenberg, V., Riedel, D. et al. (2015). Intercellular wiring enables electron transfer between methanotrophic archaea and bacteria. *Nature.* 526: 587–590.

Weinbauer, M.G. (2004). Ecology of prokaryotic viruses. *FEMS Microbiol. Rev.* 28: 127–181.

Westerholm, M., Dolfing, J., Sherry, A. et al. (2011). Quantification of syntrophic acetate -oxidizing microbial communities in biogas processes. *Environ. Microbiol. Rep.* 3: 500–505.

Williamson, S.J., Rusch, D.B., Yooseph, S. et al. (2008). The Sorcerer II global ocean sampling expedition: metagenomic characterization of viruses within aquatic microbial samples. *PLoS One* 3 (1): e1456.

Worm, P., Koehorst, J.J., Visser, M. et al. (2014). A genomic view on a syntrophic versus non-syntrophic lifestyle in anaerobic fatty acid degrading communities. *Biochim. Biophys. Acta* 1837: 2004–2016.

Wu, C., Liu, X., and Dong, X. (2006a). *Syntrophomonas erecta* subsp. *sporosyntropha* subsp. nov., a spore-forming bacterium that degrades short chain fatty acids in co-culture with methanogens. *Syst. Appl. Microbiol.* 29: 457–462.

Wu, C., Liu, X., and Dong, X. (2006b). *Syntrophomonas cellicola* sp. nov., a spore-forming syntrophic bacterium isolated from a distilled spirit fermenting cellar, and assignment of *Syntrophospora bryantii* to *Syntrophomonas bryantii* com. nov. *Int. J. Syst. Evol. Microbiol.* 56: 2331–2335.

Yan, D., Zhang, T., Su, J. et al. (2016). Diversity and composition of airborne fungal community associated with particulate matters in Beijing during haze and non-haze days. *Front. Microbiol.* 7: 487: https://doi.org/10.3389/fmicb.2016.00487.

Zerulla, K. and Soppa, J. (2014). Polyploidy in haloarchaea: advantages for growth and survival. *Front. Microbiol.* 5: 274. https://doi.org/10.3389/fmicb.2014.00274. eCollection 2014.

Further Reading

Das, S. and Mangwqni, N. (2015). Ocean acidification and marine microorganisms: responses and consequences. *Oceanologia* 57: 349–361. doi: 10.1016/j.oceano.2015.07.003.

Marxsen, J. (2016). *Climate Change and Microbial Ecology: Current Research and Future Trends.* Caister Academic Press.

Zhang, Y., Cong, J., Lu, H. et al. (2014). An integrated study to analyze soil microbial community structure and metabolic potential in two forest types. *PLoS One* 9 (4): e93773. doi: 10.1371/journal.pone.0093773.

Index

Environmental Microbiology and Microbial Ecology, First Edition. Larry L. Barton and Robert J.C. McLean.

d

f

n

Printed in the USA/Agawam, MA
December 18, 2023

857405.083